ASTRONOMY AND
ASTROPHYSICS LIBRARY

Springer
Berlin
Heidelberg
New York
Barcelona
Hong Kong
London
Milan
Paris
Singapore
Tokyo

ASTRONOMY AND ASTROPHYSICS LIBRARY

Series Editors: I. Appenzeller · G. Börner · M. Harwit · R. Kippenhahn
J. Lequeux · P. A. Strittmatter · V. Trimble

Theory of Orbits (2 volumes)
Volume 1: Integrable Systems and Non-perturbative Methods
Volume 2: Perturbative and Geometrical Methods
By D. Boccaletti and G. Pucacco

Galaxies and Cosmology
By F. Combes, P. Boissé, A. Mazure and A. Blanchard

The Solar System 2nd Edition   By T. Encrenaz and J.-P. Bibring

The Physics and Dynamics of Planetary Nebulae By G. A. Gurzadyan

Astrophysical Concepts 2nd Edition   By M. Harwit

Stellar Structure and Evolution By R. Kippenhahn and A. Weigert

Modern Astrometry By J. Kovalevsky

Astrophysical Formulae 3rd Edition (2 volumes)
Volume I: Radiation, Gas Processes and High Energy Astrophysics
Volume II: Space, Time, Matter and Cosmology
By K. R. Lang

Observational Astrophysics 2nd Edition
By P. Léna, F. Lebrun and F. Mignard

Galaxy Formation By M. S. Longair

General Relativity, Astrophysics, and Cosmology
By A. K. Raychaudhuri, S. Banerji and A. Banerjee

Tools of Radio Astronomy 2nd Edition
By K. Rohlfs and T. L. Wilson

Atoms in Strong Magnetic Fields
Quantum Mechanical Treatment and Applications
in Astrophysics and Quantum Chaos
By H. Ruder, G. Wunner, H. Herold and F. Geyer

The Stars By E. L. Schatzman and F. Praderie

Gravitational Lenses By P. Schneider, J. Ehlers and E. E. Falco

**Relativity in Astrometry, Celestial Mechanics
and Geodesy** By M. H. Soffel

The Sun An Introduction   By M. Stix

Galactic and Extragalactic Radio Astronomy 2nd Edition
Editors: G. L. Verschuur and K. I. Kellermann

Reflecting Telescope Optics (2 volumes)
Volume I: Basic Design Theory and its Historical Development
Volume II: Manufacture, Testing, Alignment, Modern Techniques
By R. N. Wilson

A. M. Fridman · N. N. Gorkavyi

Physics of Planetary Rings

Celestial Mechanics of Continuous Media

Translated by D. ter Haar
With 125 Figures and 29 Tables

 Springer

Professor Alexei M. Fridman

Institute of Astronomy RAS
Moscow State University
48 Pyatnitskaya St.
Moscow 109017, Russia, and

Sternberg Astronomical Institute
Moscow State University
Universitetskii pr. 13
Moscow 119899, Russia

Translator

Professor D. ter Haar

P.O. Box 10
Petworth, West Sussex
GU28 0RY, United Kingdom

Dr. Nikolai N. Gorkavyi

Crimean Astrophysical Observatory
Goddard Space Flight Center, Code 685
Greenbelt, MD 20771, USA

Cover picture: Voyager-1 image of Saturn and its rings taken Nov. 16, 1980, four days after its closest approach to Saturn, from a distance of 5 300 000 km (3 300 000 miles). This viewing geometry, which shows Saturn as a crescent, is never achieved from Earth. The Saturnian rings, like the cloud tops of Saturn itself, are visible because they reflect sunlight. (Data available from U.S. Geological Survey, EROS Data Center, Sioux Falls, SD)

Title of the Russian edition: *Fisika planetnych kolets: Nebesnaya mechanika sploshnoĭ sredy.*
By N. N. Gor'kavyĭ, A. M. Fridman. © Nauka, Moscow 1994

Library of Congress Cataloging-in-Publication Data:
Gor'kavyi, N. N. (Nikolai Nikolaevich) [Fizika planetnykh kolets. English] Physics of planetary rings: celestial mechanics of continuous media / A. Fridman, N. Gor'kavyi; translated by D. ter Haar. p. cm. – (Astronomy and astrophysics library, 0941-7834) Gorkavyi's name appears first on the original Russian ed. Includes bibliographical references and index. ISBN 3-540-64864-X (alk. paper) 1. Planetary rings. 2. Celestial mechanics. I. Fridman, A. M. (Aleksei Maksimovich) II. Title. III. Series. QB603.R55 G6713 1999 523.9'8–dc21 99-19567

ISSN 0941-7834
ISBN 3-540-64864-X Springer-Verlag Berlin Heidelberg New York

© Springer-Verlag Berlin Heidelberg 1999
Printed in Germany

Text keyboarding: D. ter Haar
Data conversion and Page make-up: Steingraeber Satztechnik GmbH, Heidelberg
Cover design: *design & production* GmbH, Heidelberg

SPIN: 10630441   55/3144/ba-5 4 3 2 1 0 - Printed on acid-free paper

Preface by the Translator/Editor

When the Russian edition of the present book was published it contained testimonials from a galaxy of Soviet scientists who all praised the pioneering work of the authors; they pointed out how their analysis had led them to predict the existence of several new satellites which subsequently were observed by the Voyager spacecraft. The editor of the Russian edition, Academician B.B. Kadomtsev, lists these predictions as follows:

1. The existence of a series of small satellites beyond the outer edge of the Uranian rings: Nine out of the 10 new Uranian satellites were found in that region.
2. The impossibility of satellites being formed inside the Uranian ring zones: Only one extremely small satellite was found inside the outer edge of the rings.
3. The region of localisation of not yet discovered Uranian satellites between 50 000 and 82 500 km characterised by the lower 1:2, 2:3, and 3:4 Lindblad resonances with the rings: Eight out of the 10 new satellites were found in this region – they have resonances of the predicted kind with a correlation coefficient Q between the position of the rings and the lower resonances with the satellites being as high as $Q \approx 0.84$.
4. The orbits of five satellites each of which determines simultaneously the position of two rings: Four of the discovered satellites determine the position of two rings simultaneously and their orbits coincide with the orbits of the predicted satellites to an accuracy of better than 0.5%.
5. The presence of shepherd satellites for only the single external ϵ ring: Only near the single ϵ ring were shepherd satellites found.
6. An average satellite diameter of about 100 km: The average diameter of the discovered satellites turned out to be approximately 70 km.
7. The range of the albedo α of the satellites to lie between 0.03 and 0.3: It turned out that $\alpha \approx 0.05$.

Academicians Ambartsumyan, Ginzburg, Kadomtsev, and Obukhov compared their predictions with those of Le Verrier and Adams concerning the existence of Neptune while Academician Arnold compared them with those made about atomic properties on the basis of Mendeleev's periodic system of elements.

It has been a great pleasure for me to have been involved in the preparation of the English edition of this book by Professor Fridman and Dr. Gor'kavyi and thus to make it possible for the English-speaking world to follow the detailed analysis which was the basis for their predictions. It will be seen that they have been able to tell the story of the planetary rings in such a way that, on the one hand, the romance of the subject has been fully demonstrated and, on the other hand, the scientific challenge they present us with has been met and answered. The Russian edition was published in 1994 and for the present edition the authors have updated the material to take into account the latest observational data. I am sure that this edition will be very successful.

Western readers should bear in mind that it is a well established Russian tradition to list authors in alphabetical order. This means that whereas for articles in Russian journals Dr. Gor'kavyi's name appears before Professor Fridman's, the order is reversed for articles in English journals. The reason is that in the English alphabet "F" comes before "G", whereas in the Russian alphabet "Γ" comes near the beginning of the alphabet and "Φ" near the end.

Preface to the English Edition

We are grateful to Springer-Verlag for publishing an English translation of our book. The contemporary physics of the planetary rings is based upon observations obtained by the American Voyager-1 and Voyager-2 missions. During the last twenty years tens of American and European scientists have contributed to the theory of planetary rings. The translation of "The Physics of Planetary Rings" makes it possible for English-speaking readers to judge the models and calculations which have been generated by the authors during almost thirty years mainly in the Institute for Astronomy of the Russian Academy of Sciences in Moscow and at the Simciz Observatory in the Crimea. We believe that the present book will be of interest to

a wide circle of amateur astronomers. The Saturnian rings are one of the most impressive cosmic objects which can be seen by a telescope. We hope that the presence of some mathematics in the text will not prevent the reader from learning many interesting facts about the rings of Saturn and of other planets;

students of astronomy, physics, mechanics, hydrodynamics, mathematics, and computing. We solve in our book a number of interesting scientific problems, and many other problems are just mentioned or stated so that the present book can be of great assistance for future investigators of the Universe when choosing their subject of study;

astronomers, celestial mechanicians, and those involved in the Voyager and Cassini missions to which the most specialised parts of the book have been addressed;

specialists in the theory of collisions, stability, or resonances; specialists in the theory of linear and non-linear waves, and finally anybody interested in collective processes in hydrodynamics, plasma physics, solid-state physics, or optics.

As in any field of scientific studies, our theories are often not identical to those of other authors. However, we have very carefully studied alternative models of the rings and, although not agreeing with them, we are, in fact, much obliged to many of them. In particular, the celebrated Goldreich–Tremaine model which predicted up to 18 shepherd satellites in the Uranian rings was realised only for the outermost rings – the F ring of Saturn and the ε ring of Uranus which have two shepherd satellites near them. The origin of the other eight main Uranian rings turned out to be connected, as we predicted in 1985, with Lindblad resonances with new satellites beyond the outer boundaries of the rings. A similar situation occurs for the narrow ringlets in the C ring of Saturn which are connected with the small satellites at the outer boundary of the rings that were discovered by Voyager. However, the intrinsic logic of Goldreich and Tremaine's shepherd model, as well as its validity for Saturn's F ring and Uranus' ε ring, cannot be doubted and we have shown in Chap. 11 that, without playing an important role in the formation of the other rings, the two shepherd satellites of the ε ring are of fundamental importance for the equilibrium of the whole of the present system of the Uranian rings.

We feel that the development of the physics of planetary rings is greatly indebted to these two prominent scientists, P. Goldreich and S. Tremaine, as well as to many other investigators of the rings, both observers and theorists: K. Aksness, S. Araki, R. Beebe, D. Bliss, M.S. Bobrov, P. Bodenheimer, A. Boischot, N. Borderies, J.M. Boyce, A. Brahic, F.A. Bridges, G.A. Briggs, T. Brophy, J. Burns, H. Camichel, G. Colombo, A.F. Cook, D. Cruikshank, D. Currie, J. Cuzzi, M.E. Davies, A. Dobrovolskis, L. Dones, E. Duncham, R. Durissen, J. Elliot, V.R. Eshleman, L. Esposito, B.C. Flynn, F.A. Franklin, R.G. French, T. Gold, R. Greenberg, D. Gresh, K.A. Hameen-Anttila, C. Harris, W. Hartmann, A.P. Hatzes, E.A. Holberg, C.W. Hord, W.B. Hubbard, G.E. Hunt, A.P. Ingersoll. R.J. Johnson, O.V. Khoruzhii, A.L. Lane, D.N.C. Lin, J. Lissauer, P.-Y. Longaretti, J. Lukkari, K. Lumme, H. Masursky, E. Marouf, N. Meyer-Vernet, R.L. Millis, E.D. Miner, D. Mink, D. Morrison, P. Nicholson, T. Owen, J.C.B. Papaloizou, J.B. Plescia, J. Pollack, V.L. Polyachenko, C. Porco, P. Rosen, C. Sagan, H. Salo, J.D. Scargle, A.M. Shoemaker, M.R. Showalter, F. Shu, I.G. Shuchman, B. Sicardy, K.E. Simmons, B. Smith, L.A. Soderblom, G.A. Steigman, G. Stewart, E.C. Stone, R.G. Strom, S.P. Synott, T.A. Taidakova, R.J. Terrile, W.T. Thompson, J. Trulsen, G. Tyler, T.C. Van Flandern, J. Veverka, W. Ward, S. Weidenschilling, R.A. West, H. Zebker, and many others; their work has importantly contributed to our understanding of the dynamics of the planetary rings. We also want to note the classical results of C. Hunter, A.J. Kalnajs, C.C. Lin, D. Lynden-Bell, V.S. Safronov, and A. Toomre in the physics of plane gravitating systems, which include the planetary rings. Moreover, without the successful realisation of the Voyager project the present book would have been deprived both of the observational basis of a number of the models proposed here, and also of the observational confirmation of the theoretical predictions. We should therefore add to the above list the whole group of participants in that project. We express our gratitude to JPL/NASA for their permission to use some Voyager photographs.

The English edition appears five years after the Russian one. In this edition we have

1. Undertaken the correction of a number of errors and inaccuracies.
2. Introduced a refinement of the observational data.
3. Added many explanations which now make it possible for a non-professional reader to follow the consistency of most of the conclusions. As a result it turned out that practically all chapters have been expanded, especially Chap. 11 which is devoted to a discussion of the generation, evolution, and present state of the Uranian rings. In this chapter we have, in particular,

> considered in detail the mechanism for the accumulation of the ring particles in the lower Lindblad resonances of predicted new satellites of Uranus and we discussed the conditions under which this accumulation would occur;

> calculated the balance of the angular momentum of the present Uranian rings, proving the equilibrium of the observed ensemble of rings, without requiring new "invisible" satellites inside the ring zone;

> established for each of the 10 narrow Uranian rings its "curriculum vitae" – determining where and when they were born, indicating the resonance responsible, considering the further drift evolution of the ring and the reason why it is now stable.

The English edition of this book is for the authors the final result of almost thirty years of active study of the planetary rings. We have intellectually enjoyed the study of the mechanisms behind the origin and evolution, the stability and instabilities of the broad Saturnian rings and of the narrow Uranian and Neptunian ringlets. Nature has been shown to have an enormous potential for constructing dynamic puzzles. We believe that anyone who reads the present book will agree that the planetary rings are a unique and balanced construction testifying to the quality of Nature's dynamic refinement.

O.V. Khoruzhii and V.L. Polyachenko have greatly and with great expertise assisted us in the preparation of the English edition. Professor A.A. Boyarchuk has helped us with advice for the revision of Chap. 14. We express our deep gratitude to these gentlemen.

The resonance interaction with satellites plays an important role in the dynamics and the formation of structure of the planetary rings. We had the opportunity to discuss these problems with Professor G. Contopoulos, a pioneer in modern dynamic astronomy and the theory of resonance phenomena. Professor Contopoulos' support greatly assisted us in solving the problem of an English edition of the present book.

In conclusion we wish to note the huge, skilled task accomplished by Professor ter Haar in translating and editing the English edition of our book. We do not need to remind the reader of Professor ter Haar's achievements, and his profound knowledge of various fields of physics, mathematics, and mechanics is well known. From all this it is clear that any improvement of the English edition in comparison to the Russian original is to a large extent due to its outstanding translator/editor.

April 1999   Alexei Fridman, Nikolai Gorkavyi

Preface to the Russian Edition

With the Saturnian rings, which have been known for about four centuries, the Uranian, Jovian, and Neptunian rings form a new and interesting class of objects in the Solar System; they were discovered and extensively studied between 1977 and 1989. The present book is the first in the world to contain a review of the four ring systems of the major planets and an exposition of the basis of the physics of the planetary rings. The theoretical models described here explain the origin of the rings, the stratification of the Saturnian rings, the formation and stability of the narrow Uranian ringlets, and the dynamics of the striking Neptunian "arcs".

During the past century the cosmos has become the object of studies, and even the place of work, of many people so that such an uncommon element of our cosmic home as the rings of the major planets attracts close attention both from scientists and from amateur astronomers whose poetic attention is not governed by formulae, scientific competition, or pecuniary advantages.

The authors have tried to make the contents of the present book at the same time rigorous in the scientific sense (for the professionals), as detailed as possible (for students – future astronomers and planetary physicists), and as light-hearted as possible – for the amateurs who are, of course, the best and most numerous part of mankind.

As all "ring-students" the authors are utterly convinced that the planetary rings are an important element in the understanding of the origin and the dynamics of such disc-shaped objects as satellite systems and solar systems, as well as accretion and galactic discs. They present the possibility to try out the theory of viscous discs on the unique proving ground of planetary rings – the only disc object about which we know far more than we understand. The information avalanche – of 115 thousand photographs of the work of some twelve years and the discovery of 27 new satellites – was basically due to the American Voyager-1 and Voyager-2 twins which visited the huge orange Jupiter in 1979 and the smokey golden Saturn in 1980 and 1981. Voyager-2 had a single fly-past of the azure Uranus in 1986 and of the spotty blue Neptune in August 1989.

The Voyager spacecraft have transformed the whole of planetary physics. The satellites of the major planets which should have been dead worlds with crater pockmarks turned out to be very restless provinces of the Solar System. The extremely powerful sulfuric vulcanoes on the Jovian satellite Io are surrounded by liquid lakes of red and black sulfur; the remarkable Europa – yet another Jovian satellite – looks like a cracked billiard ball and is covered by thick ice over an ocean of water; Saturn's largest satellite, Titan, is blanketed by a thick nitrogen atmosphere under which there is possibly a methane–ethane ocean; the tiny Uranian satellite Miranda – a film star from the Voyager photographs – is literally ploughed up by unexpectedly active tectonic processes; and, finally, there is the pinkish Neptunian giant satellite Triton with a small nitrogen atmosphere and splendid 10 km geyser plumes, blown by a nitrogen wind at a temperature of $-235\ °C$.

If we bear in mind that these bewitching pictures are illuminated by the Sun and also by the huge disc of the planet which takes up half the horizon of the sky of many satellites we can only envy future space travellers. It is difficult to enumerate all the surprises presented by the rings. The Saturnian rings are stratified into thousands of ringlets and the Uranian rings are compressed into narrow streams, which for some reason or other differ from circular orbits like the wheel of an old bicycle. The edge of the rings is jagged and the rings themselves are pegged down under the gravitational pressure of the satellites, bending like a ship's wake. There are spiral waves, elliptical rings, strange interlacing of narrow ringlets, and to cap it all one has observed in the Neptunian ring system three dense, bright arcs – like bunches of sausages on a transparent string. For celestial mechanics this is a spectacle as unnatural as a bear's tooth in the necklace of the English queen.

The flight of Voyager-2 has changed planetary physicists into horse racing fanatics, betting on what Voyager's TV cameras will show near the next planet. The astronomers watched with bated breath during the 1986 and 1989 fly-pasts.

In the realm of planetary rings the idea of shepherd satellites competed with that of self-organising rings. The first developed from classical considerations of the rings as accumulations of passive particles – the sheep. When order appears – some kind of structure – "shepherds" were needed, satellites positioned in the neighbourhood and actively acting on the ring using the gravitational field as a "whip".

The second idea, which the authors of the present book adhere to, was developed from the formalism of the gravitational physics of collective processes which twenty years earlier led in the dynamics of galaxies and stellar clusters to a shift in theoretical mechanics (see Fridman 1975 and Fridman and Polyachenko 1984). In the dynamics of planetary rings the physics of collective interactions was supplemented by taking collisions between particles into account. One was led to study a kinetic equation with a rather complex collision integral – because the collisions are inelastic – which later on made it possible, both by using the Chapman–Enskog method and by using the solution of the kinetic equation for a plasma in a magnetic field, to reduce it to a closed set of (hydrodynamical) moment equations. In that sense the physics of planetary rings is in many respects the celestial mechanics of continuous media studying the dynamics of a gas of inelastic particles in the field of gravitational and centrifugal forces – the approximate equilibrium of these two in the case of a continuous medium of inelastic bodies inevitably means the formation of a rotating disc around a central body. The main difference in the idea of self-organising rings is the ability of the rings to form various structures without the interference of external forces such as shepherd satellites.

A reader resolving to go on a excursion along the proving ground of the planetary rings still covered by the fragments of destroyed theoretical constructions will find in the Introduction, among many other delights, a short exposition of the long history of studies of the planetary rings, and in Chap. 2 the present-day data about all four planetary ring systems and about related discs. At the end of that chapter we give a comparative analysis of all these systems. We present in Chap. 3 those data from celestial mechanics which, in our opinion, are useful to know in order to understand the present book. We get in Chap. 4 closer to a study of the planetary rings, starting – not because it is simple, but because it is fundamental for the later models – with the most elementary study of collisions between the macroparticles which form the planetary rings – as well as a number of other disc systems. Since the present book is devoted to a realistic "natural" object whose "vitality" began long before terrestrial science split up into different disciplines, for the study of the physics of the rings we must turn to completely different methods and theories: to celestial mechanics and plasma physics, to shock theory and the theory of gases, to the physics of ice and snow, and to the theory of instabilities. We shall use analytical calculations as well as astronomical observations, and also numerical as well as actual experiments. Chapters 5 and 6 are also devoted to the dynamics of separate particles – more precisely, to problems which can be studied in the framework of models for separate particles. In Chaps. 7 to 9 we change to a study of collective and resonance phenomena in planetary rings. In Chaps. 10 to 12 we study the actual Uranian and Neptunian ring systems and the titles of their sections indicate their detailed content. We make an attempt in Chap. 13 to apply the experience obtained in studying planetary rings to the cosmogony of the Solar System. Chapter 14 describes a number of space projects and also contains some information about the scientific infrastructure.

The present book is based on the work carried out by the authors and their collaborators. The material of Sect. 13.2 and parts of Sect. 8.2 was obtained in collaboration with V.L. Polyachenko. Chapter 14 is based on work carried out with V.A. Minin as collaborator. T.A. Taĭdakova is the author of all numerical programmes used in the present book and is the coauthor of the papers on which Chap. 6 and Sects. 5.2, 10.2, 12.2, and 12.3 are based. Appendix I was written with O.V. Khoruzhiĭ and A.S. Libin and Appendices II and V with O.V. Khoruzhiĭ as coauthors.

We are grateful to V.A. Ambartsumyan, V.I. Arnol'd, V.V. Beletskiĭ, M.S. Bobrov, A.A. Boyarchuk, T.G. Brophy, B.V. Chirikov, D.G. Currie, J.N. Cuzzi, A.E. Dudorov, A.M. Dykhne, T.M. Eneev, L.W. Esposito, A.A. Galeev, V.L. Ginzburg, O.V. Khoruzhiĭ, L.V. Ksanfomaliti, G.A. Leĭkin, M.L. Lidov, J.J. Lissauer, M.Ya. Marov, V.A. Minin, A.G. Morozov, A.I. Morozov, A.M. Obukhov, V. Petrosyan, V.L. Polyachenko, V.S. Safronov, I.G. Shukhman, B.M. Shustov, R.I. Soloukhin, Yu.M. Torgashin, G.L. Tyler, A.V. Vityazev, A.V. Zasov, Ya.B. Zel'dovich, I.N. Ziglina, and our colleagues at the Astronomical Institute of the Russian Academy of Sciences, the Simeiz Observatory, the Institute for the Physics of the Earth, the Institute of Applied Mathematics, and the Space Research Institute for their interest in this work and for useful discussions.

The authors express their deep gratitude to the editor of this book, B.B. Kadomtsev, and the reader, V.B. Beletskiĭ, for their efforts to improve the text, to D.V. Bisikalo and N.M. Gaftonyuk for their participation in the preparation of the programmes and the carrying out of the numerical calculations, and also to S. Filikov, S. Kryuchkov, and L. Kokurin for their help in the computer composition of the text. The authors thank L.S. Shtirberg and the laser location group for support and providing computer time.

In conclusion we want especially to express our gratitude to the person who supported our first research on collective processes in gravitational system, the person as modest as his inheritance is rich – Mikhael Aleksandrovich Leontovich. He left behind elegant classical results in various fields of theoretical physics and a distinguished scientific school that he created. But, however large this part of his inheritance, immeasurably more valuable for us is something else – his vision of what is valuable in this world. (Einstein once remarked that Dostoevskiĭ's novels gave him more than Lorentz's papers.)

It is more than ten years since Mikhael Aleksandrovich died. And now in a time which is difficult for all of us we recall his integrity and courage, his uncompromising behaviour, his implacable opposition to hypocrisy, his sarcasm when he saw toadyish or presumptuous behaviour, and his childish joy when understanding beautiful scientific ideas whosoever they were. Whoever knew him well could only love him.

We dedicate the present book to Mikhael Aleksandrovich Leontovich – in our opinion the standard of the scientific conscience.

Table of Contents

Appendices

1 Introduction

Fool: "The reason why the seven stars are no more than seven is a pretty reason."
Lear: "Because they are not eight?"
Fool: "Yes, indeed: thou wouldst make a good fool."

W. Shakespeare, King Lear, Act I, Scene V.

1.1 Rings as Characteristic Features of Astrophysical Discs

Practically anything which rotates – and in our Universe practically every thing rotates – sooner or later becomes a disc. The flattening of rotating systems is the result of the opposite action of two basic dynamic forces – the gravitational and the centrifugal forces. The gravitational attraction tends to compress the system in all directions while the rotation, if it is sufficiently fast, prevents the compression at right angles to the rotation axis. As a result of a compromise the system is forced to flatten along the axis of rotation, the z-axis (Fig. 1.1). This is the genealogy of galactic discs, accretion discs in systems of close binary stars, proto-planetary clouds near single stars, proto-satellite discs, and planetary rings.

Exceptions are systems with a large velocity dispersion or a large pressure, such as elliptical galaxies, spherical star clusters, and the stars themselves. In those objects the pressure inhibits flattening in the z-direction. Of course, one must distinguish collisional from collisionless systems. In spherical systems with dissipation or inelastic collisions the high thermal velocities of the particles decrease through friction or shocks and this leads to the formation of a disc. This process is much slower in almost collisionless stellar systems. If the latter have a relatively large rotational angular momentum, it is in principle possible that they change into a disc, provided there are various instabilities present which develop naturally on a dynamic time scale. This means that stable systems do not evolve into a flat disc.

Elliptic galaxies and elliptic subsystems of old stars in spiral galaxies – and also the spherical halo – have a small rotational angular momentum. Their shape is caused by the anisotropy of the dispersion of the stellar velocities along the symmetry axis. The main cause for this are the stars which are rescued from compression and flattening in the z-direction through thermonuclear "doping".

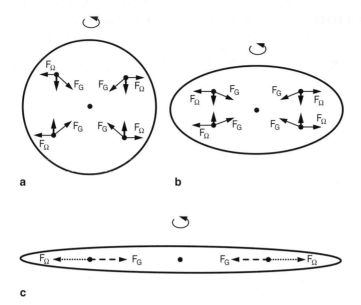

Fig. 1.1. The scenario of flattening in the z-direction of rotating gravitating systems. The fact that the gravitational force F_G and the centrifugal force F_Ω do not cancel one another leads to a compression of the system along the z-axis.

We shall be interested only in thin astrophysical discs with a ratio of the thickness h to the radius R much smaller than unity: $h/R \ll 1$. This ratio is characteristic for all galactic, circumstellar, and circumplanetary discs mentioned above, except hot and thick accretion discs which are more like a doughnut or a torus. The fact that h/R is small means that the discs considered by us are rather cold and that the pressure gradient in them is much smaller than the two main forces – the gravitational and the centrifugal forces. Hence, we can in the equilibrium equations neglect the pressure – or the velocity dispersion. However, in the linearised equations describing the perturbed quantities the pressure gradient turns out to be one of the main terms. To a first approximation there occur in the equilibrium conditions of the two main forces also such actively operating factors as thermal and viscous processes, the self-gravitation of the disc, and electromagnetic effects. The mutual cancellation of the centrifugal and gravitational forces in the disc, which releases into the dynamic arena a whole spectrum of much weaker mechanisms, means that cold thin discs are energetically evolving dynamic systems.

The fact that the ratio h/R is small is important also in another respect. We shall be interested in spatially regular structures, such as the annular stratification which appears in thin discs. The characteristic scale L of such structures is "squeezed" naturally between the basic scales of the disc – the thicknesss h and the radius R: $h < L < R$. It is clear that the smaller the ratio h/R the more characteristic scales one can fit into the range between h and R and the richer the possibilities for the formation of structures in a disc system.

The next characteristic property of the disc systems considered here is differential rotation. Circumstellar and circumplanetary discs move round a massive central body practically exactly following Kepler's law, that is, each layer of the disc of radius R rotates faster when it approaches the central body and the angular velocity increases as $R^{-3/2}$. Such discs are often called "Keplerian" discs. Flat galaxies do not have such a clearly marked relatively massive centre and the degree of differential rotation varies strongly in them. To understand the role of the differential rotation one must note that circumstellar and circumplanetary discs have an appreciable viscosity. This, in conjunction with the differential rotation, leads to the production of a constant energy source in the discs. This is a simple effect, physically speaking: a differentially rotating disc is a typical case of a large-scale shear flow. The viscosity transfers energy from this flow (that is, the orbital energy of the disc) into heat – the random motion of the particles in the disc. This heat is released in any shear (in fact, simply in any real) motion, for instance, it raises the temperature of brake shoes in a car in the braking process, the sea heats up in a storm, and one gets pleasantly warmer by rubbing one's hands.

The power of the energy source in differentially rotating discs may be very small in magnitude – the thermal power of the Saturnian rings is only of the order of 100 kW – but its action is extraordinarily important for the dynamics of the discs. The development of various instabilities in the systems, and the formation of regular spatial structures – "self-organisation" of the systems – necessarily requires that the system is "open", or, simply expressed, that it requires an energy source. If we pour oil into a frying pan, nothing whatever happens – any housewife will tells us this – until we put the frying pan on the stove, that is, supply energy. Only then can one produce in the frying pan the famous convective Bénard cells provided we follow the correct procedure.

The self-organisation of the ocean proceeds only under the action of the Sun which causes temperature gradients in it and also motions in the atmosphere. As a result we have a variety of surface waves, oceanic vortices, and undercurrents.

The fact that viscous dissipation produces a specific long-term energy source in differentially rotating discs significantly distinguishes the continuous medium of astrophysical discs from the usual gaseous or fluid medium and makes the "cosmic fluid" of the discs a highly active medium. By the celestial mechanics of a continuous medium we can understand the dynamics of a viscous fluid in the gravitational field of a massive body when there is equilibrium between the gravitational and centrifugal forces. By the way, as in any thermal engine, the differentially rotating discs have "refrigerators" where the spent heat is dumped. Most often the energy of the orbital rotation which is converted through the viscous dissipation into the energy of the random motion of the particles, afterwards through inelastic collisions of the macro-bodies is spent in heating the matter itself and the subsequent infrared emission into the surrounding space. If the disc is gaseous the inelastic collisions between the atoms and molecules directly produce the emitted radiation.

We draw attention to one more characteristic feature of astrophysical discs – they are systems with a long lifetime. It is natural for rotating systems to calculate the lifetime by the number of rotations; all observed disc systems have in this way a very considerable age: the "youngest" are the galactic systems with tens of rotations. The age in terms of rotations is an important characteristic of the dynamic maturity of a system and of the stability of the observed structures.

We have discussed the general characteristic features of astrophysical discs and we can at once note that planetary rings are typical examples of astrophysical discs. The rings are thin, differentially rotating, viscous, and long living discs consisting of inelastic macroparticles. However, we have here several times omitted the qualification "the most" as we shall discuss in the next subsection.

1.2 The Planetary Rings as Unique Disc Systems

Typical for all rings is a whole range of features distinguishing the planetary rings from other disc systems.

a) They are the thinnest of all astrophysical discs: $h/R = 10^{-6}$. According to this parameter a sheet of paper is much thicker than the planetary rings which are so thin as the very sensitive result of the scenario considered earlier of flattening in the z-direction.

b) The energetic conditions for self-organisation in the planetary rings are extremely favourable. If we disregard the inelastic collisions which cool the disc, the temperature – or the velocity dispersion of the particles in the disc – of the rings will increase at a very fast rate – faster than in other cosmic discs (see Sect. 2.8).

c) They are the most ancient disc systems, existing already about trillions (10^{12}) of rotations. In comparison with planetary rings the galaxies are one-day embryos besides a fossil just dug up with an unimaginable age of 100 million years.

d) They are the most studied kind of disc systems: the American Voyager 1 and Voyager 2 spacecraft alone transmitted in the period from 1979 to 1989 thousands of photographs of the rings of Jupiter, Saturn, Uranus, and Neptune with a resolution of a few kilometers; and this is apart from observations with ground-based and satellite observatories, probing of the rings by radio and stellar occultations, and polarisation, spectral, and other studies.

1.3 The Planetary Rings as a Proving Ground for Theorists

An obvious truth is that the less information there is about an object the more freedom the theorists have and the more varied and contradictory theories can be cultivated in the field of the sparse observational data. The enormous volume of information, obtained during the last 10 to 12 years about the four systems of planetary rings have, on the one hand, stimulated an extraordinary curiosity of researchers but, on the other hand, has placed very tight constrictions on the theoretical constructions. The American researcher Terrile has compared the difficulty of interpreting such a quantity of information with attempts to drink from a fire hose. This hyperinformation syndrome arises from the lack of balance between the observational abundance and the theoretical indigestion. Such a lack of balance precipitates the theorists into a kind of "gold fever" and forces them into a headlong construction of more and more speculative constructions to withstand the information "earthquake" caused by the Voyagers. The interest is due, on the one hand, to the great variety and unexpectedness of the dynamic effects of the rings themselves and, on the other hand, to the fact that a reliable theory of the planetary rings would become the basis of the celestial mechanics of continuous media or of celestial collective dynamics and would enable us to consider from a completely different perspective the evolution of other cosmic disc systems – first and foremost the development of our Solar system.

The planetary rings were for reseachers of planetology and mechanics, and for physicists a gift of fortune – a true proving ground for theoretical models just as a living dynosaur would be for a paleontologist. Tens of theorists from the USA, France, England, Germany, Finland, and the USSR rushed to this proving ground armed with their own specific ideas, but most often their elegant and well studied theories encountered the cold reality and quite rapidly were pulled to pieces. Particularly shattering was the devastation of a number of theoretical ideas following the fly-past by Voyager 2 of Uranus in January 1986 and of Neptune in August 1989. These dramatic events will be reflected in our book. One can easily understand the feeling of a researcher whose theoretical offspring – worked out, published, and contributed to conferences – perishes when he sees on the display screen a photograph just obtained by Voyager. Of course, this is cruel, but natural selection made it possible to distinguish that set of theoretical points of view which is able to explain the observed effects and to form the basis of the new physics of planetary rings. One must say at once that attempts to understand the rings purely on the basis of classical celestial mechanics – to represent them simply as an accumulation of passive test particles in the field of a few active gravitating centres (the satellites) – were defeated.

It turned out that the correct conception of the rings is the one which takes into account the collective properties of the particles, the presence of a particular source of energy, and the freedom to self-organise in differentially rotating viscous discs. The specific dynamics of the active medium of the planetary rings forced us to work out special methods for studying them and led to the reconsideration of a whole number of settled opinions. We note that the transition from classical mechanics to collective dynamics started to come to the physics of galaxies in the sixties (Fridman 1975; Fridman and Polyachenko 1984). The transition from N-body mechanics to a new collective gravitational physics – more precisely to a self-consistent field method: the solution of the set of Boltzmann-Vlasov equations, that is, Boltzmann's kinetic equation and the Poisson equation – was made easier by the fact that one needed the collisionless approximation for the overwhelming majority of dynamic problems in a galactic disc.

In the present book we summarise the work of theorists – these perpetual hostages to a peculiar curiosity, the stalkers in the fascinating and difficult proving ground of the planetary rings. However, before that we shall give a small historical sketch of the study of planetary rings from Galilei to the present.

1.4 Historical Journey

The study of our Solar system underwent three qualitative jumps, three epochs in which there was a cardinal change in our ideas. The first epoch is the 30 years period from 1492 to 1522 of large geographical discoveries when through the efforts of the expeditions of Christopher Columbus (1451–1506), Vasco da Gama (1469–1524) and Fernando Magellaen (1480–1521) the fundamental characteristics of our Earth – the planet of our Solar system which is the most important one for us! – were determined: its true size and the existence of a single ocean, while at the same time the large continents were discovered.

The second period started with the invention of the telescope and the third is dated by the start of cosmic flights. In describing the historical events of the telescopic era we follow the accounts of Grebenikov and Ryabov (1984) and of Silkin (1982), as well as the interesting paper by Pasha (1983). The Dutch optician Hans Lippersheij made the first telescope in 1608. Hearing this the professor of mechanics at the University of Padua, Galileo Galilei (1564–1642), constructed his own telescope and made a number of interesting discoveries: he discovered the mountains on the Moon and the phases of Venus. From the 7th to the 13th of January 1610 Galilei discovered the four large satellites of Jupiter: Io, Europe, Ganymede, and Callisto.[1] Galilei published these discoveries (together with the observation of the mountains on the Moon) in a book

---

[1] Galilei not only observed Jupiter's satellites, but he also carefully studied their motion in order that one could predict the position of the satellites with the accuracy necessary for navigators. Galilei thought that for the navigators the

appearing in the middle of March of the same year, 1610 (one can only envy such an efficiency of the publisher!) called portentously: "Stellar bulletin, disclosing important and to a high degree astonishing spectacles ...". The history of the Saturnian rings also started in 1610. Galilei, turning his telescope to Saturn in July 1610 published a coded message: "I observed a triple, removing itself from the planet". The Saturnian rings appeared in Galilei's telescope as two hazy spots at the edge of the planet. Galilei compared the "branch stars" of Saturn with obedient attendants which assisted the aged Saturn to move and always supported him each on its own side. The coded message was not for nothing: Johannes Kepler's (1571–1630) "New Astronomy ..." had already appeared in 1609 and the laws of celestial mechanics categorically denied the possibility of a fixed position of satellites next to the planet.[2] The "abnormal" Saturn continued to give trouble – in 1612 the rings turned their edge to the Earth and disappeared from the sight of the astronomers. A wise Galilei pondered "Has Saturn eaten his children?" and resolved that one should wait and the "children" would reappear. Indeed, in 1614 the Jesuit Christopher Scheiner, professor at the University of Ingolstadt, saw the "branch stars" of Saturn in his telescope, and in 1616 Galilei himself saw them, while between 1630 and 1660 Pierre Gassendi, Francesco Fontana, Giovanni-Batista Riccioli, and Johannes Hevelius observed them. And although different sketches showed ring features nobody succeeded to unravel the secret of the Saturnian ornaments, notwithstanding the fact that Gevelius observed a periodicity in the changes of the phases of the visibility.

The laurels went to the best telescope builder of that period, the Dutchman Christian Huygens (1629–1695). This young researcher discovered in 1655 the largest satellite of Saturn, Titan, and afterwards published a coded sentence about Saturn: "It is girdled by a thin flat ring, nowhere touching, inclined to the ecliptic." Huygens decoded his sentence and explained the periodicity of the vanishing of the rings in his "The System of Saturn", published in 1659. In it he also predicted future dates for the disappearance of the rings: July 1671, March 1685, and December 1700. This work turned out to have the effect of an exploding bomb in the European circles of scientists and provoked violent criticism from the orthodox. To understand the reluctance of the educated astronomers of the seventeenth century to believe in a ring, fixed to Saturn, one must take any telescope and use it to look at

---

Jupiter satellites might be more convenient than the planets because of their shorter orbital period – the satellite Io at a distance of 4.2×10^{10} cm from Jupiter has an orbital period of 42.5 hours.

[2] One should note that Galilei never read the huge "New Astronomy" book which Kepler had sent to him because of the "ponderous Latin of the author". Right from the start he treated Kepler's laws with prejudice, denying the possibility that the planets moved along ellipses. Galilei was convinced that the planets could only orbit uniformly, that is, move along circles. By referring to Kepler's book we wanted to draw attention to the level of development of the ideas about the motion of planets and satellites in 1610.

the object of our discussion. The sight of a magnificent ring shocks the un-
accustomed eye and it is in shape and by impression associated only with a
flying saucer. On the other hand, not even the most magnificent photograph
of the rings can be compared with the stunning sight of the original rings. Of
course, this may just be our subjective impression. Let us, though, return to
the seventeenth century. The rings properly disappeared in accordance with
Huygens' predictions, confirming the correctness of the great Dutchman. This
started the long history of the study of the Saturnian rings which have been
constantly excited the imaginations of researchers by their unique shape.

The Saturnian rings have been studied by such brilliant astronomers, mechanicians, and mathematicians as Galilei, Huygens, Kant, Cassini, Laplace,
Maxwell, and Poincaré. Here we shall merely briefly note the most important
events:

1664. Giuseppe Campani discovered the rings: the outer one the darker and
the inner one the brighter – the A and B rings.

1675. Giovanni Dominico Cassini (1625–1712), director of the Paris Observatory, discovered the division between the two main Saturnian rings which
was later called after him.

1715. A paper by Jacques Cassini, who directed the Paris Observatory after
his father, appeared. In it the opinion was expressed that the rings were an
"accumulation of satellites" rotating in one plane around the planet which
are too small to observe the gaps between them.

1730s. Maupertuis suggested that the rings are formed from the tails of
comets captured by Saturn. Meran and Buffon, colleagues of Maupertuis in
the Paris Academy, assumed the rings to be the remnant of the equatorial
matter of Saturn. Meran assumed that Saturn had been compressed and the
rings had remained while Buffon thought that the rings where separated from
the planet through an excess of centrifugal force.

1739. The dark interior C ring was discovered by the Englishman Thomas
Wright.

1755. In his "Allgemeine Naturgeschichte und Theorie des Himmels" (General Natural History and Theory of the Heavens) Immanuel Kant (1724–1804)
was the first to predict the existence of a fine structure of the Saturnian rings.
He envisaged the ring as a plane disk of colliding particles moving according
to Kepler's law. According to Kant it is the differential rotation which causes
the ring to break up into narrow ringlets. Kant wrote: "However little the
particles obstruct one another, if we take into account how far apart they
are in the plane of the ring, it is probable that the lagging behind of the
particles which are farther away in each of their orbits must gradually slow
down and retard the faster moving particles which are closer to the planet;
on the other hand, the latter must transmit their faster orbiting to the particles which are further away; if this interaction were not interrupted this
process would continue until all particles in the ring – both the nearer and
those farther away – would orbit at the same rate. However, in the case of

such a movement the ring would disintegrate completely. On the other hand, one need not fear that such a disruption would happen. The mechanism for creating the motion of the ring leads to a stable state for the same reasons that would lead to its disintegration; this state is reached by the ring being split into a few concentric circular bands which lose contact with each other because the gaps between them are separated. ... I expect – and this gives me great satisfaction – that the actual observations somehow will confirm my assumption."

It is opportune to comment briefly. Kant's prediction about the subdivision of the rings was confirmed only after 225 years. We note that one of the instabilities producing such a subdivision is the negative diffusion instability which, indeed, causes an effective decrease in the viscosity of the disc (see Sect. 8.2) although it does not make it vanish completely. We note that Kant was wrong in assuming that the viscosity in the disc leads in the course of the evolution to a rigid-body rotation since the result of the viscous interaction will only be a spreading of the disc in radius. Indeed, particles which are accelerated in viscous collisions go to higher orbits and acquire a slower orbital motion. We consider in Sect. 3.3 the paradox of the dynamics in a gravitational field that as the result of acceleration the velocity decreases and as the result of deceleration it increases.

Kant was too far ahead of the level of the astronomy of the eighteenth century and his work remained relatively unknown. Thirty or forty years after the publication of Kant's work Laplace (1749–1827) discussed a much more artificial model of the Saturnian rings. We now continue our discussion of the events concerning Saturn.

1760. The Scot James Short saw the Saturnian rings divided into several ringlets. This made a strong impression on the astronomers, such as Laplace.

1796. Laplace's "Exposition du Système du Monde" (Exposition of a World System) was published. Laplace proved the instability of a solid broad ring which must inevitably disintegrate in the field of the planet and he put forward the hypothesis that Saturn's rings consist of a set of narrow solid rings which enclose one another (Laplace 1789). Laplace wrote: "The rings around Saturn are continuous solid bodies with varying widths at different points of their circumference ..." The variable width of narrow rings is, according to Laplace, "necessary to maintain the ring in equilibrium round the planet since, if it were completely uniform in all its parts its equilibrium would be destroyed by even the smallest force ... and the ring would fall onto the planet."

1837. Johann-Franz Encke discovered along the outer edge of the ring a narrow division – the so-called Encke Division.

1848. The Frenchman E.Roche calculated the force of the tidal disintegration of a satellite and found that near the planet self-gravitation of the satellite cannot keep it intact. Roche expressed the hypothesis that the rings are the result of the disintegration of a large satellite near Saturn.

1851. The English astronomer William Dawes observed in the Saturnian rings a "number of narrow concentric bands, each of which was somewhat brighter than the adjacent inner band. One can clearly distinguish four bands. They look like steps leading down to the black valley between the ring and the planet." Earlier, the Englishman Henry Cater (in 1825) and the Roman priest Francesco de Vico (in 1838) had seen various bands in the rings.

1855. The director of the Harvard Observatory, William Bond, and his son John Bond noted when observing that "The inner bright ring is minutely divided into a large number of smaller ringlets." The division started at the inner edge and occupied about two thirds of the width of the ring. "At a 800 fold magnification the lines were completely clear but they weakened in the direction of the outer edge. Their form was comparable to a series of waves. The tops corresponded to the narrow rings and the bottoms to the divisions" (see Plate[3] 1). However, according to the English astronomer R.Proctor the Saturnian rings should look as shown in Plate 2 – as a comparison we give in Plate 3 the computer representation of these rings, constructed using the data from Voyager-1.

1856. Laplace's problem of the stability of a solid ring round a planet was again considered by the future creator of electrodynamics, James Clerk Maxwell (1831–1879), who received the prestigious Adams Prize of Cambridge University for a paper (Maxwell 1859) in which he showed that a narrow solid ring is stable only if one "sticks" to it a satellite which is 4.5 times as massive as the ring. Maxwell concludes: "Since the shape of the rings does not indicate anything like this or that there are the prerequisites for such a large non-uniformity in the mass, the theory that the rings are solid becomes implausible. ... Such a theory makes no sense." And finally: "The only viable system of rings reduces to an innumerable number of un-correlated particles orbiting around the planet with different velocities ... These particles may be assembled into a system of narrow rings or they may move one after the other without any particular regularity. ... The tendency of the particles to coagulate into narrow rings ... retards the disintegration process."

One of the authors of the present book recalls how Academician Budker asked a leading expert in the theory of plasma stability: "Do you know what is the first paper on collective processes?" and, after having waited so long that the person to whom he had addressed this question could fully feel the depth of his ignorance, he said: "It is Maxwell's paper on the stability of the Saturnian rings", and after a silence he added solemnly: "The great Maxwell." Indeed, this study is the first paper on the theory of collective processes written at a contemporary level: Maxwell used in his stability analysis a characteristic equation which is nowadays called the dispersion equation.

We note that Maxwell's correct inference about the meteoritic structure of the Saturnian rings was the result of an incorrect conclusion (following

---

[3] All the plates can be found together on pages 15 to 20.

Laplace) in his paper that a continuous ice ring would collapse onto the planet. This conclusion relates only to a ring with an infinite strength; Fridman, Morozov, and Polyachenko (1984; see also Gor'kavyĭ, Morozov, and Fridman 1986) have shown that a ring consisting of water ice (or any other real substance) much before collapsing disintegrates into lumps due to the small-scale bending instability in the plane of the ring.

1895. Belopol'skiĭ of the Pulkovo Observatory, Keeler of the Lick Observatory, and Delandre in Paris used the Doppler effect to measure the orbiting velocity of the rings. These measurements showed the differential rotation in accordance with Kepler's law which agreed with the many-particle model of the rings. The hypothesis of a meteoritic structure of the Saturnian rings, first expressed by Jacques Cassini, was thus observationally confirmed at the end of the nineteenth century.

During the twentieth century new data about the planetary rings were gradually piling up: the Soviet astronomer Bobrov (1910–1990) in constructing a theory about the eclipsing of the particles of the rings by each other, was the first to obtain reliable estimates about the size and density of the particles in the Saturnian rings (Bobrov 1970) and the Frenchman Camichel (1958) discovered in 1958 the puzzling effect of the azimuthal variability of the brightness of the A ring of Saturn (for more recent observational data of this effect see Thompson ct al. 1981). The first measurements of the visible thickness of the ring from the edge were made from the observations of Kiladze of the Abastumani Observatory and of Dolfuss and Focas of the Pic du Midi Observatory during the period when the Earth passed through the plane of the Saturnian rings in 1966: 1 to 3 km (Bobrov 1970). Kuiper and coworkers (Kuiper, Cruikshank, and Fink 1970) concluded in 1970 from their observations of the infrared spectrum of the Saturnian rings and a comparison with the spectrum of hoarfrost that the Saturnian rings consisted of ice.

We must also note the following events:

1970. Bobrov's "The Rings of Saturn" (Bobrov 1970) was the first book devoted to planetary rings.

At that time the researchers of the rings on the terrestrial globe could be counted on one's fingers. This gentle scientific activity ended on March 10th, 1977. On that day a new stage started in the development of the study of the outer planets; it was connected with the appearance of modern equipment and space probes.

However, before we give a brief exposition of recent years which are saturated with events we simply need to make an important observation about the results of the studies of the rings in the previous period from 1610 to 1977. Models of the rings were developed on the background of Newtonian celestial mechanics which essentially reduced to solving the problem of several gravitating point particles, usually three: the Sun, the planet and an asteroid, or the planet and two satellites, and so on. The path of science (after Newton) during the last three centuries as far as the motion of celestial bodies was concerned was, indeed, a triumphal procession of the celestial mechanics paradigm of N bodies; exact solutions were found and profound mathematical methods were developed. The model of several point particles corresponded badly only to a single entity in the Solar system – the Saturnian rings. However, it was noted by Kuhn (1977) that "they try to 'squeeze' Nature into a paradigm as into an *a priori* manufactured and rather tight small box. ... The phenomena which do not fit into that small box are often essentially pushed out of sight." After it finally had become clear that the rings were a collection of small particle-satellites it gradually became a common idea that these satellites were practically independent of one another and the dynamics of the rings was reduced to the dynamics of separate particles in the field of the planet. From the interaction between the particles one derived only a slow diffusion spreading in radius of the rings – and then the narrower the ring the faster the spreading. Thus for a long time – since Kant! – the opinion shared by observers and theoreticians had completely forgotten the possibility of subdivisions. Up to 1977 the scientists had in this way arrived at the idea that the rings were passive flat clouds of quasi-non-interacting satellites.

The new era for the rings started on the 10th of March 1977 when the narrow coal-black Uranian rings, which lag far behind each other, were discovered. This discovery was unexpected: when the 91 cm telescope of the flying Kuiper observatory (in a Boeing) was prepared to study the parameters of the Uranian atmosphere through the occultation of stars, the apparatus which was included had earlier registered nine eclipses of a star before (and later also after) it was covered by the planet (Elliot, Dunham, and Mink 1977).

Further events unrolled at a fast tempo.

On the fourth of March 1979 the American interplanetary Voyager 1 spacecraft observed transparent stony rings also round Jupiter (Jewitt 1985). Shortly after that Saturn was also studied by a series of American spacecraft: Pioneer 11 (October 1979), Voyager 1 (November 1980), and Voyager 2 (August 1981). The most impressive discovery was the observation of a hierarchical subdivision of the Saturnian rings into narrower ring structures. In January 1986 Voyager 2 studied the narrow elliptical Uranian rings. In this case the scientists were most surprised not by the rings themselves, which have been extensively studied by observers from the Earth, but by the absence of "shepherd" satellites next to the rings which, according the more popular American models, should be responsible for many oddities in the dynamics of the Uranian rings. The revenge of the "shepherd" paradigm which goes beyond the classical ideas about the passive nature of the rings and the necessity to organise the rings by the external gravitational field of the satellites had been expected near Neptune where in the middle of the eighties broken rings, or "arcs", were discovered (Brahic and Sicardy 1986). According to the common opinion at that time these were the result of the action of rather large shepherd-satellites.

In August 1989 Voyager 2 flew near Neptune and photographed the unique formation of three – according to other counts four or five – dense arcs strung like sausages on a continuous narrow transparent ring-filament. To great astonishment once again no shepherds (with a radius of more than 6 km, the limit of the resolution of the photographs by Voyager 2) were found to inhibit the spreading of the arcs along the orbit (Stone and Miner 1989). This demonstrates the level of complexity and "unexpectedness" of the phenomena observed in the rings.

In fact, during the last twenty years a new class of objects has been found and studied in the Solar system. The planetary rings turned out to be a necessary regular element of the satellite systems of the major planets. The rings demonstrate an extraordinary wealth of so far unknown collective processes and a large number of finely balanced dynamic constructions which do not confine themselves to the familiar models. The great interest in the rings was shown in 1984 by the publication of two collections of papers (Brahic 1984; Greenberg and Brahic 1984), of a total volume of about 1500 pages, with the common title "Planetary Rings". During a period of ten years the number of researchers of planetary rings increased to several hundred.

Plates

Plate 1: Layering of the Saturnian rings as observed by the American astronomers William and John Bond (Pasha 1983).

Plate 2: Structure of the Saturnian rings suggested in 1882 by the English Astronomer R.Proctor.

Plate 3: Computer representation of the Saturnian rings constructed from the 1980 data from Voyager-1.

Plate 4: Saturn's C ring (Voyager-2 photograph). In the central part of the ring one can distinguish a regular thousand-kilometer structure with a low density contrast. A few narrow ringlets in the outer and inner regions of the C ring are connected with the resonance action of satellites.

Plate 5: Outer region of the B ring (6000 km section; Voyager-2 photograph (Cuzzi et al. 1985)). The dark corner of the photograph is the Cassini division. One can clearly see a hierarchy of different scales of ring structures (from thousands to tens of kilometers). Note that these structures are not connected with resonances with satellites.

Plate 6: Composition of four images – from a Voyager-2 photograph (Smith et al. 1982) – of the inner part of the Cassini division.

Plate 7: Uranian rings – from a Voyager-2 photograph (Thompson et al. 1981). The top part is taken in back-scattered light and the bottom in forward-scattered light; the dust structures can be distinguished very clearly.

Plate 8: The faint Jovian rings (Showalter et al. 1985).

Plate 9: The two brightest Neptunian rings (Smith et al. 1989). One can see the three main arcs of the outer ring.

Plate 10: The Neptunian rings in back-scattered light; one can see the "arcs" and the traces of two satellites.

Plate 11: The Neptunian rings in forward-scattered light; one can distinguish four narrow ring details.

Plate 12: Ring occultation by Neptune. One can see separately the trail from the 1989N5R to 1989N2R rings.

Plate 13: Voyager-2 photograph of a regular chain of four – or more probably six – compact clumps positioned at distances of 0.5 to 0.8 angular degrees (550 to 900 km) from one another in the "trailing" A5 Neptunian arc (Smith et al. 1989). According to the American specialists the structure in this photograph is a blurred picture of such a chain of clumps – the blurring is connected with the orbital motion of the clumps and the uncompensated motion of the camera. The resolution of the photograph is 14 km/line and the photograph was taken in forward-scattered light with a phase angle of around 135°.

1

2

3

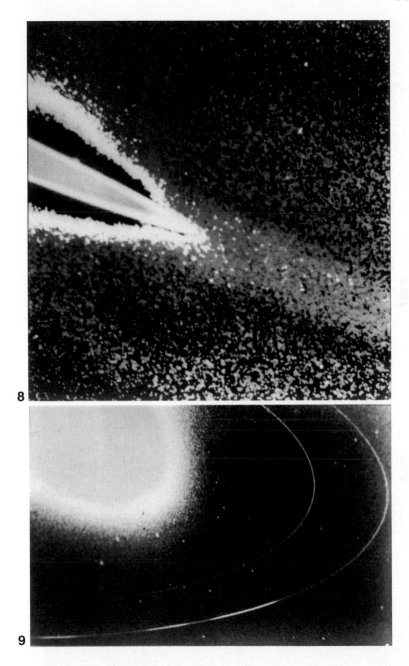

8

9

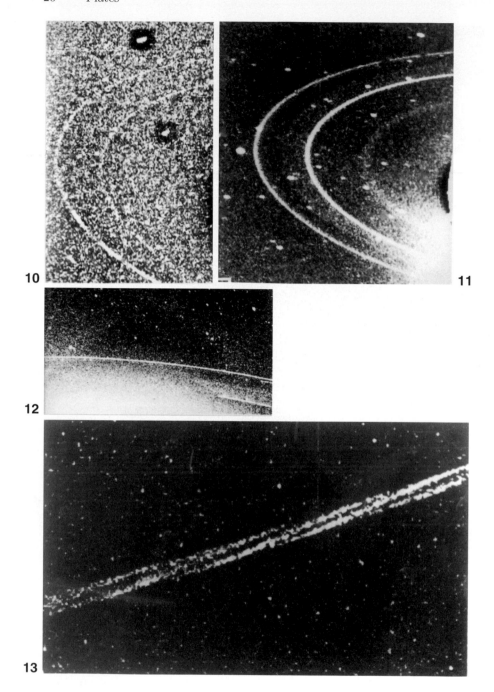

10

11

12

13

2 Observational Data

Lear: "Your eyes are in a heavy case, your purse in a light:
yet you see how this world goes."
Gloster: "I see it feelingly."
Lear: "What, art mad? A man may see how this world goes
with no eyes. Look with thine ears ..."

W. Shakespeare, King Lear, Act IV, Scene VI.

In the present chapter we shall consider the present-day data about three classes of discs: (1) discs around planets; (2) disc systems around stars; (3) galactic discs.

Discs around planets evolve into satellite systems and planetary rings. In Sects 2.1 to 2.4 we describe the system of satellites and rings of the major planets: Saturn, Uranus, Jupiter, and Neptune – in chronological order of the discovery of the rings as this enables us better to understand the evolution of the theoretical ideas about the rings.

Disc systems around stars are proto-planetary discs around single stars and accretion discs in systems of tight binary systems. The planets are formed gradually from the proto-planetary clouds and the best known but by far not the only example of this process is our own solar system. An accretion disc arises in the system of a binary star through the flow of matter from one star to the other. These discs are often hot and turbulent.

Galactic discs are huge systems in which, notwithstanding the size and the low matter density, various collective processes take place of which the spiral waves are the most impressive ones.

2.1 The Saturnian System

We show in Table 2.1 the external characteristics of the Saturnian system; in that table we give the orbital radius, eccentricity, and inclination of the orbits of the satellites and of the rings, as well as the size and albedo (reflecting power) of the satellites and rings – according to the Voyager data (Cuzzi 1983; Brahic 1984; Cuzzi et al. 1984; Esposito et al. 1984; Greenberg and Brahic 1984; Lissauer and Cuzzi 1985; Zebker, Marouf, and Tyler 1985; Esposito 1986; Ruskol 1986). We note that the large family of Saturnian satellites was enriched in 1990 by yet another satellite, Pan, positioned in the Encke

Table 2.1. The Saturnian system[a]

Satellite [ring]	Orbital radius [10^3 km]	Eccentricity [$\times 10^3$]	Inclination [degree]	Radius [width] [km]	Albedo
Phoebe	12952	163	175.3	115×110 × 105	0.06
Iapetus	3561.3	28.3	7.52	730±8	0.5/0.04
Hyperion	1481.1	104.2	0.43	175 × 120 × 100	0.2
Titan	1221.85	29.2	0.33	2575 ± 2	0.2
Rhea	527.04	1.0–0.3	0.35	765 ± 4	0.6
Dione	377.40	2.2	0.02	559 ± 5	0.55
Helene	377.40	5	0.2	18×?×?	0.6
Tethys	294.66	0.0	1.09	524 ± 5	0.8
Telesto	294.66	–	–	15 × 12 × 8	0.6
Calypso	294.66	–	–	? × 12 × 11	0.9
Enceladus	238.02	4.5	0.02	251 ± 5	1.0
[E ring]	181–483	–	–	[302 000]	?
Mimas	185.520	20.2	1.53	197 ± 3	0.7
[G ring]	170.10	–	–	[1000]	?
Janus	151.470	7	0.14	110 × 95 × 80	0.5
Epimetheus	151.420	9	0.34	70 × 58 × 50	0.5
Pandora	141.700	4.2	0.0	55 × 43 × 33	0.5
[F ring]	140.185	2.6 ± 0.6	0.0	45	?
Prometheus	139.350	2.4	0.0	70 × 50 × 37	0.5
Atlas	137.670	2	0.3	19×? × 14	0.5
[A ring]	122.17–136.78	–	–	[14610]	0.5
Pan	133.600	?	?	10	0.5?
[Cassini Division]	117.52–122.17	–	–	[4650]	0.2
[Huygens ringlet]	117.820	0.40 ± 0.17	–	[43]	?
[B ring]	92.00–117.52	–	–	[25520]	0.5
[C ring]	74.51–92.00	–	–	[17490]	0.2
[Maxwell ringlet]	87.491	0.34 ± 0.04	–	[64]	?
[Titan ringlet]	77.871	0.26 ± 0.02	–	[25]	?
[D ring]	66.97–74.51	–	–	[7450]	?

[a] Cuzzi 1983; Cuzzi et al. 1984.

Division, close to the outer edge of the Saturnian rings (Science News 1990). This satellite was confidently predicted by Cuzzi and Scargle (1985) from the wave perturbations discovered in the neighbouring A ring and in the small ringlet inside the Encke Division; from the perturbation the orbital radius was determined to be $R = 133600$ km and the radius of the satellite itself to be about 10 km (Cuzzi and Scargle 1985; Showalter et al. 1986). There are unconfirmed and less certain hypotheses about the existence of satellites also in other parts of the Saturnian rings – in the F ring, in the Cassini Division, and in the middle of the B ring (Lissauer, Shu, and Cuzzi 1981; Marouf and Tyler 1986; Cuzzi and Burns 1988; Spahn and Sponholz 1989). In 1995 the

Hubble Space Telescope discovered another four small Saturnian satellites: between the outer edge of the A ring and the co-orbital Janus-Epimetheus satellites.

During their orbital motion the Saturnian rings change the angle at which they are seen by a terrestrial observer from 0 to 26.7°. The latest transition through 0° was in 1995 after which the angle of their inclination is increasing slowly. At this moment (in 1999) we see the Saturnian rings – which are positioned in the equatorial plane of the planet – "from above" as we are to the North of Saturn's equator.

The thickness and chemical composition of the rings, the density of the rings or the fraction of the volume of the rings occupied by particles, the structure of the particle surface and the volumetric density of the particles, or the size distribution of the particles – these "intrinsic" or "invisible" characteristics of the Saturnian rings – can be determined only by constructing a quantitative theoretical model which connects the "invisible" characteristics of the rings with the "visible" brightness, spectrum, and so on. Of course, for this it is necessary that the observations of the rings are also quantitative. The first quantitative study of the Saturnian rings was carried out a hundred years ago by Müller (1893) who discovered the "opposition effect" – the fact that the brightness of the rings decreases as the phase angle of the observation (the Earth–Saturn–Sun angle) increases. Due to the orbital motion of the Earth this angle changes from 0 to 6.5°: the zero angle corresponds to the Earth being strictly between Saturn and the Sun and the maximum angle to the Earth and the Sun being at the same distance from Saturn – the Sun illuminates the rings from the side. Müller explained this effect by the shading of some particles of the rings by others. In fact, an observer positioned between the light source and the rings does not see the shadow. When he shifts slightly from the Saturn–Sun line he starts to see the shadows cast by the particles upon one another and the visible brightness of the rings decreases steeply. Seeliger (1887, 1895) developed a theory of this effect assuming cylindrical shadows of the particles. Further measurements of the dependence of the brightness of the rings on the phase angle revealed a discrepancy between the observations and Seeliger's theoretical phase curves. Bobrov (1970) constructed an improved theory using the idea of conical particle shadows – following naturally from the finite size of the Sun. This theory enabled Bobrov to estimate the fraction of space in the rings occupied by the particles to be 0.01, the size of the particles to be between 3 cm and 30 m, and the total mass of the rings to be between 10^{-6} and 10^{-9} the mass of Saturn (Bobrov 1970). These results were confirmed over thirty years by the data from spacecraft.

One other important result was the high-frequency measurement of the infrared spectrum of the rings which made it possible to conclude that the particles of the rings mainly consisted of water ice with possibly a small admixture of stony rocks (Kuiper, Cruikshank, and Fink 1970). Data about

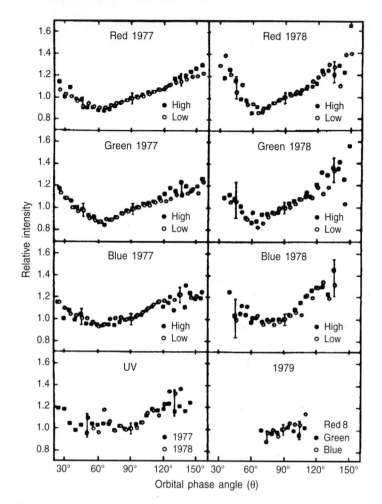

Fig. 2.1. Azimuthal asymmetry of reflected light in the A ring. The 1977 data refer to an inclination of 16.5° of the rings and the 1978 data to a −11.5° inclination. The squares and circles correspond to larger (> 1°) and smaller (< 1°) values of the angle between the Earth and the Sun.

the thermal inertia of the particles (Froidevaux, Matthews, and Neugebauer 1981; one measures through the infrared emission the temperature of the ring particles before they enter the shadow of the planet and after they leave this shadow) as well as spectrometric (Pollack 1978) and photopolarimetric (Steigmann 1984) data showed that the particles are covered by a layer of "hoarfrost" or a fine icy dust of micrometric dimensions.

A noteworthy achievement of terrestrial observations was the discovery by Camichel (1958) of the azimuthal asymmetry of the brightness of the A ring of Saturn. This puzzling effect which did not fit into the usual ideas

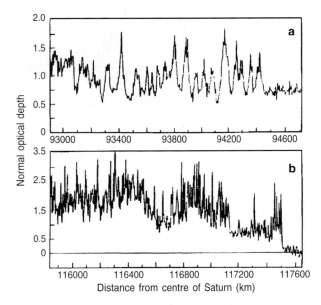

Fig. 2.2. Profile of the optical depth of (**a**) the inner and (**b**) the outer parts of the B ring (Cuzzi et al. 1984). In the inner part of the ring one can distinguish a fifty-kilometer structure and in the outer region one observes a hierarchy of structures from hundreds of kilometers down to the limit of the resolution (< 3 km).

about the rings was studied intensively during several years at the Lowell, Mauna Kea, Perth, and Cerro Tololo Observatories in the framework of a programme using the international planetary patrol network (Thompson et al. 1981). From forty thousand photographs the 400 best ones were chosen and analysed. Some of the results of these observations are shown in Fig. 2.1 (Thompson et al. 1981). If one reckons the azimuthal angle along the ring from the Saturn–Sun line in the direction in which the ring rotates (counterclockwise, if we are above Saturn's Northern hemisphere) it turns out that the brightness of the rings decreases rather rapidly with increasing azimuthal phase angle and reaches a minimum at 75° after which it increases again to a maximum at 165°; the brightness of the second half of the rings (from 180 to 360°) is symmetric with the first half (from 0 to 180°). The brightness contrast reaches 40% and depends on the inclination of the rings. There is no such asymmetry in the B ring (Camichel 1958, Thompson et al. 1981, Cuzzi et al. 1984). The asymmetry is greatest for the average inclination angles of the rings – from 6 to 16° (Cuzzi et al. 1984).

The main supply of data about Saturn's rings has come from spacecraft fly-pasts of Saturn: Pioneer-11 on 1979 September 1st, Voyager-1 on 1980 November 12th, and Voyager-2 on 1981 August 26th. These three autumn information squalls guaranteed the scientists a huge volume of data after

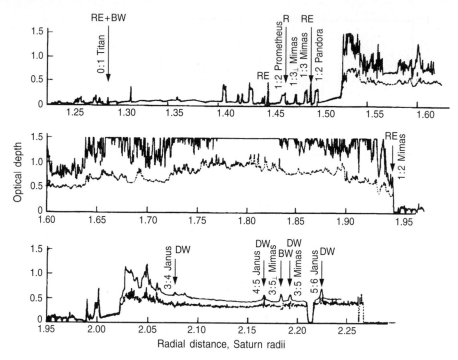

Fig. 2.3. Correlation between the positions of the narrow ringlets and resonances in the Saturnian system. R is a narrow ring, RE a narrow ring with a finite eccentricity, DW a spiral density wave, and BW a spiral bending wave. The profile of the optical depth is taken from the paper by Cuzzi et al. (1984). The dashed profile is the opacity of the rings in the 3.6 cm radio band. The resonances $1:3_\perp$ and $3:5_\perp$ indicate resonances involving vertical particle oscillations.

which there is now a pause until July 2004 when the Cassini probe should start to orbit Saturn.

The main methods of the space study of the planetary rings in the Voyager era are:

1) direct photography – the visible range of the spectrum;
2) measurement of the light of a star eclipsed by a ring – ultraviolet region;
3) radio-occultation of a Voyager signal by the rings – when the ring happens to be between the spacecraft and the Earth.

These spacecraft methods have given a spatial resolution in the photographs or the cross-sections of the rings from 10 km down to 100 m which is 100 to 10000 times better than the available possibilities of terrestrial observations. This made it possible to discover the following new effects:

a) The subdivision of the rings into separate ringlets with widths down to the limits of the resolution (Cuzzi et al. 1984). We show in Plates 4 and

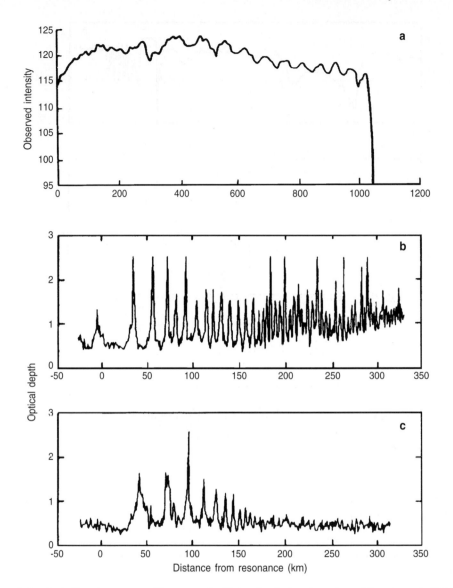

Fig. 2.4. Density waves ("wave trains") in the Saturnian rings (Cuzzi et al. 1984). (**a**) The linear wave in the Cassini Division which is connected with the 0:1 nodal resonance with Iapetus (a resonance is called nodal if the precession frequency of the particles in the ring, due to the asphericity of the field of the planet, is the same as the orbital frequency of a satellite, which is usually very far away); (**b**) the nonlinear extended spiral wave in the B ring due to Janus (Lindblad 1:2 resonance); (**c**) strongly damped wave due to Mimas (3:5 resonance) in the A ring.

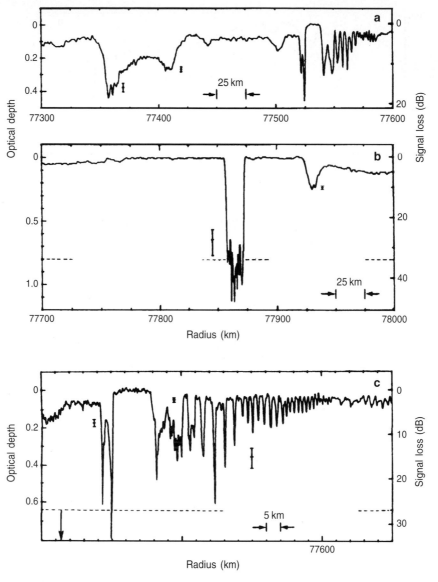

Fig. 2.5. Opacity of a section of the C ring in the 3.6 cm radio band. (**a**) Bending wave due to the 1:0 nodal resonance with Titan (Marouf, Tyler, and Rosen 1986); (**b**) narrow eccentric ringlet near the 1:0 resonance with Titan (Marouf, Tyler, and Rosen 1986); (**c**) Bending wave due to Titan (close-up); the position of the resonance is indicated by an arrow (Rosen and Lissauer 1988).

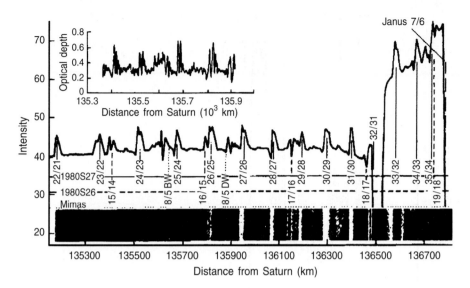

Fig. 2.6. Spiral waves at the outer edge of the A ring (Cuzzi et al. 1984). The spiral waves are excited by even high-order resonances. Mimas gives rise to two kinds of waves – bending waves (BW) propagating from the point of resonance towards the planet, and density waves (DW) propagating away from the planet. Waves of the first kind are caused by a resonance between the orbital frequency of the satellite and the frequency of the vertical oscillations of the particles in the ring and waves of the second kind are caused by a resonance between the orbital frequency of the satellite and the frequency of the radial oscillations of the particles.

5 and Fig. 2.2 thin divisions of the C and B rings. The A ring is more uniform but it may have a small-scale subdivision of hundreds of meters. The existence of such a radial structure in Saturn's rings is connected with intrinsic evolutionary processes. One can observe in the B ring a change in the albedo of the particles connected with the radial ring structure. The difference in the particle albedo in the density minima and maxima may reach 50% – from 0.6 to 0.4, from the albedo of snow to the reflectivity of stone (Cuzzi et al. 1984).

b) Still another class of spatial structures – spiral waves – are in continuous rings caused by outer satellite resonances. A large number of them have been observed in Saturn's rings, especially in the A ring (Fig. 2.3; Cuzzi et al. 1984). Spiral waves can be split into density waves (Fig. 2.4) and bending waves, which deflect particles from the equatorial plane (Fig. 2.5; Holberg, Forrester, and Lissauer 1982, Gresh et al. 1986, Marouf, Tyler, and Rosen 1986, Esposito, Harris, and Simmons 1987, Rosen and Lissauer 1988, Rosen 1989). The strongest spiral waves – with an appreciable amplitude and extension – are caused by lower-order resonances (1:2, 2:3, 3:4, 4:5, 5:6, 3:5, 6:7) but high-order resonances (such as 32:33) also excite small waves (Fig. 2.6). The appearance of the Cassini Division is connected with the ear-

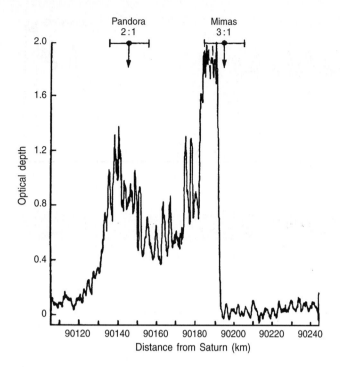

Fig. 2.7. Narrow ringlet in the C ring "squeezed" between two resonances due to Pandora (2:1) and Mimas (3:1) (Cuzzi et al. 1984).

lier existence of a spiral wave due to the 2:1 Mimas resonance (Goldreich and Tremaine 1978b, Fridman et al. 1996; see also Appendix V).

c) The existence of eccentric rings. Outside the outer edge of the A ring there is the narrow elliptical F ring with near it two "shepherd" satellites close by: Prometheus and Pandora. In the C ring and the Cassini Division some narrow, dense, and eccentric ringlets have been observed which are mainly connected with resonances with outer satellites (see Figs. 2.3, 2.5, 2.7, and 2.8, where one can see that the edge of the B ring is perturbed in the $m = 2$ mode and the Huygens ringlet and a broader ring (BAND 1) next to it by an elliptical mode ($m = 1$), and Plate 6; Smith et al. 1982, Cuzzi et al. 1984, Porco, Nicholson, et al. 1984, Marouf, Tyler, and Rosen 1986, Porco and Nicholson 1987, Flynn and Cuzzi 1989). Sometimes resonance ring structures do not have an appreciable eccentricity. It is possible that the formation of the narrow Maxwell ring is connected with a 1:2 resonance with one of the new Saturnian satellites which were discovered in 1995 (Fig. 2.9). We note that the 7:6 Janus resonance forces the outer edge of the A ring to oscillate in the $m = 7$ mode and the 2:1 Mimas resonance drives the $m = 2$ mode of the outer edge of the B ring (see Fig. 2.10; Porco, Danielson, et al. 1984). We give in Table 2.2 data about the main resonance structures in Saturn's rings.

Table 2.2. Resonance structures in the Saturnian rings[a]

Satellite	Resonance[*]	Structure	Radius [$\times 10^3$ km]	Width of the structure [km]	Figure
Iapetus	1:0	DW	120.976	1000	2.4a
Titan	$1:0_\perp$	BW	77.515	90	2.5
	1:0	RE	77.871	25	2.5
Mimas	$3:1_\perp$	R	88.702	2	2.3
	3:1	RE	90.198	60	2.7
	2:1	RE	117.553	43	2.8
	$5:3_\perp$	BW	131.900	160	2.3
	5:3	DW	132.340	170	2.4
	$8:5_\perp$	BW	135.642	46	2.6
	8:5	DW	135.875	60	2.6
Janus	2:1	DW	96.249	333	2.4
	4:3	DW	125.269	266	2.3
	5:4	DW	130.702	175	2.3
	6:5	DW	134.264	122	2.3
Pandora	2:1	RE	90.171	60	2.7
Prometheus	2:1	R	88.714	20	2.3

[a] Holberg, Forrester, and Lissauer 1982; Porco et al. 1984; Gresh, Rosen, Tyler, and Lissauer 1986; Esposito, Harris, and Simmons 1987; Porco and Nicholson 1987; Rosen and Lissauer 1988; Rosen 1989; Flynn and Cuzzi 1989.
[*] Here and henceforth we have a twofold notation of the Lindblad resonances – for instance, 2:1 or 1:2.
Note. R indicates a narrow ring, RE a narrow ring with eccentricity, DW a spiral density wave, BW a spiral bending wave, 5:3 a resonance of the frequency of the radial oscillations of the particles with the orbital frequency of the satellite, $5:3_\perp$ a resonance of the frequency of the vertical oscillations of the particles with the orbital frequency of the satellite, 1:0 a nodal resonance of the orbital frequency of the satellite with the precession frequency of the particle orbits in the aspherical planetary gravitational field.

d) "Spokes". In the B ring one can also observe shortlived (no longer than 3 to 4 hours) structures which extend along the radius and are situated on the corotation radius where the orbital period is equal to the rotational period of the planet. These clusters are called "spokes", they show rigid-body rotation, and their appearance is connected with the action of the planetary magnetic field on the charged dust. In backscattered light the spokes are dark and in forward scattered light bright which indicates that they consist of small grains. The optical depth of the grains forming the spokes is 1 to 2% of the optical depth of the ring (Doyle, Dones, and Cuzzi 1989).

e) Characteristic features of the particle size spectrum. Data from the radio-occultation of Saturn's rings by Voyager signals which have been processed in detail by Tyler's group at Stanford have given us important information about the sizes of the particles in the rings (Marouf, Tyler, and

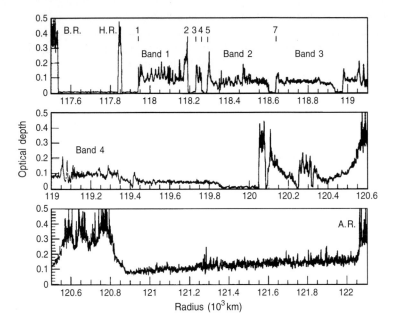

Fig. 2.8. Optical depth profile (Flynn and Cuzzi 1989) of the Cassini Division. B.R.: edge of the B ring, A.R.: Edge of the A ring, H.R.: elliptical Huygens ringlet.

Eshleman 1982, Zebker 1984, Zebker, Marouf, and Tyler 1985, Tyler 1987). It was shown that the particle sizes varied from the very small to tens of meters. The particle size spectrum $n(a) = na^{-q}$ has an index $q = 2.8$ to 3.4 in the range of sizes up to the maximum $a_{max} = 5$ to 10 m. The number density of particles with radii larger than the maximum one decreases steeply with an index $q = 5$ to 6. The main mass of the rings is contained in meter-size particles (Fig. 2.10; Zebker, Marouf, and Tyler 1985). It was possible to determine the surface density and the thickness of the rings from an analysis of their radio-occultation. We give the main physical characteristic of Saturn's rings in Table 2.3. The density data following from resonance theories are obtained from the theoretical dependence of the length of the spiral wave on the surface density of the disc (see, for instance, Eq. (A5.63)).

The Voyager and Pioneer data have given us an estimate about the mass of the rings as being 5×10^{-8} times Saturn's mass which is close to the mass of a satellite such as Mimas. In recent fly-pasts of spacecraft the new D, E, F, and G rings as well as ten new satellites (including Helene) were discovered.

In conclusion we shall give a brief survey of the properties of Saturn's satellites (Hunten et al. 1984, Burns 1986, Morrison 1986, Ruskol 1986, Veverka et al. 1986). They are practically all bright and consist of water ice. One may note the following features of each of them – we consider the satellites starting from the one closest to the planet.

Table 2.3. Main physical characteristics of the Saturnian rings

Part of ring	Thickness [m][a,b]	Average optical depth[c]	Surface density [g/cm²]		Maximum size of particles[b] [m]
			Radio occultation[b]	Resonance theories[d]	
C ring:					
interior part	5	0.08	3.2 ± 1.8	1.5	4.5
exterior part	5	0.12	4.3 ± 2.5	1–5	2.4 –5.3
B ring:					
interior part	5–10	1.21	–	70 ± 4	
central part	5–10	1.76		(99 ± 6)	
exterior part		1.84			
Cassini Division	20	0.12	18.8 ± 0.5	16 ± 3	7.5
A ring:					
interior part	10–30	0.70	34 ± 6	40 ± 2.5	5.4*
exterior part	10–30	0.57	24.4 ± 7.2	37 ± 2.0	8.9–11.2
G ring	10^5	10^{-6}–10^{-4}	–	–	
E ring	10^7	10^{-6}–10^{-5}	–	–	

[a] Esposito 1986.
[b] Zebker, Marouf, and Tyler 1985.
[c] Cuzzi, Lissauer, Esposito, et al. 1984.
[d] Rosen 1989.
* For the middle part of ring A.

1) The recently discovered satellite, Pan, in the Encke Division has a radius of about 10 km and is noteworthy because it causes a series of spiral waves in nearby sections of the ring which made it possible to "compute" its location (Cuzzi and Scargle 1985; Showalter, Cuzzi, et al. 1986)

2) Atlas is situated at the outer edge of the Saturnian A ring.

3) Prometheus is the inner "shepherd" of the F ring.

4) Pandora is the outer "shepherd" of the F ring.

These shepherd satellites cause strong wave perturbations in the F ring which gives the illusion of an interwoven ring. This apparent interweaving of the ring into a "plait" was at the time the reason of violent disputes.

5–6) Janus and Epimetheus which are co-orbital satellites – that is, they move practically along the same orbit – are in a 1:1 resonance with each other. Among co-orbital satellites they are unique in that they have comparable masses. Librating around their equilibrium positions they periodically change places – first one then the other satellite is closest to the planet (Lissauer, Goldreich, and Tremaine 1985).

7) In contrast to the previous six, Mimas was discovered from the Earth and has a spherical shape. The Herschel crater is situated on it; it has a size

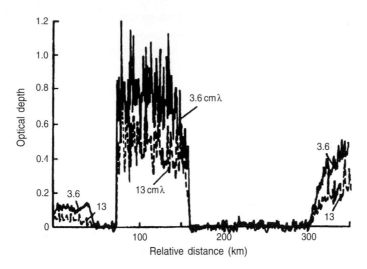

Fig. 2.9. Narrow elliptical Maxwell ring in the outer region of the C ring (Cuzzi et al. 1984). It is unique in that it is not connected with a resonance (see Fig. 2.3). The optical depth was measured in the 3.6 and 13 cm radio bands.

of 130 km which is one third of the diameter of the satellite itself. A slightly larger meteorite could simply break up the satellite into separate parts.

8) Enceladus has an uniquely high albedo (1.0), the brightest body in the solar system. It is probably totally covered by a thin layer of a geologically recent hoarfrost. It is the geologically most active Saturnian satellite. The energy source for this high activity is so far unknown. It is assumed that Enceladus is the source of the matter for the dusty, tenuous E ring.

9) Tethys is noteworthy because of the Odysseus crater which has a 400 km diameter – two fifth of the diameter of the satellite – and the giant Ithaca canyon, which stretches for more than 270° across the satellite, is nearly 3000 km long. There are two small co-orbital satellites, Telesto and Calypso, in the L_4 and L_5 Lagrangian points – 60° along the orbit, in front of and behind Tethys. Telesto and Calypso were discovered through Voyager data. Tethys is connected with Mimas by a 1:2 resonance.

10) Dione has, like Tethys, a small co-orbital satellite, Helene – 60° in front of Dione. Helene was detected in 1980 by the terrestrial observations of J.Lecacheux and coworkers; its existence was confirmed by data from the Voyager-1 mission. Dione is connected with Enceladus by a 1:2 resonance.

11) Rhea is a densely cratered satellite which is geologically less active than Dione.

12) Titan is the largest Saturnian satellite, the second largest satellite in the solar system – it is slightly smaller in size and mass than the Jovian satellite Ganymede. It has a large atmosphere with dense mists and clouds of aerosols. The atmospheric pressure at the surface is 1.6 atmos and the

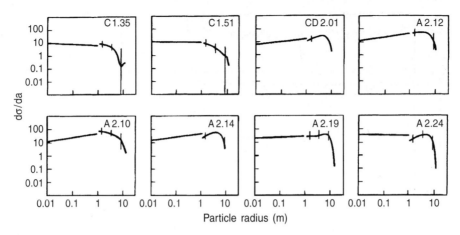

Fig. 2.10. Particle mass spectrum in eight regions of the Saturnian rings: C ring, A ring, and Cassini Division (Zebler, Marouf, and Tyler 1985). The numbers indicate the distance from the planet in Saturnian radii (60330 km); σ is the surface density in kg/m², a the particle size in meters. We assume that the particle volume density is 0.9 g/cm³. A distribution law with $q = 3$ corresponds to a horizontal line, since $dn/da \propto a^{-q}$, that is, $d\sigma/da \propto a^{3-q}$.

temperature 94 K. Titan's colour is reddish brown with seasonal changes. Theoretical models indicate the existence of a surface methane-ethane ocean of a few kilometers depth. It is the only satellite in the solar system the surface of which cannot be observed by the usual optical means. Titan's atmosphere and surface will be studied by direct probing during the Cassini mission.

13) Hyperion is a dark, irregularly shaped satellite with a chaotic intrinsic rotation – the rotational period changes by tens of percents in a week. Hyperion is connected with Titan by a 4:3 resonance.

14) Iapetus is noteworthy for the sharp asymmetry of the albedo of the hemispheres – from 0.04 to 0.5. Its surface is heavily cratered.

15) Phoebe is the farthest removed and darkest Saturnian satellite, orbiting the planet in the opposite direction. Most researchers feel that Phoebe is a captured satellite.

2.2 The Uranian System

The history of the Uranian rings started on 10th of March 1977 when several groups of researchers observed the occultation of the star SAO 158687 by the Uranian rings (Elliot, Dunham, and Mink 1977; Millis, Wasserman, and Birch 1977; Bhattacharyya and Kuppuswamy 1977). The best data were obtained by the 91 cm telescope of the flying Kuiper Observatory (Elliot, Dunham, and Mink 1977). Subsequently more than 200 occultations of stars by Uranus were

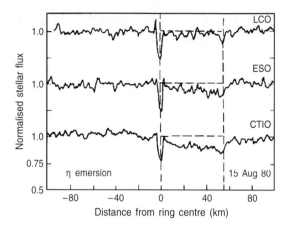

Fig. 2.11. Records of the eclipse of a star by the η ring of Uranus, obtained by three observatories in 1980 (Elliot and Nicholson 1984). The broad component can be seen.

observed as a result of which nine rings were revealed and their properties determined. Elliot and French from MIT and independently Nicholson from Cornell and their collaborators have been the most active terrestrial observers of the Uranian rings.

The Uranian rings are a set of narrow dense rings (Plate 7) which are often elliptical and inclined to the equatorial plane (Elliot and Nicholson 1984). We give the characteristics of the Uranian system in Table 2.4. The rings have sharp edges and precess in the aspherical field of the planet as a single body. In Figs. 2.11 and 2.12 we show the profile of two of the rings. Part of the rings has a variable width – maximal in the apocentre and minimal in the pericentre (Fig. 2.13). A characteristic feature is the arrangement of narrow dense rings near the lower-order resonances with the outer satellites (see Sect. 10.2). We note that one can in the widest α and ε rings clearly see a radial structure of kilometric dimensions (see Fig. 2.12). The region between the dense rings is filled with fine dust with an optical depth of 0.001 to 0.0001. This fine dust is distributed inhomogeneously and forms a number of ring structures (see Plate 7 and Fig. 2.14) The appreciable density of the outer Uranian atmosphere causes a fast drift of the dust components towards the planet and worsens the problem of the stability of the rings (Broadfoot et al. 1986). The Uranian rings are very black with an albedo of about 3%, the infrared spectrum of the rings corresponds most closely to carbonaceous chondrites (Elliot and Nicholson 1984; Lane et al. 1986). The size of the particles in the rings varies from centimeters to several meters.

Voyager-2 in its fly-past of Uranus on 24th January 1986 gave more precise data about the nine main rings and discovered a number of weaker rings – 1986U1R, 1986U2R, and so on. The main discovery of Voyager was the detection of ten new Uranian satellites (from Puck to Cordelia; see Table 2.4).

Table 2.4. The Uranian system[a]

Satellite [ring]	Orbital radius [km]	Eccentricity [$\times 10^3$]	Inclination [degree]	Radius [width] [km]	Albedo [optical depth]
Oberon	582596 ± 71	0.8	0.10	775 ± 10	0.24
Titania	435844 ± 86	2.2	0.14	805 ± 5	0.28
Umbriel	265969 ± 48	5.0	0.36	595 ± 10	0.19
Ariel	191239 ± 57	3.4	0.31	580 ± 5	0.40
Miranda	129783 ± 66	2.7	4.22	242 ± 5	0.34
Puck	86006 ± 25	(0.1)	0.3	77 ± 3	0.07
Belinda	75256 ± 29	(0.1)	0.0	34 ± 4	–
Rosalind	69942 ± 26	(0.1)	0.3	29 ± 4	
Portia	66090 ± 34	(0.2)	(0.1)	55 ± 6	–
Juliet	64350 ± 27	0.6	(0.1)	42 ± 5	–
Desdemona	62675 ± 24	(0.2)	0.2	29 ± 3	–
Cressida	61776 ± 27	(0.2)	0.0	33 ± 4	–
Bianca	59172 ± 26	0.9	(0.2)	22 ± 3	–
Ophelia	53794 ± 39	10.1	(0.1)	16 ± 2	–
[ε ring]	51149	7.94	0.000	[20/96]	[1.2/4]
[1986U1R]	50023	(0.0)	(0.0)	[1–2]	[0.1]
Cordelia	49771 ± 17	0.5	0.1	13 ± 2]	–
[δ ring]	48299	(0.04)	(0.002)	[3/9]	[0.3/0.4]
[γ ring]	47627	(0.10)	(0.006)	[1/4]	[1.3/2.3]
[η ring]	47176	(0.00)	(0.001)	[1–2]	[0.1–0.4]
[β ring]	45661	0.44	0.005	[7/12]	[0.2]
[α ring]	44718	0.76	0.015	[7/12]	[0.3/0.4]
[ring 4]	42571	1.06	0.032	[2–3]	[0.3]
[ring 5]	42235	1.90	0.054	[2–3]	[0.5–0.6]
[ring 6]	41837	1.01	0.062	[1/3]	[0.2/0.3]
[1986U2R]	37–39.5×10^3	–	–	[2500]	[0.001–0.0001]
Unreliable details of the ring system[b]					
[ring 1]	50660 ± 30	–	–	[16]	[0.1]
[arc 1]	41760 ± 30	–	–	[2]	[0.2]
[arc 2]	41470 ± 30	–	–	[4]	[0.2]
[arc 3]	38430 ± 50	–	–	[2]	[0.2]
[ring 2]	38280 ± 50	–	–	[1]	[0.2]

[a] Elliot and Nicholson 1984, Lane et al. 1986, Broadfoot et al. 1986, Smith et al. 1986, French et al. 1986, Owen and Synnot 1987, French et al. 1988, Thomas et al. 1989.
[b] Smith et al. 1984.
Note. The quantities in square brackets and separated by a stroke indicate variable widths (or optical depths); the quantities in round brackets are only known with a poor accuracy. The semiaxes of the main rings are determined with an accuracy of a few hundred meters.

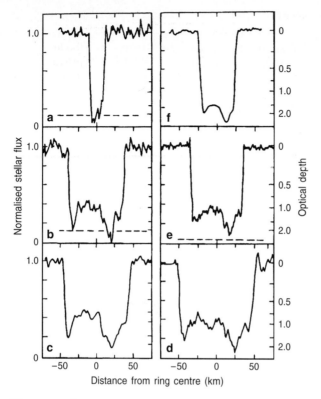

Fig. 2.12. Some cross-sections of the ε ring with a variable width – from 20 to 96 km (Elliot and Nicholson 1984). The position of these cross-sections on the general map of the ε ring is shown in Fig. 2.13.

We note that by tradition the new Uranian satellites are given the names of characters from Shakespeare's plays. All new satellites are black as coal; on a photograph of the largest of them, Puck, one sees a cratered surface. We shall briefly list some features of the five large Uranian satellites (Smith et al. 1986).

1) Miranda is a small Uranian satellite with many signs of an unexpected violent geological past. Scientists have carefully calculated the balance of Miranda's energy budget to explain such an acitivity. In January 1986 Voyager-2 obtained excellent pictures of Miranda – with a resolution down to 1 km – from which the NASA specialists were able to construct even a stereoscopic picture of the exotic relief of Miranda where one could distinguish vast furrow-like regions, reminding one of ploughed fields. The region where the furrows joined at an angle was unofficially called a "chevron".

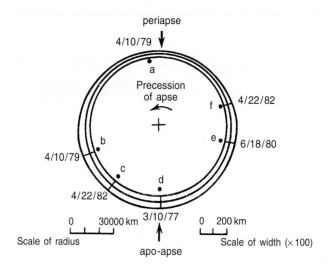

Fig. 2.13. The ε ring with an eccentricity and a variable width (the scales of the width and of the radius are shown in the figure; Elliot and Nicholson 1984). We have indicated the dates of the stellar occultations by the ring.

2) Ariel the brightest Uranian satellite is geologically active and has clear signs of ancient vulcanic activity.
3) Umbriel is the darkest satellite with a faceless strongly cratered surface.
4) Titania, together with Oberon, forms the pair of the two largest Uranian satellites. It has a network of large fractures and there are signs of ancient vulcanic activity.
5) Oberon has an old, strongly cratered surface. There is little evidence of large tectonic fractures.

We note a characteristic feature of the system of Uranian satellites: since Uranus rotates "lying on its side" the system of satellites, which is situated in the equatorial plane, rotates "on edge" – like a Ferris wheel in an amusement park. At this moment (1997) Uranus has turned its North pole towards the Earth and the line of sight of a terrestrial observer is almost perpendicular to the orbital plane of the satellites which gives us maximally favourable conditions for observing the satellites and rings – through the stellar occultation method.

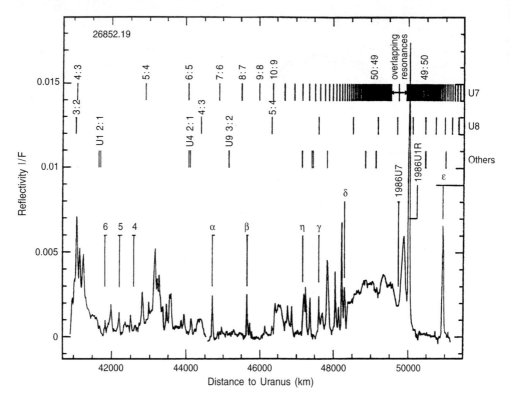

Fig. 2.14. Dust distribution in the Uranian ring system (Cuzzi, private communication). The position of the resonances with the 1986U7, 1986U8, and other satellites are shown.

2.3 The Jovian System

The first fly-past studies of the Jovian system were carried out by the American spacecraft Pioneer-10 and Pioneer-11 in 1973/4 and Voyager-1 and Voyager-2 in March and July 1979. The Jovian rings were discovered in March 1979 in the television survey by Voyager-1 – the indirect detection of the Jovian rings from the Pioneer data in 1974 remained unacknowledged. The Voyager-2 data greatly supplemented the information about the Jovian rings (Brahic 1982, Showalter et al. 1985; see Plate 8). We give the characteristics of the rings in Table 2.5. The spectrum of the rings corresponds to rocky material (Brahic 1982). On the outer edge of the main ring there are two small satellites. The amount of reflected light indicates the presence in the main ring of large particles although on the whole the Jovian rings consist of dust (Jewitt 1985). The magnetic field of the planet plays the main role in the dynamics of the dust particles. The particles from the main ring pass to the tenuous ring (pale veil) under the action of braking forces, especially the Poynting–Robertson effect. This is often a countereffect, forcing the particles to drift towards the planet; this is connected with the deceleration of the orbital velocity of the particles – loss of their angular momentum – due to their interaction with the solar wind. A particle moving even strictly at right angles to the light flux in its own frame of reference undergoes a drag due to the counterflow of radiation – just as one always has rain in the face when moving. Particles with a diameter of less than 0.4 μm can pass into the halo under the action of the magnetic field – provided the particle is slightly charged.

Table 2.5. The Jovian system[a]

Satellite	Orbital radius [10^3 km]	Eccentricity [$\times 10^3$]	Inclination [degree]	Radius [km]	Albedo
Sinope	23700	275	153	20	–
Pasiphae	23500	378	148	35	–
Carme	22600	207	163	22	–
Ananke	21200	169	147	15	–
Elara	11737	207	28	40	0.03
Lysithea	11720	107	29	20	–
Himalia	11480	158	28	90	0.03
Leda	11094	148	27	8	–
Callisto	1883	7	0.281	2400	0.2
Ganymede	1070	0.6/1.5	0.195	2631	0.4
Europa	670.9	10.1/0.1	0.470	1569	0.6
Io	421.6	4.1/0.01	0.040	1815	0.6
Thebe	221.9	15	0.8	? × 55 × 45	0.05–0.1
Amalthea	181.3	3	0.40	135 × 82 × 75	0.06
Adrastea	128.98	0	0	12.5 × 10 × 7.5	0.05–0.1
Metis	127.96	4	0	? × 20 × 20	0.05–0.1

Ring	Radius [10^3 km]	Width [10^3 km]	Thickness [km]	Optical depth
The gossamer ring	123–210	87	< 4000	10^{-7}
The bright ring	122.8–129.2	6.4	< 30	3×10^{-5}
The faint ring	71.4–122.8	51.4	< 1000	7×10^{-6}
Halo	71.4–123	51.6	> 10000	5×10^{-6}

[a] Jewitt 1985, Lissauer and Cuzzi 1985.

Little is known either about the three inner Jovian satellites – Metis, Adrastea, and Thebe, which were discovered by interplanetary spacecraft or about the small outer satellites with orbits which are strongly inclined and are very eccentric. We restrict ourselves to the features of the largest satellites (for details see the review by Jewitt (1985)).

1) Amalthea is a dark-red satellite of irregular shape covered by craters. It consists of refractory rock.

2) Io is one of the four Galilean satellites of Jupiter. There are several (nine during the Voyager fly-pasts) active volcanoes on reddish-orange Io which throw out plumes up to several hundreds of kilometers. Sulphur or sulphur dioxide is the active substance of the volcanoes. Different parts of Io's surface are heated up to 650 K. There are no impact craters on Io because of the intensive volcanic rejuvenation of the surface. Io is the source of an orbital plasma torus of ionised oxygen and sulphur atoms and of neutral clouds of sodium and potassium atoms. Io's density is 3.53 g/cm^3. The source of the energy for Io's volcanic activity is the dissipation of the oscillations of the tidal bulge at a height of about 8 km – oscillations in the amplitude of the bulge and its motion along the surface. The bulge itself is caused by the action of Jupiter and the oscillations are connected with the small eccentricity of Io's orbit. Io's orbital eccentricity which vanishes due to the tidal interaction between Io and Jupiter is sustained by a 1:2 resonance with Europa which in turn is in a 1:2 resonance with Ganymede. All three satellites – Io, Europa, and Ganymede – are interconnected through a three-frequency Laplace resonance.

3) Europa is a bright satellite with a smooth, icy, and young surface covered by a global network of bright and dark narrow greatly extended strips. The drops in height are less than 100 m and there are practically no craters. There seems to be a water ocean under the ice core and then a massive silicate kernel. The density of the satellite is 3.04 g/cm^3.

4) Ganymede is the largest satellite in the solar system. It consists of a silicate-ice mixture with a density of 1.93 g/cm^3. Impact craters and furrowed regions are characteristic for its surface.

5) Callisto is the second largest Jovian satellite. Its density is 1.83 g/cm^3. Its ancient surface is completely saturated with meteoritic craters.

2.4 The Neptunian System

After the detection in 1977 of the Uranian rings through stellar occultation similar investigations started to be made also for possible Neptunian rings. Nicholson and Jones (1980) communicated in 1980 that on 21st August 1978 they had observed the occultation of a star by a Neptunian ring at a distance of one and a half times the planet's radius. On the 10th and 24th May 1981 two stars passed through the Neptunian system; on 10th May six telescopes from 0.6 to 3 m were pointed to observe this and on 24th May the 4 meter telescope from Cerro Tololo, but no occultation occurred (Elliot et al. 1981). These contradictory results caused a lively polemic (Elliot 1984, Maddox 1985, Thompson 1987). The first reliable data appeared after the observation of a stellar occultation on 22nd July 1984 (Hubbard et al. 1985, 1986). The data from various observations are given in Table 2.6. Some of the data from

Table 2.6. Groundbased observations of the Neptunian rings

Ref.[*]	Radius [km]	Width [km]	Optical depth[a]	Radius[a] [km]	Width[a] [km]	Optical depth[a]
				41680 or 41860	< 4	> 0.3
1, 2	54200 ± 50	15	–	53730–54270	9.1	0.14
3	62600 ± 160 or	8	0.9	54060–58930	25	0.075
	63700 ± 120	8	2.0			
4	64190–64400	–	–	60000 ± 9000	80	> 0.7
1, 5, 6	66000 ± 3000	15	0.12	62580 or 64010	8	0.12–0.3
7	69880 ± 300	9.0	0.17	67000 ± 4000	15	0.14
	70020 ± 300	5.1	0.13			
	70050 ± 300	5.3	0.1			
2	74000	17	0.1			
5	75000	–	–			

[a] According to Lissauer's survey.
[*] References: 1: Brahic and Sicardy 1986; 2: Sicardy, Brahic, Bouchet, et al. 1985; 3: Covault, Glass, French, and Elliott 1986; 4: Pandey and Mahra 1987; 5: Hubbard, Brahic, Bouchet, et al. 1985; 6: Hubbard, Brahic, Sicardy, et al. 1986; 7: Cooke, Nicholson, Matthews, and Elias 1985.

the 1985 IAU Circular No 4100 which were subsequently not confirmed have been omitted. Lissauer has given a more complete selection of terrestrial observations in a contribution to the Institute of Terrestrial Physics in the summer of 1988. The peculiar feature of the Neptunian rings is the absence of a second occultation so that one started to talk about the Neptunian rings as "arcs". Moreover, the observation of a star at the same time by two telescopes indicated that the density of an arc drops abruptly over a distance of a few hundred kilometers along the orbit – one telescope found the occultation of the star while the other at another spot on the Earth did not observe it.

We give in Table 2.7 the data of the Neptunian system obtained in August 1989 by Voyager-2 (IAU 1989 Circulars 4824, 4830, and 4867, Lane et al. 1989, Smith et al. 1989, Stone and Miner 1989, Porco 1991, Ferrari and Brahic 1994). We show photographs of the rings in Plates 9 to 12. It is is clear that all the rings are continuous, but on the outer ring there are three bright segments. The observation of the occultation of stars by the dense part of the outer ring shows that it contains a condensation of the width by 15 km, surrounded by a diffuse halo of dust to 50 km. A chain of separate bunches has been detected inside the segments (Plate 13) at distances of several hundred kilometers from one another. These bunches can be seen both in reflected and in scattered light.

Information about the six small inner satellites discovered by Voyager 2 on its fly-past of Neptune is given in Table 2.7. Very little is known about Nereid, the outermost Neptunian satellite. Triton is the largest Neptunian

Table 2.7. The Neptunian system[a]

Satellite [ring]	Orbital radius [10^3 km]	Eccentricity [$\times 10^3$]	Inclination [degree]	Radius [width] [km]	Albedo [optical depth]
Nereid	5510	750	27.6	170 ± 25	0.14–0.035
Triton	354.8	< 0.5	158.5	1353 ± 3	0.6–0.9
Proteus	117.6	?	< 1	200 ± 10	0.06
Larissa	73.6	?	< 1	95 ± 10	0.056
Adams ring	62.9	0.47[b]	0.06	[15–50]	[0.01– 0.1][c]
Galatea	62.0	?	< 0.1[d]	79 ± 12	–
[1989N4R]	53.2–59	–	–	[5800]	[0.0001]
Le Verrier ring	53.20	–	–	[100]	[0.01]
Despina	52.5	?	< 1	75 ± 15	0.054
Thalassa	50.0	?	< 1	40 ± 8	–
Naiad	48.0	?	4.5	27 ± 8	–
[1989N3R]	41.9	–	–	[1700]	[0.0001]

[a] Stone and Miner 1989; Smith, Soderblom, Banfield, et al. 1989; Porco 1991.
[b] Eccentricity of the clumps in the arcs.
[c] The optical depth of the continuous ring is 0.01 and that of the arcs 0.1.
[d] Inclination relative to the Adams ring is $i = 0.02$ to $0.01°$.

Table 2.7a. Characteristics of the individual Neptunian arcs in the Adams ring[a]

Designation	Number of observed clumps	Angle of Centre of arc [degree]	Length of arc [degree]	Optical thickness of arc
A1/C	1	285.03 ± 0.05	1.0 ± 0.1	0.04
A2/L	2	277.73 ± 0.05	4.0 ± 0.1	$\cong 0.1$
A3/E	1	265.4 ± 0.5	1.0 ± 0.5	0.08
A4/E	1	262.6 ± 0.5	3.0 ± 0.5	0.13
A5/F	11	251.88 ± 0.05	9.6 ± 0.1	$\cong 0.1$

[a] Porco 1991; Ferrari and Brahic 1994.

satellite and one of the two satellites in the solar system – the other being Titan – having a nitrogen atmosphere; it was studied by Voyager which detected on the pinkish surface of Triton a very rich set of geological structures and – completely unexpectedly – active geysers with plumes of many kilome-

Fig. 2.15. Evolution of a proto-planetary disc according to the popular Shmidt-Safronov model (Safronov 1969).

ters which were dispersed over long distances by the wind in the rarefied atmosphere. The energy source for the volcanic activity is apparently solar heating caused by a greenhouse effect in the depth of the transparent nitrogen ice.

2.5 The Solar System

Our solar system according to our present information consists of the Sun, nine planetary systems – with 65 satellites and four ring systems – the asteroid belt, and comets – with a possible reservoir (the Oort cloud) at the periphery. We have given in Table 2.8 the orbital characteristics (Marov 1987). We remind ourselves that the six planets closest to the Sun, up to and including

Saturn, can be seen with the naked eye and have been known since time immemorial. Uranus was discovered by Herschel in 1781 using a telescope. Neptune's existence was predicted by Adams and Le Verrier from an analysis of the perturbations of Uranus's motion and was detected by Galle in 1846, using Le Verrier's data. Pluto was detected by Tombaugh in 1930.

Although it is now clear that planets have been formed from a gaseous dusty disc rotating round the Sun (Fig. 2.15; Safronov 1969, Gehrels 1978, Safronov and Vityazev 1983, Hayashi, Nakazawa, and Nakagawa 1985) there are still many nebulous parts in the problem of the formation and evolution of such a disc and they are the cause of polemics. Most researchers are inclined to accept Safronov's model (Safronov 1969, Safronov and Vityazev 1983) which assumes a cold gaseous dusty disc with a small mass (a few percent of the solar mass) the formation of which was possible due to the low angular momentum of the initial proto-solar cloud. The basic evolutionary mainspring of the process for the formation of the planets is the appearance of planetesimals – bodies of asteroid size – and their accretional enlargement (that is, the sticking together in collisions) to planetary sizes. In this stage the proto-planetary disc is similar to the planetary rings. Observations in recent years have given persuasive evidence for the existence of proto-planetary discs around other single stars, for instance, near β Pictoris (Smith and Terrile 1984, Beust et al. 1989). A cold gaseous dusty disc with a size of a few hundred or thousand astronomical units seems to be a characteristic and, perhaps, universal feature of single star systems. During 1995 and 1996 astronomers discovered massive planets near about ten stars – including stars in binary systems – which conclusively indicates the widespread existence of planetary systems in our Universe.

2.6 Accretion Discs

Most stars are binaries. The evolution of the components in close binary systems often leads to one of the stars increasing in size and filling its Roche lobe. This means that matter from the surface of such a star starts to flow out in the form of a jet – mainly through the inner Lagrangian point – into the attraction sphere of the second star. The non-vanishing angular momentum of the outflowing matter leads to the formation of an accretion disc around the second star (Fig. 2.16; Cherepashchuk 1981, Pringle 1981, Hayley and Starr 1988). An analysis of the luminosity curve of a binary star makes it possible to establish accurately the geometry of the binary system and to detect whether there is a disc in it (Cherepashchuk 1981). Such an analysis made it possible in the forties and fifties of the present century to discover accretion discs in several binary systems. Theorists afterwards convincingly showed that the existence of such a disc is inevitable for a number of scenarios of the evolution of stars in binary systems – when a star fills the Roche lobe.

Table 2.8. The planets of the solar system

Planet	Semi-major axis [a.u.]	Eccentricity	Orbital inclination [degree]	Equatorial radius [km]	Equatorial inclination [degree]	Rotational period	Number of satellites [in 1999]	Density [g/cm^3]
Mercury	0.387	0.206	7.0	2439	< 30	58.6 d	–	5.44
Venus	0.723	0.007	3.4	6051.5	177	243 d	–	5.24
Earth	1	0.017	0	6378	23.5	23.9 h	1	5.52
Mars	1.524	0.093	1.8	3394	25.2	24.6 h	2	3.95
Asteroids	$\approx$ 2.7	$\approx$ 0.14	$\approx$ 9.5	< 500	–	$\cong$5–10 h	> 1	$\approx$ 3.5
Jupiter	5.203	0.048	1.3	71398	3.1	9.9 h	16	1.33
Saturn	9.539	0.056	2.5	60330	26.4	10.2 h	22	0.70
Uranus	19.182	0.047	0.8	25400	98	17.24 h	17	1.15
Neptune	30.058	0.009	1.8	24750	29	16.05 h	8	1.55
Pluto	39.439	0.250	17.2	1100	94	6.4 h	1	2.1

Note. 1 a.u. $= 1.496 \times 10^8$ km.

Numerical modeling is a powerful method for analysing the formation process of accretion discs (Hayley and Starr 1988). The shape of the accretion disc depends on the balance between heating and cooling. The disc is heated by the incident jet of matter – which possesses a large amount of kinetic energy – the viscous dissipation of the orbital energy, and the radiation of the stars in the system and it is cooled by emission from its surface. Efficient cooling leads to a thin disc. If the heating source turns out to be more powerful the disc becomes thick ("plump") and like a torus (Hayley and Starr 1988).

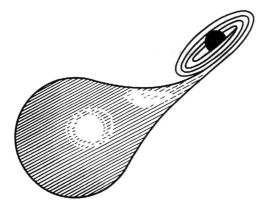

Fig. 2.16. Accretion disc formed by matter flowing between components in a binary system.

Accretion discs exist not only near stars but also near black holes (Shakura and Sunyaev 1973) which may be components of binary systems. There are also more exotic variants of accretion discs of stars themselves which surround massive black holes in the nuclei of active galaxies. An accretion disc in such a system is a powerful source of radiation, processing an appreciable fraction of the rest mass of the matter falling onto the black hole into quanta and it is one of the most efficient energy sources of our Universe.

We note that the viscosity of gaseous accretion discs in binary systems is very large, due to turbulence (Fridman and Ozernoy 1991). The dynamics of thin accretion discs is also interesting because spiral waves can be excited in them – due to the resonance action of a neighbouring star.

2.7 Galactic Discs

The history of galaxies, like the history of planetary rings, starts in 1610 and is connected with the name of Galileo. We shall relate this history, following Mitton (1976).

In 1610 in his "Stellar Herald" Galileo described his astonishing discoveries: "The Milky Way is an accumulation of countless stars". However, the fast development of the science of galaxies started in an unusual way. Halley published in 1705 a book about comets where he expressed the opinion that the Halley comet – as it was called later – had a periodic motion. After this book a real hunt for comets started which often looked like hazy spots in the sky. Then one discovered fixed nebulosities which clearly were no comets. In order that these fixed "pseudocomets" should not get under the feet of the hunters for real comets, Messier and Meshain collected a catalogue of 103 of such nebulosities and this became the important first catalogue of galaxies. In 1750 the Englishman Wright in his booklet "Original Theory or New Hypothesis of the Universe" correctly interpreted the Milky Way as clearly a picture of our own disc-shaped galaxy "from the inside" and the bright nebulae as accumulations of faraway stars like our own galaxy. A few years later Kant developed similar ideas – the island universes model – indicating that the disc shape of the Milky Way and of the far nebulae may be determined by rotation. Herschel made an important contribution to the science of nebulae when he divided the observed nebulae into galactic, that is, intergalactic, nebulae – planetary nebulae and gas clouds in our galaxy – and extragalactic nebulae which are other galaxies. Parsons, third Count Rosse, started in 1845 observations using the 180 cm telescope, the largest of the time and constructed by Lord Rosse himself, and made important discoveries: in the M51 galaxy – this means the 51st object in Messier's catalogue – he observed a spiral structure. This discovery greatly excited – and still excites – the astronomers. This persistent interest is connected not only with the aesthetic thrill when one sees such grandiose spirals: the theory of spiral waves is still far from complete (Fridman 1975, Fridman and Polyachenko 1984).

Notwithstanding impressive observational successes there was a large divergence of opinions as far as understanding the nature of the galaxies was concerned. Controversies went on about the distances to the extragalactic nebulae and about whether they were gaseous or stellar objects. There were difficulties in constructing a model of our Milky Way: right up to the twenties it was assumed that the Sun was at its centre.

Shapley from Harvard constructed a model of our galaxy which is close to reality in size and in the position of the Sun which is an ordinary star at the edge of the galaxy. Hubble, using the largest telescope of his time – the 2.5 m Mount Wilson telescope – distinguished separate stars in the large spiral Andromeda nebula M31 thus confirming the hypothesis that these nebulae are huge conglomerations of stars. In 1923 Hubble found in M31 several

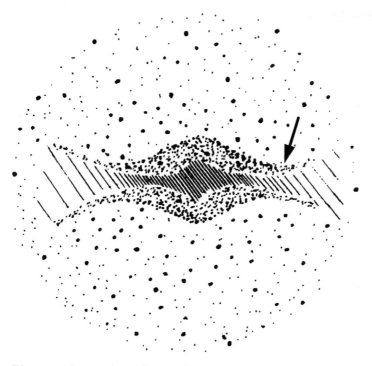

Fig. 2.17. View of our Galaxy from the edge. The disc-shaped component of gas, dust, and stars is embedded in a spherical halo of old stars and star clusters. The position of the Sun is indicated by an arrow.

pulsating variable stars – Cepheids – which made it possible to determine the distance from M31 – more than a million light years! The Universe was suddenly thrown open in front of a stunned mankind in all its grandeur.

We note that Hubble became the creator of the present-day classification of galaxies and in 1929 discovered the recession of the galaxies, that is, the expansion of the Universe.

From the whole range of galaxies we are interested only in those which have a disc-shaped component. In those discs one often meets with spirals which indicates the wealth of dynamics of such systems. A typical spiral galaxy, like our own galaxy, is a set of discs of stars, gas, and dust, embedded in a spherical halo (Fig. 2.17). Collective processes play an important role in the dynamics of the stellar and gaseous discs (Fridman 1975, 1986, Fridman and Polyachenko 1984). In the study of wave processes in the stellar subsystem one uses the collisionless approximation; the dynamics of the gaseous disc is described by hydrodynamic equations (Fridman 1990) where viscosity plays an important role (Fridman 1989). Galactic discs differ greatly from the discs considered earlier by us in a relatively slow rotation and a significantly larger size. In the next section we shall discuss the comparative characteristics of the various discs.

2.8 Comparative Analysis

Before we start to construct theoretical models we give a comparative analysis of planetary discs and other disc systems, based mainly on observational data given in the previous sections (see also Goldreich and Tremaine 1982, Gor'kavyĭ and Fridman 1985b, 1990, Gresh 1989, Fridman and Gor'kavyĭ 1992).

2.8.1 Primary and Secondary Rings

All the ring structures around the planets can be divided into two classes: "primary" and "secondary" rings (Gor'kavyĭ and Fridman 1985a, Fridman and Gor'kavyĭ 1992). The first class of structures are dense rings with rather large particles – up to several meters in size. These rings have a rather long life-time, which may clearly be comparable to cosmological times. Certainly the classical A, B, and C rings of Saturn, the nine dense Uranian rings, the primary Neptunian rings – including the Adams ring with its segments, and the main Jovian ring with the large particles belong to the primary class. The second kind of rings are the rarefied gaseous dusty rings which need a constant influx of matter in order to exist for a long time. Such rings cannot be observed from Earth because they are transparent and are usually detected by interplanetary spacecraft. To this second class belong the E ring – with the satellite Enceladus as the source of matter – and G ring of Saturn, the dusty Uranian rings, which are positioned between the dense rings and are connected with the sweeping up of small grains from these rings, the gaseous torus on the orbit of Io, which is the most volcanically active Jovian satellite, and the rarefied Jovian rings – the halo and the gossamer and faint rings.

One observes a characteristic feature in all ring systems: while secondary rings may occur at any distance from the planet – depending of the position of the "maternal" matter source – the outer radius of the primary rings is sharply limited and equal to about two radii of its "own" planet: 1.8 for Jupiter, 2.3 for Saturn, 2.0 for Uranus, and 2.5 for Neptune (Fig. 2.18). The band of satellites starts, as a rule beyond the limit of the primary rings, but in a narrow "boundary" zone of the ring there may also be satellites "mixed in". An exception is the Neptunian system where inside the zone of primary rings there are successively three satellites (see Fig. 2.18). The secondary rings consist of particles of micrometric or submicrometric size. The primary Saturnian rings contain particles of sizes ranging from micrometers to 10 to 20 meters and meter-size bodies form the main mass of the rings. The absence

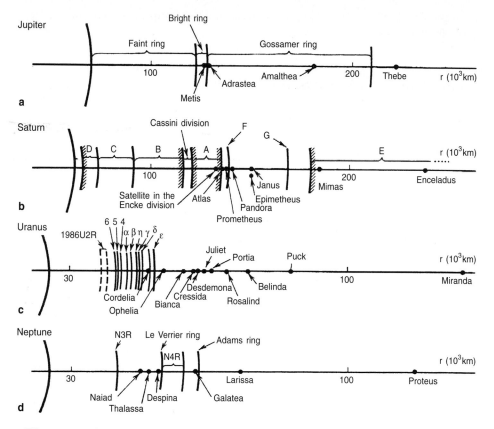

Fig. 2.18a–d. Relative position of the rings and nearest satellites in the systems of the four giant planets.

of larger-size particles is an important feature of the planetary rings and is also observed in the Uranian system (Gresh 1989). The optical depth of the rings is determined by both meter-size and smaller particles.

2.8.2 Density Distribution in the Systems of the Giant Planets

We show in Fig. 2.19 the assumed density distribution in the discs around Jupiter, Saturn, Uranus, and Neptune. The method of constructing this distribution is simple: the mass of a satellite is "smeared out" in the ring zone between the middle distances to the nearest satellites. We have combined the masses of Atlas, Prometheus, and Pandora, and of Janus and Epimetheus.

The Uranian and Neptunian systems turned out to be different in principle as regards the mass of the proto-disc in the ring zone. The peak of the surface density (60 g/cm^2) in the inner region of the Uranian system lies in

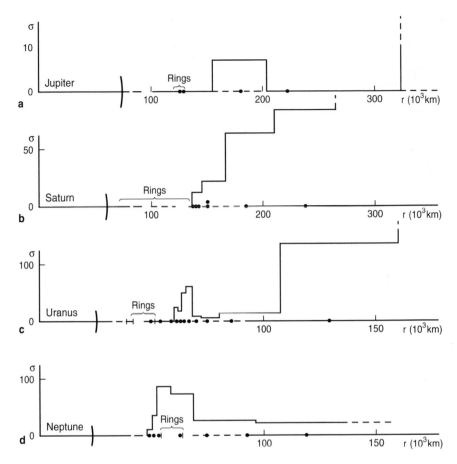

Fig. 2.19a–d. Reconstruction of the surface density in the systems of the giant planets and in the solar system in the proto-stage. The density of the rings has been neglected. The volume density of the satellites is assumed to be 1 g/cm^3.

the Portia zone, after which the density systematically decreases with decreasing orbital radius and in the zone of the shepherd-satellites Ophelia and Cordelia, bounding the ring zone, has reached 1 g/cm^2. The current density of matter in the ring zone is approximately still another order lower. In contrast to the Uranian rings, the Neptunian rings are positioned in the zone of maximum density in the disc which indicates completely different conditions of formation. In particular, the density in the Neptunian proto-disc was close to the critical density at which the disc is subject to the Jeans instability. This may be connected with the peculiarities of the Neptunian system: the presence of satellites in the zone of the primary rings and the existence of a compaction in one of the rings.

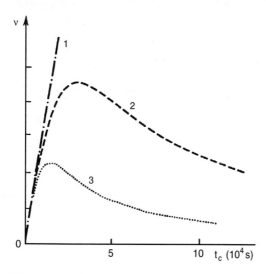

Fig. 2.20. The kinematic viscosity of a rotating system as function of the mean free flight time. *1:* $\Omega = 0$; *2:* $\Omega = 3 \times 10^{-4} \text{ s}^{-1}$; *3:* $\Omega = 6 \times 10^{-4} \text{ s}^{-1}$.

2.8.3 Dissipation in a Disc System

Slightly anticipating later discussions we write down from the equations of viscous hydrodynamics (Landau and Lifshitz 1987) the equation for the increase in the temperature T_0 per unit mass of a differentially rotating viscous disc:

$$\frac{3}{2} \frac{\partial T_0}{\partial t} = \nu \left(r \frac{d\Omega_0}{dr} \right)^2, \tag{2.1}$$

where ν is the kinematic viscosity. Since T_0 is the temperature (strictly speaking, the energy per unit mass) we have $2T_0 = c_s^2$ where c_s is the sound velocity. In Eq. (2.1) and henceforth the index "0" will indicate unperturbed quantities which are independent of the angle φ. Equation (2.1) has been written down just for those quantities and we have neglected the thermal conductivity. Dividing the left- and right-hand sides by c_s^2 and integrating (2.1) over the time from 0 to t we find

$$c_s^2(t) = c_s^2(0) \exp \left[\frac{4}{3} \int_0^t S(t)\, dt \right],$$

where

$$S(t) \equiv \frac{\nu}{c_s^2} \left(r \frac{d\Omega_0}{dr} \right)^2. \tag{2.2}$$

The meaning of $S(t)$ follows from the fact that

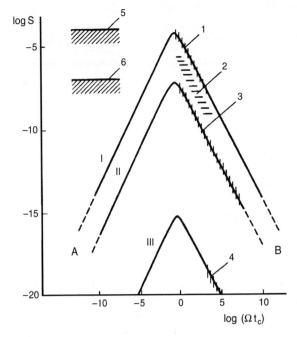

Fig. 2.21. The dissipative function S for various disc systems. *I*: $\Omega = 10^{-4}$ s^{-1}; *II*: $\Omega = 10^{-7}$ s^{-1}; *III*: $\Omega = 10^{-15}$ s^{-1}. Region *1*: planetary rings; *2*: proto-satellite discs; *3*: proto-planetary planetesimal disc; *4*: galactic disc of stars; *5* and *6*: maximally turbulised accretion discs with angular frequencies of 10^{-4} and 10^{-7} s^{-1}. *A*: region of classical gas dynamics; *B*: region of celestial mechanics.

$$\tfrac{4}{3} S(t) \equiv \frac{4}{3} \frac{\nu}{c_s^2} \left(r \frac{d\Omega_0}{dr} \right)^2 = \frac{1}{t_{ch}},$$

where t_{ch} is a characteristic time over which the disc temperature increases by a factor e due to the viscous dissipation of the energy of the orbital motion. The dissipative index S of the system reflects whether it is energetically possible for the system to self-organise, that is, to reach spatial structure. To use Prigogine's language (Nicolis and Prigogine 1980) S is the openness index of the system. The viscosity occurring in S depends in a complex manner on the collision frequency in the disc and we derive this function in Chap. 7. We note merely a basic difference between the viscosity of a normal gas and the many-particle medium of the disc: the viscosity of a gas increases with increasing mean free flight time, that is, with increasing rarefaction of the gas – a well known paradox (Balescu 1975). The same is true for the viscosity of the disc as long as the mean free flight time is considerably shorter than the period of rotation and each particle collides several times during a single revolution. In discs where the free flight time is comparable with or longer than the period of rotation the situation is radically changed – the viscosity

Table 2.9. Characteristic parameters of disc systems

System	Size R [km]	Oblateness h/R	Surface density [g/cm^2]	Degree of dissipation S^*	Age [number of revolutions]	Angular velocity [s^{-1}]
Planetary ring	10^5	10^{-6}	10	10^{-6}–10^{-5}	10^{12}	10^{-4}
Protosatellite disc:						
gas component	10^6	0.1	10^4–10^6	10^{-6}–10^{-5}	10^8–10^{11}	10^{-6}–10^{-5}
dust component	10^6	10^{-6}–10^{-4}	10^1–10^2	10^{-8}–10^{-5}	10^8–10^{11}	10^{-6}–10^{-5}
Protoplanetary disc:						
gas component	5×10^9	0.1	10^2–10^3	10^{-8}–10^{-7}	10^7–10^9	10^{-8}–10^{-6}
dust component	5×10^9	10^{-5}–10^{-3}	10	10^{-15}–10^{-8}	10^7–10^9	10^{-8}–10^{-6}
Cold accretion disc	10^7	< 0.1	0.01	10^{-5}–10^{-4}	> 100	10^{-4}
Galactic stellar disc	3×10^{18}	0.1	0.01	10^{-20}–10^{-18}	10–100	10^{-15}

* Gaseous systems are assumed to be maximally turbulent.

starts to decrease steeply with increasing rarefaction of the disc (Fig. 2.20). Such a major difference is connected with the fact that a particle in a non-rotating medium moves along straight trajectories, which are deflected only in collisions and transfer momentum over long distances. In a rotating rarefied medium the particles move under the action of the planetary field along curved orbits – in a rotating system of coordinates this looks like the motion of a particle along a closed ellipse (see Chap. 3). In exactly the same way the viscosity of a plasma changes across a magnetic field – because the particle trajectories circle around the magnetic field lines along Larmor circles. Of course, under such conditions the ability of a particle to transfer momentum decreases steeply and there remains for it nothing else but patiently to await collisions with a closely approaching particle.

After what we have just said the shape of the dependence of the quantity S on the parameter Ωt_c (t_c is the free flight time) shown in Fig. 2.21 for various disc systems becomes understandable. Figure 2.21 shows that the planetary discs are champions in dissipativity and only accretion discs with maximum turbulisation can compete with them. In this respect the rings are as far as the collision frequency is concerned in a boundary region – between celestial mechanics and classical hydrodynamics, confirming that it is justified to apply the term "celestial mechanics of a continuous medium" to the dynamics of planetary rings.

2.8.4 Table of the Parameters of Disc Systems

We give in Table 2.9 the main characteristics of disc systems: size, oblate-ness, dissipativity, and the rotational parameters. The information given in the table needs no comments and confirms the idea expressed earlier that the rings are recordholders in flattening, life-time (expressed in rotational peri-ods), and in dissipativity. As to the latter index the rings may be beaten by cold accretion discs.

With this we end the review of observational data of planetary rings and similar disc systems. The remainder of the book is devoted to theoretical models of the planetary rings.

3 Celestial Mechanics Minimum

If a scientist does not understand a problem, he writes down many formulae, but as soon as he has understood it, there remain at most two.

Niels Bohr

Earl of Gloster: "When shall I come to the top of that same hill?"
Edgar: "You do climb up it now: look, how we labour."
Earl of Gloster: "Methinks the ground is even."

W. Shakespeare, King Lear, Act IV, Scene VI.

In the present chapter we shall give those results from celestial mechanics which we shall use in the present book. This makes our exposition more self-contained and makes it unnecessary for the reader to turn to textbooks on celestial mechanics.

3.1 Basic Equations

Let us consider two equations: Newton's Second Law of Motion and the Newtonian Law of Universal Gravitation:

$$\boldsymbol{F} = m\boldsymbol{a}, \tag{3.1}$$

$$\boldsymbol{F} = -\frac{GMm}{R^3} \boldsymbol{R}. \tag{3.2}$$

Bearing in mind that the acceleration satisfies the relation $\boldsymbol{a} = d^2\boldsymbol{R}/dt^2$ we get from (3.1) and (3.2) the vector equation of motion of a particle in the gravitational field of a spherical body,

$$\frac{d^2\boldsymbol{R}}{dt^2} = -\frac{GM}{R^3} \boldsymbol{R}, \tag{3.3}$$

where $\boldsymbol{R}$ is the radius vector of a test particle relative to a central body of mass M. From (3.3) we get the following set of equations in the Cartesian system of coordinates x', y', z':

$$\frac{d^2 x'}{dt^2} = -\frac{GM}{R^3} x', \tag{3.4}$$

$$\frac{d^2 y'}{dt^2} = -\frac{GM}{R^3} y', \tag{3.5}$$

$$\frac{d^2 z'}{dt^2} = -\frac{GM}{R^3} z', \tag{3.6}$$

where $R = \sqrt{x'^2 + y'^2 + z'^2}$. It is convenient when we study rotating discs to use a system of coordinates rotating with an angular velocity Ω – in general, corresponding to a rotation vector $\boldsymbol{\Omega}$. We shall take the z-axis along the axis of rotation. We write the set (3.4) to (3.6) in a rectangular rotating frame by using the formulae

$$x' = x \cos \Omega t - y \sin \Omega t, \tag{3.7}$$

$$y' = x \sin \Omega t + y \cos \Omega t, \tag{3.8}$$

$$z' = z. \tag{3.9}$$

Substituting (3.7) into (3.4) and (3.8) into (3.5) we find

$$\left(\ddot{x} - 2\Omega \dot{y} - \Omega^2 x \right) \cos \Omega t - \left(\ddot{y} + 2\Omega \dot{x} - \Omega^2 y \right) \sin \Omega t$$
$$= -\frac{GM}{R^3} \left(x \cos \Omega t - y \sin \Omega t \right), \tag{3.10}$$

$$\left(\ddot{x} - 2\Omega \dot{y} - \Omega^2 x \right) \sin \Omega t + \left(\ddot{y} + 2\Omega \dot{x} - \Omega^2 y \right) \cos \Omega t$$
$$= -\frac{GM}{R^3} \left(x \sin \Omega t + y \cos \Omega t \right). \tag{3.11}$$

We multiply (3.10) by $\cos \Omega t$ and (3.11) by $\sin \Omega t$, and add the resulting equations. After that we multiply (3.10) by $-\sin \Omega t$ and (3.11) by $\cos \Omega t$, and again add the resulting equations. Finally, substituting (3.9) into (3.6), we get the set of equations we are looking for:

$$\ddot{x} - 2\Omega \dot{y} - \Omega^2 x = -\frac{GM}{R^3} x, \tag{3.12}$$

$$\ddot{y} + 2\Omega \dot{x} - \Omega^2 y = -\frac{GM}{R^3} y, \tag{3.13}$$

$$\ddot{z} = -\frac{GM}{R^3} z, \tag{3.14}$$

where now $R = \sqrt{x^2 + y^2 + z^2}$. The set (3.12) to (3.14) can be written in vector form as follows:

$$\frac{d\boldsymbol{V}}{dt} + 2[\boldsymbol{\Omega} \wedge \boldsymbol{V}] + [\boldsymbol{\Omega} \wedge [\boldsymbol{\Omega} \wedge \boldsymbol{R}]] = \frac{\boldsymbol{F}}{m}, \tag{3.15}$$

where $\boldsymbol{V} = d\boldsymbol{R}/dt$. The term $2[\boldsymbol{\Omega} \wedge \boldsymbol{V}]$ describes the Coriolis force and $[\boldsymbol{\Omega} \wedge [\boldsymbol{\Omega} \wedge \boldsymbol{R}]]$ the centrifugal force.

Let us consider the equations of motion in cylindrical coordinates. We use the equations

$$x' = r\cos\varphi, \qquad y' = r\sin\varphi, \qquad z' = z \tag{3.16}$$

to change from the Cartesian coordinates x', y', z' to the cylindrical coordinates r, φ, z. Substituting these equations into (3.4) to (3.6) we get

$$\left[\ddot{r} - r(\dot{\varphi})^2\right]\cos\varphi - (2\dot{r}\dot{\varphi} + r\ddot{\varphi})\sin\varphi = -\frac{GM}{R^3}r\cos\varphi, \tag{3.17}$$

$$\left[\ddot{r} - r(\dot{\varphi})^2\right]\sin\varphi + (2\dot{r}\dot{\varphi} + r\ddot{\varphi})\cos\varphi = -\frac{GM}{R^3}r\sin\varphi, \tag{3.18}$$

where now $R = \sqrt{r^2 + z^2}$. Proceeding with these last two equations as we did with (3.10) and (3.11) we get the following equations of motion in cylindrical coordinates in an inertial frame:

$$\frac{d^2r}{dt^2} - r\left(\frac{d\varphi}{dt}\right)^2 = -\frac{GM}{R^3}r, \tag{3.19}$$

$$\frac{1}{r}\frac{d}{dt}\left(r^2\frac{d\varphi}{dt}\right) = 0, \tag{3.20}$$

$$\frac{d^2z}{dt^2} = -\frac{GM}{R^3}z, \tag{3.21}$$

Substituting (3.16) into (3.12) to (3.14) we get similarly the set of equations of motion in cylindrical coordinates in a rotating frame:

$$\frac{d^2r}{dt^2} - r\left(\frac{d\varphi}{dt}\right)^2 - 2\Omega r\frac{d\varphi}{dt} - r\Omega^2 = -\frac{GM}{R^3}r, \tag{3.22}$$

$$\frac{1}{r}\frac{d}{dt}\left(r^2\frac{d\varphi}{dt}\right) + 2\Omega\frac{dr}{dt} = 0, \tag{3.23}$$

$$\frac{d^2z}{dt^2} = -\frac{GM}{R^3}z. \tag{3.24}$$

We note that to find the equations of motion of a test particle in the gravitational field of several bodies we must substitute on the right-hand sides of the equations we obtained the sum of the forces exerted by the N attractive centres. For instance, in the case of three attractive bodies (say, a planet of mass M and two satellites of masses m_1 and m_2) (3.12) and (3.13) will look as follows:

$$\ddot{x} - 2\Omega\dot{y} - \Omega^2 x = -\frac{GM}{R^3}x + \frac{Gm_1}{R_1^3}(x_1 - x) + \frac{Gm_2}{R_2^3}(x_2 - x), \tag{3.25}$$

$$\ddot{y} + 2\Omega\dot{x} - \Omega^2 y = -\frac{GM}{R^3}y + \frac{Gm_1}{R_1^3}(y_1 - y) + \frac{Gm_2}{R_2^3}(y_2 - y), \tag{3.26}$$

where R_1 and R_2 are the distances from the test particle to the two satellites with coordinates $\boldsymbol{r}_1$ and $\boldsymbol{r}_2$.

3.2 Solution for a Single Point Particle

Let us consider the equations of motion (3.19) to (3.21) in an inertial frame for the case of cylindrical coordinates in the two-dimensional case ($z = 0$). We write (3.19) in the form

$$\ddot{r} - r\dot{\varphi}^2 = -\frac{GM}{r^2}. \tag{3.27}$$

From (3.20) we get the Law of Areas, or the conservation of angular momentum:

$$r^2\dot{\varphi} \equiv L = \text{constant}. \tag{3.28}$$

From (3.28) we get the following relations:

$$(r\dot{\varphi})^2 = \frac{L^2}{r^2}, \qquad dt = \frac{r^2}{L}\, d\varphi. \tag{3.29}$$

Writing $U = 1/r$ and using (3.29) we can reduce (3.27) to the form

$$U^2 L \frac{d}{d\varphi}\left\{ U^2 L \frac{d}{d\varphi}\frac{1}{U}\right\} - U^3 L^2 = -GMU^2. \tag{3.30}$$

After some simple transformations (3.30) takes the form

$$\frac{d^2 U}{d\varphi^2} + U = \frac{GM}{L^2}, \tag{3.31}$$

where we have used the fact that L is a constant of the motion so that $\partial L/\partial\varphi = 0$. The general solution of (3.31) is

$$U = \frac{GM}{L^2} + A\cos(\varphi - \varphi_0), \tag{3.32}$$

where A and φ_0 are integration constants. We introduce the notation

$$p \equiv \frac{L^2}{GM}, \qquad e \equiv \frac{AL^2}{GM}. \tag{3.33}$$

After changing back from U to r we get from (3.32) the equation of a conical section in polar coordinates:

$$r = \frac{p}{1 + e\cos(\varphi - \varphi_0)}. \tag{3.34}$$

By solving the two-body problem we have thus found the trajectories of a test particle in the gravitational field of a point particle or a sphere. These trajectories are described by (3.34) and can be classified according to the magnitude of the eccentricity e: for $e = 0$ the trajectory is a circle, for $0 < e < 1$ it is an ellipse, for $e = 1$ a parabola, and for $e > 1$ a hyperbola

(Fig. 3.1). If we choose the polar coordinates r and f such that SB $= r$, $f = \angle$ASB, (3.34) becomes

$$r = \frac{p}{1 + e \cos f} = \frac{a(1 - e^2)}{1 + e \cos f}. \tag{3.35}$$

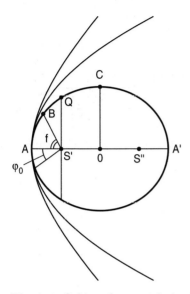

Fig. 3.1. Orbits of a particle in the gravitational field of a central body.

We have introduced here the major semi-axis:

$$a \equiv \tfrac{1}{2} \text{AA}',$$

the minor semi-axis:

$$b \equiv \text{OC} = a\sqrt{1 - e^2}, \tag{3.36}$$

and the (focal) parameter ($2p$ is called the latus rectum of the orbit)

$$p \equiv \text{SQ} = a(1 - e^2). \tag{3.37}$$

The quantity p is called the parameter of the conical section. The point A is called the pericentre and A$'$ the apocentre. In the pericentre we have

$$r = a(1 - e), \tag{3.38a}$$

and in the apocentre

$$r = a(1 + e). \tag{3.38b}$$

Let us consider the problem of the orbital period of a particle. We note that in (3.28) the constant L is equal to twice the areal velocity. As the area of the ellipse is equal to $\pi a b$ and this area is swept through in one period T, we have

$$\frac{2\pi a b}{T} = L,$$

or, using Eq. (3.36) for the minor semi-axis,

$$2\pi a^2 \sqrt{1 - e^2} = LT.$$

From (3.33) and (3.37) for the parameter p we find

$$\frac{L^2}{GM} = a(1 - e^2), \tag{3.39}$$

and using (3.39) to eliminate L from (3.38) we get for the orbital period

$$T = 2\pi \sqrt{\frac{a^3}{GM}}. \tag{3.40}$$

If we introduce the orbital frequency $\Omega = 2\pi/T$ we can rewrite this extremely important relation in the form

$$\Omega = \sqrt{\frac{GM}{a^3}}. \tag{3.41}$$

For circular orbits we have $r = a$ and the frequency therefore decreases according to the law $\Omega \propto r^{-3/2}$. This relation is known as Kepler's Third Law and discs rotating in agreement with this law are known as "Keplerian discs".

Let us consider the problem of the velocity of a particle in an elliptical orbit. This velocity is directed along the tangent to the ellipse; it has a radial component $\dot{r}$ and a transverse component $r\dot{f}$. Hence we have

$$V^2 = \dot{r}^2 + r^2 \dot{f}^2. \tag{3.42}$$

Differentiating (3.35) with respect to t and using the law of areas (3.28) to eliminate $\dot{f}$ and (3.35) to eliminate r we find

$$\dot{r} = \frac{L}{p} e \sin f. \tag{3.43}$$

Moreover, from (3.28) and (3.35) we get

$$r\dot{f} = \frac{L}{r} = \frac{L}{p}(1 + e \cos f). \tag{3.44}$$

Squaring (3.43) and (3.44) and adding the results we get the following expression for V^2:

$$V^2 = \frac{L^2}{p^2}\left(1 + 2e\cos f + e^2\right), \tag{3.45}$$

or

$$V^2 = \frac{L^2}{p^2}\left[2(1 + e\cos f) - (1 - e^2)\right]. \tag{3.46}$$

Using (3.35) we get

$$V^2 = \frac{2L^2}{rp} - \frac{L^2}{p^2}\left(1 - e^2\right). \tag{3.47}$$

Since $L^2/GM = p = a(1 - e^2)$ we have

$$V^2 = GM\left(\frac{2}{r} - \frac{1}{a}\right). \tag{3.48}$$

In the pericentre where $r = a(1 - e)$ the velocity V has its maximum:

$$V_{\mathrm{pc}}^2 = \frac{GM}{a}\frac{1 + e}{1 - e}, \tag{3.49}$$

and in the apocentre where $r = a(1+e)$ the particle velocity has its minimum:

$$V_{\mathrm{ac}}^2 = \frac{GM}{a}\frac{1 - e}{1 + e}. \tag{3.50}$$

For circular orbits we have $a = r$ and from (3.48) we get the well known relation

$$V^2 = \frac{GM}{r}. \tag{3.51}$$

We have thus considered the motion of particles in the gravitational field of a spherical body: we have found the trajectories, the rotational period in a closed orbit, and the velocity in an elliptical orbit. All these solutions referred to a non-rotating, or inertial, frame of reference. Let us now consider the motion of a particle round a planet in a rotating frame. We write down the equations of motion (3.12) and (3.13) for the two-dimensional case:

$$\ddot{x} - 2\Omega\dot{y} - \Omega^2 x = -\frac{GM}{(x^2 + y^2)^{3/2}}x, \tag{3.52}$$

$$\ddot{y} + 2\Omega\dot{x} - \Omega^2 y = -\frac{GM}{(x^2 + y^2)^{3/2}}y. \tag{3.53}$$

Let the y-axis be directed along the orbit in the direction of the motion and the x-axis along the radius from the planet (Fig. 3.2). If the particle is moving along a circular orbit with angular velocity $\Omega = \sqrt{GM/x_*^3}$, it will be at rest in the point $(x_*, 0)$ in a frame of reference which is rotating with the same velocity. Let us now consider a small deviation from the circular

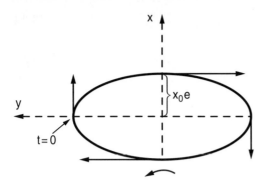

Fig. 3.2. Trajectory of the epicyclic motion of a particle in the rotating frame of reference. The planet is at the bottom; the orbital motion and the rotation of the frame of reference are both counterclockwise.

orbit – and hence from the point $(x_*, 0)$: $x = x_* + x'$, $x' \ll x_*$, $y \ll x_*$. If we use the fact that $y \ll x_*$ we can write (3.52) and (3.53) in the form

$$\ddot{x} - 2\Omega\dot{y} = x\left(\Omega^2 - \frac{GM}{x^3}\right), \tag{3.54}$$

$$\ddot{y} + 2\Omega\dot{x} = y\left(\Omega^2 - \frac{GM}{x^3}\right). \tag{3.55}$$

Putting $x = x_* + x'$ and linearising, we get

$$\ddot{x}' - 2\Omega\dot{y} = 3\Omega^2 x', \tag{3.56}$$

$$\ddot{y} + 2\Omega\dot{x}' = 0. \tag{3.57}$$

Differentiating (3.56) and using (3.57) to eliminate $\ddot{y}$ we find

$$\dddot{x}' + \Omega^2\dot{x}' = 0, \tag{3.58}$$

and hence

$$\dot{x}' = C_1 \cos\Omega t + C_2 \sin\Omega t, \tag{3.59}$$

$$x' = \frac{C_1}{\Omega}\sin\Omega t - \frac{C_2}{\Omega}\cos\Omega t + C_3. \tag{3.60}$$

Determining the integration constants in such a way that $x' = 0$ for $t = 0$ and $t = \pi/\Omega$, we have $C_2 = C_3 = 0$. According to (3.35) we have in the apocentre $x = (1 + e)x_*$ and hence $x' = ex_*$. We thus find that $C_1 = \Omega e x_*$. The final result is now

$$\dot{x}' = \Omega e x_* \cos\Omega t. \tag{3.61}$$

From (3.61) and (3.57) we find an equation for y:

$$\ddot{y} = -2\Omega^2 e x_* \cos \Omega t, \tag{3.62}$$

whence

$$\dot{y} = -2\Omega e x_* \sin \Omega t + C_4, \tag{3.63}$$

and

$$y = 2e x_* \cos \Omega t + C_4 t + C_5. \tag{3.64}$$

Let us have $y = 0$ for $t = \pi/2\Omega + n\pi/\Omega$, or $t = T/4 + nT/2$, with $n = 0, 1, 2, \ldots$. We then have $C_4 = C_5 = 0$ and from (3.60) and (3.64):

$$x' = e x_* \sin \Omega t, \tag{3.65}$$

$$y = 2e x_* \cos \Omega t. \tag{3.66}$$

Squaring (3.65) and (3.66) and adding we have the equation of an ellipse:

$$x'^2 + \tfrac{1}{4} y^2 = e^2 x_*^2.$$

Therefore, if the particle moves in the inertial system along an elliptical orbit with a small eccentricity c, in the rotating frame its trajectory traces an ellipse with a ratio of $\frac{1}{2}$ of the axes (see Fig. 3.2). This ellipse is called an epicycle, by analogy with the kindred, but hypothetical Ptolemaean epicycles. One sees easily from the equations (3.61) and (3.63) for the velocities that the maximum of V_y is twice the maximum of V_x and this explains that the epicycle is stretched out along the y-axis:

$$V_{x\,\text{max}} = \Omega e x_*, \qquad V_{y\,\text{max}} = 2\Omega e x_*.$$

When the eccentricity decreases to become zero, the epicycle is compressed into a point, while keeping the ratio of its axes unchanged. The period of rotation along the epicyclic is, of course, equal to the period of rotation of the frame of reference, or of the point particle itself around the planet.

Let us now consider the case when the epicycle is shifted along the x-axis by an amount x_0. We determine the integration constants in Eq. (3.60) from the condition that $x' = x_0$ for $t = 0$ and π/Ω, so that $C_2 = 0$, $C_3 = x_0$. Equation (3.64) must satisfy the condition that $y = 0$ for $t = 0$ so that $C_5 = 0$. From (3.63) we see that the integration constant C_4 is equal to $\dot{y}$ at time $t = 0$. To determine C_4 we look at (3.56). We find from (3.61) that at time $t = 0$ we have

$$\ddot{x}' = -\Omega^2 e x_* \sin \Omega t = 0.$$

Since we have $x' = x_0$ at $t = 0$ we find from (3.56) that

$$C_4 = \dot{y} = -\tfrac{3}{2}\Omega x_0.$$

The trajectory of the particle when the epicycle is shifted will thus be described by the equations

$$x' = ex_* \sin \Omega t - x_0, \tag{3.67a}$$

$$y = 2ex_* \cos \Omega t - \tfrac{3}{2}\Omega x_0 t. \tag{3.67b}$$

We see from this that the epicycle drifts along the Y-axis with a drift velocity $-\tfrac{3}{2}\Omega x_0$.

3.3 Main Perturbing Factors

We shall now consider thin disc systems, in most cases around a massive centre – such as a planet or a star. This means that in most cases the trajectories of the particles are well described by Kepler ellipses and all factors perturbing the particle orbits are small. We may thus assume that the particle orbits are ellipses the elements of which, such as the major semi-axis and the eccentricity, change slowly under the effect of the perturbing factors. Plasma physicists would say that the orbit drifts, experts of celestial mechanics would argue that the elements of the orbit osculate. The problem of describing the particle orbits in the case of small perturbations reduces to finding the equations which describe the change in the elements of an elliptical orbit as a function of the perturbing factors. In the present section we obtain a set of equations for the osculating (drifting) elements and we consider the effect on the particle trajectory of such perturbing factors as the non-sphericity of the gravitational field of a planet, aerodynamic friction, interaction with the solar radiation, and collisions between the particles.

3.3.1 Equations for the Osculating Orbital Elements

In our derivation of these equations we follow standard textbooks (see, for instance, Brouwer and Clemence 1961 or Roy 1978). We write (3.3) in the form

$$\frac{d^2 \boldsymbol{R}_0}{dt^2} = \nabla U_0, \tag{3.68}$$

where

$$U_0 = \frac{GM}{R_0}. \tag{3.69}$$

Let us now add to the potential energy U_0 a small perturbing potential energy U':

$$\frac{d^2 \boldsymbol{R}}{dt^2} = \nabla \left(U_0 + U' \right). \tag{3.70}$$

The solution of (3.70) will give us the coordinates of the body moving along the elliptical orbit as function of the parameters of the orbit and of the time:

$$x, y, z = f_k(a, e, i, \varphi, \varpi, \varepsilon, t), \qquad k = 1, 2, 3, \tag{3.71}$$

where $a, e, i, \varphi, \varpi, \varepsilon$ are the slowly changing parameters of the elliptical orbit: the major semi-axis, the eccentricity, the inclination, the length of the ascending node, the argument of the periapse, and the time of the perihelion passage.

The inclination i is the angle at which the orbital plane intersects the base plane – for which one chooses usually the equatorial plane of the planet. The particle moving along an inclined orbit intersects the base plane in two points – the ascending and the descending nodes. The angle between the co-ordinate axis in the base plane and the radius vector of the particle when it passes through the ascending node is called the length of the ascending node. The sum of the length of the ascending node and the angle between the radius vectors of the particle as it passes through the ascending node and the pericentre of the orbit is called the argument of the periapse. These characteristics together with a and e uniquely determine the position of the orbit in space. The time of perihelion passage enables us to determine the position of the particle on this trajectory.

Let us denote the parameters of the orbit by c_i; let us assume that at a particular, given time t_0 they have the values c_{0i} which satisfy (3.68), that is,

$$x_0, y_0, z_0 = f_k(c_{0i}, t); \qquad i = 1, \ldots, 6; \qquad k = 1, 2, 3. \tag{3.71a}$$

We now look for solutions $f_k(c_i, t)$ such that the c_i at t_0 satisfy (3.71a). If we now assume that the parameters of the orbit c_i ($i = 1, \ldots, 6$) depend on the time, we get by differentiating (3.71)

$$\frac{dx}{dt} = \frac{\partial x}{\partial t} + \sum_{i=1}^{6} \frac{\partial x}{\partial c_i} \frac{dc_i}{dt}. \tag{3.72}$$

However, in the "zeroth approximation" (3.71a) we have

$$\frac{d\mathbf{R}_0}{dt} = \frac{\partial \mathbf{R}_0}{\partial t}, \tag{3.73}$$

and

$$\sum_{i=1}^{6} \frac{\partial x_0}{\partial c_i} \frac{dc_{0i}}{dt} = 0, \tag{3.74}$$

with similar equations for y and z. Differentiating the x-component of (3.72) we find

$$\frac{d^2 x}{dt^2} = \frac{\partial}{\partial t}\left(\frac{dx_0}{dt}\right) + \sum_{i=1}^{6} \frac{\partial}{\partial c_i}\left(\frac{dx_0}{dt}\right) \frac{dc_i}{dt}, \tag{3.75}$$

However, according to (3.73) we have

$$\frac{dx_0}{dt} = \frac{\partial x_0}{\partial t}, \tag{3.76}$$

and hence

$$\frac{d^2 x}{dt^2} = \frac{\partial^2 x_0}{\partial t^2} + \sum_{i=1}^{6} \frac{\partial \dot{x}_0}{\partial c_i} \frac{dc_i}{dt}. \tag{3.77}$$

From (3.69) and (3.70) we have

$$\frac{\partial^2 x_0}{\partial t^2} = \frac{\partial U_0}{\partial x}, \tag{3.78}$$

$$\frac{d^2 x}{dt^2} = \frac{\partial U_0}{\partial x} + \frac{\partial U'}{\partial x}. \tag{3.79}$$

From (3.77) to (3.79) it follows that

$$\sum_{i=1}^{6} \frac{\partial \dot{x}_0}{\partial c_i} \frac{dc_i}{dt} = \frac{\partial U'}{\partial x}. \tag{3.80}$$

We get similar equations for y and z. We now drop the subscript "0" and use (3.74) and (3.80) and a Lagrange transformation to determine the required equations for the dc_i/dt (Brouwer and Clemence 1961). We multiply the six equations (3.74) and (3.80) successively by

$$-\frac{\partial \dot{x}}{\partial c_j}, \quad -\frac{\partial \dot{y}}{\partial c_j}, \quad -\frac{\partial \dot{z}}{\partial c_j}, \quad \frac{\partial x}{\partial c_j}, \quad \frac{\partial y}{\partial c_j}, \quad \frac{\partial z}{\partial c_j},$$

respectively. For $j = 1$ we have

$$-\left(\sum_{i=1}^{6} \frac{\partial x}{\partial c_i} \frac{dc_i}{dt} \right) \frac{\partial \dot{x}}{\partial c_1} = 0,$$

$$-\left(\sum_{i=1}^{6} \frac{\partial y}{\partial c_i} \frac{dc_i}{dt} \right) \frac{\partial \dot{y}}{\partial c_1} = 0,$$

$$-\left(\sum_{i=1}^{6} \frac{\partial z}{\partial c_i} \frac{dc_i}{dt} \right) \frac{\partial \dot{z}}{\partial c_1} = 0,$$

$$\left(\sum_{i=1}^{6} \frac{\partial \dot{x}}{\partial c_i} \frac{dc_i}{dt} \right) \frac{\partial x}{\partial c_1} = \frac{\partial U'}{\partial x} \frac{\partial x}{\partial c_1},$$

$$\left(\sum_{i=1}^{6} \frac{\partial \dot{y}}{\partial c_i} \frac{dc_i}{dt} \right) \frac{\partial y}{\partial c_1} = \frac{\partial U'}{\partial y} \frac{\partial y}{\partial c_1},$$

$$\left(\sum_{i=1}^{6} \frac{\partial \dot{z}}{\partial c_i} \frac{dc_i}{dt} \right) \frac{\partial z}{\partial c_1} = \frac{\partial U'}{\partial z} \frac{\partial z}{\partial c_1},$$

$$\tag{3.81}$$

Adding the six equations (3.81) and repeating the exercise another five times with $j = 2, \ldots, 6$ we get the following set of equations:

$$\sum_{j=1}^{6} [c_i, c_j] \frac{dc_j}{dt} = \frac{\partial U'}{\partial c_i}, \qquad i = 1, 2, 3, 4, 5, 6, \tag{3.82}$$

where we have introduced the following Lagrange brackets:

$$\begin{aligned}
[c_i, c_k] &= \frac{\partial \dot{x}}{\partial c_k} \frac{\partial x}{\partial c_i} - \frac{\partial x}{\partial c_k} \frac{\partial \dot{x}}{\partial c_i} + \frac{\partial \dot{y}}{\partial c_k} \frac{\partial y}{\partial c_i} \\
&\quad - \frac{\partial y}{\partial c_k} \frac{\partial \dot{y}}{\partial c_i} + \frac{\partial \dot{z}}{\partial c_k} \frac{\partial z}{\partial c_i} - \frac{\partial z}{\partial c_k} \frac{\partial \dot{z}}{\partial c_i}.
\end{aligned} \tag{3.83}$$

We have used here the fact that U' does not depend on the velocity so that

$$\frac{\partial U'}{\partial x} \frac{\partial x}{\partial c_1} + \frac{\partial U'}{\partial y} \frac{\partial y}{\partial c_1} + \frac{\partial U'}{\partial z} \frac{\partial z}{\partial c_1} = \frac{\partial U'}{\partial c_1}. \tag{3.84}$$

From (3.83) it is clear that

$$[c_i, c_i] = 0, \qquad [c_i, c_k] = -[c_k, c_i]. \tag{3.85}$$

Hence, there are only 15 unknown Lagrange brackets; each of them depends only on the equations of the elliptical motion and does not depend explicitly on the time (Brouwer and Clemence 1961). In Brouwer and Clemence's book Whittaker's method for evaluating the Lagrange brackets is described in detail and we therefore give at once the results:

$$\begin{aligned}
[\varepsilon, a] &= \tfrac{1}{2}\Omega a, \\
[\varpi, a] &= -\tfrac{1}{2}\Omega a \left[1 - \sqrt{1 - e^2}\right], \\
[\varphi, a] &= -\tfrac{1}{2}\Omega u \sqrt{1 - e^2}(1 - \cos i), \\
[\varpi, e] &= -\frac{\Omega a^2 e}{\sqrt{1 - e^2}}, \\
[\varphi, e] &= \frac{\Omega a^2 e}{\sqrt{1 - e^2}}(1 - \cos i), \\
[\varphi, i] &= -\Omega a^2 \sqrt{1 - e^2} \sin i.
\end{aligned} \tag{3.86}$$

Remember that $[p, q] = -[q, p]$; all other brackets are equal to zero. Substituting (3.86) into the set (3.82) we get

$$[\varepsilon, a] \, \frac{da}{dt} = \frac{\partial U'}{\partial \varepsilon},$$

$$[\varpi, a] \, \frac{da}{dt} + [\varpi, e] \, \frac{de}{dt} = \frac{\partial U'}{\partial \varpi},$$

$$[\varphi, a] \, \frac{da}{dt} + [\varphi, e] \, \frac{de}{dt} + [\varphi, i] \, \frac{di}{dt} = \frac{\partial U'}{\partial \varphi},$$

$$[a, \varepsilon] \, \frac{d\varepsilon}{dt} + [a, \varpi] \, \frac{d\varpi}{dt} + [a, \varphi] \, \frac{d\varphi}{dt} = \frac{\partial U'}{\partial a},$$

$$[e, \varpi] \, \frac{d\varpi}{dt} + [e, \varphi] \, \frac{d\varphi}{dt} = \frac{\partial U'}{\partial e},$$

$$[i, \varphi] \, \frac{d\varphi}{dt} = \frac{\partial U'}{\partial i}.$$

(3.87)

Solving (3.87) we find

$$\frac{da}{dt} = \frac{2}{\Omega a} \frac{\partial U'}{\partial \varepsilon},$$

$$\frac{de}{dt} = -\frac{\sqrt{1-e^2}}{\Omega a^2 e} \left[1 - \sqrt{1-e^2} \right] \frac{\partial U'}{\partial \varepsilon} + \frac{\sqrt{1-e^2}}{\Omega a^2 e} \frac{\partial U'}{\partial \varpi},$$

$$\frac{di}{dt} = -\frac{\tan \frac{1}{2} i}{\Omega a^2 \sqrt{1-e^2}} \left[\frac{\partial U'}{\partial \varepsilon} + \frac{\partial U'}{\partial \varpi} \right] - \frac{1}{\Omega a^2 \sqrt{1-e^2} \sin i} \frac{\partial U'}{\partial \varphi},$$

$$\frac{d\varepsilon}{dt} = -\frac{2}{\Omega a} \frac{\partial U'}{\partial a} + \frac{\sqrt{1-e^2}[1-\sqrt{1-e^2}]}{\Omega a^2 e} \frac{\partial U'}{\partial e} + \frac{\tan \frac{1}{2} i}{\Omega a^2 \sqrt{1-e^2}} \frac{\partial U'}{\partial i},$$

(3.88)

$$\frac{d\varpi}{dt} = \frac{\sqrt{1-e^2}}{\Omega a^2 e} \frac{\partial U'}{\partial e} + \frac{\tan \frac{1}{2} i}{\Omega a^2 \sqrt{1-e^2}} \frac{\partial U'}{\partial i},$$

$$\frac{d\varphi}{dt} = \frac{1}{\Omega a^2 \sqrt{1-e^2} \sin i} \frac{\partial U'}{\partial i}.$$

In order to make (3.88) more convenient to use we introduce the following transformations. We decompose the perturbing acceleration into components S, T, W with S the radial component directed along the orbital radius of the particle motion, T the transverse component lying in the orbital plane at right angles to S and making with the velocity vector an angle less than 90°, and W the component at right angles to the orbital plane (it is positive if it is directed to the North of the orbital plane). The derivatives of U' can be expressed in terms of S, T, W as follows (Brouwer and Clemence 1961, Roy 1978):

$$\frac{\partial U'}{\partial a} = \frac{R}{a} S,$$

$$\frac{\partial U'}{\partial e} = -aS \cos f + R \sin f \left(\frac{1}{1-e^2} + \frac{a}{R} \right) T,$$

$$\frac{\partial U'}{\partial i} = RW \sin u,$$

$$\frac{\partial U'}{\partial \varphi} = -2RT \sin^2 \tfrac{1}{2} i - RW \cos u \sin i,$$

$$\frac{\partial U'}{\partial \varepsilon} = \frac{aeS \sin f}{\sqrt{1-e^2}} + \frac{Ta^2 \sqrt{1-e^2}}{R},$$

$$\frac{\partial U'}{\partial \varpi} = -\frac{\partial U'}{\partial \varepsilon} + RT,$$

(3.89)

where $u = \varpi - \varphi + f$ and f is the true anomaly. Substituting (3.89) into (3.88) we get

$$\frac{da}{dt} = \frac{2}{\Omega \sqrt{1-e^2}} \left(Se \sin f + \frac{p}{R} T \right),$$

$$\frac{de}{dt} = \frac{\sqrt{1-e^2}}{\Omega a} \left[S \sin f + T(\cos E + \cos f) \right],$$

$$\frac{di}{dt} = \frac{WR \cos u}{\Omega a^2 \sqrt{1-e^2}},$$

$$\frac{d\varepsilon}{dt} = \frac{e^2}{1 - \sqrt{1-e^2}} \frac{d\varpi}{dt} + 2\sqrt{1-e^2} \sin^2 \tfrac{1}{2} i \frac{d\varphi}{dt} - \frac{2R}{\Omega a^2} S,$$

$$\frac{d\varpi}{dt} = \frac{\sqrt{1-e^2}}{\Omega a e} \left[-S \cos f + T \left(1 + \frac{R}{p} \right) \sin f \right] + 2 \sin^2 \tfrac{1}{2} i \frac{d\varphi}{dt},$$

$$\frac{d\varphi}{dt} = \frac{WR \sin u}{\Omega a^2 \sqrt{1-e^2} \sin i},$$

(3.90)

where we have used the standard connection between the eccentric anomaly E and the true anomaly f:

$$\cos E = \frac{R}{a} \cos f + e.$$

If we decompose the perturbing acceleration into components, T' and N, which are tangential to and normal to the orbit of the body instead into the radial and transverse components S and T, the set of equations becomes (Subbotin 1968):

$$
\begin{aligned}
S\sqrt{1 + 2e\cos f + e^2} &= eT'\sin f - (1 + e\cos f)N, \\
T\sqrt{1 + 2e\cos f + e^2} &= (1 + e\cos f)T' + eN\sin f, \\
\frac{da}{dt} &= \frac{2a(2a - R)}{RV}T', \\
\frac{de}{dt} &= \frac{2(e + \cos f)}{V}T' - \frac{R\sin f}{aV}N, \\
\frac{d\varpi}{dt} &= \frac{2\sin f}{eV}T' + \frac{1}{V}\left[2 + \frac{R\cos f}{ae}\right]N + 2\sin^2 \tfrac{1}{2}i\,\frac{d\varphi}{dt}, \\
\frac{d\varepsilon}{dt} &= -\frac{2\sqrt{1 - c^2}}{V}\left(\frac{e\sin f}{1 + e\cos f}T' - N\right) \\
&\quad + \frac{e^2}{1 + \sqrt{1 - e^2}}\frac{d\varpi}{dt} + 2\sqrt{1 - e^2}\sin^2 \tfrac{1}{2}i\,\frac{d\varphi}{dt},
\end{aligned}
\tag{3.91}
$$

where $V = \sqrt{GM(1/R - 2/a)}$.

3.3.2 Satellite Orbit in the Field of an Aspherical Planet

We can in the case of an aspherical planet with axial symmetry express the potential energy at an external point as follows (Brouwer and Clemence 1961, Roy 1978):

$$
U = \frac{GM}{R}\left[1 - \sum_{n=2}^{\infty} J_n \left(\frac{R_{\text{pl}}}{R}\right)^n P_n(\sin \Phi)\right], \tag{3.92}
$$

where the J_n are constants, R_{pl} is the radius of the planet, the P_n are Lagrange polynomials, and Φ is the angle between the equator of the planet and the radius vector of the satellite. Let us consider an orbit with zero inclination so that $\Phi = 0$, $\sin \Phi = 0$ and the first two Lagrange polynomials become

$$
\begin{aligned}
P_2(\sin \Phi) &= \tfrac{1}{2}(3\sin^2 \Phi - 1) = -\tfrac{1}{2}, \\
P_3(\sin \Phi) &= \tfrac{1}{2}(5\sin^3 \Phi - 3\sin \Phi) = 0.
\end{aligned}
\tag{3.93}
$$

We can then write the perturbing potential function U', up to the first two terms, in the form

$$
U' = U - \frac{GM}{R} = \frac{GM}{2R}J_2\left(\frac{R_{\text{pl}}}{R}\right)^2. \tag{3.94}
$$

We rewrite the set (3.88) for the case considered here in the following form:

$$\frac{da}{dt} = \frac{2}{\Omega a} \frac{\partial U'}{\partial I},$$

$$\frac{de}{dt} = \frac{1 - e^2}{\Omega a^2 e} \frac{\partial U'}{\partial I} - \frac{\sqrt{1 - e^2}}{\Omega a^2 e} \frac{\partial U'}{\partial \varpi},$$

$$\frac{di}{dt} = \frac{\cot i}{\Omega a^2 \sqrt{1 - e^2}} \frac{\partial U'}{\partial \varpi} - \frac{1}{\Omega a^2 \sin i \sqrt{1 - e^2}} \frac{\partial U'}{\partial \varphi},$$

$$\frac{d\varphi}{dt} = \frac{1}{\Omega a^2 \sin i \sqrt{1 - e^2}} \frac{\partial U'}{\partial i}, \tag{3.95}$$

$$\frac{d\varpi}{dt} = -\frac{\cot i}{\Omega a^2 \sqrt{1 - e^2}} \frac{\partial U'}{\partial i} + \frac{\sqrt{1 - e^2}}{\Omega a^2 e} \frac{\partial U'}{\partial e},$$

$$\frac{dI}{dt} = \Omega - \frac{1 - e^2}{\Omega a^2 e} \frac{\partial U'}{\partial e} - \frac{2}{\Omega a} \frac{\partial U'}{\partial a},$$

where the mean anomaly I takes the place of the variable ε through the relation

$$I = \Omega t + \varepsilon. \tag{3.96}$$

If we average U' over I, that is, eliminate the short-period oscillations of frequency Ω as follows:

$$\overline{U'} = \frac{1}{2\pi} \int_0^{2\pi} U' \, dI,$$

we get from (3.94) (Roy 1978):

$$\overline{U'} = \frac{GM J_2 R_{\text{pl}}^2}{2a^3 \sqrt{(1 - e^2)^3}}. \tag{3.97}$$

From (3.95) we find the averaged parameters of the satellite orbit (for $i_0 = 0$):

$$\bar{a} = a_0, \qquad \bar{e} = e_0, \qquad \bar{i} = i_0,$$

$$\bar{\varpi} = \varpi_0 + \frac{3J_2 R_{\text{pl}}^2}{p^2} \overline{\Omega} t,$$

$$\overline{\Omega} = \Omega \left[1 + \frac{3J_2 R_{\text{pl}}^2}{2p^2} \sqrt{1 - e^2} \right], \tag{3.98}$$

$$\bar{I} = I + \overline{\Omega} t,$$

where again $p = a(1 - e^2)$.

The rotational velocity of a satellite round a planet thus increases due to the aspherical harmonic J_2 of the gravitational field and the orbit itself precesses in the same direction as the velocity:

$$\frac{d\bar{\varpi}}{dt} = 3J_2 \frac{R_{\text{pl}}^2}{p^2} \overline{\Omega}. \tag{3.99}$$

We have neglected here a factor $1 - e^2$, assuming the eccentricity of the satellite orbit to be small: $e \ll 1$. The precession of an elliptical orbit in an aspherical field plays an important role in the dynamics of rings and we shall therefore often encounter it in what follows.

3.3.3 Effect of Aerodynamic Friction on the Orbit of a Satellite

The aerodynamic friction force per unit mass is given by the following expression:

$$F = \frac{1}{2m} C_D A \varrho V_r^2, \tag{3.100}$$

where C_D is the aerodynamic drag coefficient, $A = \pi a^2$ is the cross-section of the satellite, and V_r is the velocity of the satellite of mass m relative to the atmosphere with density ϱ. The aerodynamic coefficient C_D is for a dense atmosphere close to unity, and for a rarefied atmosphere – when the mean free path is much larger than the size of the satellite – we have $C_D \simeq 2$.

Let us consider the set (3.91). The aerodynamic friction force will be directed along the tangent to the orbit, if we assume that the atmosphere does not rotate, $F = -T'$, and the normal component will be equal to zero. We then get from (3.91) the following set of equations for the major semi-axis, the eccentricity, and the pericentre length:

$$
\begin{aligned}
\frac{da}{dt} &= -\frac{2\sqrt{1 + e^2 + 2e \cos f}}{\Omega\sqrt{1 - e^2}} F, \\
\frac{de}{dt} &= -\frac{2\sqrt{1 - e^2}}{\Omega a} \frac{e + \cos f}{\sqrt{1 + e^2 + 2e \cos f}} F, \\
\frac{d\varpi}{dt} &= -\frac{\sqrt{1 - e^2}}{\Omega a e} \frac{2 \sin f}{\sqrt{1 + e^2 + 2e \cos f}} F.
\end{aligned}
\tag{3.101}
$$

The aerodynamic friction does not change the inclination of the orbit. The main effect of the aerodynamic friction is a decrease of the major semi-axis:

$$\frac{da}{dt} \approx -\frac{2}{\Omega} F, \tag{3.102}$$

where we have assumed that $e \ll 1$. If we bear in mind that for a circular orbit we have

$$V = \sqrt{\frac{GM}{a}},$$

we find

$$\frac{dV}{dt} = -\tfrac{1}{2}\Omega \frac{da}{dt}, \tag{3.103}$$

or, substituting (3.102) into (3.103),

$$\frac{dV}{dt} = F. \tag{3.104}$$

Equation (3.104) expresses a well known paradox: the velocity of the satellite increases as a result of damping – in this case aerodynamic friction, but the same will be true for any other kind of damping (Moulton 1914, Balk 1965, Abalakin, Aksenov, Grebenikov, and Ryabov 1971, Nazarenko and Skrebyshevskiĭ 1981). If the eccentricity is small we have

$$\frac{de}{dt} = -\frac{2}{\Omega a} F \cos f. \tag{3.105}$$

This means that the change in the eccentricity is basically of a short-periodic nature. One could have reached the same conclusion also for the change in the pericentre length. Aerodynamic friction is one of the most important factors of the dynamics of small grains in planetary rings, especially in the Uranian system.

3.3.4 The Poynting–Robertson Effect

This is the phenomenon of the damping of particles due to the solar radiation. This effect is connected with the fact that the speed of light is finite so that for a particle moving across a light beam the light quanta fly slightly towards it – just as rain always hits the face of a runner. One must distinguish the heliocentric and the planetocentric Poynting–Robertson effect: in the first case the solar radiation decelerates a body rotating round the Sun and in the second case particles moving round a planet are slowed down (Burns et al. 1979). The nature of the braking on the orbit of a particle can be seen from the results of the previous section. The characteristic rate at which the orbital radius is decreased can be written in the following form (Goldreich and Tremaine 1982):

$$\frac{da}{dt} = -a\frac{5S}{2mc^2}\sigma Q, \tag{3.106}$$

where m is the mass of a particle of radius r_0, c is the light velocity, $\sigma = \pi r^2$, σQ is the scattering cross-section, and S is the solar energy flux. For a simple estimate one can give a simple formula for the characteristic time for the reduction of the orbit (Goldreich and Tremaine 1982):

$$t_{\mathrm{PR}} \cong \left(\frac{1}{a}\frac{da}{dt}\right)^{-1} \approx 10^9 \text{ year} \left[\frac{\varrho}{\text{g cm}^{-3}}\right]\left[\frac{r_0}{\text{cm}}\right]\left[\frac{a_0}{10 \text{ a.u.}}\right]^2. \tag{3.107}$$

One sees that for the outer planets (from Saturn onwards) the Poynting–Robertson effect is effective for particles of radius of about a centimeter or smaller. At the same time one can discard the solar radiation as an effect for satellite-particles up to several meters moving round Mercury or Venus. This may explain the fact that these planets not only do not have any rings, but also do not have any satellites.

3.3.5 Collisions and Particle Orbits

Let the particle, which we are looking at, acquire a velocity due to a collision with another particle, that is, let its three velocity components in the radial, transverse, and z-directions increase:

$$\Delta V_S = S\Delta t; \qquad \Delta V_T = T\Delta t; \qquad \Delta V_W = W\Delta t. \tag{3.108}$$

From (3.90) we find for the changes in the orbital elements due to collisions:

$$\Delta a = \frac{2}{\Omega\sqrt{1-e^2}}\left(e\Delta V_S \sin f + \frac{p}{R}\Delta V_T\right),$$

$$\Delta e = \frac{\sqrt{1-e^2}}{\Omega a}\left[\Delta V_S \sin f + (\cos E + \cos f)\Delta V_T\right],$$

$$\Delta i = \frac{R\cos u}{\Omega a^2\sqrt{1-e^2}}\,\Delta V_W,$$

$$\Delta\varphi = \frac{R\sin u}{\Omega a^2\sqrt{1-e^2}\sin i}\,\Delta V_W, \tag{3.109}$$

$$\Delta\varpi = \frac{\sqrt{1-e^2}}{\Omega a e}\left[-\Delta V_S \cos f\right.$$

$$\left. + \left(1 + \frac{R}{p}\right)\Delta V_T \sin f\right] + 2\sin^2 \tfrac{1}{2}i\,\Delta\varphi,$$

$$\Delta\varepsilon = \frac{e^2}{1-\sqrt{1-e^2}}\,\Delta\varpi + 2\sqrt{1-e^2}\sin^2 \tfrac{1}{2}i\,\Delta\varphi - \frac{2R}{\Omega a^2}\,\Delta V_S,$$

where f and E are the true and the eccentric anomalies.

It is clear from the first equation of the set (3.109) that the major semi-axis of the particle orbit remains unchanged, if the collision is at right angles to the plane of the orbit. The major semi-axis a of an orbit with a small eccentricity hardly depends on the radial increment of the velocity and the most efficient way of changing a is a tangential collision:

$$\Delta a \approx \frac{2}{\Omega}\,\Delta V_T. \tag{3.110}$$

In the case of an orbit with a small eccentricity the eccentricity depends on both the normal and the tangential perturbations:

$$\Delta e \approx \frac{\Delta V_S \sin f + 2\Delta V_T \cos f}{\Omega a}. \tag{3.111}$$

However, the tangential collisions are twice as effective as the radial perturbations. The increase – or decrease – in the eccentricity depends on the point in the orbit where the particle is when it collides. The eccentricity of an orbit which has a small eccentricity changes appreciably through a tangential collision when the particle is in the pericentre or the apocentre. Radial perturbations affect the eccentricity most strongly in the middle points of the orbit where $r = a$.

4 Elementary Particle Dynamics
I Rigid Body Collisions

Edgar: "Hadst thou been aught, but gossamer, feathers, air, so many fathom down precipitating, thoud'st shiver'd like an egg: but thou dost breathe; ..."

W. Shakespeare, King Lear, Act IV, Scene VI.

One fundamental problem of the physics of planetary rings is to find the mechanical characteristics of a typical particle. It is impossible to understand either the origin of the rings or the dynamics of the collective processes without a detailed study of the properties of the particles. In the present section we shall therefore study the physics of collisions between rigid bodies which are the simplest and most popular model of the particles in the planetary rings. Collisions between rigid bodies are, however, interesting also for many other fields of astrophysics; they are an important evolutionary springboard for modern cosmogonical models. The formation of the planets and satellites started with collisions and the coalescence of solid particles (Safronov 1969, Safronov and Vityazev 1983). The surface structure of the large bodies in the solar system is in many respects caused by meteoritic impacts.

The range of the observed collisional velocities is from a milimeter per second (for planetary rings) to tens of kilometers per second (for meteorites). One may consider a collision to be a low-velocity (high-velocity) one if $V < V_{cat}$ ($V > V_{cat}$) where V_{cat} is the velocity needed for a catastrophic break-up in which the body ceases to be whole and splits up into separate parts. For ice we have $V_{cat} = 9$ m/s and for basalt $V_{cat} = 37$ m/s (Hartmann 1978).

High-velocity impacts of bodies on targets and the accompanying formation of craters have been studied both experimentally and theoretically (Ishlinskiĭ and Chernyĭ 1981, Hartmann 1985). We shall be interested in low-velocity impacts, the main characteristic of which is the restitution coefficient

$$q(V) = \frac{V'}{V}, \tag{4.1}$$

where V and V' are, respectively, the velocities of the body before and after the collision.

4.1 Theoretical Models

A large number of authors of theoretical models of the Saturnian rings use a smooth ice sphere as a typical ring particle (Goldreich and Tremaine 1978a, Borderies, Goldreich, and Tremaine 1984, Bridges, Hatzes, and Lin 1984, Thompson 1984, Kerr 1985, Araki and Tremaine 1986, Hatzes, Bridges, and Lin 1988, McDonald et al. 1989). Bridges and coworkers have even studied the restitution coefficient of such a sphere experimentally – in vacuo at low temperatures and with characteristic impact velocities V of between 0.1 and 10 mm/s. They found that the restitution coefficient of a sphere colliding with a plate was close to unity for velocities of $V = 0.1$ mm/s and decreased with increasing collisional velocity. This behaviour is in accordance with that of $q(V)$ for any smooth bodies in the well studied "everyday" range of impact velocities of $V = 1$ to 6 m/s (Goldsmith 1960) corresponding to the velocity with which objects hit the ground when falling from the height of a man.

We shall follow papers by one of us (Gor'kavyĭ 1985a, 1989b) to study the collisions of hypothetical smooth particles and we shall show that due to mutual collisions they are inevitably covered by loose snowlike regolith which qualitatively changes their elastic properties and makes theoretical models based upon the assumption of smooth particles inapplicable to real rings.

4.1.1 Some Relations from the Theory of the Collisions of Smooth Spheres

Hertz constructed in 1880 a comprehensive theory of collisions of ideally elastic spheres. From Hertz's theory we obtain the following expression for the maximum energy of elastic deformation of colliding identical smooth spheres (see, for instance, Landau and Lifshitz 1970):

$$U_{\max} = 5.4 \left(1 - \mu_{\mathrm{P}}^2\right) \frac{P_{\max}^2}{E} R_{\mathrm{c}}^3, \tag{4.2}$$

where $R_{\mathrm{c}} \approx \pi a (1 - \mu_{\mathrm{P}}^2) P_{\max}/2E$ is the maximum radius of the zone of elastic contact, $\mu_{\mathrm{P}} = 0.36$ is the Poisson coefficient, a the radius of the sphere, $P_{\max} = 5 \times 10^7$ to 10^8 dyne/cm^2 the breaking point (maximum) tension of ice, and $E = 10^{11}$ dyne/cm^2 the Young modulus (Bogorodskiĭ and Gavrilo 1980).

We find the critical collisional velocity – at which the ice starts to deform irreversibly in the contact zone – by putting the kinetic energy $\frac{1}{2}mV^2$ equal to $U_{\max}$:

$$V_{\mathrm{cr}} \approx 3.2 \frac{(1 - \mu_{\mathrm{P}}^2)^2}{\sqrt{\varrho}} \frac{P_{\max}^{5/2}}{E^2} \sim 4.5 \times 10^{-3} \text{ to } 2.5 \times 10^{-2} \text{ cm/s}, \tag{4.3}$$

where $\varrho = 0.9$ g/cm^3. For $V > V_{cr}$ ice starts to break up, but the break-up can be either elastobrittle (like, for instance, in a collision of glass objects) or elastoplastic (like the deformation of lead spheres).

Let us consider the problem of the nature of the break-up of ice spheres in the rings. The data from many terrestrial experiments show that the kind of break-up of ice depends on the rate of deformation: elastobrittle break-up dominates if the rate of deformation $d\varepsilon/dt > 10^{-3}$ to 10^{-2} s^{-1}, where ε is the relative deformation (Bogorodskiĭ and Gavrilo 1980). For sufficiently slow deformations, $d\varepsilon/dt < 10^{-3}$ s^{-1}, ice suffers elastoplastic break-up. The data refer to terrestrial conditions; at temperatures around 70 K ice is more brittle, that is, it suffers brittle break-up at lower deformation rates.

Let us estimate the deformation rate for the hypothetical smooth particles in the Saturnian rings. The collision time (in seconds) of spheres which are not breaking up – from the start of the contact to the maximum depression of the surface – for $V \cong 0.1$ cm/s is according to Hertz's theory (Landau and Lifshitz 1970)

$$\tau_{\text{coll}} \approx 4.6 \left[\frac{(1 - \mu_{\text{P}}^2)\varrho}{\sqrt{V}E} \right]^{2/5} a = 0.025a. \tag{4.4}$$

Here a is the particle radius in meters. It is clear that for real materials the time of non-destructive contact – from the start of the collision to the moment of break-up – is even shorter.

For the maximum depression of the surface – the absolute deformation of the sphere at the start of the break-up – we find from Hertz's theory:

$$H_{\text{max}} \approx 10.3 \left(1 - \mu_{\text{P}}^2 \right)^2 \frac{P_{\text{max}}^2}{E^2} a. \tag{4.5}$$

In collisions between smooth spheres basically a contact region of radius R_{el} is deformed (Landau and Lifshitz 1970). This is clear from (4.2) if we bear in mind that $P_{\text{max}}^2/2E$ is the elastic deformation energy per unit volume and that the volume of the contact zone is approximately $\frac{4}{3}\pi R_{\text{el}}^3$. From this we can estimate the maximum deformation of ice – at the moment the break-up tension P_{max} is reached:

$$\varepsilon_{\text{max}} \sim \frac{H_{\text{max}}}{R_{\text{el}}} \sim 5.7 \frac{P_{\text{max}}}{E} \sim (3 \text{ to } 6) \times 10^{-3}. \tag{4.6}$$

This estimate is in agreement with experimental data about the break-up of ice under terrestrial conditions (Bogorodskiĭ and Gavrilo 1980): $\varepsilon_{\text{max}} \sim 3 \times 10^{-3}$. Knowing the magnitude of the relative deformation of ice and the break-up time (4.4) we can estimate the rate of deformation:

$$\frac{d\varepsilon}{dt} \sim \frac{\varepsilon_{\text{max}}}{\tau_{\text{coll}}} \geqslant \begin{cases} 1.2 \times 10^{-2} \text{ s}^{-1} & \text{for } a = 10 \text{ m,} \\ 1.2 \times 10^{-1} \text{ s}^{-1} & \text{for } a = 1 \text{ m.} \end{cases} \tag{4.7}$$

This means that the large ice particles in the Saturnian rings when colliding with characteristic velocities of 1 mm/s will break up as brittle bodies with the formation of small particles – icy dust particles – in the contact zone (Mellor 1980).

4.1.2 Break-Up of Ring Particles (Estimates)

The formation of small particles in the contact zone is characteristic for the break-up under impact of brittle materials (Cherepanov 1974). In the break-up under contact of brittle spheres the kinetic energy of the collision changes initially into elastic deformation energy and after that – when the critical stress is reached – basically into the surface energy of the fragments which are formed – although partially the collisional energy is spent on the friction between the fragments and the heating of the material. The break-up region is localised in the region of maximum deformation with a characteristic size R_{el}. One can get a lower estimate of the size of the microfragments by assuming that the whole of the collisional energy is spent on the formation of a new surface of microgranules – for instance, in the shape of cubes: $r_0 \simeq 6\gamma_{\mathrm{s}}/\varepsilon_{\mathrm{V}} \simeq 10^{-5}$ to 10^{-4} cm, where $\gamma_{\mathrm{s}} \simeq 10^2$ erg/cm^2 is the surface energy of ice and $\varepsilon_{\mathrm{V}} \simeq 10^7$ to 10^8 erg/cm^3 the experimentally determined break-up energy of a unit volume of ice (Bogorodskiĭ and Gavrilo 1980). We note that it is more natural to determine the break-up energy in terms of the characteristic size of the fragments $\varepsilon_{\mathrm{V}} \simeq 6\gamma_{\mathrm{s}}/r_0$, rather than the other way round. In the case of the break-up of monolithic ice the size of the microfragments corresponds to the sizes of the most stable micrometer blocks in the ice structure. When the characteristic size of the fragments increases, the magnitude of ε_{V} decreases.

A smooth surface of ice particles thus starts to be covered through collisions by a layer of finely broken up ice. Let us estimate the mass Δm of the broken up ice after a single collision of particles of mass m and a velocity V between around 0.1 and 0.6 cm/s, assuming that the whole of the collisional energy is spent on the break-up:

$$\frac{\Delta m}{m} \sim \frac{\varrho V^2}{2\varepsilon_{\mathrm{V}}} \sim 5 \times 10^{-11} \text{ to } 2 \times 10^{-8}. \tag{4.8}$$

Assuming a constant break-up rate we get a characteristic time for the complete break-up of ice particles (for $t_{\mathrm{c}} = 10$ hr)

$$t_{\mathrm{br}} \sim \frac{m t_{\mathrm{c}}}{\Delta m} \sim 2.5 \times 10^8 \text{ to } 7 \times 10^5 \text{ years}, \tag{4.9}$$

where t_{c} is the mean free flight time of the particles in the rings. Hence we can conclude that the particles in the Saturnian rings are most likely lumps of finely broken up ice or snow. However, it is possible that the ice particles stop being broken up after their surface has accumulated a thin loose layer of finely broken up ice which absorbs the collisional energy and protects the monolith from break-up. The conclusion that there is at least a surface layer of loose regolith on the ring particles is confirmed both by theoretical arguments (Hartmann 1984; Gor'kavyĭ 1985a) and by observations (see Sect. 2.1).

However, we have been rather hasty in calling the regolith layer "loose": it is well known that terrestrial snowdrift after a few days turns into a very firm mass – the snow particles freeze together. How will the cosmic "snow-drift" behave near Saturn? The problem is whether ice particles can coalesce under the conditions characteristic for the Saturnian rings. It is well known that there are several mechanisms for coalescence of small grains: viscous flow, sublimation – evaporation of the ice and subsequent condensation of the vapour – and volume and surface diffusion. These processes have been intensively studied for the physics of snow (Andersen and Benson 1985) and for powder metallurgy (Geguzin 1969). The formulae applicable for snow experts and metallurgists are exactly the same, but in the first case the process of coalescence is called regelation and in the second case it is called sintering. Those are the mechanisms which are responsible for the hardening of terrestrial loose snow, but they are practically impotent for the snowdrifts near Saturn. This is connected with the fact that such characteristics of ice as the viscosity or the diffusion coefficient are strongly temperature dependent (Friedson and Stevenson 1983) and at temperatures between 70 and 100 K the viscosity turns out to be so large and the diffusion coefficient so small that they prevent the metamorphosis of the ice grains. Clark, Fanale and Zent (1985) came to the conclusion that micrometric ice grains are stable near Saturn during the period the solar system has existed.

The rate of coalescence of micrometric ice grains at temperatures around 70 K is thus extraordinarily small and we can neglect the fusion of the grains in the contact zone during a collision because of the low kinetic energies of the colliding particles; hence, the surface regolith is a loose medium weakly bound by self-adhesion forces.

There now arises the obvious question as to in how far the mechanical impact properties of particles covered by a thin layer of regolith are similar to those of smooth spheres. This problem is much more wide-ranging than the problematics of the planetary rings: practically all geologically weakly active bodies in the solar system – such as the Moon, Phobos, or the asteroids – are covered by loose regolith and the planetesimals may possibly have had the same coating at the time when the planets were formed.

4.1.3 Model of Collisions Between Particles Covered by Regolith

Gor'kavyĭ (1985a) has proposed a three-stage model for the collisions of monolithic bodies covered by a thin layer of loose regolith. At sufficiently low impact velocities the whole of the particle energy goes into an irreversible deformation of the surface layer – formation of a dent – and the collision will be practically completely inelastic. When the impact velocity increases only part of the kinetic energy will be spent on disturbing the surface layer and the remainder of the energy goes into the potential energy of the elastic deformation of the monolithic nucleus. At a certain impact velocity the tension in the monolith or in the regolith grains becomes larger than its critical value and the monolith and/or the particles of the loose surface layer start to break up.

We denote the energy spent on the inelastic deformation of the regolith by ΔE, the energy of the elastic deformation of the monolith by U, and the critical value of U before the monolith breaks up by U_{cr}. We shall consider how the collision process changes when the particle velocity increases.

A. *Stage of Completely Inelastic Impact.* The kinetic energy is spent on the irreversible compression of the loose surface regolith:

$$\tfrac{1}{2}mV^2 < \Delta E, \qquad U = 0, \qquad q(V) = 0.$$

B. *Stage of Elastic Deformation of the Monolith Nucleus.* Part of the kinetic energy of the particles, remaining after the compression of the regolith, is spent on a reversible deformation of the monolith:

$$\Delta E < \tfrac{1}{2}mV^2 < \Delta E + U_{\max},$$

$$U = \tfrac{1}{2}mV^2 - \Delta E, \qquad q(V) = \sqrt{1 - \frac{2\Delta E}{mV^2}}. \tag{4.10}$$

C. *Break-Up Stage.* The collisional energy is so large that the monolith nucleus starts to deform irreversibly or that the regolith grains start to break up:

$$\Delta E + U_{\max} < \tfrac{1}{2}mV^2, \qquad U = U_{\max}, \qquad q(V) = \sqrt{\frac{2U_{\max}}{mV^2}}. \tag{4.11}$$

The natural next stage of the collision is the rebound – the transformation of the reversible deformation energy U or $U_{\max}$ back into kinetic energy. The quantities ΔE and $U_{\max}$ may depend on V, but less strongly than V^2. We shall consider this function $E(V)$ for a smooth particle.

4.1.4 Restitution Coefficient of a Smooth Particle

For $\Delta E = 0$ there remain only two collisional stages, B and C, and one unknown quantity: $U_{\max}$. The particles do not break up when $V < V_{\mathrm{cr}}$ and we have $q = 1$. An irreversible local deformation of the contact zone starts for $V > V_{\mathrm{cr}}$.

We can estimate $U_{\max}$ after the break-up has started by assuming that (4.2) is approximately true if R_c is the radius of the contact zone at break-up. Let a spherical segment of base radius R_s and height l be broken up in the contact zone; the kinetic energy will then be spent on the break-up of a segment of volume $\pi a l^2$ and on the elastic deformation of the particle (Gor'kavyĭ 1989b):

$$\tfrac{1}{2}mV^2 \;=\; \varepsilon_V \pi a l^2 + 5.4 \left(1 - \mu_P^2\right) \frac{P_{max}^2}{E}\, R_s^3 . \tag{4.12}$$

The second term on the right-hand side is the elastic deformation energy U_{max}. Hence, we have from (4.11)

$$q \;=\; \sqrt{\frac{10.8\,(1 - \mu_P^2)\,P_{max}^2\,(2al)^{3/2}}{mV^2 E}}. \tag{4.13}$$

Eliminating the quantity l from (4.12) and (4.13) we find the following equation for $q(V)$ (Gor'kavyĭ 1989b):

$$q^2 + V^{2/3}\left(\frac{q}{L}\right)^{8/3} \;=\; 1, \tag{4.14}$$

with

$$L \;=\; 2.32\, \frac{\sqrt{(1 - \mu_P^2)}\,P_{max}}{\varrho^{1/8} E^{1/2} \varepsilon_V^{3/8}}.$$

One can call L the "elasticity parameter": the larger L, the higher the restitution coefficient.

From (4.14) we find (Gor'kavyĭ 1985a, 1989b)

$$q(V) \;=\; \begin{cases} 1 & \text{as } V \to 0, \\ L V^{-0.25} & \text{for } V \gg L^4. \end{cases} \tag{4.15}$$

We note that Johnson (1989) has recently given an expression similar to (4.15) and that an equation close in spirit to (4.12) has been discussed by Tabor (1948). The fact that $q(V) \propto V^{-0.25}$ means that $U_{max} \propto V^{3/2}$ when $V \gg L^4$.

4.2 Experimental Data

4.2.1 Comparison Between the Smooth Particle Model and the Experimental Data

The solutions of (4.14) are in good agreement with the experimental data of Hatzes, Bridges, and Lin (1988) for $L = 1.15$ and those of Bridges, Hatzes, and Lin (1984) for $L = 0.35$. These values of L are obtained for $\mu_P = 0.36$, $\varrho = 0.9$ g/cm^3, $E = 10^{11}$ dyne/cm^2, $\varepsilon_V = 10^8$ erg/cm^3 for a breaking strength of ice equal to $P_{max} = 1.7 \times 10^8$ and 5×10^7 dyne/cm^2, respectively.

We show in Fig. 4.1 the theoretical function $q(V)$ for smooth particles (using (4.14); qualitatively these functions were derived by Gor'kavyĭ (1985a)); the experimental data are also indicated. Bridges et al. (1984) gave the following approximate expressions:

$$q = \begin{cases} 1 \quad \text{for } V < V_{\text{cr}} \approx 8 \times 10^{-3} \text{ cm/s}, \\ (0.32 \pm 0.02) \, V^{-0.234 \pm 0.008} \quad \text{for } 1.5 \times 10^{-2} < V < 5 \text{ cm/s}. \end{cases} \tag{4.16}$$

Hatzes et al. (1988) obtained for ice particles with a very smooth surface the relation

$$q(V) = 0.82 \, V^{-0.047}, \tag{4.16a}$$

where V is in cm/s. A comparison of the theoretical points with the experimental curves of Fig. 4.1 and the relations (4.15) and (4.16) show that the suggested analytical model satisfactorily describes the basic behaviour of the collisions between smooth particles. We note that Borderies et al. (1984) proposed to describe the collisions between smooth particles, using Andrews's theory (1930) which is based upon an assumption about plastic deformation of the particles and which gives the following relations:

$$q = \begin{cases} 1 \quad \text{for } V < V_{\text{cr}}, \\ 1.4 \, \sqrt{\dfrac{V_{\text{cr}}}{V}} \quad \text{for } V \gg V_{\text{cr}}. \end{cases} \tag{4.17}$$

Expressions (4.17) do not agree with the experimental data as far as the power-law dependence is concerned.

As alternative estimates of the break-up of particles in a single collision we note two relations proposed by Goldreich and Tremaine (1978a) (V is expressed in cm/s and a in m).

$$\left(\frac{\Delta m}{m}\right)_1 = 3 \times 10^{-8} \left(\frac{V}{10^{-2}}\right)^3 a^{3/2}, \tag{4.18a}$$

$$\left(\frac{\Delta m}{m}\right)_2 = 10^{-9} \left(\frac{V}{10^{-2}}\right)^{6/5}. \tag{4.18b}$$

Hartmann (1984) has criticised these expressions, comparing them with his experimental data (Hartmann 1978) which show that $\Delta m/m$ depends solely on the kinetic energy density $\frac{1}{2}\varrho V^2$, that is, $\Delta m/m \propto V^2$, and is independent of the size of the particles. Borderies et al. (1984) apply yet another estimate of the broken up mass:

$$\frac{\Delta m}{m} \sim \left(\frac{R_{\text{el}}}{a}\right)^3 \sim \left(\frac{P_{\text{max}}}{E}\right)^3 = \text{const } (V), \tag{4.19}$$

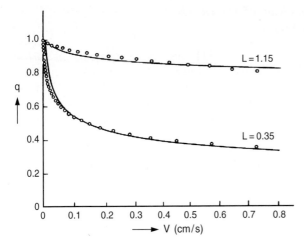

Fig. 4.1. The restitution coefficient of smooth particles as function of the velocity. The full-drawn curves give the experimental curves (Bridges et al. 1984 and Hatzes et al. 1988) and the circles the theoretical results (Gor'kavyĭ 1985a, 1989b).

which also does not agree with Hartmann's experimental data. We note that expression (4.8) is free of the deficiencies pointed out by Hartmann (1984) and agrees with his experimental data. Hartmann also noted that expressions (4.18) proposed by Goldreich and Tremaine (1978a) lead to incorrect estimates for the rate of catastrophic particle break-up. Indeed, for $\Delta m/m \simeq \frac{1}{2}$ we get from (4.18a) and (4.18b), respectively,

$$V_{\text{cat},1} \sim 2.5 \text{ to } 25 \text{ cm/s} \qquad \text{for} \qquad a = 1 \text{ to } 0.01 \text{ m},$$
$$V_{\text{cat},2} \sim 2 \times 10^7 \text{ cm/s}. \tag{4.20}$$

Both estimates are far from the experimental value (Hartmann 1978, 1984) of $V_{\text{cat}} \sim 9 \times 10^2$ cm/s. Expression (4.19) is completely unable to describe the catastrophic break-up of the particles. For $\Delta m/m \simeq \frac{1}{2}$ we get from expression (4.8)

$$V_{\text{cat}} \sim \sqrt{\frac{\varepsilon_V}{\varrho}} \sim 3.2 \times 10^3 \text{ to } 10^4 \text{ cm/s}. \tag{4.21}$$

We have used here $\varepsilon_V \sim 10^7$ to 10^8 erg/cm^3. The estimate (4.21) agrees qualitatively with Hartmann's experiments. One must take into account that in a catastrophic particle break-up the size of the fragments increases and as a result the magnitude of ε_V must decrease. Indeed, for some break-up of the matter when $V < V_{\text{cat}}$ the size of the largest fragment is limited by the dimensions of the impact zone. In the case of a catastrophic break-up of the particles, when $V > V_{\text{cat}}$, the largest fragment can reach the size of half of that of the particle. If we put $\varepsilon_V \sim 10^6$ erg/cm^3 into (4.21) we get $V_{\text{cat}} \sim 10^3$ cm/s, in excellent agreement with experiments.

4.2.2 Restitution Coefficient of Particles Covered by a Regolith Layer

It is clear from the relations (4.9) to (4.11) that in the presence of surface regolith, absorbing part of the collisional energy ($\Delta E \neq 0$), the scenario of the collision changes qualitatively (see Fig. 4.2c). For low velocities the restitution coefficient is zero (stage A), $q(V) = 0$; with increasing velocity $q(V)$ also begins to increase (stage B); when V increases further the restitution coefficient must again decrease (stage C).

As far as we know, the fact that the restitution coefficient increases with increasing velocity had not been experimentally noted before. Panovko (1977) remarked in his book, where he obtained a similar dependence for a formally constructed discrete model of a collision with a parallel coupling of an elastic element and a dry friction element, that "the model is questionable since ... the restitution coefficient increases with increasing velocity, whereas experimentally one sees clearly the opposite tendency."

To check theoretical relations such as (4.9) to (4.11) an experiment was carried out to measure the restitution coefficient of a metallic sphere falling on a massive plate, covered by a layer of loose regolith (Gor'kavyǐ 1989b). The diameter of the sphere was 5 cm and its mass 0.5 kg. The impact velocity varied from 1 to 6 m/s.

In the first case the experiment was carried out with a massive stone slab of solid Crimean limestone. Dry fine sea sand served as regolith. Figure 4.2a shows the experimental results for the restitution coefficient.

In the second case a $7 \times 48 \times 48$ cm concrete slab on the flat ground was the substrate. Dry cement dust was the regolith. The data obtained are shown in Fig. 4.2b.

The restitution coefficient of a sphere hardly depends on the velocity when there is no regolith – one may note some decrease of q from 0.5 to 0.4 in the 3 to 6 m/s region for a concrete slab. It can be seen from Fig. 4.2 that the presence of regolith fundamentally changes the picture: $q = 0$ for $V < 1$ m/s after which q increases steeply. This is in complete accordance with the first two stages of the model described by relations (4.9) to (4.11). Moreover, in two cases (the data in Fig. 4.2b for regolith layers of 1 and 2 mm) one can notice also the third stage – when the restitution coefficient again starts to decrease. The experimental data of Figs. 4.2a and b correspond to the following analytical expressions:

a) Stone Slab:

for a 1 mm regolith layer:

$$q = \begin{cases} 0 \text{ for } 0 < V < 1.5 \text{ m/s}, \\ \sqrt{1 - \dfrac{1.129}{V^{0.3}}} \text{ for } 1.5 < V < 5 \text{ m/s}; \end{cases} \tag{4.22}$$

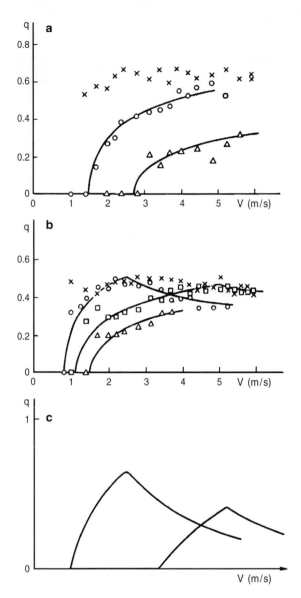

Fig. 4.2. The restitution coefficient of particles covered by regolith as function of the velocity. (**a, b**) experimental data. The restitution coefficient of a metallic sphere was measured as function of the impact velocity. A sphere with a 5 cm diameter falls on a plate covered by a thin layer of loose regolith. (**a**) massive slab of solid Crimean limestone and dry fine sea sand serving as regolith; (**b**) a 7 × 48 × 48 cm concrete slab was lying on the flat ground and the regolith was dry cement dust; the crosses refer to a clean plate, the circles to a 1 mm regolith layer, the squares to a 2 mm layer, and the triangles to a 3 mm layer; (**c**) theoretical curves (Gor'kavyĭ 1985a).

for a 3 mm regolith layer:

$$
q = \begin{cases} 0 \ \text{for } 0 \ < \ V \ < \ 2.75 \ \text{m/s}, \\ \sqrt{1 - \dfrac{1.141}{V^{0.13}}} \ \text{for } 2.75 \ < \ V \ < \ 6 \ \text{m/s}. \end{cases}
\tag{4.23}
$$

b) Concrete Slab:

for a 1 mm regolith layer:

$$
q = \begin{cases} 0 \ \text{for } 0 \ < \ V \ < \ 0.8 \ \text{m/s}, \\ \sqrt{1 - \dfrac{0.946}{V^{0.25}}} \ \text{for } 0.8 \ < \ V \ < \ 2.5 \ \text{m/s}, \\ \dfrac{0.787}{\sqrt{V}} \ \text{for } 2.5 \ < \ V \ < \ 5.5 \ \text{m/s}; \end{cases}
\tag{4.24}
$$

for a 2 mm regolith layer:

$$
q = \begin{cases} 0 \ \text{for } 0 \ < \ V \ < \ 1.15 \ \text{m/s}, \\ \sqrt{1 - \dfrac{1.021}{V^{0.15}}} \ \text{for } 1.15 \ < \ V \ < \ 5 \ \text{m/s}, \\ \dfrac{0.721}{V^{0.3}} \ \text{for } 5 \ < \ V \ < \ 6 \ \text{m/s}; \end{cases}
\tag{4.25}
$$

for a 3 mm regolith layer:

$$
q = \begin{cases} 0 \ \text{for } 0 \ < \ V \ < \ 1.5 \ \text{m/s}, \\ \sqrt{1 - \dfrac{1.047}{V^{0.11}}} \ \text{for } 1.5 \ < \ V \ < \ 4 \ \text{m/s}. \end{cases}
\tag{4.26}
$$

All qualitative features of the model described by relations (4.9) to (4.11) (Gor'kavyǐ 1985a, 1989b) have been confirmed. Let us note some details: the inelastic deformation energy of the surface layer depends quite strongly on the collisional velocity: $\Delta E \propto V^{1.7}$ to $V^{1.9}$ which may be connected with the fact that the regolith grains are broken up more strongly when the impact velocity increases. Theoretically one expected that ΔE would be proportional to the volume of the compressed part of the regolith, that is, $\Delta E \propto h^2$, where h is the thickness of the regolith layer. Experimentally one finds that the dependence $\Delta E(h)$ is weaker and closer to linear: $\Delta E \propto h$; ΔE is here defined as the quantity equal to the energy of the sphere at the point where the graph in Fig. 4.2 starts to rise, that is, where stages A and B of the collision join. This may be connected with the observed sweeping-out of part of the regolith from under the sphere at the moment of impact, which diminishes the volume of the deformed loose material. That effect is probably connected with a compression air wave arising at the lower part of the surface of the incident sphere (Gor'kavyĭ 1989b). The fact that there exists an air wave sweeping the regolith away from under the incident body has been confirmed in Hartmann's experiments (1978) in which a 14 g sphere fell with a velocity of 5.58 m/s on a thick layer of regolith. The experiment was carried out both under normal atmospheric pressure and in vacuo. The dispersion of the regolith in the atmosphere was three orders of magnitude larger than the ejection of regolith from the impact zone in vacuo.

Let us estimate the thickness of the regolith layer which is free to absorb the main fraction of the impact energy and qualitatively to change the characteristics of the colliding particles. Let the energy spent on the deformation of the layer be equal to $\Delta E = E_V \pi a h^2$, where E_V is the deformation energy of unit volume of regolith. We assume here that there is no sweeping-away and that $\Delta E \propto h^2$. Equating ΔE to the kinetic energy $\frac{2}{3}\varrho \pi a^3 V^2$ we get

$$\frac{h}{a} \sim \sqrt{\frac{2\varrho}{3E_V}}\, V \sim 0.2 \times 10^{-3}\, V, \tag{4.27}$$

where V is in cm/s. In the experiments E_V reached the considerable values of $\sim 5 \times 10^7$ erg/cm^3 for cement dust, and of $\sim 2 \times 10^8$ erg/cm^3 for sand. These values will clearly be appreciably smaller for low velocities. For the planetary ring particles we get $h/a \sim 10^{-5}$ to 10^{-4} if we put $\varrho \sim 1$ g/cm^3, $E_V \sim 10^6$ to 10^8 erg/cm^3, $V \sim 0.1$ cm/s. From this it follows that a 10 meter planetary ring particle becomes completely inelastic already for millimeter thicknesses of regolith. For a bazalt sphere with $\varrho \sim 3$ g/cm^3 and $V \sim 500$ cm/s, incident onto regolith with $E_V \sim 5 \times 10^7$ erg/cm^3, the impact will be practically inelastic for $h/a \sim 0.1$. This is in agreement with the corresponding experiments of Hartmann's (1978).

The theoretical discussion and the experimental data show that smooth particles cannot exist in the planetary rings – they are inevitably covered by finely crushed ice formed through collisions, and even a thin layer of regolith changes the elastic properties of the particles qualitatively: the restitution coefficient is close to zero for low velocities and increases with increasing collisional velocity. The energy balance in the models of non-gravitating particles used by various authors (Goldreich and Tremaine 1978a, Bridges, Hatzes, and Lin 1984, Borderies, Goldreich, and Tremaine 1984, Araki and Tremaine 1986, Hatzes, Bridges, and Lin 1988) is consistent only if the following two conditions are satisfied: (1) the restitution coefficient decreases when the velocity increases (see Chap. 8); (2) the quantity q is larger than some critical value $q_{cr} = 0.67$. The analysis carried out above showed the incompatibility of these conditions with the model of real particles covered by a regolith layer. Indeed, in the first place, the function $q(V)$ does not decrease monotonically but is complicated (see Fig. 4.2). Secondly, the condition $q > 0.67$ is hardly feasible when there is regolith present. We are thus led to the conclusion that for the particles in the planetary rings the smooth particle model is inapplicable. We note that this conclusion must clearly also be taken into account in models for the accretional growth of satellites and planets which also use smooth particle models (Greenberg et al. 1978). The effect of regolith can appreciably increase the efficiency of the sticking together of planetesimals and other particles in a proto-satellite cluster.

5 Elementary Particle Dynamics
II Ring Cosmogony

> Earl of Gloster: "O ruin'd piece of nature! This great world shall so wear out to nought."
>
> *W. Shakespeare, King Lear, Act IV, Scene VI.*

5.1 Tides or Collisions?

As soon as the Saturnian rings had been discovered the problem arose as to how they had originated. The first suggestions about the causes of the existence of the rings have already been touched upon in the Introduction: hypotheses about the capture of cometary tails (Maupertuis), separation of the rings when the planet was compressed (Meran), or the detaching of the rings from a planet because the centrifugal force was too strong (Buffon). We note that all these hypotheses in some way or other solved the problem of the origin of the matter making up the rings but completely ignored the problem of the dynamics: in what respect does the evolution of the matter of the rings differ from the evolution of the matter from which the satellites are formed? Why does the matter of rings remain uniformly spread along the orbit?

In the present chapter we shall consider more modern theories about the origin of the rings and the restrictions which they impose on the mechanical properties of the particles. The problem about the causes of the existence of planetary rings can be split into the following points:

1) Whence comes the matter from which the rings are formed?
2) Why are the particles limited in size, or, why do the rings not agglomerate into satellites?
3) What determines the outer limit of the rings?

In what follows we shall understand by rings solely the primary rings with relatively large particles (see Sect. 2.8).

5.1.1 Discussion of the Traditional Point of View That the Region of the Primary Rings Is the Roche Zone

In 1849 Roche considered the balance of tidal forces and the self-gravitation of a satellite and put forward the hypothesis that the Saturnian rings would be the result of the break-up of a large body through the tidal pull near the planet. Let us briefly dwell upon the nature of the tidal forces.

The planet attracts different parts of the satellite with different forces: the closer a point of the satellite is to the planet, the larger the acceleration (Fig. 5.1),

$$g = \frac{GM}{R^2}.\tag{5.1}$$

The gravitational acceleration opposes the centrifugal one,

$$f = \Omega^2 R,\tag{5.2}$$

which, in contrast, increases with increasing distance from the planet – since all parts of the satellite orbit with the same angular velocity, provided we assume a synchronous rotation with the satellite always showing the same side to the planet, like the Moon to the Earth.

Assuming for the sake of simplicity that the satellite has a uniform density ϱ and a spherical shape, we can find the condition that its initial shape will

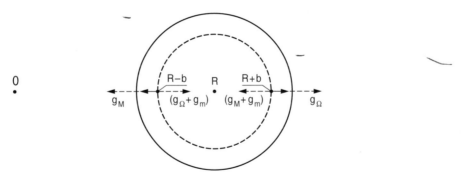

Fig. 5.1. Nature of the tidal forces acting on a satellite. The tidal force arises due to the difference in the way the centrifugal and the gravitational forces, which act at each point of a satellite orbiting a planet, depend on the distance. The result is a tidal force which tends to break up the satellite. The planet of mass M is situated at the point O; R is the orbital radius of (the centre of mass of) the satellite and $R \pm b$ are arbitrary points on the straight line OR and its extension. According to Newton's theorem, at the points $R \pm b$ only that part of the mass of the satellite, indicated by $m(b)$, will act which lies inside the dashed surface. The acceleration at each point is indicated by the dashed arrows where g_Ω is the centrifugal acceleration and g_M the gravitational acceleration in the field of the planet, while $g_{m(b)}$ is the gravitational acceleration caused by the central part of the satellite within the constant radius b. The resulting acceleration is indicated by the full-drawn arrows.

as a result of the forces acting upon it start to be deformed into a cucumber aimed at the planet. Let there be a planet of mass M at the point O and let the centre of the satellite be at the point R. For any pair of interior points of the satellite $R \pm b$ lying on the line OR and its extension (Fig. 5.1) we find for the resulting acceleration:

$$g_\Omega + g_M + g_{m(b)} = \Omega^2(R \pm b) - \frac{GM}{(R \pm b)^2} \mp \frac{Gm(b)}{b^2}$$

$$= \pm \left(3\Omega^2 - \tfrac{4}{3}\pi G\varrho\right) b, \tag{5.3}$$

where g_Ω is the centrifugal acceleration, g_M the gravitational acceleration in the field of the planet, $g_{m(b)}$ the gravitational acceleration due to the central part of the mass of the satellite inside the (constant) radius b, $m(b) \equiv \tfrac{4}{3}\pi\varrho b^3$, $b \leqslant a$, and a is the radius of the satellite. The upper (lower) signs in (5.3) refer to the point $R + b$ ($R - b$). In deriving this formula we have used Newton's theorem that the mass of spherical layers outside the points $R \pm b$ will not produce any resultant forces.

From (5.3) it follows that if the condition

$$3\Omega^2 > \tfrac{4}{3}\pi G\varrho \tag{5.4}$$

is satisfied, the resulting acceleration in the points $R \pm b$ is directed away from the sphere $b = \text{const}$. Hence, condition (5.4) is the necessary and sufficient condition for the break-up of a so-called "fluid" satellite, that is, one which is without intermolecular cohesive forces.

From (5.4) we get an outer limit R_R for the Roche zone inside which a self-gravitating liquid satellite is broken up by the tidal force:

$$R_R = \alpha \left(\frac{M}{\varrho}\right)^{1/3}. \tag{5.5}$$

In our estimate we have $\alpha \approx 0.9$; Roche calculated $\alpha = 1.5$. Of course, Roche's model is very attractive and answers all three questions given at the beginning of the present chapter: the material of the broken-up satellite – which happened to be in the danger zone – can be used for the construction of the rings; the tidal forces prevent the existence of satellites in the ring zone; and the boundary of the rings is given by (5.5) – at that radius the tidal forces are sufficiently decreased to become comparable with the self-gravitation forces.

After Roche's model had existed peacefully for a hundred years it started to encounter trouble: Jeffreys (1947) showed that molecular cohesion is for small satellites more important than self-gravitation and found an additional condition for break-up:

$$P_{max} < 1.68\varrho a^2 \Omega^2, \tag{5.6}$$

where P_{max} is the tensile strength of the material of a satellite of density ϱ and radius a, and Ω is the angular velocity of the orbital motion. We can easily understand the origin of this condition by equating the tidal force which breaks the satellite up,

$$F = \tfrac{1}{2}mf = \tfrac{2}{3}\pi\varrho a^3 f,$$

to the force actually necessary for the break-up of a satellite with a given tensile strength:

$$F' = \pi a^2 P_{\max}.$$

Hence we get the condition $P_{\max} < 2\varrho a^2 \Omega^2$, which is close to (5.6). It follows from (5.6) that a body of ice with a tensile strength of 10^7 dyne/cm^2 will not be broken up in the Saturnian ring zone, if its radius a is less than 200 km. Thus, the ring particles which are much smaller cannot be the result of a tidal break-up of a large satellite: fragments of a satellite are already stable against tidal break-up. Roche's catastrophic hypothesis was therefore gradually replaced by a condensation model according to which the rings are the remnants of a circumplanetary proto-satellite cloud. The tidal forces began to emerge in the role of a factor preventing the accretional growth of the particles and the formation of satellites – without it being exactly clear whether this is through preventing accretion or through breaking up the growing particles.

Davis et al. (1984) showed that the accretional growth of the particles in the rings is not forbidden and they considered the balance between the accretional growth (sticking together) of particles and their breaking up by tidal forces when they reach their maximum size of 10 m. One finds easily from (5.6) that 10 m size particles are efficiently broken up when the tensile strength of the material of the particles is 10^{-2} dyne/cm^2: this is four to five orders of magnitude lower than the tensile strength of loose terrestrial snow (Voĭtkovskiĭ 1977) or of loose lunar regolith (Jaffe 1967). In order to avoid having to assume abnormal mechanical properties of the particles, Davis et al. (1984) suggested that the maximum size of the ring particles reached kilometers and that observers simply had not noticed these bodies. We then get from (5.6) an altogether realistic tensile strength of the particles of 10^2 dyne/cm^2. However, the majority of observers feels that the upper limit of the particle sizes had been reliably determined (Cuzzi et al. 1984). Since the model of tidal break-up requires for 10 m size particles an unrealistically low tensile strength, one must conclude that the tidal action on large ring particles is weak. This conclusion brings us back to the three problems stated at the beginning of this chapter.

5.1.2 Collisional Break-Up of Particles in Grazing Collisions

Let us consider a completely different mechanism for limiting both the size of the ring particles and the radius of the region in which the rings can occur. Gor'kavyĭ and Fridman (1985; Gor'kavyĭ and Taĭdakova 1989; Fridman and

Gor'kavyĭ 1992) have studied the collisional break-up of ring particles. The growth of the particles will be limited if (a) the particles are broken up and (b) the fragments are spread out beyond the gravitational pull of the particles.

The splitting up of a particle is a problem of how much energy there is in the break-up. We shall in what follows estimate the amount of energy per unit volume in the break-up for which the collisions between particles will be efficient in breaking down the material. The problem of the spreading out of the material from the sphere of action of the particle is more complicated. According to Safronov (1972) the particle velocity v in a proto-planetary disc with a power-law mass distribution, $dN(m) = Cm^{-q}\,dm$, is determined by the mass m and the radius a of the largest particle: $v = \sqrt{Gm/\Theta a}$, where Θ is the Safronov parameter which for $q < 2$ is a few units so that v turns out to be less than the characteristic velocity $v \simeq \sqrt{Gm/a}$ – the escape (primary cosmic) velocity at the surface of a large particle. In this case the fragments cannot overcome the gravitational attraction of the body and, notwithstanding the break-up, the planetesimals grow through mutual collisions. How can particle collisions limit the accretional growth of the particles in the rings? Gor'kavyĭ and Fridman (1985) suggested that around a planet there exist two physically different zones:

a) an inner zone of the cosmogonically primary rings where the fragments of the particles do not return to the "parent" particles in the collisions;
b) the outer zone of the satellites where the break-up is inefficient – the fragments rapidly return to the "parent" particles.

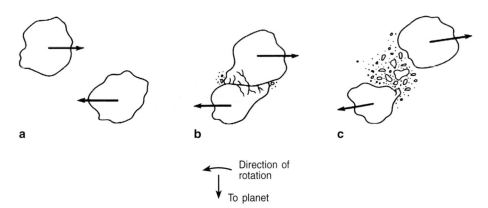

Fig. 5.2a–c. Collision of two loose particles in the field of the planet, pictured in the rotating reference frame.

The enhanced destructibility of the particles in the rings is connected with the large magnitude of the shear velocities due to the differential rotation of the rings. Quasi-grazing collisions have the largest relative velocities in which case the semi-major axes of the colliding particles differ approximately by twice the radius of the particles (Fig. 5.2). Assuming the velocity of the fragments relative to the particles which are breaking up in the collisions to be comparable to the shear velocity of the particles and equating the latter to the characteristic velocity of these particles – the escape velocity at their surface, $\sqrt{Gm/a}$ – we get an expression for the limit of the rings (Gor'kavyĭ and Fridman 1985):

$$\Omega a \approx \sqrt{\frac{Gm}{a}}, \tag{5.7}$$

or

$$R_c \approx \beta \left(\frac{M}{\varrho}\right)^{1/3}. \tag{5.8}$$

Here m and M are, respectively, the masses of the particle and of the planet, and β (≈ 1) is a numerical coefficient.

The radius of the Roche zone is also described by a similar formula ($\beta = 1.5$) which is not surprising since in both cases we consider the balance between the self-gravitation of a particle and effects connected with the gravitational force exerted by the planet – with tidal forces or with shear velocities. Of course, the numerical coefficients may differ.

We shall follow Gor'kavyĭ and Taĭdakova (1989) and consider more rigorously the problem of the motion of the fragments in the gravitational field of two colliding particles and the planet.

5.2 Dynamics of Particle Fragments in the Four-Body Problem

For our calculations we shall make the following assumptions:

a) The mass of the fragments is negligibly small and they do not collide with one another.
b) After the collision the large particles move in one plane along elliptical orbits.
c) The mass of the large particles is small as compared to the planetary mass.
d) The gravitational field of the large particles is spherically symmetric.

A note about the terminology: we shall call the small particles formed in the break-up: "fragments", particulates, or test particles, and we shall call the large gravitating particles simply particles or bodies.

We write the dynamical equations of a separate fragment in the rotating x, y, z coordinate system where the x-axis is directed along the radius and the y-axis in the direction of the orbital motion of the particles (Fig. 5.3)

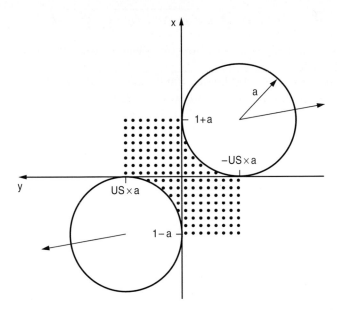

Fig. 5.3. Separation of colliding particles in the rotating (x, y) system of coordinates at time $t = 0$. The system rotates counterclockwise, and the planet is situated below the figure; a is the radius of the particle, $2USa$ is the initial distance along the y-axis between the centres of the particles (here we have $US = 1$).

while the origin is at the centre of the planet. We introduce the following units: the unit of length is the radius R of the orbit, the unit time is $T/2\pi$ with T the rotational period of the coordinate system, and the unit mass is the mass of the planet. In that system of units the angular velocity of the system of coordinates and the gravitational constant are equal to unity. The set of equations has the form (Roy 1978):

$$\ddot{x} = 2\dot{y} + x + \frac{\partial}{\partial x}\psi(x, y, z), \tag{5.9}$$

$$\ddot{y} = -2\dot{x} + y + \frac{\partial}{\partial y}\psi(x, y, z), \tag{5.10}$$

$$\ddot{z} = \frac{\partial}{\partial z}\psi(x, y, z). \tag{5.11}$$

Here

$$\psi = \frac{1}{r_1} + \frac{m_2}{r_2} + \frac{m_3}{r_3}$$

is the gravitational potential of the planet and the two large particles of masses m_2 and m_3; r_1, r_2, and r_3 are, respectively, the distances to the planet and to the two large particles:

$$r_1 = \sqrt{x^2 + y^2 + z^2}, \qquad r_{2,3} = \sqrt{(x - x_{2,3})^2 + (y - y_{2,3})^2 + z^2},$$

where $x_2(t), y_2(t), x_3(t), y_3(t)$ ($z_2 = z_3 \equiv 0$) are the given coordinates of the centres of the large particles.

Gor'kavyǐ and Taǐdakova (1989; Taǐdakova 1990) have solved the set (5.9) to (5.11) numerically using an implicit second-order-accuracy method which has been described, for instance, by Potter (1975). To evaluate the derivatives we write (5.9) to (5.11) in the form

$$\ddot{x} = 2\dot{y} + x + F_x, \tag{5.12}$$

$$\ddot{y} = -2\dot{x} + y + F_y, \tag{5.13}$$

$$\ddot{z} = F_z. \tag{5.14}$$

$$\boldsymbol{F} = -\left\{ \frac{\boldsymbol{r}}{r_1^3} + \frac{m_2(\boldsymbol{r} - \boldsymbol{r}_2)}{r_2^3} + \frac{m_3(\boldsymbol{r} - \boldsymbol{r}_3)}{r_3^3} \right\}, \tag{5.15}$$

where $\boldsymbol{r}$, $\boldsymbol{r}_2$, $\boldsymbol{r}_3$, and $\boldsymbol{F}$ are vectors with components $\{x, y, z\}$, $\{x_2, y_2, z_2\}$, $\{x_3, y_3, z_3\}$, and $\{F_x, F_y, F_z\}$, respectively.

We follow Gor'kavyǐ and Taǐdakova (1989) and write (5.12) to (5.15) in the form ($\boldsymbol{v} = \{v_x, v_y, v_z\}$)

$$\frac{dv_x}{dt} = 2v_y + x + F_x, \tag{5.16}$$

$$\frac{dv_y}{dt} = -2v_x + y + F_y, \tag{5.17}$$

$$\frac{dv_z}{dt} = F_z, \tag{5.18}$$

$$\frac{d\boldsymbol{r}}{dt} = \boldsymbol{v}. \tag{5.19}$$

We can use the transformation

$$\hat{x} = x + iy, \qquad \hat{v} = v_x + iv_y, \qquad \hat{F} = F_x + iF_y \tag{5.20}$$

to write (5.16) to (5.19) in the form

$$\frac{d\hat{x}}{dt} = \hat{v}. \tag{5.21}$$

$$\frac{d\hat{v}}{dt} = -2i\hat{v} + \hat{x} + \hat{F}, \tag{5.22}$$

We have thus obtained a set of ordinary time-dependent differential equations ((5.18), the z component of (5.19), and relations (5.21) and (5.22)). The implicit second-order-accuracy method gives for equations of the form

$$\frac{df}{dt} + \varPhi(f, t) = 0, \tag{5.23}$$

with given initial conditions $f(t_0) = f_0$ the solution

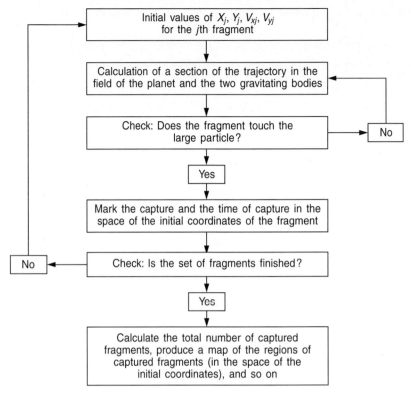

Fig. 5.4. Block diagram for calculating the motion of fragments in the four-body problem.

$$f^{[n+1]} = f^{[n]} - \tfrac{1}{2}\left(\varPhi^{[n]} + \varPhi^{[n+1]}\right)\Delta t. \tag{5.24}$$

We note that in the case of interest to us the function $\varPhi$ depends on the space coordinates for (5.18) and (5.22). Applying the scheme (5.24) directly to (5.18) and (5.22) leads to complicated algebraic equations. To simplify matters we shall evaluate the spatial coordinates at half-integral time values $n + \tfrac{1}{2}$. Doing this in applying the scheme (5.24) to (5.18), (5.19), (5.21), and (5.22), and going back from the complex to real quantities we get the following set of difference equations (Taĭdakova 1990):

$$
\begin{aligned}
v_x^{[n+1]} &= \frac{1}{1+\Delta t^2} \left\{ v_x^{[n]}(1 - \Delta t^2) + 2v_y^{[n]}\Delta t \right. \\
&\quad \left. + \left(x^{[n+0.5]} + F_x^{[n+0.5]} \right)\Delta t + \left(y^{[n+0.5]} + F_y^{[n+0.5]} \right)\Delta t^2 \right\}, \\
v_y^{[n+1]} &= \frac{1}{1+\Delta t^2} \left\{ v_y^{[n]}(1 - \Delta t^2) - 2v_x^{[n]}\Delta t \right. \\
&\quad \left. + \left(y^{[n+0.5]} + F_y^{[n+0.5]} \right)\Delta t + \left(x^{[n+0.5]} + F_x^{[n+0.5]} \right)\Delta t^2 \right\}, \\
v_z^{[n+1]} &= v_z^{[n]} + F_z^{[n+0.5]}\Delta t, \\
\boldsymbol{r}^{[n+1]} &= \boldsymbol{r}^{[n]} + \tfrac{1}{2}\left(\boldsymbol{v}^{[n+1]} + \boldsymbol{v}^{[n]} \right)\Delta t,
\end{aligned}
\tag{5.25}
$$

where

$$
\begin{aligned}
\boldsymbol{r}^{[n+0.5]} &= \boldsymbol{r}^{[n]} + \tfrac{1}{2}\boldsymbol{v}^{[n]}\Delta t, \\
\boldsymbol{F}^{[n+0.5]} &= \boldsymbol{F}(x^{[n+0.5]}, y^{[n+0.5]}, z^{[n+0.5]}, t^{[n+0.5]}).
\end{aligned}
$$

The function $\boldsymbol{F}$ is given by (5.15).

We show in Fig. 5.3 the initial uniform grid of fragments in the x, y-plane between the particles; this is an idealised picture of the break-up shown in Fig. 5.2. The initial velocity of the fragments corresponds to a circular Kepler motion. We show in Fig. 5.4 the block diagram of our programme. The calculations show that the trajectories of the separate fragments depend on the positions at $t = 0$. We show in Fig. 5.5 by different symbols the initial positions of those fragments which in the course of two orbital revolutions will again be captured by the large bodies. The particulates which are not captured start their motion in those regions where there are no symbols. One can divide all fragments into four basic classes:

1) Fast accreting fragments: those which practically at once – during the first one third of an orbital revolution – return to the particles, that is hit their surface. The regions of the initial positions of such fragments are indicated in Fig. 5.5c by the number *1*.
2) Slowly accreting fragments: those which are captured by the particles after a longer time – more than a third of an orbital revolution. The regions of such fragments are indicated in Fig. 5.5c by the number *2*.
3) Uncaptured advancing fragments: those moving to orbits with semi-major axes larger than $R+a$ or less than $R-a$, as a result of which the fragments move away from the place where the collision took place faster than the large bodies and can no longer be captured by them even after many orbital revolutions. When drawing Fig. 5.5 we assumed circular motion of the large bodies. The regions of the advancing fragments are indicated in Fig. 5.5c by the number *3*.
4) Lagging uncaptured fragments: they have new orbits with semi-major axes in the range from $R - a$ to $R + a$ so that they lag behind the large particles. The region of the initial positions of this class of fragments is indicated by the number *4* in Fig. 5.5c.

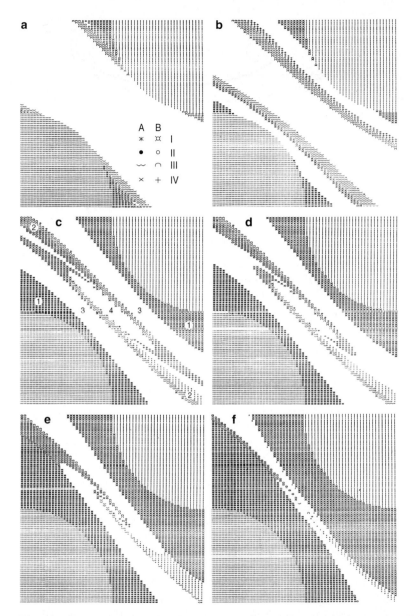

Fig. 5.5. Various regions for the initial positions of captured fragments. The different symbols correspond to different fragment life-times (as fractions of an orbital revolution) before they hit the surface of one of the particles before the completion (*I*) of one third of an orbital revolution; (*II*) of two thirds; (*III*) of one orbital revolution; (*IV*) of two orbital revolutions. The A (B) symbols refer to fragments captured by the upper (lower) particle which is vertically (horizontally) hatched; (**a–f**) correspond to α values of 0.65 (**a**), 0.81 (**b**), 1.05 (**c**), 1.09 (**d**), 1.29 (**e**), and 1.61 (**f**), respectively; *1* to *4* are the main fragment classes: fast accreting, slowly accreting, uncaptured advancing, and lagging uncaptured fragments.

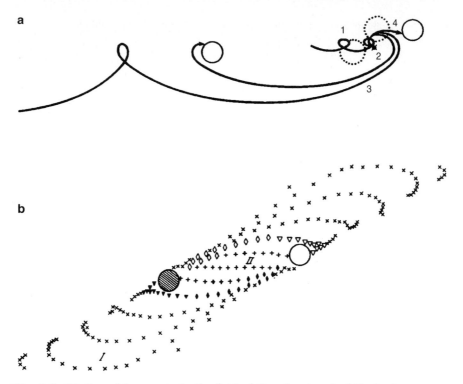

Fig. 5.6. Motion of fragments in the field of the planet and of the broken-up particles. (**a**) The trajectories *1* to *4* of a separate fragment at different distances from
the planet. The initial positions relative to the large bodies are the same; *1* to
4 correspond to $\alpha = 0.81$ (*1*), 1.05 (*2*), 1.09 (*3*), and 1.29 (*4*). (**b**) the shape of
a cluster of fragments after the completion of two thirds of an orbital revolution
($\alpha = 1.05; US = 1$). (*I*) uncaptured advancing particles; (*II*) lagging uncaptured
particles; the rest of the symbols indicate fragments captured within two orbital
revolutions. The full drawn and open symbols refer to fragments hitting the shaded
or the open particle, respectively.

The sequence from Fig. 5.5a to Fig. 5.5f corresponds to a decrease of the
effect of the gravitational field of the planet and an increase of the gravitational effect of the particles. This can be obtained by increasing either R,
the distance from the planet, or the ratio m/M of the particle mass to the
planetary mass. The calculations show that the dependence of the dynamical
scenario of the separation of the fragments on these quantities can be reduced
to dependence on a single parameter, $\alpha = R(\varrho/M)^{1/3}$ (see (5.8)). Figs. 5.5a
to f were obtained for $\alpha = 0.65$; 0.81; 1.05; 1.09; 1.29; and 1.62, respectively.
The outer limit of the planetary rings corresponds, according to the data of
Sect. 6.4, to $\alpha = 0.82 \pm 0.05$.

In Fig. 5.6 we show the trajectories of separate fragments. The initial
values of the coordinates were chosen in the same point of Figs. 5.5b to e.

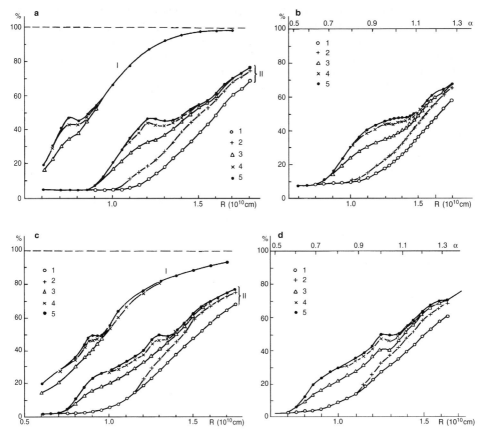

Fig. 5.7. Percentage of captured particles as function of the orbital radius and the elapsed time. Fraction of particles captured in the first one third of an orbital revolution (*1*), in the first two thirds of an orbital revolution (*2*), in the first orbital revolution (*3*), in the first four thirds of an orbital revolution (*4*), and within the first two orbital revolutions (*5*). (**a**) two-dimensional case, $US = 0.5$, particle density $\varrho = 0.9$ g/cm^3 (*I*), 0.2 g/cm^3 (*II*). (**b**) three-dimensional case, $US = 0.5$, the initial thickness of the cluster of fragments in the z-direction is equal to the radius a of the large body. We consider only a particle density $\varrho = 0.2$ g/cm^3, corresponding to the lower scale of distances. The upper scale gives the values of the parameter α. (**c**) two-dimensional case, $US = 1$, $\varrho = 0.9$ g/cm^3 (*I*), 0.3 g/cm^3 (*II*). (**d**) three-dimensional case, $US = 1$, $\varrho = 0.3$ g/cm^3. The upper scale gives the values of the parameter α.

$$\frac{\sigma \Omega}{4\varrho}\, t_{\mathrm{a}} \cong 0.1 a_{\mathrm{cr}}. \tag{5.37}$$

Substituting expression (5.35) for σ into (5.37) we get an equation where apart from q we have the unknown $a_{\max}$, as an integration limit, and t_{a}, the time it takes to accrete. Since the differential density $\sigma(a)$ decreases fast for $a > a_{\mathrm{cr}}$ we can for the upper limit of integration take $a_{\max} \cong a_{\mathrm{cr}}$. Starting from the stationarity condition for the spectrum we may assume that the time t_{a} it takes a particle to grow through accretion from 0 to a_{cr} must be equal to the time it takes a particle to break down from a_{cr} to 0 through collisions.

We can estimate the latter time as the free flight time of a particle of radius a_{cr} in a medium of the same particles:

$$t_{\mathrm{a}} \sim t_{\mathrm{c}} \sim \frac{1}{2\Omega\tau(a_{\mathrm{cr}})}. \tag{5.38}$$

Using (5.35) and substituting (5.38) into (5.37) we get after some simple manipulations

$$\left(\frac{a_{\mathrm{cr}}}{a_0}\right)^q = 0.68(4+q)\frac{n(a_{\mathrm{cr}})}{n_0 a_{\mathrm{cr}}}. \tag{5.39}$$

We have used here the fact that the characteristic size of a particle of the accreting medium is $l \cong a_{\max}$, which corresponds to the assumption that $q > -4$. We can now find q from (5.39) and data following from observations, $n(a_{\mathrm{cr}}) \cong 10^{-4}$ to 10^{-3} m^{-2}, $a_{\mathrm{cr}} = 5$ m, $a_0 = 0.01$ m, and $n_0 \cong 3 \times 10^3$ m^{-2}/m. The dimensionality [m^{-2}/m] of the function $n(a)$ indicates the number of particles per square meter in the plane of the rings as function of the particle size, measured in meters. The result is $q = -(2.7$ to $3.1)$ which agrees with the mass spectrum obtained from the radio-occultation of the rings.

We note that in September 1989 Longaretti (1989) arrived at a model which is the same as the one considered earlier by us (Gor'kavyǐ and Fridman 1985). Longaretti studied the collisional break-up of particles – he estimated the tidal break-up to be negligible – and arrived at the conclusion that the Saturnian rings consisted of loose particles and that the model of smooth hard spheres leads to an unrealistic size distribution law. Longaretti also obtained the same distribution law for loose particles as we have found, namely (5.33).

The mechanism of collisional break-up therefore limits the size of the particles in planetary rings and determines the outer limit of the rings. If we bear in mind that most researchers consider the rings as the relict remnant of a circumplanetary proto-satellite disc we have obtained an answer to the three questions put at the beginning of Sect. 5.1. We have not used any anomalous assumptions. The model considered here makes it possible to estimate the mechanical characteristics of the particles which are agglomerates of loose matter – in the case of the Saturnian rings snow – and also to explain the observed particle size distribution spectrum.

6 Elementary Particle Dynamics
III Wave, Photometric, and Other Effects

"What is your substance, whereof are you made,
That millions of strange shadows on you tend?
Since every one hath, every one, one shade,
And you, but one, can every shadow lend."

W. Shakespeare, Sonnet 53.

In Chap. 5 we solved the problem of the origin of the rings by studying the behaviour of particles in the field of a planet and two gravitating bodies. However, one can also use the same method to study a series of other problems and we shall do that in the present chapter.

6.1 A Satellite in a Differentially Rotating Disc

Large bodies in a disc of small particles act on the particles through their gravitational fields, and they also can absorb disc particles by collecting them on their surface. The gravitational action can be a resonance action – in the case when the ratio of the orbital periods of the satellite and of the particle are the ratio of integers. We shall consider this kind of interaction between satellites and particles in Chaps. 9 and 10. Here, we shall be interested in the behaviour of particulates in the disc in the immediate vicinity of a satellite or a large particle. We shall study the dynamics of particles in the field of a planet and a single satellite using the model suggested in Sect. 5.2, dropping the "spare" satellite and expanding the field of the initial positions of the test particles.

It is well known from classical mechanics that the lowest energy of particles moving in a central field of force occurs when they are moving along purely circular orbits. Any additional energy source, whether an internal source, such as the energy of the thermal motion of particles (see, for instance, Fig. 2 in a paper by Fridman et al. (1981)), or an external source, such as the gravitational interaction of particles with a satellite (see Fig. 6.1), will perturb the circular motion of the particles causing them to oscillate around the unperturbed circular trajectory in the plane of the orbit. This is called epicyclic motion (see, for instance, Fridman and Polyachenko 1984).

In the case when the relative effect of a satellite is large the amplitude of the epicyclic oscilations will also be large, sharply "turning the particles round" in the rotating frame of reference; see Fig. 6.1.

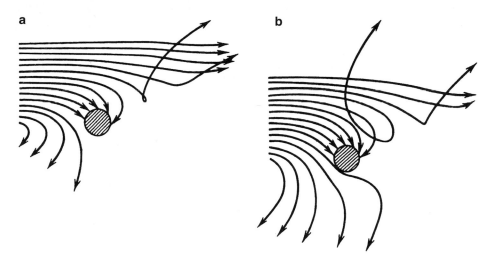

Fig. 6.1. Perturbation by a satellite of purely circular particle trajectories around a planet. In (**a**) the relative effect of the satellite is less than in (**b**). One can clearly see the repulsive regions of epicyclic oscillations.

The particles perturbed by the satellite follow epicyclic trajectories and form a spiral density wave in the plane of the disc (see Fig. 6.2). This kind of spiral wave has, in fact, been observed in the A ring at the edge of the Encke gap which indicates the existence of an invisible satellite in that gap (Goldreich and Tremaine 1982). A similar spiral wave due to the "shepherd" satellites in Saturn's F ring gives the illusion of a three-dimensional inter-weaving of a thin ring (Goldreich and Tremaine 1982). Of course, this is a very simple model of a spiral wave and it neglects viscous and inelastic interactions between the particles, which lead to appreciable changes in its shape and the size of the region over which it propagates. We note that the wavelength of the spiral wave is very simply connected with the distance S from the satellite: $\lambda = 3\pi S$. This relation can be obtained by elementary arguments, as follows.

Let us change to a frame of reference which rotates with the satellite with an angular velocity $\boldsymbol{\Omega}_\mathrm{s}$, with the origin as before at the centre of the planet. In this frame the velocity of a particle at a distance $\boldsymbol{r}_\mathrm{p}$ from the origin is given by the equation (Landau and Lifshitz 1976)

$$\boldsymbol{u}(\boldsymbol{r}_\mathrm{p}) \; = \; \boldsymbol{v}_\mathrm{p}(\boldsymbol{r}_\mathrm{p}) - [\boldsymbol{\Omega}_\mathrm{p} \wedge \boldsymbol{r}_\mathrm{p}]. \tag{6.1}$$

If we now assume that the satellite and the particle rotate in the $z = 0$-plane along circular orbits with radii r_s and $r_s + s$, respectively, the azimuthal component of the relative velocity of the particle will be equal to

$$u_\varphi = (v_p)_\varphi - \Omega_s r_p = (\Omega_p - \Omega_s)\, r_p \cong \left.\frac{\partial \Omega}{\partial r}\right|_{r=r_s} s = -\tfrac{3}{2}\Omega_s s. \quad (6.2)$$

We have used here the Kepler orbit law: $\Omega \sim r^{-3/2}$. In the same rotating frame of reference the perturbed particle moves, as the result of the interaction with the satellite, along an epicyclic with a rotational period $T = 2\pi/\Omega_s$. If we bear in mind that the wavelength is $\lambda = |u_\varphi T|$ we obtain the required relation $\lambda \cong 3\pi s$.

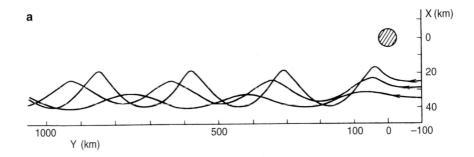

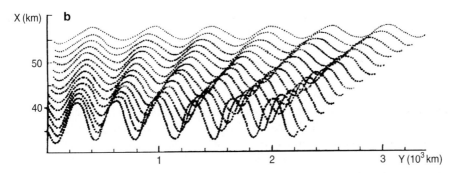

Fig. 6.2. Perturbation of the trajectories of particles in a disc by the field of a satellite. **(a)** Strong perturbation close to the satellite – it is clear how fast the particle trajectories become entangled. **(b)** Weaker perturbation; a spiral wave is formed in the disc.

Some disc particles are captured by the satellite in the accretion process – we do not consider the unlikely case when a particle hits the satellite and irreversibly rebounds from it. We shall call that region of the disc from which

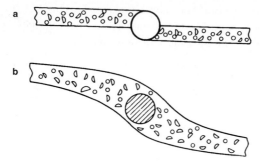

Fig. 6.3. Feeding zone of a satellite. (**a**) Case of a non-gravitating satellite; the capture is determined solely by the differential rotation of the disc. (**b**) Feeding zone of a gravitating satellite.

the particles are absorbed by the satellite its feeding zone. A satellite without any gravitating field will have a feeding zone determined solely by the differential rotation of the disc particles (Fig. 6.3a). The gravitational field of the satellite increases the feeding zone (Fig. 6.3b).

6.2 Two Large Bodies in a Disc of Small Particles

Let us consider the behaviour of small disc particles in the vicinity of two large gravitating bodies which come close to one another along their orbits. This is an altogether realistic situation – in fact, a large number of large particles in the ring will get close to one another undergoing quasi-grazing collisions and one can consider the assembly of small particles around these bodies as a set of test particles. We shall study this situation in the framework of the same numerical model – with small variations – making the same assumptions. This problem has been partially touched upon in Sect. 5.2 but there we were interested in the dynamics of a cluster of fragments in a very localised region – between two particles which are moving away from one another. We shall now consider the problem in a somewhat wider context. We show in Fig. 6.4 the feeding zones of two bodies moving towards one another along parallel orbits in a differentially rotating disc. The two bodies also move in circular Kepler orbits. The complicated twisting of the feeding zones demonstrates the perturbation of the medium by the gravitational field of the bodies. Generally speaking, this is one additional way in which large bodies interact in the disc – indirectly through perturbing the accreting medium.

In Fig. 6.5 we show a picture of the feeding zones in the period after the particles have already encountered one another and when they move away from each other continuing to shake up the "troughs" of their neighbour – in Fig. 6.5 we show only one perturbed zone, the other being positioned symmetrically. The evolution of that part of this zone which is close to the

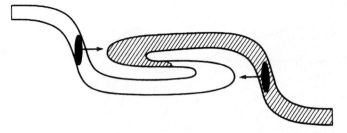

Fig. 6.4. Deformation of the feeding zones of two gravitating bodies in a disc of particles before their grazing encounter. The picture has been compressed in the y-direction.

particles themselves has, as function of the parameter α, been considered in Sect. 5.2. To widen the field of our "vision" to the left (along the path of the lower particle) we compress the scale of the picture in the y-direction – along the orbit. This picture of the feeding zone is given in Fig. 6.6. The feeding zone is distorted because the lower particle, moving to the left, causes a spiral wave in the medium of the small particles; it lies above it and moves to the right, overtaking the second particle – bear in mind that the action is taking place in the rotating frame of reference. Such a "shaking up" of the feeding zone increases the fraction of the test particles in the disc which are captured by 35 to 40%. The size of the feeding zone in the x-direction is increased by about a factor 2. This means that the rate at which the particles in the disc grow through accretion is increased.

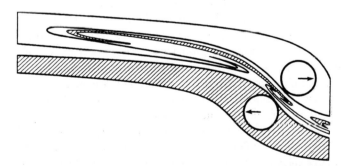

Fig. 6.5. Feeding zones of two gravitating bodies after their encounter. The scale has not been distorted.

We draw attention to the complex geometry of the feeding zones between the two particles (see Fig. 6.5) and on the boundary of the bananalike regions of uncaptured particles inside the feeding zone (see Fig. 6.6). On this boundary the feeding zones of both particles together with the regions of un-

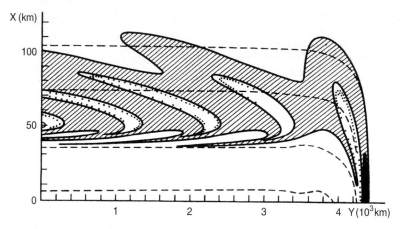

Fig. 6.6. Feeding zones of two gravitating bodies after their encounter. The picture has been compressed by a factor 20 along the y-axis.

captured particles are strongly entangled. We shall in what follows consider the interesting dynamics of the test particles from this "mixed-up" region.

6.3 Wanderer Particles in the Four-Body Problem

We show in Fig. 6.7 "mixed-up" regions with ever increasing resolution, that is, for a denser initial grid of the fragments and a longer calculation period. To fix the ideas we have indicated by figures the initial coordinates of fragments which are captured by one of the particles and by symbols the initial positions of the fragments hitting the other particle. From Fig. 6.7 it is clear that at the edge of the "figure" region there are several layers of a symbol region and on the edge of a symbol layer one can find a microscopic number region – registering a group of fragments with much longer life-times. Such a fractal structure of the boundary can clearly be traced practically ad infinitum. The entanglement of the feeding zones and the zones of uncaptured fragments is explained by the presence of wanderer fragments which are alternately accelerated and decelerated in the field of the large particles moving away from one another, travelling between them along rather complicated trajectories. We show in Fig. 6.8 three trajectories of fragments which differ extremely little in their initial coordinates but this insignificant deviation rapidly increases – one can say that the trajectories are unstable – and the fate of the three neighbouring fragments turns out to be very dissimilar. We can understand the further entanglement of the zones if we assume that trajectory *3* in Fig. 6.8 does not end on a particle but passes near it – as in Fig. 6.9. The wanderer particle decelerating in the field of one of the large particles rushes to overtake the other one. The fate of a "wanderer" near a

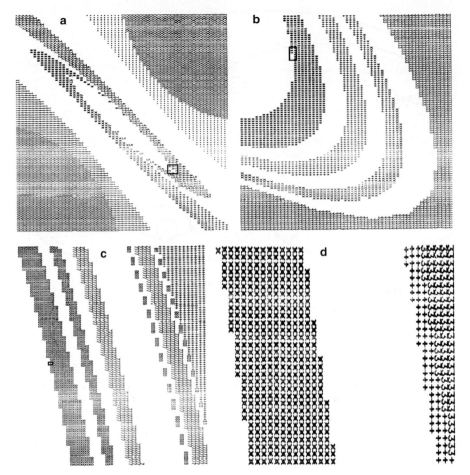

Fig. 6.7. Fractal structure of the boundary of the feeding zone. Each of the figures (**b–d**) gives an enlargement of the indicated section of the boundary of the preceding figure (**a–c**, respectively).

satellite which it has overtaken can also develop in different ways depending on completely negligible shifts in the initial coordinates. One can say that the wanderer particles are the stochastic component of our dynamic and basically deterministic system. However, the fraction of wanderer particles is very small and decreases after each revolution.

We note that in each successive approach of a fragment to a large particle the eccentricity of the orbit of the fragment increases – the eccentricity is proportional to the distance from the maximum, epicentre, and the minimum, pericentre, radius of the orbit. This is a fundamental effect lying at the basis of the whole dynamics of differentially rotating discs: the transformation of part of the energy of the orbital motions into chaotic motions

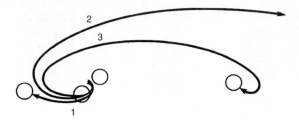

Fig. 6.8. Three trajectories with completely different fates due to a very slight difference in the initial coordinates.

Fig. 6.9. Trajectory of a wanderer particle which gets close to two satellites which move away from one another. The numbers indicate the time when it approaches the satellites in their orbits.

(read: increasing the eccentricity) when the particles collide – or come into each other's gravitational vicinity. NASA has been using this effect of the increase in kinetic energy of a particle when it interacts with bodies which orbit around a central mass. Because of financial difficulties NASA chooses for an interplanetary probe those trajectories which compensate for the insufficient power of the spacecraft by the free acceleration in the gravitational field of the planets. This was the basis of the flights of the Pioneers and the Voyagers in the seventies and the eighties and the same effect was used by Galileo to reach Jupiter in 1995.

6.4 Azimuthal Brightness Asymmetry of the Saturnian Rings

It is forty years ago that Camichel (1958) discovered the striking effect of the azimuthal brightness asymmetry in the Saturnian A ring. Nowadays there exist a number of high-quality observations of the brightness asymmetry of the rings both from the Earth (Thompson et al. 1981) and from the interplanetary Voyager spacecraft (Franklin et al. 1987). A number of hypotheses have been put forward (see Brahic's review article (1982)) to explain this phenomenon; these are based on assumptions about the synchronous rotation of the particles or about their asymmetric shape in the form of extended ellipsoids directed at a small angle to their orbit, or about an asymmetric surface albedo.

The synchronous rotation of particles is unrealistic from the point of view of collisional dynamics, and the inclined arrangement of ellipsoidal bodies is unstable. The most preferable model is therefore the one (Franklin et al. 1987) according to which the brightness asymmetry of the rings is connected with the spiral waves caused by the gravitational effect of the large particles. There are no quantitative estimates of the contribution of this effect to the azimuthal brightness asymmetry of the rings.

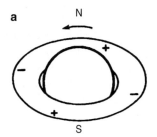

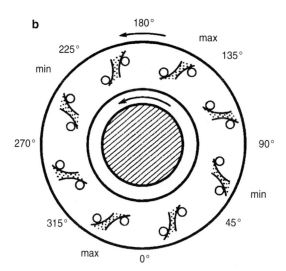

Fig. 6.10. Observed asymmetry in the brightness of the Saturnian A ring (**a**) and position of clusters of dust and small particles relative to the observer and the Sun, which lies below the figure (**b**). It is clear that in the regions of minimum brightness the dust cloud is shielded by the large particle – or projected upon it which also means no contribution to the brightness of the rings. In the region of maximum brightness the occultation of the clouds is insignificant.

Gor'kavyĭ and Taĭdakova (1989) have considered a mechanism for the azimuthal brightness asymmetry which is connected with a process which is fundamental for the planetary rings, namely the collisional break-up of large loose particles. The cluster of broken up material (see Fig. 5.7b) is a good reflector of the light from the Sun but the large particles moving away from each other after a collision almost completely shield the cloud of small particles both from the Sun and from the observer in well determined phases of the orbital angles (Fig. 6.10). This is the main factor for the occurrence of an azimuthal brightness asymmetry of the Saturnian rings: one sees more fragment-clusters in the bright parts of the rings and fewer clusters in the dark parts. We have taken into account that the collisional break-up of large particles is practically the only source of small particles and dust which give the main contribution to the brightness of the rings. Since the time for the existence of a cluster of fragments – until it is scattered by neighbouring particles – is equal to a few hours like the mean free flight time of a small particle in the rings one may assume that most small particles which determine the optical depth and the brightness of the rings are bunched together in the form of extended clouds formed in the collisions of large particles.

Gor'kavyĭ and Taĭdakova (1989) considered the azimuthal brightness asymmetry of the Saturnian rings for the case when the observer is on one line with the Sun. This corresponds closely to the case of the terrestrial observations (Camichel 1958, Thompson et al. 1981). The evaluation of the evolution of the fragment cluster was carried out in the three-dimensional variant – taking into account the thickness of the fragment cloud in the z-direction – by the same method as in the preceding sections. We give in Fig. 6.11 the block diagram of our calculations. Since there is in each region a cloud in a different stage of its expansion and, hence, with a different contribution to the asymmetry, we consider several shapes of the clouds with times of their development from the initial moment to the completion of one revolution (Fig. 6.12). For each shape of the fragment cluster one can find a projection on the plane at right angles to the line from Saturn to the Sun (observer). The plane is divided into 40 cm by 40 cm squares which gives the effective size of a single cloud. The area of the projection decreases when the small particles are occulted by one another and by the large bodies. The area of the projection of the large particles is neglected since these bodies produce merely a symmetric brightness background. The projection of the clouds in different stages of development are summed for each value of the angle. The result we obtain for the way the area of projection depends on the orbital azimuthal angle and other parameters can directly be compared with observational curves since the area of projection is proportional to the optical depth and hence determines the brightness of the rings.

We show in Fig. 6.13 the calculated curves for the area of projection of the clouds. The shape of the curve depends strongly on the density of the large particles, that is, on the capacity of the latter to scatter a cloud of

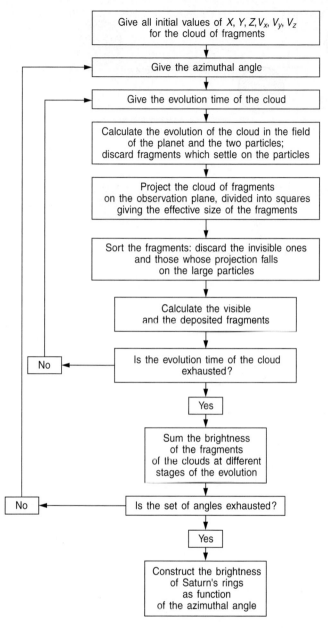

Fig. 6.11. Block diagram for the evaluation of the brightness of the Saturnian rings as function of the azimuthal angle.

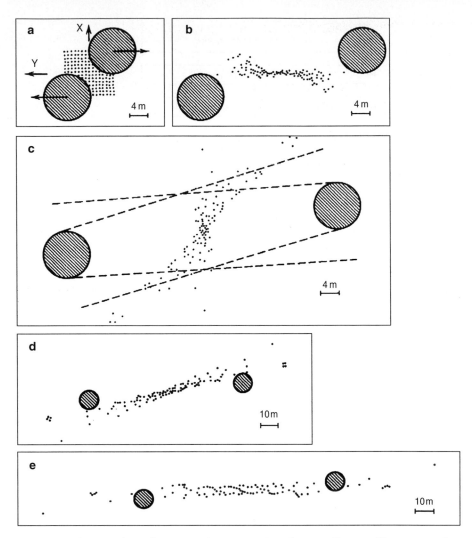

Fig. 6.12. Shape of the fragment cluster, produced in a collision of large particles, at different times: $t = 0$ (**a**), after one third of a revolution (**b**), after five ninth of a revolution (**c**), after seven ninth of a revolution (**d**), after one revolution (**e**); in Fig. 6.12c we have drawn the lines of the projection of the large particles in the two brightness minima of the dust cloud.

small fragments. The best agreement with observations occurs for particles with densities around 0.15 g/cm^3 (Fig. 6.13 – for an initial distance between the centres of the particles in the y-direction of $2a$) or around 0.1 g/cm^3 – with a distance between the centres of the particles of $1.5a$. In that case we took the thickness of the cluster in the z-direction to be a and the number of fragments in a single cluster around 1500; a collision changes the velocity of the large particles: they receive on top of the circular velocity additional velocities; for the particle which is further away these are $\Delta v_x = 0.25a\Omega$ and $\Delta v_y = 0.25a\Omega$ and for the nearer particles the same but with opposite signs. This variant corresponds well with observations: the minimum occurs in the neighbourhood of 65°, the maximum in the region of 160°. We show in Fig. 6.13 the data from measurements of the asymmetry (Thompson et al. 1981, Cuzzi et al. 1984); the calculations also give the observed asymmetry in the steepness of the wings – the curve increases from the minimum more steeply in the small angle region – as well as the presence of a plateau or minimum near 80° to 85° (see Fig. 6.13).

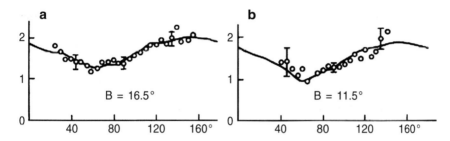

Fig. 6.13a,b. Theoretical curves for the area of the fragment clusters (in units of cells in the plane of projection). B is the inclination of the rings, the density of the large particles is 0.15 g/cm^3, and the circles are the experimental data (Thompson et al. 1981).

It is clear from the calculations that when one sums the contributions from the clusters "with a long life-time" (of the order of a revolution) the brightness minimum shifts from 60° to 85°. The fact that the observed minimum is close to 60° indicates that the clusters have a short life-time. This is connected with three processes: the scattering of the cluster under the influence of the background particles, the settling of fragments on the surfaces of massive bodies, and the fact that the fragments stick together which leads to a diminution of the optical depth and of the brightness of the cluster. The increase in the asymmetry when the inclination of the rings decreases (Thompson et al. 1981, Cuzzi et al. 1984) is an important observational fact. This dependence of the asymmetry on the angle of inclination also shows up in the calculations (see Fig. 6.13) where the cases a and b correspond to angles equal to 16.5° and 11.5°. One can easily understand this dependence:

large particles efficiently shield the cluster of fragments from the observer only near the plane of the rings and do not hide the fragments from view "from above". Of course, for extremely small angles of inclination the asymmetry again decreases until it disappears completely – at such angles the A ring becomes opaque and uniformly bright – the same cause also explains that the B ring which is dense and bright in reflected light shows no asymmetry: one cannot see the bright cluster against the bright background.

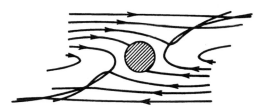

Fig. 6.14. Asymmetry caused by a separate body in a disc of particles.

Gor'kavyĭ and Taĭdakova (1989) have studied a possible contribution to the asymmetry from spiral waves caused by the gravitational action of the large particles. Indeed, a large particle causes in the layer of surrounding small particles an asymmetry in their distribution – due to capture and gravitational perturbations (Fig. 6.14). They considered around a large particle a circle with a diameter of 5 to 10 times that of the particle itself and considered the projections of about 1200 small particles distributed in that region. For a density of the central particle of 0.1 g/cm^3 and a ring thickness of 30 m there was no asymmetry larger than 2.5%. No significant asymmetry was found either for other particle densities (up to 3 g/cm^3) or thicknesses of the layer of small particles. The brightness asymmetry of the rings is thus basically connected with the interaction between the particles and the clusters of the broken-up matter.

The model of the asymmetry considered here connects the exterior, integral characteristics of the rings with interior processes on the level of individual particles. The study of the azimuthal brightness asymmetry makes it possible to get a very reliable determination of the density of the ring particles and to calculate the coefficient β in (5.8) which connects the density of the particles with the outer radius of the planetary rings: $\beta = 0.77$ to 0.88. Further elaboration of theoretical models and an accumulation of observations of the azimuthal brightness asymmetry of the Saturnian rings may present us with a powerful new method for studying the planetary rings from a distance.

Having determined the coefficient β in (5.8) we can find the particle density from the radius of the rings – understood to be the mass of the particle divided by its volume – in other rings:

Table 6.1. Comparison of the characteristics of the planetary rings

Planet	Density of the planet [g/cm³]	Radius of the rings [R_{pl}]	Density*		Albedo	Composition
			Particles in the ring [ϱ_s] [g/cm³]	Matter in the ring [g/cm³]		
Jupiter	1.33	1.81	$\simeq 4$ 0.4 to 0.6	3.6	0.1?	stone
Saturn	0.69	2.27	1 0.1 to 0.15	0.9	0.6	ice
Uranus	1.15	1.95	$\simeq 3$ 0.3 to 0.45	2.6	0.05	carbonaceous chondrite
Neptune	1.55	2.54	$\simeq 2$ 0.2 to 0.3	1.7	0.05	carbonaceous chondrite and ice

* By "density of the particles" we mean the average density of the individual particles which is given in relative units (with the density in the Saturnian rings being taken as unit) and in g/cm³; by "density of the matter in the ring" we mean the density of the material – ice, stone, and so on, from which the rings are made – in the monolithic state.

$$\varrho = \left(\frac{\beta}{R_r}\right)^3 M_p, \tag{6.1}$$

where R_r is the outer radius of the primary rings and M_p the mass of the planet. This does so far not tell us anything about the density of the material of which the particle is made itself, since we do not know the porosity of the particle. Data about the Saturnian rings enable us to estimate this porosity as there are unambiguous indications that the rings consist of ice – water ice. We can thus for the matter of the Saturnian rings take a density of 0.9 g/cm³. From the results of the present chapter follows a particle density in the Saturnian rings of 0.1 to 0.15 g/cm³. The porosity of the particles in the Saturnian rings is thus about 85%. If we assume that the porosity of the material in different planetary rings is about the same we can estimate from (6.1) the density of the matter of the other rings of the large planets since we now know the outer radii R_{cr} of the dense – primary – rings. We give in Table 6.1 data about the planetary rings obtained using these assumptions. We note that these estimates about the matter density are in good agreement with other data – about the spectrum, the albedo, and so on. It should be mentioned that estimates about the density of the particles themselves, for instance, in the Saturnian rings (0.1 to 0.15 g/cm³) until recently were disagreeing with the data from other authors which determined the particle density from the estimate of the total density in the rings – using the wavelength of the spiral waves – and from the spectrum of the particles found from the radio-occultation of the rings. However, in a recent paper by radio-occultation specialists (Rosen *et al.* 1991) these data have been reconsidered and for the particles in the C ring they found $\varrho = 0.38^{+0.43}_{-0.23}$ which at the lower limit is the same as our estimate.

7 Collective Dynamics of Disc Particles

I Formalism

> Bel-Affris: "When the wall came nigh, it changed into a line
> of men ... . Every man of them flung his javelin ..."
>
> *G.B.Shaw, Caesar and Cleopatra, Act I.*

7.1 Transport Theories for Macroparticles

One can understand the reasons why the brightness of the Saturnian A ring shows an azimuthal asymmetry by considering the dynamics of the separate particles. However, the layering of the Saturnian rings and the occurrence of other spatial structures in the planetary rings are caused by collective processes and it is natural to study those in the framework of a hydrodynamic model where the "gas" of the colliding macroparticles is described in the same way as an ordinary molecular gas. The results of Chaps. 4 and 5 show that one can take as a typical ring particle a practically completely inelastic loose meter-size sphere. One must here take into account the gravitational field of such particles; this plays an important role in collision processes and in the break-up of large bodies and the motion of the small fragments. It is impossible to speak about the applicability of hydrodynamics to planetary rings without indicating the characteristic sizes and time scales of the processes which are described; they must be significantly larger than the mean free path and the mean free flight time of a particle, respectively. We shall show in Chap. 8 that these inequalities are satisfied for the large-scale processes in which we are interested.

Let us introduce the fundamental concept of a particle distribution function. Each particle in the gas is characterised by three coordinates, x, y, z, and three velocity components, v_x, v_y, v_z. These six quantities, considered as coordinates form a six-dimensional space in which the gaseous medium occupies some volume. The particle density in the six-dimensional coordinate-velocity space is the particle distribution function $f(\boldsymbol{r}, \boldsymbol{v}, t)$. Its evolution is described by the kinetic equation

$$\frac{\partial f}{\partial t} + \left(\boldsymbol{v} \cdot \frac{\partial f}{\partial \boldsymbol{r}} \right) + \left(\frac{d\boldsymbol{v}}{dt} \cdot \frac{\partial f}{\partial \boldsymbol{v}} \right) = C(f), \qquad \frac{d\boldsymbol{v}}{dt} = \frac{\boldsymbol{F}}{m}, \tag{7.1}$$

where $\boldsymbol{F}$ is the total force acting upon a particle and $C(f)$ is the collisional term which describes the evolution of the distribution function due to collisions between the particles. The actual form of $\boldsymbol{F}$ and $C(f)$ depends on the kind of system and its characteristics.

Once we know the distribution function $f(\boldsymbol{r}, \boldsymbol{v}, t)$ we can obtain the usual average characteristics of the gas:

1) the density,

$$n(\boldsymbol{r}, t) = \int f(\boldsymbol{r}, \boldsymbol{v}, t) \, d^3\boldsymbol{v}, \tag{7.2}$$

2) the average (hydrodynamic) velocity,

$$\boldsymbol{V}(\boldsymbol{r}, t) = \frac{1}{n} \int \boldsymbol{v} \, f(\boldsymbol{r}, \boldsymbol{v}, t) \, d^3\boldsymbol{v}, \tag{7.3}$$

3) and the temperature,

$$T(\boldsymbol{r}, t) = \frac{1}{n} \int \tfrac{1}{3} m (\boldsymbol{v} - \boldsymbol{V})^2 f(\boldsymbol{r}, \boldsymbol{v}, t) \, d^3\boldsymbol{v}. \tag{7.4}$$

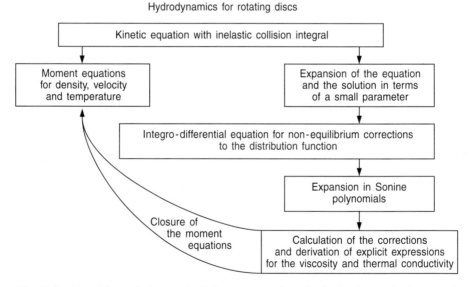

Fig. 7.1. Algorithm of the method for constructing the hydrodynamic (moment) equations for a rotating flat system of inelastically colliding particles.

The classical methods of kinetic theory, developed by Chapman, Enskog, and other authors (Chapman and Cowling 1953, Hirschfelder et al. 1954, Braginskii 1958, 1965, Shkarovskiĭ et al. 1969, Ferziger and Kaper 1972, Alekseev 1982), made it possible to obtain from the kinetic equation (7.1) a closed system of moment equations for $n, \boldsymbol{V}, T$ or the set of equations of viscous hydrodynamics. We show in Fig. 7.1 the algorithm of a Chapman–Enskog type of method. The set of moment equations obtained from (7.1) turns out to be open because the n-th order moment equation ($n = 0, 1, 2$) involves an $n+1$-st order moment which contains the unknown transport coefficients – the viscosity, thermal conductivity, and so on. The derivation of these coefficients and the closure of the set of moment equations turns out to be one of the most complicated problems in transport theory. There exist various methods for expanding the kinetic equations and, hence, the distribution function, in terms of a small parameter. One obtains for the first non-equilibrium correction to the distribution function an integro-differential equation; the solution of that equation makes it possible explicitly to evaluate the viscosity and thermal conductivity, that is, to determine them as functions of n, T, and the other ring parameters.

In order to construct the hydrodynamics of planetary rings one must obtain from the kinetic equation for inelastic gravitating particles a set of moment equations and close it through using kinetic theory to evaluate the viscosity and thermal conductivity coefficients. The construction of a transport theory of rotating discs of inelastic particles is made complicated by two factors – rotation and inelasticity. Rotation means that between collisions a particle in the disc will move not along a straight line, as in the case of the usual medium, but follow an epicyclic (see Chaps. 3 and 6), whether at rest or drifting. Inelasticity somewhat changes the form of the moment equations and makes the problem of whether the distribution function is stationary more complicated. Let us begin by trying to deal with the rotation.

To do this we first of all formulate the Larmor theorem for a particle in a gravitational field (Landau and Lifshitz 1975). The equation of motion of such a particle can be written in the same form as the equation of motion of a charged particle in an electric and magnetic field (Landau and Lifshitz 1975).

This analogy makes it possible to apply well established methods from plasma physics to a gravitating medium (Fridman and Polyachenko 1984) and we shall do so in what follows. In the present chapter we shall follow the important papers by Braginskii (1958, 1965) and construct the hydrodynamics of a disc of gravitating elastic particles, rotating as a rigid body, which takes into account all basic features of a rotating system and afterwards we shall generalise it to the case of a differentially rotating disc of inelastic particles (Gor'kavyĭ 1986b).

7.1.1 The Larmor Theorem for a Particle in a Gravitational Field

In Chap. 3 we used Newton's second law to obtain the equations of motion of a particle. There is also another way of obtaining the equations of motion – using the Lagrangian equations (Landau and Lifshitz 1976),

$$\frac{d}{dt}\frac{\partial L}{\partial \boldsymbol{v}} = \frac{\partial L}{\partial \boldsymbol{r}}, \tag{7.5}$$

where L is the Lagrangian of the system under consideration. For a particle moving in a potential field U the Lagrangian is the difference between the kinetic and the potential energy:

$$L = \tfrac{1}{2}m\boldsymbol{v}^2 - U. \tag{7.6}$$

In a frame of reference rotating with a constant angular velocity $\boldsymbol{\Omega}$ the Lagrangian of a particle moving in the gravitational field Ψ_G has the form (Landau and Lifshitz 1976)

$$L = \tfrac{1}{2}m(\boldsymbol{v}+\boldsymbol{W})^2 - m\Psi_G, \qquad \boldsymbol{W} \equiv [\boldsymbol{\Omega}\wedge\boldsymbol{r}], \tag{7.7}$$

where $\boldsymbol{v}$ is the particle velocity relative to the rotating frame of reference.

Following Landau and Lifshitz (1976) for the calculation of the derivatives occurring in the Lagrangian equation of motion we write down the total differential of L:

$$dL = m\left(\{\boldsymbol{v}+[\boldsymbol{\Omega}\wedge\boldsymbol{r}]\}\cdot d\boldsymbol{v}\right)$$
$$+ m\left(\left\{[\boldsymbol{v}\wedge\boldsymbol{\Omega}]+[[\boldsymbol{\Omega}\wedge\boldsymbol{r}]\wedge\boldsymbol{\Omega}]-\frac{\partial\Psi_G}{\partial\boldsymbol{r}}\right\}\cdot d\boldsymbol{r}\right).$$

Substituting that into (7.5) we obtain the well known equation of motion (Landau and Lifshitz 1976)

$$m\frac{d\boldsymbol{v}}{dt} = 2m[\boldsymbol{v}\wedge\boldsymbol{\Omega}]+m[\boldsymbol{\Omega}\wedge[\boldsymbol{r}\wedge\boldsymbol{\Omega}]]-\frac{\partial\Psi_G}{\partial\boldsymbol{r}}.$$

We can write the Coriolis acceleration in the form

$$2[\boldsymbol{v}\wedge\boldsymbol{\Omega}] = [\boldsymbol{v}\wedge\operatorname{curl}\boldsymbol{W}].$$

We can finally write the equation of motion in the following form:

$$\frac{d\boldsymbol{v}}{dt} = -\nabla\left(\Psi_G - \tfrac{1}{2}\boldsymbol{W}^2\right) + [\boldsymbol{v}\wedge\operatorname{curl}\boldsymbol{W}]. \tag{7.8}$$

Introducing the notation

$$\boldsymbol{E} \equiv -\nabla\left(\Psi_G - \tfrac{1}{2}\boldsymbol{W}^2\right), \tag{7.9}$$
$$\boldsymbol{h} \equiv \operatorname{curl}\boldsymbol{W}, \tag{7.10}$$

we get the equation

$$\frac{d\boldsymbol{v}}{dt} = \boldsymbol{E} + [\boldsymbol{v} \wedge \boldsymbol{h}]. \tag{7.11}$$

The kinetic equation for the system under consideration now takes the form

$$\frac{\partial f}{\partial t} + \left(\boldsymbol{v} \cdot \frac{\partial f}{\partial \boldsymbol{r}}\right) + \left(\{\boldsymbol{E} + [\boldsymbol{v} \wedge \boldsymbol{h}]\} \cdot \frac{\partial f}{\partial \boldsymbol{v}}\right) = \hat{C}, \tag{7.12}$$

where $\hat{C}$ is the collision integral. Equations (7.11) and (7.12) are the same as those for a charged particle in an electromagnetic field in a system of units where the particle charge e, its mass m, and the velocity of light in vacuo c are equal to unity: $e = m = c = 1$. Of course, (7.9) and (7.10) satisfy the first pair of the stationary Maxwell equations

$$\operatorname{curl} \boldsymbol{E} = 0, \qquad \operatorname{div} \boldsymbol{h} = 0.$$

In fact, the analogy which we have just noted is already contained in the Larmor theorem (Landau and Lifshitz 1975) with two inessential differences. The first difference lies in that instead of the case of a particle moving in the field of a gravitational centre Larmor studied the motion of a charged particle in the electric field of a fixed particle with the opposite sign of charge. This difference is unimportant as the Coulomb and Newtonian forces are the same, apart from constant factors. The second difference lies in that in the equation of motion (7.8) Larmor dropped the quadratic term $\frac{1}{2}\boldsymbol{W}^2$. This difference is also unimportant as by treating the difference $\Psi_G - \frac{1}{2}\boldsymbol{W}^2$ as the electrostatic potential – and $\boldsymbol{W}$ as the vector potential – we achieve that expressions (7.9) and (7.10) describe the electric and magnetic field strengths.

The analogy we have just found enables us to apply plasma physics methods to gravitating media (Fridman 1975, Fridman and Polyachenko 1984, Gor'kavyĭ 1986b).

7.1.2 Derivation of the Moment Equations

We can now obtain from (7.12) the moment equations or the transport equations using the usual methods. To do this we multiply (7.12) by some function of the velocities and integrate over the velocities. Let us write down some relations we need for what follows (Braginskii 1965):

$$\int \Phi(\boldsymbol{v}) \frac{\partial f}{\partial t} d^3\boldsymbol{v} = \frac{\partial}{\partial t} \int \Phi(\boldsymbol{v}) f\, d^3\boldsymbol{v} = \frac{\partial}{\partial t} n \langle \Phi \rangle, \tag{7.13}$$

$$\int \Phi(\boldsymbol{v}) v_i \frac{\partial f}{\partial x_i} d^3\boldsymbol{v} = \frac{\partial}{\partial x_i} \int \Phi(\boldsymbol{v}) v_i f\, d^3\boldsymbol{v} = \frac{\partial}{\partial x_i} n \langle \Phi(\boldsymbol{v}) v_i \rangle, \tag{7.14}$$

$$\int \Phi(\boldsymbol{v}) E_i \frac{\partial f}{\partial v_i} d^3\boldsymbol{v} = -E_i \int \frac{\partial \Phi(\boldsymbol{v})}{\partial v_i} f\, d^3\boldsymbol{v} = -E_i n \left\langle \frac{\partial \Phi}{\partial v_i} \right\rangle, \tag{7.15}$$

$$\int [\boldsymbol{v} \wedge \boldsymbol{h}]_i \Phi(\boldsymbol{v}) \frac{\partial f}{\partial v_i} \, d^3\boldsymbol{v} \;=\; -\int [\boldsymbol{v} \wedge \boldsymbol{h}]_i \frac{\partial \Phi(\boldsymbol{v})}{\partial v_i} \, f \, d^3\boldsymbol{v}$$

$$= -n \left\langle [\boldsymbol{v} \wedge \boldsymbol{h}]_i \frac{\partial \Phi}{\partial v_i} \right\rangle , \tag{7.16}$$

where the pointed brackets indicate averaging over the distribution function and where $x_i = x, y, z$, $v_i = v_x, v_y, v_z$ for $i = 1, 2, 3$, respectively. Equations (7.15) and (7.16) have been derived assuming that $\Phi f \to 0$ as $v_i \to \infty$. If we now take $\Phi(\boldsymbol{v}) = 1$, expressions such as (7.15) or (7.16) vanish and, using (7.13) and (7.14), we get by integrating (7.12) the equation of continuity

$$\frac{\partial n}{\partial t} + \left(\frac{\partial}{\partial \boldsymbol{r}} \cdot n\boldsymbol{V} \right) \;=\; \int \hat{C} \, d^3\boldsymbol{v}. \tag{7.17}$$

We now multiply (7.12) successively by $\Phi = m\boldsymbol{v}$ and $\Phi = \frac{1}{2}m\boldsymbol{v}^2$. Using (7.13) to (7.16) we get after integration

$$\frac{\partial}{\partial t} nmV_i + \sum_j \frac{\partial}{\partial x_j} \left(nm \langle v_i v_j \rangle \right) - nm \left\{ \boldsymbol{E} + [\boldsymbol{V} \wedge \boldsymbol{h}] \right\}_i = \int mv_i \hat{C} \, d^3\boldsymbol{v}, \tag{7.18}$$

$$\frac{\partial}{\partial t} nm \langle \tfrac{1}{2} v^2 \rangle + \sum_j \frac{\partial}{\partial x_j} \left(nm \langle \tfrac{1}{2} v^2 v_j \rangle \right) - nm(\boldsymbol{E} \cdot \boldsymbol{V}) = \int \tfrac{1}{2} mv^2 \hat{C} \, d^3\boldsymbol{v}. \tag{7.19}$$

We now introduce the random velocity which is equal to the difference between the total particle velocity and its hydrodynamic average:

$$\boldsymbol{v}_1 \;=\; \boldsymbol{v} - \boldsymbol{V}. \tag{7.20}$$

Using the fact that $\langle \boldsymbol{v}_1 \rangle = 0$, we then get the following relations:

$$\langle v_i v_j \rangle \;=\; V_i V_j + \langle v_{1i} v_{1j} \rangle , \tag{7.21}$$

$$\langle v^2 \rangle \;=\; V^2 + \langle v_1^2 \rangle , \tag{7.22}$$

$$\langle v^2 v_i \rangle \;=\; \langle v_1^2 v_{1i} \rangle + \langle v_1^2 \rangle V_i + V^2 V_i + 2 \left(\langle v_{1i} \boldsymbol{v}_1 \rangle \cdot \boldsymbol{V} \right). \tag{7.23}$$

We can use the set (7.20) to (7.23) to transform the moment equations (7.17) to (7.19) to a more familiar form. We use in the equation of motion the continuity equation, and when transforming the energy equation both the equation of motion and the continuity equation are used. We note an important fact: when the number of particles is conserved in the collisions, the first moment of the collisional term vanishes:

$$\int \hat{C} \, d^3\boldsymbol{v} \;=\; 0. \tag{7.24}$$

This is not the case when we are dealing with chemical reactions or in the case of other particle sinks and sources. If momentum is conserved in the collisions – and this is, of course, true even for completely inelastic particles – the second moment of the collisional term also vanishes:

$$\int m\boldsymbol{v}\,\hat{C}\,d^3\boldsymbol{v} \;=\; 0. \tag{7.25}$$

The right-hand side of the equation of motion may be non-vanishing when there are cross-interactions – when in a two-component medium there is friction between particles of different kinds.

   In the case of collisions between perfectly elastic particles the energy conservation law is satisfied so that

$$\int \tfrac{1}{2}m\boldsymbol{v}^2\,\hat{C}\,d^3\boldsymbol{v} \;=\; 0. \tag{7.26}$$

For inelastic particles – and all macroparticles happen to be inelastic – this integral does not vanish and we shall take this deciding factor for the dynamics of planetary rings into account in the next section. Assuming for the present the validity of all conservation laws we get from (7.17) to (7.19) the following set of moment equations (Braginskii 1958, 1965):

$$\frac{\partial n}{\partial t} + \mathrm{div}(n\boldsymbol{V}) \;=\; 0, \tag{7.27}$$

$$
\begin{aligned}
mn\frac{dV_i}{dt} \;-\; &-\frac{\partial p}{\partial x_i} - \frac{\partial \pi_{ik}}{\partial x_k} \\
&+ mn\left[-\nabla\left\{\Psi_G - \tfrac{1}{2}\boldsymbol{W}^2\right\} + [\boldsymbol{V} \wedge \mathrm{curl}\,\boldsymbol{W}]\right]_i,
\end{aligned}
\tag{7.28}
$$

$$\tfrac{3}{2}n\frac{dT}{dt} + p\,\mathrm{div}\boldsymbol{V} \;=\; -\,\mathrm{div}\boldsymbol{q} - \pi_{ik}\frac{\partial V_i}{\partial x_k}, \tag{7.29}$$

where $\boldsymbol{q}$ is the heat flux vector, π_{ik} the viscous stress tensor, and

$$
\frac{d}{dt} \;=\; \frac{\partial}{\partial t} + (\boldsymbol{V}\cdot\nabla), \qquad \boldsymbol{q} \;=\; mn\left\langle \tfrac{1}{2}v_1^2\boldsymbol{v}_1\right\rangle,
$$

$$\pi_{ik} \;=\; mn\left\langle v_{1i}v_{1k} - \tfrac{1}{3}\boldsymbol{v}_1^2\delta_{ik}\right\rangle, \tag{7.30}$$

$$p \;=\; nT, \qquad n \;=\; \int f^{(0)}\,d^3\boldsymbol{v}, \qquad \boldsymbol{V} \;=\; \frac{1}{n}\int \boldsymbol{v}f^{(0)}\,d^3\boldsymbol{v},$$

$$T \;=\; \frac{1}{n}\int \tfrac{1}{3}m\boldsymbol{v}_1^2 f^{(0)}\,d^3\boldsymbol{v}. \tag{7.31}$$

We have used here standard notations; p is the pressure and T the temperature in energy units.

7.1.3 Integro-differential Equation
for the Non-equilibrium Correction to the Distribution Function

We now write the kinetic equation in the following form

$$\hat{D}f \;=\; \frac{1}{\varepsilon}\,\hat{K}f, \tag{7.32}$$

where ε is a formal small parameter. We look for the distribution function in the following form:

$$f = f^{(0)} + \varepsilon f^{(1)} + \varepsilon^2 f^{(2)} + \dots \ . \tag{7.33}$$

Substituting (7.33) into (7.32) and equating the coefficients of the same powers of ε on both sides of the equation we get an infinite chain of equations:

$$\left(\hat{K}f\right)^{(0)} = 0, \tag{7.34}$$

$$\left(\hat{D}f\right)^{(0)} = \left(\hat{K}f\right)^{(1)}, \tag{7.35}$$

and so on.

The meaning of this set of equations is clear: first one retains in the kinetic equation the main terms – which determine the "basic features" of the equilibrium distribution function – and then one looks for the corrections to the equilibrium distribution function caused by the existence of other terms in the kinetic equation. If the Maxwell distribution function satisfies (7.34) one can use the Chapman–Enskog method (Chapman and Cowling 1953) to construct a transport theory.

Let us change in (7.12) to the random velocities, that is, introduce a new system of coordinates:

$$t_1 = t, \qquad \boldsymbol{r}_1 = \boldsymbol{r}, \qquad \boldsymbol{v}_1(\boldsymbol{r},t) = \boldsymbol{v} - \boldsymbol{V}(\boldsymbol{r},t).$$

In terms of the new coordinates we have

$$\frac{\partial f(\boldsymbol{r}_1,\boldsymbol{v}_1,t_1)}{\partial t} = \frac{\partial f}{\partial t_1}\frac{\partial t_1}{\partial t} + \left(\nabla_1 f \cdot \frac{\partial \boldsymbol{r}_1}{\partial t}\right) + \left(\nabla_{\boldsymbol{v}_1} f \cdot \frac{\partial \boldsymbol{v}_1}{\partial t}\right)$$
$$= \frac{\partial f}{\partial t} - \left(\frac{\partial \boldsymbol{V}}{\partial t} \cdot \nabla_{\boldsymbol{v}_1} f\right),$$

$$(\boldsymbol{v} \cdot \nabla f(\boldsymbol{r}_1,\boldsymbol{v}_1,t_1)) = \frac{\partial f}{\partial t_1}(\boldsymbol{v} \cdot \nabla t_1) + (\nabla_1 f \cdot (\boldsymbol{v} \cdot \nabla)\boldsymbol{r}_1) + (\nabla_{\boldsymbol{v}_1} f \cdot (\boldsymbol{v} \cdot \nabla)\boldsymbol{v}_1)$$
$$= (\boldsymbol{v} \cdot \nabla) f - (\boldsymbol{v} \cdot \nabla)(\boldsymbol{V} \cdot \nabla_{\boldsymbol{v}_1}) f,$$

$$\left(\frac{d\boldsymbol{v}}{dt} \cdot \nabla_{\boldsymbol{v}} f(\boldsymbol{r}_1,\boldsymbol{v}_1,t_1)\right) = \frac{\partial f}{\partial t_1}\left(\frac{d\boldsymbol{v}}{dt} \cdot \nabla_{\boldsymbol{v}} t_1\right) + \left(\nabla_1 f \cdot \left(\frac{d\boldsymbol{v}}{dt} \cdot \nabla_{\boldsymbol{v}}\right)\boldsymbol{r}_1\right)$$
$$+ \left(\nabla_{\boldsymbol{v}_1} f \cdot \left(\frac{d\boldsymbol{v}}{dt} \cdot \nabla\right)\boldsymbol{v}_1\right) = \left(\frac{d\boldsymbol{v}}{dt} \cdot \nabla_{\boldsymbol{v}_1}\right) f,$$

where $\nabla \equiv \partial/\partial\boldsymbol{r}$, $\nabla_1 \equiv \partial/\partial\boldsymbol{r}_1$, $\nabla_{\boldsymbol{v}} \equiv \partial/\partial\boldsymbol{v}$, and $\nabla_{\boldsymbol{v}_1} \equiv \partial/\partial\boldsymbol{v}_1$.

Substituting these expressions into (7.1) we find

$$\frac{\partial f}{\partial t} + (\boldsymbol{v} \cdot \nabla) f + \left(\frac{d\boldsymbol{v}}{dt} \cdot \nabla_{\boldsymbol{v}_1}\right) f - \left(\left[\frac{\partial \boldsymbol{V}}{\partial t} + (\boldsymbol{v} \cdot \nabla)\boldsymbol{V}\right] \cdot \nabla_{\boldsymbol{v}_1}\right) f = \hat{C}.$$

Now using (7.20) and (7.30) to replace $\boldsymbol{v}$ by $\boldsymbol{V} + \boldsymbol{v}_1$ and $\partial/\partial t$ by $d/dt - (\boldsymbol{V} \cdot \nabla)$ we get

$$\frac{df}{dl} + (\boldsymbol{v}_1 \cdot \nabla) f + \left(\left[\frac{d\boldsymbol{v}}{dt} - \frac{d\boldsymbol{V}}{dt} \right] \cdot \nabla_{\boldsymbol{v}_1} \right) f - ((\boldsymbol{v}_1 \cdot \nabla) \boldsymbol{V} \cdot \nabla_{\boldsymbol{v}_1}) f \ = \ \hat{C}.$$

Using first (7.11) and then (7.20), we finally arrive at (Braginskii 1965):

$$\frac{df}{dt} + (\boldsymbol{v}_1 \cdot \nabla f) - \left(\left[\nabla \left(\Psi_G - \tfrac{1}{2} \boldsymbol{W}^2 \right) + [\operatorname{curl} \boldsymbol{W} \wedge \boldsymbol{V}] + \frac{d\boldsymbol{V}}{dt} \right] \nabla_{\boldsymbol{v}_1} f \right)$$
$$- \left((\boldsymbol{v}_1 \cdot \nabla) \boldsymbol{V} \cdot \nabla_{\boldsymbol{v}_1} \right) f + ([\boldsymbol{v}_1 \wedge \operatorname{curl} \boldsymbol{W}] \cdot \nabla_{\boldsymbol{v}_1} f) \ = \ \hat{C}, \tag{7.36}$$

We now write (7.36) in a form similar to (7.32):

$$\hat{D} f \ = \ \hat{C} - ([\boldsymbol{v}_1 \wedge \operatorname{curl} \boldsymbol{W}] \cdot \nabla_{\boldsymbol{v}_1} f). \tag{7.37}$$

The right-hand side of (7.37) vanishes if $f^{(0)}$ is the Maxwell distribution:

$$f^{(0)} \ = \ n \left(\frac{m}{2\pi T} \right)^{3/2} e^{-m\boldsymbol{v}_1^2/2T}, \tag{7.38}$$

or

$$\ln f^{(0)} \ = \ \ln n - \tfrac{3}{2} \ln T - \frac{m\boldsymbol{v}_1^2}{2T} + \text{const.} \tag{7.39}$$

In fact, the Coriolis term $([\boldsymbol{v}_1 \wedge \operatorname{curl} \boldsymbol{W}] \cdot \nabla_{\boldsymbol{v}_1} f)$ vanishes whenever $f^{(0)}$ is a spherically symmetric function of the velocity, and the collision integral $\hat{C}$ vanishes if $f^{(0)}$ is the Maxwell distribution.

In the zeroth approximation – when we neglect all non-equilibrium corrections to the distribution function and this leads to the vanishing of all viscous stresses and heat currents – the set of moment equations (7.27 – 7.29) has the form

$$\frac{dn}{dt} = -n(\nabla \cdot \boldsymbol{V}),$$
$$\frac{d\boldsymbol{V}}{dt} = \frac{\boldsymbol{F}}{m} - \frac{1}{mn} \nabla(nT), \tag{7.40}$$
$$\frac{dT}{dt} = -\tfrac{2}{3} T(\nabla \cdot \boldsymbol{V}).$$

Substituting (7.38) into the right-hand side of (7.37), and using (7.39) and (7.40) we get, after symmetrisation,

$$(\hat{D} f)^{(0)} \ = \ f^{(0)} \left[\left(\frac{m\boldsymbol{v}_1^2}{2T} - \frac{5}{2} \right) (\boldsymbol{v}_1 \cdot \nabla \ln T) \right.$$
$$\left. + \frac{m}{2T} \left(v_{1i} v_{1k} - \tfrac{1}{3} \boldsymbol{v}_1^2 \delta_{ik} \right) \hat{W}_{ik} \right], \tag{7.41}$$

$$L = \tfrac{1}{2}\left(\dot{r}^2 + r^2\dot{\varphi}^2\right) - \Psi(r, \varphi).$$

Let us evaluate the partial derivatives of that function with respect to r and $\dot{r}$:

$$\frac{\partial L}{\partial r} = r\dot{\varphi}^2 - \frac{\partial \Psi}{\partial r} = \frac{v_\varphi^2}{r} - \frac{\partial \Psi}{\partial r}, \qquad \frac{\partial L}{\partial \dot{r}} = \dot{r} = v_r;$$

substituting these expressions into the radial Lagrangian equation, we find the radial acceleration:

$$\frac{dv_r}{dt} = \frac{v_\varphi^2}{r} - \frac{\partial \Psi}{\partial r}.$$

Similarly, after finding the derivatives of L with respect to φ and $\dot{\varphi}$,

$$\frac{\partial L}{\partial \varphi} = -\frac{\partial \Psi}{\partial \varphi}, \qquad \frac{\partial L}{\partial \dot{\varphi}} = r^2\dot{\varphi},$$

we find the azimuthal Lagrangian equation of motion:

$$\frac{d}{dt}\left(r^2\dot{\varphi}\right) = -\frac{\partial \Psi}{\partial \varphi},$$

or

$$\frac{d}{dt}\left(rv_\varphi\right) = v_r v_\varphi + r\frac{dv_\varphi}{dt} = -\frac{\partial \Psi}{\partial \varphi},$$

whence follows the following expression for the azimuthal acceleration:

$$\frac{dv_\varphi}{dt} = -\left(\frac{v_r v_\varphi}{r} + \frac{1}{r}\frac{\partial \Psi}{\partial \varphi}\right).$$

Finally, the kinetic equation in the cylindrical laboratory frame of reference K has the form

$$\frac{\partial f}{\partial t} + v_r\frac{\partial f}{\partial r} + \frac{v_\varphi}{r}\frac{\partial f}{\partial \varphi} + \left[\frac{v_\varphi^2}{r} - \frac{\partial \Psi}{\partial r}\right]\frac{\partial f}{\partial v_r}$$

$$-\left[\frac{v_r v_\varphi}{r} + \frac{1}{r}\frac{\partial \Psi}{\partial \varphi}\right]\frac{\partial f}{\partial v_\varphi} = \hat{C}. \tag{7.78}$$

We now change to a frame of reference K_1 which rotates with respect to the frame K with an angular velocity $\Omega(r)$. The distribution function f in the frame K will then in the frame K_1 correspond to a function F which depends on new variables which we give an index "1":

$$t_1 = t, \quad r_1 = r, \quad \varphi_1 = \varphi - \Omega(r)t, \quad v_{r1} = v_r, \quad v_{\varphi1} = v_\varphi - \Omega(r)r. \tag{7.79}$$

We give the expressions for the partial derivatives in terms of the new coordinates:

$$\frac{\partial f}{\partial t} = \frac{\partial F}{\partial t_1}\frac{\partial t_1}{\partial t} + \frac{\partial F}{\partial r_1}\frac{\partial r_1}{\partial t} + \frac{\partial F}{\partial \varphi_1}\frac{\partial \varphi_1}{\partial t} + \frac{\partial F}{\partial v_{r1}}\frac{\partial v_{r1}}{\partial t} + \frac{\partial F}{\partial v_{\varphi 1}}\frac{\partial v_{\varphi 1}}{\partial t}$$

$$= \frac{\partial F}{\partial t_1} - \Omega\frac{\partial F}{\partial \varphi_1},$$

$$\frac{\partial f}{\partial r} = \frac{\partial F}{\partial t_1}\frac{\partial t_1}{\partial r} + \frac{\partial F}{\partial r_1}\frac{\partial r_1}{\partial r} + \frac{\partial F}{\partial \varphi_1}\frac{\partial \varphi_1}{\partial r} + \frac{\partial F}{\partial v_{r1}}\frac{\partial v_{r1}}{\partial r} + \frac{\partial F}{\partial v_{\varphi 1}}\frac{\partial v_{\varphi 1}}{\partial r}$$

$$= \frac{\partial F}{\partial r_1} - \Omega' t\frac{\partial F}{\partial \varphi_1} - (\Omega + r\Omega')\frac{\partial F}{\partial v_{\varphi 1}},$$

$$\frac{\partial f}{\partial \varphi} = \frac{\partial F}{\partial t_1}\frac{\partial t_1}{\partial \varphi} + \frac{\partial F}{\partial r_1}\frac{\partial r_1}{\partial \varphi} + \frac{\partial F}{\partial \varphi_1}\frac{\partial \varphi_1}{\partial \varphi} + \frac{\partial F}{\partial v_{r1}}\frac{\partial v_{r1}}{\partial \varphi} + \frac{\partial F}{\partial v_{\varphi 1}}\frac{\partial v_{\varphi 1}}{\partial \varphi}$$

$$= \frac{\partial F}{\partial \varphi_1},$$

$$\frac{\partial f}{\partial v_r} = \frac{\partial F}{\partial t_1}\frac{\partial t_1}{\partial v_r} + \frac{\partial F}{\partial r_1}\frac{\partial r_1}{\partial v_r} + \frac{\partial F}{\partial \varphi_1}\frac{\partial \varphi_1}{\partial v_r} + \frac{\partial F}{\partial v_{r1}}\frac{\partial v_{r1}}{\partial v_r} + \frac{\partial F}{\partial v_{\varphi 1}}\frac{\partial v_{\varphi 1}}{\partial v_r}$$

$$= \frac{\partial F}{\partial v_{r1}}.$$

$$\frac{\partial f}{\partial v_\varphi} = \frac{\partial F}{\partial t_1}\frac{\partial t_1}{\partial v_\varphi} + \frac{\partial F}{\partial r_1}\frac{\partial r_1}{\partial v_\varphi} + \frac{\partial F}{\partial \varphi_1}\frac{\partial \varphi_1}{\partial v_\varphi} + \frac{\partial F}{\partial v_{r1}}\frac{\partial v_{r1}}{\partial v_\varphi} + \frac{\partial F}{\partial v_{\varphi 1}}\frac{\partial v_{\varphi 1}}{\partial v_\varphi}$$

$$= \frac{\partial F}{\partial v_{\varphi 1}},$$

where $\Omega' = d\Omega/dr$ and where we have used the fact that $\varphi_1 = \varphi_1(r,t)$ and $v_{\varphi 1} = v_{\varphi 1}(r)$.

Substituting these expressions for the derivatives into (7.78) and using (7.79), we obtain the kinetic equation in the rotating frame in the following form:

$$\frac{\partial F}{\partial t_1} + v_{r1}\frac{\partial F}{\partial r_1} + \left[\frac{v_{\varphi 1}}{r_1} - v_{r1}\Omega' t\right]\frac{\partial F}{\partial \varphi_1} + \left[\frac{v_{\varphi 1}^2}{r_1} + 2\Omega(r_1)v_{\varphi 1}\right.$$

$$\left. + \Omega^2(r_1)r_1 - \frac{\partial\Psi}{\partial r_1}\right]\frac{\partial F}{\partial v_{r1}} - \left[\frac{v_{r1}v_{\varphi 1}}{r_1} + 2v_{r1}\Omega(r_1)\right.$$

$$\left. + r_1\Omega'(r_1)v_{r1} - \frac{1}{r_1}\frac{\partial\Psi}{\partial \varphi_1}\right]\frac{\partial F}{\partial v_{\varphi 1}} = \hat{C}(f,f), \tag{7.80}$$

where $\hat{C}$ is the collision integral including inelastic collisions. Trulsen (1971, 1972) has obtained $\hat{C}$ for the case of non-gravitating inelastic spheres and Shukhman (1984) has obtained it for finite size spheres with spin.

In the case when we keep the spatial coordinates fixed and change to a frame of reference which rotates only in velocity space, we have instead of (7.79) the relations

$$t_1 = t, \quad r_1 = r, \quad \varphi_1 = \varphi, \quad v_{r1} = v_r, \quad v_{\varphi 1} = v_\varphi - \Omega(r)r. \tag{7.81}$$

In this frame of reference one needs change in (7.80) only the factor of the derivative $\partial F / \partial \varphi_1$. As a result the kinetic equations in terms of the new variables (7.81) has the form

$$
\frac{\partial F}{\partial t_1} + v_{r1} \frac{\partial F}{\partial r_1} + \left[\frac{v_{\varphi 1}}{r_1} + \Omega \right] \frac{\partial F}{\partial \varphi_1} + \left[\frac{v_{\varphi 1}^2}{r_1} + 2\Omega(r_1) v_{\varphi 1} \right.
$$

$$
+ \left. \Omega^2(r_1) r_1 - \frac{\partial \Psi}{\partial r_1} \right] \frac{\partial F}{\partial v_{r1}} - \left[\frac{v_{r1} v_{\varphi 1}}{r_1} + 2 v_{r1} \Omega(r_1) \right.
$$

$$
+ \left. r_1 \Omega'(r_1) v_{r1} - \frac{1}{r_1} \frac{\partial \Psi}{\partial \varphi_1} \right] \frac{\partial F}{\partial v_{\varphi 1}} = \hat{C}(f, f). \tag{7.82}
$$

No solutions are known of the stationary equations (7.80) or (7.82). One checks easily that the distribution functions corresponding to these solutions are, in general, anisotropic. To see this, it is sufficient to consider the simplest limiting case, $\Omega t_c \to \infty$, when there have been many revolutions around the planet during a single particle collision time. In the leading perturbation theory order in the small parameter $(\Omega t_c)^{-1}$ we find, using the equilibrium condition $\Omega^2 r = \partial \Psi / \partial r$, from (7.80) and (7.82) – dropping the index "1":

$$
\left\{ 2\Omega v_\varphi \frac{\partial}{\partial v_r} - \frac{\kappa^2}{2\Omega} v_r \frac{\partial}{\partial v_\varphi} \right\} F^{(0)} = 0, \tag{7.83}
$$

where $\kappa^2 \equiv 4\Omega^2 \left[1 + \frac{1}{2} r \Omega_0' / \Omega \right]$ is the square of the epicyclic frequency. We shall look for the solution $F^{(0)}$ of (7.83) in the form of the anisotropic Schwarzschild distribution function (Schwarzschild 1907):

$$
F^{(0)} = A \exp \left[-\frac{v_r^2}{2c_r^2} - \frac{v_\varphi^2}{2c_\varphi^2} \right], \tag{7.84}
$$

where c_r and c_φ are the radial and azimuthal particle velocity dispersions. Substituting $F^{(0)}$ from (7.84) into (7.83) we find the well known relation between c_r and c_φ (see, for instance, Fridman and Polyachenko 1984):

$$
c_\varphi = \frac{\kappa}{2\Omega} c_r.
$$

For a Keplerian disc we have $\kappa / 2\Omega = 0.5$ and, hence,

$$
\frac{c_\varphi}{c_r} = 0.5. \tag{7.85}
$$

We obtained this relatively modest amount of velocity anisotropy under the condition $\Omega t_c \to \infty$ which corresponds to the largest possible case of anisotropy. We shall see below that in planetary discs $\left(\Omega t_c \right)_{max} \cong 10$, corresponding to an even larger ratio of c_φ / c_r, that is, practically an isotropic case. This means that in planetary discs the velocity distribution function is close to Maxwellian and this can be justifiably used in our calculations, taking into consideration the accuracy of the Voyager observations.

If we forget about the accuracy of the Voyager observations and require mathematical rigour we have, of course, in the general case a more complicated anisotropic function than the Schwarzschild distribution and this raises serious problems for the applicability of the Chapman–Enskog method.

Goldreich and Tremaine (1978a) used a Gaussian function for the anisotropic distribution function. As a result the scalar energy equation became a tensor equation. The set of transport equations was closed by dropping in the energy equation a third rank tensor which in the isotropic case corresponds to the heat flux vector. We shall show later that the heat flux plays an important role for the stability of the rings so that this simplification is undesirable. On the other hand, the set of equations which Goldreich and Tremaine (1978a) had obtained remains very cumbersome and ill suited for an analysis of collective processes in the rings. For instance, the change from a scalar to a tensor energy equation increases by a factor 30 the number of terms in the dispersion equation of the linear oscillations – which is even without this quite complicated (see Sects. 8.1 and 8.4 and Appendix III).

There have been many studies of the dynamics of rings (Trulsen 1971, 1972, Goldreich and Tremaine 1978a, Brahic 1984, Bridges et al. 1984, Greenberg and Brahic 1984, Shukhman 1984, Stewart et al. 1984, Borderies et al. 1985, Shu and Stewart 1985, Araki and Tremaine 1986, Spaute and Greenberg 1987, Araki 1988, Greenberg 1988, Hameen-Anttila et al. 1988, Brophy and Esposito 1989, Brophy et al. 1990, Gor'kavyĭ and Fridman 1990) and of proto-planetary discs (Safronov 1972, Safronov and Vityazev 1983, Hayashi et al. 1985, Polyachenko and Fridman 1988, Stewart and Wetherill 1988) and they all failed to find any collective instabilities connected with an anisotropy in the thermal velocities. There is therefore a desire to obtain a set of transport equations which is convenient for a study of the stability of planetary rings but which neglects the anisotropy of the distribution function. We remind ourselves that such a desire does not come into conflict with the accuracy of the observations (see our remarks about an estimate of the maximally possible anisotropy (7.84)). Moreover, in as far as the restitution coefficient of the snow particles of the Saturnian rings is close to zero, the gravitational interactions between the particles play the role of elastic collisions. If we neglect the effect of the planet on the gravitational cross-section of the particles we can show that the frequency of gravitational collisions is 4.4 to 6.0 times larger than the frequency of contact collisions (Stewart and Kaula 1980, Fridman et al. 1992; see also Chap. 8) Hence we can approximately write $\hat{C}(f^{(0)}, f^{(0)}) \approx \hat{C}_G(f^{(0)}, f^{(0)}) \approx 0$, where $\hat{C}_G$ is the integral of the elastic (in the case of the planetary rings the gravitational) collisions. The solution of this equation is the Maxwell distribution function and this makes it possible to use the modification of the Chapman–Enskog method described earlier. We note the following features.

1) To obtain the moment equations from (7.82) we must take into account that energy is not conserved in collisions between inelastic par-

ticles. This means that in the energy equation there appears an extra term – the third moment of the collision integral of the inelastic particles, $\sigma E^- = \int \frac{1}{2}mv_1^2\hat{C}_* \, d^3v$, where $\hat{C}_*$ is the integral of the inelastic (in the case of the rings the contact) interactions. The same term also appears in the energy equation in the zeroth approximation (see the third equation of the set (7.40)).

2)   When we derive from (7.82) the equation for the corrections to the stationary distribution function we get not only vector and tensor terms, as before (see (7.41)), but also scalar ones – due to the term $\hat{C}_*(f^{(0)}, f^{(0)})$ in the first order of perturbation theory, since $\hat{C}_*(f^{(0)}, f^{(0)}) \ll \hat{C}_G(f^{(0)}, f^{(0)})$, and also because of the term σE^- in the system energy equation (7.40). The general form of the correction to the distribution function will have the following form:

$$\Psi = \Psi_0 + (\boldsymbol{\Psi} \cdot \boldsymbol{v}_1) + \Psi_{ik}\left(v_{1i}v_{1k} - \tfrac{1}{3}v_1^2\delta_{ik}\right). \tag{7.86}$$

Since the terms in (7.86) are linearly independent we can for each kind of correction – scalar, vector, or tensor – obtain a separate equation which can be solved independently. We can evaluate the vector and tensor corrections in the same way as before. The integral equation for the scalar correction is similar to the equation for the correction caused by the transfer of energy between components in a non-equilibrium plasma (Ferziger and Kaper 1976). As in plasma theory it is not necessary to solve this equation since the scalar correction caused by the inelasticity of the collisions does not affect the expressions for the vector and tensor corrections which determine the viscosity and thermal conductivity coefficients. The transport coefficients of a differentially rotating disc of gravitating particles are thus also given by (7.57), (7.58), and (7.71) to (7.75).

Let us compare the expressions for the shear viscosity obtained in the present chapter with those obtained by Goldreich and Tremaine (1978a) and Stewart and Kaula (1980). We have already noted that Goldreich and Tremaine constructed a transport theory for smooth non-gravitating inelastic spheres using an anisotropic distribution function. Stewart and Kaula used, like us (Gor'kavyĭ 1986b, Fridman et al. 1992), a Maxwell distribution and considered inelastic gravitating particles. We note that Stewart and Kaula followed the work in plasma theory by Shkarovskiĭ et al. (1969) while our calculations were based on Braginskii's work (1958, 1965).

Unfortunately it is not possible to compare the expressions for the thermal conductivity as Stewart and Kaula studied only a stationary energy balance in a viscous differentially rotating disc and did not calculate the thermal conductivity while Goldreich and Tremaine were forced to neglect the heat fluxes in order to close the set of moment equations.

The viscosity coefficient was obtained by Goldreich and Tremaine as a function of the optical depth. We write it in terms of the free flight time using the relation (Goldreich and Tremaine 1982) $\tau = 2\Omega t_c$ and form the following dimensionless function of $x \equiv \Omega t_c$:

$$\frac{\nu_{GT}\Omega}{T} = \frac{0.92x}{4x^2 + 1}.$$ (7.87)

We note that t_c for hard spheres is analogous to the free flight time for gravitating points if we take as the effective radius of the "sphere" the size of the zone of influence of the gravitational attraction.

We change in the first approximation of the viscosity coefficient obtained by Stewart and Kaula the free flight time in accordance with the definition (7.76); this gives the following function:

$$\frac{\nu_{SK}\Omega}{T} = \frac{0.83x}{(1.95x)^2 + 1}.$$ (7.88)

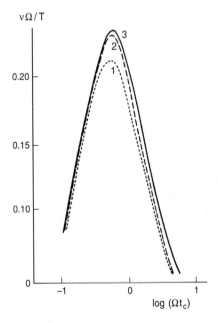

Fig. 7.2. Comparison of the viscosity coefficients obtained by Stewart and Kaula (*1*), by Goldreich and Tremaine (*2*), and by Gor'kavyǐ (*3*).

We obtain from (7.71) to (7.75) an expression for the kinematic shear viscosity coefficient: $\nu_1 = \eta_1/mn$. This formula is exact for frequent collisions, $\Omega t_c \ll 1$, or for a weakly differentially rotating disc, $|\mathrm{curl}\,\boldsymbol{W}| \approx 2\Omega$. For discs with a Kepler rotation, $|\mathrm{curl}\,\boldsymbol{W}| \approx 0.5\Omega$, and arbitrary collision frequency we neglect the anisotropy of the random velocities; this may introduce a possible error of a factor of the order of unity into the expression for the free flight time which depends on the random velocity. We shall assume in what follows that $x = \Omega t_c$ in (7.72) to (7.74) and we shall then have a good asymptotic behaviour in the limit of the classical case. As a result we get the expression

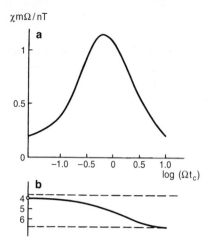

Fig. 7.3. The quantity $\chi m\Omega/nT$ as function of Ωt_c (**a**) and the Eucken ratio as function of Ωt_c (**b**).

$$\frac{\nu_1 \Omega}{T} = \frac{(4.8x^2 + 2.23)x}{16x^4 + 16.2x^2 + 2.33}. \tag{7.89}$$

The functions (7.87) to (7.89) are shown in Fig. 7.2; one can see the excellent agreement of all three expressions for the viscosity notwithstanding the differences in their derivation.

From (7.57) and (7.58) we get the following function for the thermal conductivity coefficient – in the rotation plane of the disc:

$$\frac{\chi m\Omega}{T} = \frac{(2x^2 + 2.645)x}{x^4 + 2.7x^2 + 0.677}. \tag{7.90}$$

This function is shown in Fig. 7.3. We show also the ratio of the viscosity and thermal conductivity coefficients – the Eucken coefficient – as function of Ωt_c:

$$\frac{\chi m}{\eta_1} = \frac{(2x^2 + 2.645)(16x^4 + 16.2x^2 + 2.33)}{(4.8x^2 + 2.23)(x^4 + 2.7x^2 + 0.677)}. \tag{7.91}$$

From (7.91) we find

$$\frac{\chi m}{\eta_1} = \begin{cases} 4.08 \text{ for } x \to 0, \\ 5.5 \text{ for } x = 1, \\ 6.67 \text{ for } x \to \infty, \end{cases} \tag{7.92}$$

It is clear that in the limit of frequent collisions ($x \to 0$) the coefficient (7.91) is close to the classical Eucken ratio $\frac{15}{4} = 3.75$ (in our units).

8 Collective Dynamics of Disc Particles
II Stability Analysis

> "If you throw a pebble in the water, look at the circles produced by it; otherwise the throwing will be an empty gesture."
>
> *Koz'ma Prutkov.*

8.1 General Dispersion Equation

Imagine a completely smooth, mirror-like surface of a quiet lake. A pebble falling in the water produces a small wave – in scientific terms, a perturbation – which rapidly dies out. However, if at that moment the wind is blowing, the wave will not be damped, but be amplified, to become – depending on the strength of the wind – a small ripple or a breaker wave. If we look at this situation with the eyes, not of a poet, but of a scientist we can say that if there is no wind the equilibrium state of the system (the water surface) is stable (the perturbations die out) while the wind causes an instability (growth of the initial perturbations). The instability in turn produces a new structural state of the system – a regular wave sequence. If we are dealing with a system which has not yet been studied in detail, then we may ask the question whether a given spatially uniform state of the system is stable. First of all, it would be well to ascertain whether there is an energy source in the system without which it would be impossible to produce any structure. However, the existence of such a source does not necessarily mean an instability or the appearance of a new structural state; hence we must study the dynamics of the system in detail. We may assume that we know the equations which more or less completely describe our system. How can we then examine such a system for stability?

We have already described one way to determine the stability of our system – we must throw a "pebble" in the "water" of the system we are studying, that is, artificially excite a wave perturbation in the system and watch it to see whether it grows. If the perturbation dies out the system is stable with respect to that kind of perturbation, but if the wave amplitude increases the system is, as a rule, unstable. It is much more complicated to follow the evolution of the initial wave to a new equilibrium state – one needs a non-linear analysis. For the moment we shall restrict ourselves to the linear case – to small wave amplitudes.

First of all we shall study the stability of rotating viscous discs. Many authors (Kumar 1960, Lynden-Bell and Pringle 1974, Mishyrov et al. 1976, Morozov et al. 1985) have studied the linear oscillations of such discs in the framework of ordinary hydrodynamics. Usually these studies – with the exception of the work by Morozov et al. (see below) – are applicable only to uniformly rotating discs, whatever the authors state. Differentially rotating viscous discs which are described by the usual equations of hydrodynamics are non-stationary as far as the temperature is concerned, that is, they rapidly heat up so that it is incorrect to study the instability of linear oscillations in such discs – there is no "surface" into which one can throw a "pebble". Safronov (1969) was, as far as we know, the first to indicate this. Nonetheless up to now there exists the general opinion that the transition from uniform to differential rotation is trivial (see, for instance, Mishyrov et al. 1976). To make it clear that there is a fundamental difference between the two cases we shall start by considering the linear oscillations of a uniformly rotating disc of elastic particles. This also serves as a useful exercise in stability analysis through the example of a relatively simple system.

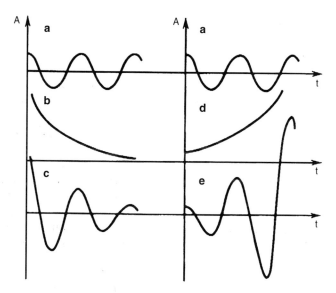

Fig. 8.1. Various regimes. (**a**) Stable oscillations: Re $\omega \neq 0$, Im $\omega = 0$; (**b**) Aperiodic damping: Re $\omega = 0$, Im $\omega < 0$; (**c**) Damped oscillations: Re $\omega \neq 0$, Im $\omega < 0$; (**d**) Aperiodic instability: Re $\omega = 0$, Im $\omega > 0$; (**e**) Unstable oscillations: Re $\omega \neq 0$, Im $\omega > 0$.

We make a general remark: when studying disc systems one uses "planar" hydrodynamic equations. This means that one uses the equations of two-dimensional hydrodynamics in the plane of the disc. It is then possible, in particular, to get rid of the temperature dependence of the volume density of the particles. The surface density of the particles, on the other hand, does not depend on the temperature. Indeed, the thickness of the disc depends as follows on the random particle velocity: $h \approx V/\Omega$, that is, it is determined by the height or the distance over which a particle with velocity V can remove itself from the rotational plane of the disc before the z-component of the gravitational attraction of the planet pulls it back (see Sect. 1.1).

Hence, when the temperature of the disc is increased in some point its thickness also increases and the volume density of the particles decreases. This makes the usual density a not very convenient characteristic. At the same time the surface density of the particles – the number of particles per unit disc area – remains constant during temperature oscillations. If one integrates the equation of continuity over the z-coordinate we obtain an equation for the surface density. Obtaining the other transport equations is not so trivial (see Appendices I and II). The equations obtained are suitable for studying low-frequency processes with characteristic spatial scales along x and y much larger than the thickness of the disc.

8.1.1 Stability of a Uniformly Rotating Disc

Let us write down the equations of two-dimensional hydrodynamics for axially symmetric perturbations in the non-rotating frame of reference and cylindrical coordinates – we give in Appendices I and II the conditions for the validity of these equations when there is no viscosity:

$$
\begin{aligned}
&\frac{\partial \sigma}{\partial t} + \frac{1}{r}\frac{\partial}{\partial r}\left(r\sigma V_r\right) = 0, \\
&\frac{\partial V_r}{\partial t} + V_r \frac{\partial}{\partial r} V_r - \frac{V_\varphi^2}{r} = -\frac{1}{\sigma}\frac{\partial P}{\partial r} - \frac{\partial \psi_G}{\partial r} \\
&\qquad + \tfrac{4}{3}\nu \frac{\partial}{\partial r}\frac{1}{r}\frac{\partial}{\partial r} r V_r + \frac{2}{3\sigma}\frac{\partial \nu\sigma}{\partial r}\left[2\frac{\partial V_r}{\partial r} - \frac{V_r}{r}\right], \\
&\frac{\partial V_\varphi}{\partial t} + V_r \frac{\partial V_\varphi}{\partial r} + \frac{V_r V_\varphi}{r} = \frac{1}{\sigma r^2}\frac{\partial}{\partial r}\left[r^3 \nu\sigma \frac{\partial}{\partial r}\frac{V_\varphi}{r}\right], \\
&\frac{3}{2}\left[\frac{\partial T}{\partial t} + V_r \frac{\partial T}{\partial r}\right] + T\frac{1}{r}\frac{\partial}{\partial r} r V_r = \frac{1}{\sigma r}\frac{\partial}{\partial r}\left(\chi_T r \frac{\partial T}{\partial r}\right) \\
&\qquad + \nu\left[r\frac{\partial}{\partial r}\frac{V_\varphi}{r}\right]^2 + \tfrac{4}{3}\nu\left\{\left(\frac{\partial V_r}{\partial r}\right)^2 + \frac{V_r^2}{r^2} - \left(\frac{\partial}{\partial r}V_r\right)\frac{V_r}{r}\right\}.
\end{aligned}
$$

(8.1)

Here σ is the surface density, which one can obtain by multiplying the volume density by the thickness of the disc: $\sigma = \varrho_r h$, ν is the kinematic viscosity coefficient which is equal to the dynamic viscosity divided by the surface density, and χ_T is the heat conductivity coefficient. We now introduce a perturbation: we assume that any quantity A, such as the density, the velocity components, or the temperature, consists of the sum of a stationary term A_0 and a small perturbation term A_1 so that $A = A_0 + A_1$ with $A_1 \ll A_0$. Of course, the stationary value of some quantity may be equal to zero, as is, for instance, the case for the radial component of the velocity. We write the perturbed quantity as a monochromatic wave:

$$A_1 = \hat{A}e^{-i\omega t + ikr}, \tag{8.2}$$

where $\hat{A}$ is the initial wave amplitude or Fourier component, ω the frequency of the oscillations, $k = 2\pi/\lambda$ the wavenumber, and λ the wavelength. We shall in what follows in most cases use the compact notation $\gamma \equiv -i\omega$. If as a result of our analysis it becomes clear that γ is a purely imaginary quantity, that is, that the frequency is real, this means that our wave propagates along the disc without changes – the system is stable. If γ is real and negative it follows from (8.2) that the wave is damped – the system is again stable and in that case γ is the damping rate. If, however, γ is real and positive the wave is amplified and in that case γ is the growth rate. The situations described here are illustrated in Fig. 8.1.

We substitute the following expressions into (8.1):

$$
\begin{aligned}
\sigma &= \sigma_0 + \hat{\sigma}e^{\gamma t + ikr}, & V_r &= \hat{V}_r e^{\gamma t + ikr}, \\
V_\varphi &= \Omega r + \hat{V}_\varphi e^{\gamma t + ikr}, & T &= T_0 + \hat{T}e^{\gamma t + ikr},
\end{aligned}
\tag{8.3}
$$

where the stationary quantities are designated by a zero subscript and the amplitude of the perturbations by a $\wedge$ sign. Splitting off the equation of dynamic equilibrium between the centrifugal and gravitational forces,

$$\Omega^2 r = \frac{GM}{r^2}, \tag{8.4}$$

using the Poisson equation for the perturbed gravitational potential (Fridman 1975, Fridman and Polyachenko 1984),

$$\frac{\partial \hat{\psi}_G}{\partial r} = -2\pi iG\hat{\sigma}, \tag{8.5}$$

and also neglecting non-linear terms – quadratic in the amplitude – we get the following set of equations:

$$\gamma\hat{\sigma} + ik\sigma_0\hat{V}_r = 0, \tag{8.6}$$

$$\gamma\hat{V}_r - 2\hat{V}_\varphi\Omega = i\frac{2\pi G\sigma_0 - kc^2}{\sigma_0}\hat{\sigma} - ik\hat{T} - \tfrac{4}{3}\nu k^2\hat{V}_r, \tag{8.7}$$

$$\gamma\hat{V}_\varphi + \frac{\kappa^2}{2\Omega}\hat{V}_r = -\nu k^2\hat{V}_\varphi, \tag{8.8}$$

$$\tfrac{3}{2}\gamma\hat{T} + ikc^2\hat{V}_r = -\chi k^2\hat{T}. \tag{8.9}$$

where κ is the epicyclic frequency, $\kappa^2 = 4\Omega^2[1 + \frac{1}{2}r(\Omega'/\Omega)]$; for a Kepler disc we have $\kappa^2 = \Omega^2$.

Assuming the masses of all the particles to be the same and equal to unity we have introduced the isothermal sound velocity through the relation $c^2 \equiv T_0$. The coefficient $\frac{3}{2}$ in (8.9) means that the medium considered by us behaves like a monatomic gas with $c_V = \frac{3}{2}$. In the same equation we introduced the quantity $\chi = \chi_T/\sigma$ which, apart from a factor c_P, corresponds to the thermal conductivity.

Moreover, we have used the equation of state (A3.11):

$$P = \sigma T. \tag{8.10}$$

A comparison of the stability criteria for different states of the medium shows that they are not wildly different from one another, possibly for the same reason that the polytropic index of real media differs only slightly from that of the polytrope of a perfect gas, $\frac{5}{3}$. We shall therefore use here the equation of state of a perfect gas $P_V = nT$ where P_V is the usual pressure for unit volume of the medium and n the number of particles per unit volume. If we take the mass of each particle to be equal to unity we can rewrite the equation of state in the form $P_V = \varrho T$, where ϱ is the volume density. We shall assume that the temperature T manages to equilibrate during the time of the process along the rotational axis of the thin disc, that is, that T is independent of the z coordinate. In that case we get by integrating the three-dimensional equation of state along z the two-dimensional equation of state (8.10) which was used to calculate the acceleration $\sigma^{-1}\partial P/\partial r$ in (8.7).

This system is obtained in the short-wavelength or WKB approximation, that is, we assumed that the wavelength is shorter than the characteristic lengths of change of the stationary parameters of the disc or – if the disc is assumed to be uniform in its characteristics – shorter than the radius of the disc. This assumption makes it possible to write down the following inequalities:

$$k^2\hat{A} \gg \frac{k}{r}\hat{A} \gg \frac{\hat{A}}{r^2}, \qquad kr \gg 1,$$

and to drop the corresponding small terms. It releases us at the same time from taking into account the boundary conditions – in fact, one considers the disc essentially to be infinite.

The set (8.6) to (8.9) is a system of algebraic equations. Calculating the determinant of this system we get the following algebraic equation, the so-called dispersion relation:

$$\gamma^4 + [\tfrac{7}{3}\nu k^2 + \tfrac{2}{3}\chi k^2]\gamma^3 + [\omega_0^2 + \nu k^2(\tfrac{4}{3}\nu k^2 + \tfrac{14}{9}\chi k^2)]\gamma^2$$
$$+ \left[\nu k^2(\tfrac{5}{3}k^2c^2 - 2\pi G\sigma_0 k) + \tfrac{2}{3}\chi k^2\left(\omega_*^2 - \tfrac{4}{3}\nu^2 k^4\right)\right]\gamma$$
$$+ \tfrac{2}{3}\chi k^2\nu k^2(k^2c^2 - 2\pi G\sigma_0 k) = 0, \tag{8.11}$$

where

$$\omega_0^2 = \tfrac{5}{3}k^2c^2 - 2\pi G\sigma_0 k + \kappa^2, \tag{8.12}$$

$$\omega_*^2 = k^2c^2 - 2\pi G\sigma_0 k + \kappa^2. \tag{8.13}$$

Kumar (1960) obtained a similar equation, in a somewhat different geometry. Let us study (8.11) and separate the different branches of oscillations. Neglecting dissipative effects – or considering only high-frequency oscillations – we get the dispersion relation of the linear oscillations of a self-gravitating rotating disc with pressure:

$$\omega^2 = \tfrac{5}{3}k^2c^2 - 2\pi G\sigma_0 k + \kappa^2. \tag{8.14}$$

These oscillations can be subject to the Jeans instability. Safronov (1960) was the first to obtain the conditions for this instability for a disc. Let us consider a disc which is Jeans stable, assuming that the self-gravitation is much smaller than the terms characterising the contributions from the pressure or the rotation. We shall now be interested in low-frequency oscillations with $\gamma \sim \chi k^2 \ll \Omega$. For such oscillations we get from (8.11) damped temperature oscillations:

$$\gamma \sim -\tfrac{2}{3}\chi k^2. \tag{8.15}$$

If the smallness of γ is of another order of magnitude, $\gamma \sim \nu k^4 c^2/\Omega^2$, $k^2 c^2 \ll \Omega^2$, we get a solution describing the dynamics of diffusion (viscous) oscillations:

$$\gamma \sim -\nu k^2 \frac{k^2 c^2 - 2\pi G\sigma_0 k}{4\Omega^2}. \tag{8.16}$$

If the condition

$$k^2 c^2 - 2\pi G\sigma_0 k < 0 \tag{8.17}$$

is satisfied we have $\gamma > 0$, and a dissipative instability occurs (Kumar 1960, Lynden-Bell and Pringle 1974, Mishyrov et al. 1976) which may be called a secular instability. This means that the system under consideration will stratify into ringlets with a wavelength for which the instability is the strongest – that is, for which the derivative $\partial\gamma/\partial k$ is equal to zero. Other wavelengths do not have enough time to grow. For a disc which is close to the Jeans instability ($\omega_0^2 \to 0$) the scenario changes appreciably. Morozov et al. (1985) have shown that the system has an interesting solution: $\gamma \sim (\nu k^2 \kappa^2)^{1/3}$ – the instability is much stronger.

So far we have used the usual Navier–Stokes equations but in principle we could have used the transport equations obtained in Chap. 7. One needs only take the case of uniform rotation and neglect the inelasticity of the particles. From the five viscosity coefficients obtained in Chap. 7 there remain only three in the two-dimensional case: ν_0, ν_1, and ν_3, and we can write the equations of motion in the form:

$$\frac{\partial V_r}{\partial t} + V_r \frac{\partial}{\partial r} V_r - \frac{V_\varphi^2}{r} = -\frac{1}{\varrho} \frac{\partial P}{\partial r} - \frac{\partial \psi_G}{\partial r}$$

$$+ \nu_3 \frac{\partial}{\partial r} \left[\frac{1}{r} \frac{\partial}{\partial r}(rV_\varphi) \right] + (\tfrac{1}{3}\nu_0 + \nu_1) \frac{\partial}{\partial r} \left[\frac{1}{r} \frac{\partial}{\partial r}(rV_r) \right], \tag{8.18}$$

$$\frac{\partial V_\varphi}{\partial t} + V_r \frac{\partial}{\partial r} V_\varphi + \frac{V_r V_\varphi}{r} = \nu_1 \frac{\partial}{\partial r} \left[\frac{1}{r} \frac{\partial}{\partial r}(rV_\varphi) \right] - \nu_3 \frac{\partial}{\partial r} \left[\frac{1}{r} \frac{\partial}{\partial r}(rV_r) \right]. \tag{8.19}$$

(Note that in the classical limit $\Omega t_c \to 0$, $\nu_3 \to 0$, $\nu_1 \to \nu_0$, and we obtain the usual Navier–Stokes equations.)

As before we linearise them and change to Fourier coefficients, and we obtain the following equations:

$$\gamma \hat{V}_r - 2\Omega \hat{V}_\varphi = -\frac{ikc^2}{\sigma_0} \hat{\sigma} - \frac{ik}{m} \hat{T} + 2\pi iG\hat{\sigma} - (\tfrac{1}{3}\nu_0 + \nu_1)k^2 \hat{V}_r - \nu_3 k^2 \hat{V}_\varphi, \tag{8.20}$$

$$\gamma \hat{V}_\varphi + 2\Omega \hat{V}_r = -\nu_1 k^2 \hat{V}_\varphi + \nu_3 k^2 \hat{V}_r. \tag{8.21}$$

There are two terms with $\hat{V}_\varphi$ in (8.20), but taking into account the inequality $\nu_3 k^2 \ll 2\Omega$ we can neglect the term with ν_3. A similar situation occurs in (8.21) and the viscosity coefficient ν_3 there leads only to a slight correction to the term $2\Omega \hat{V}_r$. The difference between ν_0 and ν_1 is also unimportant since the viscosity in the radial equation for the oscillations does not occur in the expression for the dissipative growth rates which depend mainly on the shear viscosity in the equation for the azimuthal velocity component and the thermal conductivity (see (8.15) and (8.16)). Taking these circumstances into account in what follows we shall, in order to simplify the calculations, neglect the viscosity coefficient ν_3 and replace $\tfrac{1}{3}\nu_0 + \nu_1$ by $\tfrac{4}{3}\nu$.

8.1.2 A Differentially Rotating Disc of Inelastic Particles

In this subsection we shall obtain a general dispersion relation for the linear oscillations of disc systems with inelastic colliding particles. The main system of transport equations have, if we take into account what we have said earlier, the form

$$\frac{\partial \sigma}{\partial t} + \frac{1}{r} \frac{\partial}{\partial r}(r\sigma V_r) + \frac{1}{r} \frac{\partial}{\partial \varphi}(\sigma V_\varphi) = N^+(\sigma, T) - N^-(\sigma, T), \tag{8.22}$$

$$\frac{\partial V_r}{\partial t} + V_r \frac{\partial V_r}{\partial r} + \frac{V_\varphi}{r} \frac{\partial V_r}{\partial \varphi} - \frac{V_\varphi^2}{r} = -\frac{1}{\sigma} \frac{\partial P}{\partial r} - \frac{\partial \psi_G}{\partial r}$$

$$+ \tfrac{4}{3}\nu \frac{\partial}{\partial r} \frac{1}{r} \frac{\partial}{\partial r} rV_r + \frac{4}{3\sigma} \frac{\partial \nu\sigma}{\partial r} \left[\frac{\partial V_r}{\partial r} - \frac{V_r}{2r} \right] - \frac{2}{3\sigma r} \frac{\partial}{\partial r} \left[\nu\sigma \frac{\partial V_\varphi}{\partial \varphi} \right]$$

$$- \frac{4\nu}{3r^2} \frac{\partial V_\varphi}{\partial \varphi} + \frac{1}{r\sigma} \frac{\partial}{\partial \varphi} \left\{ \nu\sigma \left[\frac{1}{r} \frac{\partial V_r}{\partial \varphi} + r \frac{\partial}{\partial r} \frac{V_\varphi}{r} \right] \right\}, \tag{8.23}$$

$$\frac{\partial V_\varphi}{\partial t} + V_r \frac{\partial V_\varphi}{\partial r} + \frac{V_\varphi}{r} \frac{\partial V_\varphi}{\partial \varphi} + \frac{V_r V_\varphi}{r} = -\frac{1}{\sigma r} \frac{\partial P}{\partial \varphi} - \frac{1}{r} \frac{\partial \psi_G}{\partial \varphi}$$

$$+ \frac{1}{\sigma r^2} \frac{\partial}{\partial r} \left[r^3 \nu \sigma \frac{\partial}{\partial r} \frac{V_\varphi}{r} \right] + \frac{1}{\sigma r} \frac{\partial}{\partial r} \left[\nu \sigma \frac{\partial V_r}{\partial \varphi} \right] + \frac{\nu}{r^2} \frac{\partial V_r}{\partial \varphi}$$

$$+ \frac{1}{r \sigma} \frac{\partial}{\partial \varphi} \left\{ \tfrac{4}{3} \nu \sigma \left[\frac{1}{r} \frac{\partial V_\varphi}{\partial \varphi} + \frac{V_r}{r} - \frac{1}{2} \frac{\partial V_r}{\partial r} \right] \right\}, \qquad (8.24)$$

$$\frac{3}{2} \left[\frac{\partial T}{\partial t} + V_r \frac{\partial T}{\partial r} + \frac{V_\varphi}{r} \frac{\partial T}{\partial \varphi} \right] + T \left[\frac{1}{r} \frac{\partial}{\partial r} r V_r + \frac{1}{r} \frac{\partial V_\varphi}{\partial \varphi} \right] = \frac{1}{\sigma r} \frac{\partial}{\partial r} \left[\chi_T r \frac{\partial T}{\partial r} \right]$$

$$+ \frac{1}{\sigma r^2} \frac{\partial}{\partial \varphi} \left[\chi_T \frac{\partial T}{\partial \varphi} \right] + \nu \left(r \frac{\partial}{\partial r} \frac{V_\varphi}{r} \right)^2 + 2\nu \left[\left(\frac{\partial V_r}{\partial r} \right)^2 + \left(\frac{V_r}{r} \right)^2 \right]$$

$$+ 2\nu \left[\left(\frac{1}{r} \frac{\partial V_\varphi}{\partial \varphi} \right)^2 + \frac{2 V_r}{r^2} \frac{\partial V_\varphi}{\partial \varphi} + \frac{1}{2r^2} \left(\frac{\partial V_r}{\partial \varphi} \right)^2 + \frac{\partial V_r}{\partial \varphi} \frac{\partial}{\partial r} \frac{V_\varphi}{r} \right]$$

$$- \tfrac{2}{3} \nu \left[\left(\frac{1}{r} \frac{\partial}{\partial r} r V_r \right)^2 + \frac{2}{r^2} \left(\frac{\partial}{\partial r} r V_r \right) \frac{\partial V_\varphi}{\partial \varphi} + \left(\frac{1}{r} \frac{\partial V_\varphi}{\partial \varphi} \right)^2 \right] - E^-. \quad (8.25)$$

In the continuity equation we have taken into account that the density in the disc may change not only through diffusive motion of the particles in the disc but also through external (or non-diffusive) fluxes of matter, for instance, when matter is accreted from a dust-gas cloud in the proto-stage – that is, we must admit the possiblity of an exchange of matter with a matter bath: the function $N^+(\sigma, T)$ describes an increase and the function $N^-(\sigma, T)$ a decrease in the mass of the disc. These terms are analogous to the terms describing chemical reactions in a gaseous medium and their existence means that the first moment of the collision integral does not vanish. We assume that these terms are so small that they do not substantially affect the other transport equations, but they are important for the equation of continuity notwithstanding their being small since if they occur that equation will depend on the temperature. The equation of motion and the energy equation depend on the temperature and the density also when they are not present so that it is not important in their case to take the very "slow" chemical reactions into account. Formally (8.22) to (8.25) are the same as the equations of viscous hydrodynamics: there is only an extra term describing the inelasticity of the particles and the viscosity has become a more complex function. We note that Stewart et al. (1984) arrived at the conclusion that it is possible to use equations of this form to study the instability of the rings but they did not analyse such equations in detail.

For a correct consideration of the linear oscillations of a viscous differentially rotating disc it is necessary that the characteristic times for changes in the stationary state are much longer than the characteristic times of the oscillations considered. A viscous differentially rotating disc is non-stationary because of the viscous transfer of angular momentum in the disc from the inner, faster rotating regions to the outer zones with a slower rotation. This leads to a slow diffusive spreading of the disc with a characteristic time which we can obtain from (8.24):

$$t_{\mathrm{d}} \sim \frac{r^2}{\nu} \ll \frac{1}{\nu k^2}, \qquad kr \gg 1. \tag{8.26}$$

In the short-wavelength approximation the characteristic time for this non-stationarity is thus much longer than the time for the viscous evolution of the perturbations under consideration. If we neglect inelasticity the temperature in the disc is non-stationary:

$$\frac{\partial T}{\partial t} = \nu (r\Omega')^2. \tag{8.27}$$

Hence, if we take the thickness of the disc $h \approx c/\Omega$ into account we get the time for the heating of the disc:

$$t_U \sim \frac{c^2}{\nu (r\Omega')^2} \sim \frac{h^2}{\nu}. \tag{8.28}$$

The wavelengths considered in the disc must thus be much larger than the thickness of the disc and this means that the characteristic time of this non-stationarity is shorter than the characteristic times of all dissipative processes considered – these times are comparable only in the limit as $kh \to 1$. This is the reason why it is wrong to consider the stability of a viscous differentially rotating disc in the framework of ordinary hydrodynamics. An exception is the paper by Morozov et al. (1985) in which linear oscillations of the disc are studied near the boundary of the Jeans instability so that $kh \sim 1$ and the characteristic time for this instability, $t_{\mathrm{m}} \sim (\nu k^2 \kappa^2)^{-1/3}$, in the case of low viscosity,

$$\nu \ll \kappa h^2,$$

turns out to be much shorter than the time (8.28) for heating the disc up. In the general case the stationarity of the temperature will be guaranteed by the cooling of the disc due to the inelastic collisions between the particles. We can write the energy balance in the form

$$\nu (r\Omega')^2 = E^- \qquad \text{or} \qquad E^+ = E^-. \tag{8.29}$$

We shall later on consider the energy balance in detail but now we make one important remark: because of the presence of differential rotation the shear viscosity coefficient can no longer be taken outside the spatial derivative signs. It must experience a perturbation at the same time as the other terms due to its temperature and density dependence. Generally speaking the terms involving the viscosity should be perturbed also in a disc which is rotating uniformly but one can easily show that they in that case lead only to non-linear terms which we drop in the linear approximation. In a differentially rotating disc there occurs the quantity $r\Omega'$ which completely changes the situation so that the perturbation of the viscosity in the equations of motion must be written in the form

$$\frac{\partial(\widehat{\nu\sigma})}{\partial r} = \mathrm{i}k\frac{\partial\nu\sigma_0}{\partial\sigma_0}\hat{\sigma} + \mathrm{i}k\sigma_0\frac{\partial\nu}{\partial T_0}\hat{T}. \tag{8.30}$$

In the energy equation there appear extra terms connected with the perturbation of the viscosity coefficient in the term $\nu(r\Omega')^2$, which describes the dissipation of the orbital rotation into kinetic energy of the random motion, and also connected with the perturbation of the term $E^-(\sigma, T)$ describing the inelastic energy losses:

$$\widehat{\sigma E^-} = \frac{\partial E^-\sigma_0}{\partial\sigma_0}\hat{\sigma} + \sigma_0\frac{\partial E^-}{\partial T_0}\hat{T}. \tag{8.31}$$

Taking all this into account we can write down a set of linearised transport equations in the WKB approximation (see Appendix III) for perturbations of the form $e^{\gamma+\mathrm{i}kr+\mathrm{i}m\varphi}$ – we assume here that the perturbations may be non-axisymmetric and may depend on φ:

$$(\gamma+\mathrm{i}m\Omega)\hat{\sigma} + \mathrm{i}k\sigma_0\hat{V}_r = \left[\frac{\partial N^+}{\partial\sigma_0} - \frac{\partial N^-}{\partial\sigma_0}\right]\hat{\sigma} + \left[\frac{\partial N^+}{\partial T_0} - \frac{\partial N^-}{\partial T_0}\right]\hat{T},$$

$$(\gamma+\mathrm{i}m\Omega)\hat{V}_r - 2\Omega\hat{V}_\varphi = \mathrm{i}\frac{2\pi G\sigma_0 - kc^2}{\sigma_0}\hat{\sigma} - \mathrm{i}k\hat{T} - \tfrac{4}{3}\nu k^2\hat{V}_r, \tag{8.32}$$

$$(\gamma+\mathrm{i}m\Omega)\hat{V}_\varphi + \frac{\kappa^2}{2\Omega}\hat{V}_r = -\mathrm{i}k\beta\hat{\sigma} - \mathrm{i}k\alpha\hat{T} - \nu k^2\hat{V}_\varphi,$$

$$\tfrac{3}{2}(\gamma+\mathrm{i}m\Omega)\hat{T} + \mathrm{i}kc^2\hat{V}_r = -\chi k^2\hat{T} - \Delta E_\sigma\hat{\sigma} - \Delta E_T\hat{T} - \mathrm{i}k\mu\hat{V}_\varphi,$$

where to simplify the equations we have introduced the notation

$$\alpha = \frac{\partial\nu(-r\Omega')}{\partial T_0}, \quad \beta = \frac{1}{\sigma_0}\frac{\partial\nu\sigma_0(-r\Omega')}{\partial\sigma_0}, \quad \mu = 2\nu(-r\Omega'),$$

$$\Delta E_\sigma = \frac{1}{\sigma_0}\left[\frac{\partial E^-\sigma_0}{\partial\sigma_0} - \frac{\partial E^+\sigma_0}{\partial\sigma_0}\right], \quad \Delta E_T = \frac{\partial E^-}{\partial T_0} - \frac{\partial E^+}{\partial T_0}. \tag{8.33}$$

Putting the determinant of the system (32) equal to zero we get the dispersion relation of the linear oscillations of a differentially rotating disc of inelastic particles – neglecting non-diffusive fluxes, that is, putting $N^+ = N^- = 0$:

$$(\gamma + im\Omega)^4 + (\gamma + im\Omega)^3[\tfrac{2}{3}(\chi k^2 + \Delta E_T) + \tfrac{7}{3}\nu k^2]$$

$$+ (\gamma + im\Omega)^2[\tfrac{4}{3}\nu^2 k^4 + \omega_0^2 + \tfrac{14}{9}\nu k^2(\chi k^2 + \Delta E_T) + \tfrac{2}{3}k^2\alpha\mu]$$

$$+ (\gamma + im\Omega)\left[\nu k^2(\tfrac{5}{3}k^2 c^2 - 2\pi G\sigma_0 k) + \tfrac{2}{3}(\tfrac{4}{3}\nu^2 k^4 + \omega_*^2)(\chi k^2 + \Delta E_T)\right.$$

$$\left. + \tfrac{8}{9}\alpha\mu\nu k^4 + \tfrac{4}{3}k^2 c^2\Omega\alpha - \tfrac{2}{3}k^2\sigma_0\Delta E_\sigma - \frac{\kappa^2}{3\Omega}k^2\mu + 2k^2\beta\Omega\sigma_0\right]$$

$$+ \tfrac{2}{3}\left[(\chi\nu k^4 + \nu k^2\Delta E_T + k^2\alpha\mu)(k^2 c^2 - 2\pi G\sigma_0 k)\right.$$

$$\left. + \sigma_0 k^4(2\chi\beta\Omega - \beta\mu - \Delta E_\sigma\nu) + 2k^2\sigma_0\Omega(\beta\Delta E_T - \alpha\Delta E_\sigma)\right] \;=\; 0, \quad (8.34)$$

where ω_0 and ω_* are given by (8.12) and (8.13). We shall study (8.34) in the following sections.

8.2 Analysis of the Axisymmetric Oscillations of a Disc; Instabilities Causing the Small- and Medium-Scale Structure of the Rings

We start by considering at an elementary physical level those instabilities which can develop for axisymmetric oscillations of a differentially rotating disc of inelastic particles. We shall show that the small-scale structure of the rings is the result of the development of gravitational and diffusion instabilities, while the quasi-secular instability produces the medium-scale structure of the rings.

8.2.1 Gravitational Instability

If we neglect all dissipative effects in the dynamics of the rings the dispersion relation (8.34) becomes the equation for the Jeans (gravitational) oscillations:

$$\hat{\omega}^2 \;=\; \omega_0^2 \;=\; \tfrac{5}{3}k^2 c^2 - 2\pi G\sigma_0 k + \kappa^2, \qquad \hat{\omega} \equiv \omega - m\Omega. \qquad (8.35)$$

It can be seen from (8.35) that the thermal motion has a stabilising effect on the Jeans instability which therefore develops most strongly in a "cold" disc. Since the sound velocity c is a measure of the thermal motion and as it is proportional to the semi-thickness h of the disc, we may conclude that the smaller the thickness of the disc, the more stable it is – provided the other parameters are not changed. Hence, the dimensionless destabilising factor is r/h where r is the radius of the disc, and the dimensionless stabilising factor is M/m where M and m are, respectively, the masses of the central body and of the disc. The meaning of the stabilising factor M/m consists in that when it increases the relative influence of the central body increases. When the attractive force between any particle and the central mass exceeds the force of mutual attraction between the particles, the system is stable, for the same reason that a point orbiting in a central field is stable – we neglect here all interactions except gravitation. In other words, the system turns out to be unstable if the destabilising factor exceeds the stabilising one, that is, if

$$\frac{r}{h} > \frac{M}{m}, \quad \text{or} \quad Q < 1, \quad Q \equiv \frac{M}{m}\frac{h}{r}. \tag{8.36}$$

The parameter Q is called the Toomre stability coefficient. More precisely, the value of the coefficient Q_T introduced by Toomre in 1964 is $Q_T \equiv c\kappa/3.36G\sigma_0$. However, if we use the condition for equilibrium of a disc in the field of a central body (Fridman and Khoruzhiĭ 1999),

$$h \simeq \sqrt{\frac{2}{\gamma}}\frac{c}{\Omega_0}, \quad m \simeq \pi r^2 \sigma_0,$$

where γ is the adiabatic index, as well as the rotational law for Kepler motion, $\Omega^2 = GM/r^3$, $\kappa = \Omega_0$, we find

$$Q_T \simeq \frac{\pi}{3.36}\sqrt{\frac{\gamma}{2}}Q.$$

We see that the numerical coefficient of Q is here of the order of unity so that $Q \approx Q_T$.

The instability condition for the disc in the form (8.36) is valid for very-short-wave perturbations with a wavelength $\lambda \simeq h$. In that case the disc is divided into rings of width $d \simeq h$. However, it follows from our considerations that if $\lambda \simeq d > h$ we must in condition (8.36) replace h by d: the larger the width of the rings, the more difficult it is to satisfy the instability criterion $Q(d) = Md/mr < 1$. According to the results of the processing of the data from Voyager-2 (Smith et al. 1982, Lane et al. 1986) for Saturn's B ring we have $Q \approx 2$, that is, the B ring is close to the limit of gravitational instability. This result is obtained assuming the ring particles to be monolithic. The high porosity of these particles ($\simeq 85\%$; see Chap. 2) may make Q appreciably larger. In what follows we shall consider possible dissipative instabilities of the rings which develop in a gravitationally stable disc.

8.2.2 Energy (Thermal) Instability

In the rotating frame of reference the random motion of the disc particles is similar to the motion of molecules in a gas. In our model the random motion of the particles is maintained by the mutual gravitational perturbations of the particles, but there are also energy sources such as the non-local viscosity connected with the finite size of the particles as well as perturbations from the satellites – this means that the energy of the orbital motion of the viscous differentially rotating disc is transformed into the energy of the random

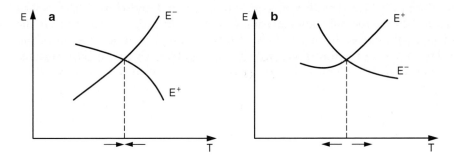

Fig. 8.2. Stable (**a**) and unstable (**b**) energy balance.

motion – "thermal" energy. On the level of individual particles this process was illustrated by our numerical calculations in Sect. 6.3.

The inelasticity of the particles makes it impossible for the random velocities to increase without bound. The balance between the influx and outflow of random motion energy can, like any balance, turn out to be unstable – like, for instance, the imbalance of a financial budget which causes economic crises. If in the cooling of some section of the rings the energy influx, which depends on the temperature of the medium, increases, the ring returns to its original temperature, but if the influx diminishes the ring suddenly cools and changes into a lower energy state (Fig. 8.2). The energy instability is thus an instability of the stationary equilibrium and it does not produce any structure. We shall consider other dissipative instabilities, assuming that the disc is energetically stable.

8.2.3 Negative Diffusion Instability

Let there be in the disc a cosinusoidal perturbation of the surface density: $\sigma \simeq \sigma_0 \cos kx$ (Fig. 8.3). Let us consider the region 1 $(0 < x < x_0)$ where the density σ_1 is increased and the region 2 $(x_0 < x < x_1)$ where we have a lowered density σ_2. At the boundary, in the point x_0, the density is unchanged. Through unit length of the boundary between 1 and 2 there is a flow of matter equal to $\sigma_1 V_1 - \sigma_2 V_2$ where V_1 and V_2 are the diffusion velocities proportional to the average thermal velocities in the regions 1 and 2, respectively. An instability occurs when the particle density in region 1 increases due to the inflow of particles from region 2, that is, when $\sigma_1 V_1 - \sigma_2 V_2 < 0$. Since $\sigma_1 > \sigma_2$ this condition for instability is satisfied, for instance, when $V \simeq \sigma^{\alpha-1}$ with $\alpha < 0$. This means that the particle velocity must decrease when the disc density increases. This is possible in the case of inelastic particles: when the density of the medium increases the collision frequency and the outflow of kinetic energy increases – hence, the random velocity of the particles decreases. The boundary between the regions 1 and 2 corresponds to a point of inflection of the function $\sigma(x)$, that is, $\partial^2 \sigma(x)/\partial x^2 = 0$ in the point x_0. It is clear from Fig. 8.3 that $\partial^2\sigma(x)/\partial x^2 < 0$ in region *1* and $\partial^2\sigma(x)/\partial x^2 > 0$ in region *2*. From the diffusion equation $\partial\sigma/\partial t = D\partial^2\sigma/\partial x^2$ it follows that the density will increase in region *1*: $\partial\sigma_1/\partial t > 0$, while it will decrease in region *2*: $\partial\sigma_2/\partial t < 0$, if the diffusion coefficient is negative, $D < 0$, in both regions. It is now clear why the instability discussed here is called the "negative diffusion instability". If we write the modulation of the density σ as a cosine wave with an exponentially increasing amplitude, $\simeq \sigma_0 e^{\gamma t}\cos kx$ we find from the diffusion equation that $\gamma = k^2|D|$, that is, the growth rate of the instability turns out to be greatest for short wavelengths.

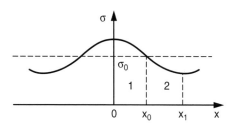

Fig. 8.3. Cosinusoidal density perturbation in the disc.

8.2.4 Analysis of the Dispersion Equation

Let us consider radial oscillations ($m = 0$) of the disc in the case when there are no non-diffusive (external) flows of matter ($N^+ = N^- = 0$). If the disc is Jeans-stable we get for low-frequency perturbations, $\gamma \simeq \nu k^2 \ll \Omega$, the following dispersion equation:

$$\gamma^2\omega_0^2 + \gamma\left[\tfrac{2}{3}\omega_*^2(\chi k^2 + \Delta E_T) + \nu k^2(\tfrac{5}{3}k^2c^2 - 2\pi G\sigma_0 k)\right.$$
$$+ \tfrac{4}{3}k^2c^2\Omega\alpha - \tfrac{2}{3}k^2\sigma_0\Delta E_\sigma - \frac{\kappa^2}{3\Omega}k^2\mu - 2k^2\beta\Omega\sigma_0\right]$$
$$+ \tfrac{2}{3}\left[(\chi\nu k^4 + \nu k^2\Delta E_T + k^2\alpha\mu)(k^2c^2 - 2\pi G\sigma_0 k)\right.$$
$$+ \sigma_0 k^4(2\chi\beta\Omega - \beta\mu - \nu\Delta E_\sigma) + 2k^2\sigma_0\Omega(\beta\Delta E_T - \alpha\Delta E_\sigma)\right] = 0. (8.37)$$

The criterion for a dissipative instability will be that the free term in (8.37) is negative since in a gravitationally stable disc we have $\omega_0^2 > 0$ and in an energetically stable disc for sufficiently long wavelengths the positive term $\tfrac{2}{3}\omega_*^2\Delta E_T$ dominates in the coefficient of γ. Equation (8.37) has two roots for long wavelengths, $kh \ll 1$, which describe the dynamics of the temperature perturbations,

$$\gamma \approx -\tfrac{2}{3}\Delta E_T, \tag{8.38}$$

and the dynamics of the diffusion oscillations,

$$\gamma \approx -Dk^2, \tag{8.39}$$

where

$$D = \sigma_0 \left[\beta - \alpha \frac{\Delta E_\sigma}{\Delta E_T} \right] \frac{2\Omega}{\kappa^2}. \tag{8.40}$$

We shall call the quantity D the diffusion coefficient. From the work by Ward (1981) and Lin and Bodenheimer (1981), who were the first to find the diffusion instability and to obtain (8.39), it follows that the disc becomes unstable when D goes through zero and becomes negative. We show from the general equation (8.37) that this is not the case.

Let us consider the case where the diffusion is small and positive. If $D \to 0$ we get from (8.37) for $kh \ll 1$:

$$\gamma = - \frac{\nu k^2}{\omega_*^2} \left[F k^2 c^2 - 2\pi f G \sigma_0 k \right], \tag{8.41}$$

where

$$f = 1 + \frac{\partial \nu}{\partial T_0} \frac{2(r\Omega')^2}{\Delta E_T},$$

$$F = f + \frac{\chi}{\nu c^2} \frac{\partial \nu \sigma_0}{\partial \sigma_0} \frac{2\Omega(-r\Omega')}{\Delta E_T} - \frac{2}{c^2} \frac{\partial \nu \sigma_0}{\partial \sigma_0} \frac{(r\Omega')^2}{\Delta E_T}$$

$$- \frac{1}{\Delta E_T c^2} \left[\frac{\partial E^-}{\partial \sigma_0} \sigma_0 - \frac{\partial E^+}{\partial \sigma_0} \sigma_0 \right]. \tag{8.42}$$

We note that we can consider dissipative oscillations only when the energy balance is stable, $\Delta E_T > 0$, in which case the temperature perturbations are damped.

Equation (8.41) describes an instability – which we call quasi-secular because its similarity with expression (8.16) – with a wavelength $\lambda_0 \simeq c^2/G\sigma_0$ (if F and $f \simeq 1$) for which it is growing fastest. One proves easily from (8.37) that this instability starts for a positive diffusion coefficient when

$$0 < D < \frac{\nu(2\pi G \sigma_0 f)^2}{\kappa^2 c^2 F}. \tag{8.43}$$

Let us consider the negative diffusion case. If $D < 0$ the diffusion instability starts; the criterion for this instability was established by Ward (1981) and Lin and Bodenheimer (1981) in the form $d(\nu\sigma)/d\sigma < 0$. From this criterion we can deduce only the fact that the instability sets in but it is impossible to determine the region of the most unstable wavelengths, that is, the characteristic stratification scale of the disc. One can use (8.37) to calculate the boundary of the diffusion instability and the wavelength of the most unstable waves. Indeed, in the long-wavelength limit the instability growth rate is $\gamma \simeq \nu k^2$. On the other hand, it follows from (8.37) that for the shortest wavelengths the term $k^6 \nu \chi c^2$ will always dominate and this stabilises the diffusion instability for $kh \simeq 1$ or $\lambda \simeq 2\pi h$ (Gor'kavyĭ 1986b). One sees this easily from the following estimate:

$$2\sigma_0 \Omega k^2 (\beta \Delta E_T - \alpha \Delta E_\sigma) \sim \frac{7\nu^2 \Omega^4 k^2}{c^2}. \tag{8.44}$$

From Sect. 7.2 we get the Eucken relation in the region $\Omega t_c \simeq 1$: $\chi \simeq 5\nu$ whence we find for the "stabilising" term: $\chi \nu c^2 k^6 \simeq 5\nu^2 c^2 k^6$. This term is comparable with the "unstable" term for $k \simeq h^{-1} \simeq \Omega/c$. The stabilisation of the diffusion instability for $kh \simeq 1$ means that the instability growth rate is a maximum for wavelengths $\lambda > 2\pi h$, or $\lambda \sim 10h$ (Gor'kavyĭ 1986b). We compare the destabilising and stabilising terms in Fig. 8.4. We note that the problem of the characteristic stratification scale of the disc was discussed by Stewart et al. (1984) who reached the conclusion that it was difficult to solve this problem in the framework of Goldreich and Tremaine's transport theory (1978a) because of the complexity of the moment equations and the neglect of the heat fluxes.

8.2.5 Criteria for the Diffusion and Energy Instabilities for Non-gravitating Smooth Spheres

For a disc of smooth particles the energy sources and sinks can be written in the form (Goldreich and Tremaine 1978a)

$$E^+ = \frac{c^2}{\Omega} \frac{\tau}{1 + \tau^2} (r\Omega')^2, \tag{8.45}$$

$$E^- = (1 - q^2) c^2 \Omega \tau, \tag{8.46}$$

and the energy balance equation takes the form

$$1 - q^2 = \frac{0.6}{1 + \tau^2}. \tag{8.47}$$

The criterion of the energy instability, $\Delta E_T < 0$, becomes the inequality

$$\frac{d(1 - q^2)}{d(V^2)} < 0, \qquad \text{or} \qquad \frac{dq}{dV} < 0. \tag{8.48}$$

The negative diffusion instability starts in an energetically stable disc if

$$\tau > \tau_{\rm cr} \equiv \sqrt{\frac{V^2}{1 - q^2} \frac{\partial(1 - q^2)}{\partial(V^2)}}. \tag{8.49}$$

If we use Bridges, Hatzes, and Lin's (1984) experimental data we have $q \propto V^{-0.25}$ and from (8.49) it follows that $\tau_{\rm cr} \approx 0.5$. The velocity dispersion is determined from the intersection point of the balance (8.47) and the experimental function $q(V)$ and is close to 0.5 mm/s. Hatzes, Bridges, and Lin's

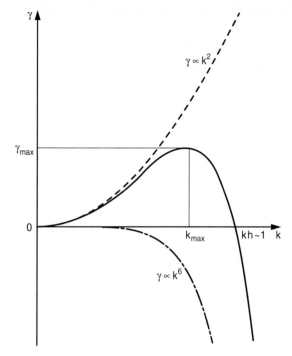

Fig. 8.4. The various terms in the dispersion equation describing the diffusion instability as functions of the wavenumber k. The term with k^6 stabilises the instability
at short wavelengths.

experimental data (1988) correspond to a smaller τ_{cr} and a much larger –
clearly unrealistic – V. However, it follows from the results of Chap. 4 that
the restitution coefficient of the snowy particles of the Saturnian rings is close
to zero and, hence, the balance (8.47) is not satisfied – the inequality is possible only, provided $q > 0.63$. Even, if the balance (8.47) were possible – the
regolith layer is for some reason very thin and only just dusts the polished
side of the ice sphere – we have $dq/dV < 0$ and the balance is according to
(8.48) unstable.

8.2.6 Energy and Diffusion Instabilities
in a Model of Gravitating Particles

The random velocity of gravitating inelastic particles in a differentially rotating disc increases when the particles approach each other closely, interacting
only gravitationally and not touching one another. If the particles come in
contact with one another in collisions the random velocity decreases.

We shall estimate the characteristic mean free flight times both in the
case when the particles come into contact, t_{cont}, and in the case of close

gravitational approaches, t_g and we shall after that compare them with one another.

1) Characteristic flight time for contact collisions.

It is well known from mechanics that in three-dimensional space the characteristic flight time for contact collisions is

$$t_{\text{cont}} \approx \frac{1}{V \sigma_{\text{cont}} n},$$

where V is the mean relative velocity, n the number of particles in unit volume, and σ_{cont} the cross-section for contact collisions.

Let us estimate the cross-section for contact collisions between particles with masses m_1 and m_2 and radii a_1 and a_2, attracting one another according to Newton's law. We show in Fig. 8.5a the case of a maximally large impact parameter ϱ_{max} when at the moment of the collision the velocity vectors of the particles are parallel to their common tangential planes – "grazing contact". The impact parameter ϱ thus varies in contact collisions from 0 to ϱ_{max}, and the cross-section σ_{cont} for such contact collisions can be found by integrating $d\sigma_{\text{cont}} = 2\pi \varrho d\varrho$ from 0 to ϱ_{max}. As a result we get the obvious result $\sigma_{\text{cont}} = \pi \varrho_{\text{max}}^2$.

In order to evaluate ϱ_{max} we change from the problem of the interaction between two particles in the laboratory frame of reference (Fig. 8.5a) to the motion of a particle with the reduced mass $\mu = m_1 m_2/(m_1 + m_2)$ in a centre of force moved to the centre of mass of the two particles (Fig. 8.5b). The trajectory of the particle with the reduced mass μ is at all times given by the function $\boldsymbol{r}(t)$ where $\boldsymbol{r} = \boldsymbol{r}_1 - \boldsymbol{r}_2$ with $\boldsymbol{r}_1$ and $\boldsymbol{r}_2$ the coordinates of the centres of the two particles in the laboratory frame of reference. Clearly we have $r_{\text{min}} = a_1 + a_2$.

In the centre of mass frame we have conservation of the total energy of the system,

$$\tfrac{1}{2}\mu v_\infty^2 = \tfrac{1}{2}\mu v^2(r_{\text{min}}) - \frac{G m_1 m_2}{r_{\text{min}}},$$

as well as conservation of the angular momentum relative to the origin,

$$\mu \varrho_{\text{max}} v_\infty = \mu r_{\text{min}} v(r_{\text{min}}).$$

Eliminating $v(r_{\text{min}})$ between these two equations we find

$$\varrho_{\text{max}}^2 = (a_1 + a_2)^2 \left[1 + 2 \frac{G(m_1 + m_2)}{(a_1 + a_2)v_\infty^2} \right].$$

Hence we find

$$\sigma_{\text{cont}} = \pi(a_1 + a_2)^2 \left[1 + 2 \frac{G(m_1 + m_2)}{(a_1 + a_2)v_\infty^2} \right].$$

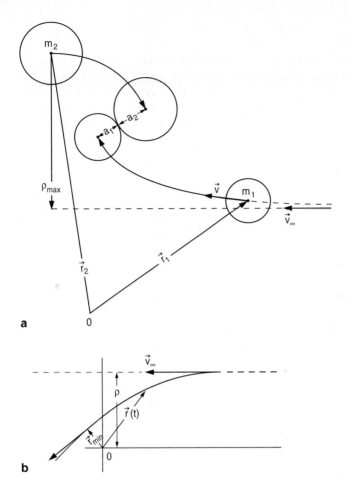

Fig. 8.5. Deriving an expression for the cross-section of contact collisions. (**a**) Case of a maximally large impact parameter ϱ_{max} when at the moment the collision takes place the velocity vectors of the particles are parallel to their common tangential planes: "grazing contact" in the laboratory frame of reference. (**b**) Motion of a particle with the reduced mass in a centre of force at the centre of mass of the particles.

In the particular case where the two colliding particles are the same ($m_1 = m_2 = m$, $a_1 = a_2 = a$) we have

$$\sigma_{cont} = 4\pi a^2 \left(1 + 2\,\frac{Gm}{aV^2}\right),$$

where V is the mean relative velocity of the particles. We can therefore write the expression for the characteristic free flight time for contact collisions in the form

$$t_{\text{cont}} = \frac{1}{4\pi a^2 nV[1 + 2(Gm/aV^2)]}.$$

If we use again the condition of equilibrium along the z-coordinate in the disc in the field of a large central mass (Fridman and Khoruzhiĭ 1999), $h \simeq \sqrt{(2/\gamma)}(v_z/\Omega)$ and introduce the optical depth τ of the disc in the form $\tau = 2\pi a^2 nh$ – bearing in mind that v_z is the velocity dispersion along z, γ is the adiabatic index, and h the semi-thickness of the disc – we can finally write the expression for t_{cont} in the form

$$t_{\text{cont}} \approx \frac{1}{\zeta \Omega \tau (1 + x^{-1})}, \tag{8.50}$$

where $\zeta \equiv 2\sqrt{(\gamma/2)}(V/v_z)$, $x \equiv aV^2/2Gm$. To determine the ratio V/v_z we calculate the average value of the relative velocity

$$\langle V \rangle = \langle |\boldsymbol{v}_1 - \boldsymbol{v}_2| \rangle = \frac{1}{2\pi} \int_0^{2\pi} \sqrt{v_1^2 + v_2^2 - 2v_1 v_2 \cos \varphi} \, d\varphi \approx \frac{4}{\pi} v,$$

where we have assumed that $|\boldsymbol{v}_1| = |\boldsymbol{v}_2| = v$. On the other hand, we have $v_z \approx v/\sqrt{3}$ and $\gamma = 1$ since the equation of state $P = \sigma T$ which we use in the present chapter is valid only for a disc which is isothermal along z. Hence we have $V/v_z \approx 4\sqrt{3}/\pi$ and hence $\zeta \approx 3.1$ and $x \approx 0.8av^2/Gm$.

2) Characteristic flight time for gravitational collisions

We use Braginskii's formula (Braginskii 1965) for the characteristic free flight time for collisions between ions with the usual changes to go from electrostatics to gravostatics: $Z^2 e^2 \to Gm^2$, where Z is the charge number and e the charge of an electron. As a result we find

$$t_{\text{g}} \approx \frac{v^3}{4\sqrt{3\pi}\Lambda G^2 m^2 n},$$

where Λ is the Coulomb logarithm. Using the relations found earlier we can rewrite the expression for t_{g} as follows:

$$t_{\text{g}} \approx \frac{x^2}{\psi \Omega \tau}, \tag{8.51}$$

where $\psi \approx 1.5\sqrt{\gamma}\Lambda$.

3) Derivation of the energy balance equation.

We write the shear viscosity coefficient in a simple form (Goldreich and Tremaine 1978a):

$$\nu = \frac{sTt_{\text{c}}}{(b\Omega t_{\text{c}})^2 + 1}, \tag{8.52}$$

where $s = 0.9$, $b = 2$. In deriving (8.52) Goldreich and Tremaine used the "analogous expression for the viscosity of a plasma in a magnetic field"

(Spitzer 1962). However, whereas we understand in plasma physics by t_c the free flight time for Coulomb collisions which corresponds to the free flight time of gravitational collisions t_g for ring particles, Goldreich and Tremaine obtained (8.52) for contact collisions of non-gravitating inelastic spheres which can correspond only to small particles. The closeness of the behavior of the function $\nu(\Omega t_c)$ in Goldreich and Tremaine's paper to the analogous functions when the gravitational interaction between the particles is taken into account, which is demonstrated in Fig. 7.2, makes it possible for us to use (8.52) for particles of all sizes. It then turns out that the numerical values $s = 0.9$ and $b = 2$ are universal and we shall continue to denote by t_g and t_{cont} the free flight times between gravitational and contact collisions, respectively.

If in a differentially rotating disc all radial – as well as all azimuthal – gradients are negligibly small, except $\Omega' \equiv d\Omega/dr$, we can reduce the energy balance equation (8.25) to the following form:

$$\frac{3}{2}\frac{\partial T}{\partial t} - \nu(r\Omega')^2 + \frac{\partial E^-}{\partial t} = 0.$$

The last term on the left-hand side is the energy loss per unit mass due to inelastic collisions. If this were absent, the disk would "heat up" rapidly: the random velocities of the particles, and thus also the thickness of the disc, would increase. The energy source is just the differential rotation. Since, in contrast to the contact collisions which are inelastic, the gravitational collisions are elastic they tend to equalise the angular velocities in neighbouring radial regions. As a result the rotational energy is transformed partly into peculiar motions of the particles. According to the above energy balance equation a differentially rotating disc can be in a stationary state if

$$\nu(r\Omega')^2 = \frac{\partial E^-}{\partial t}.$$

The right-hand side of this equation can be written as $\Delta E / t_{cont}$ where $\Delta E = v_a^2 - v_b^2$ is the change in the kinetic energies of the two colliding particles with the indices "a" and "b" denoting the velocities before and after the collision, respectively, and where v is the three-dimensional random velocity. Since the energy loss is possible only through inelastic collisions, the right-hand side of the last equation depends only on the typical free flight time t_{cont} between contact collisions. If we introduce the "inelasticity coefficient" $q = v_b/v_a$ we can write the energy balance equation in a stationary differentially rotating disc in the form

$$\nu\left(r\frac{d\Omega}{dr}\right)^2 = \frac{(1-q^2)v^2}{t_{cont}}.$$

Substituting here expression (8.52) for ν we find

$$s\frac{Tt_c}{(b\Omega t_c)^2 + 1}(r\Omega')^2 = \frac{3(1-q^2)T}{t_{cont}}.$$

We shall show in what follows that if the optical depth is small, $\tau \simeq (\Omega t_c)^{-1} \ll 1$, and also when it is large, $\tau \gg 1$, and for small values of q^2, the free flight times between gravitational collisions turn out to be at least five times smaller than the free flight times between contact collisions. Hence, we can on the left-hand side of the last equation replace t_c by t_g. The final form of the energy balance equation thus becomes

$$s \frac{T t_g}{(b\Omega t_g)^2 + 1} (r\Omega')^2 = \frac{3(1 - q^2)T}{t_{\text{cont}}}. \tag{8.53}$$

If we use (8.50) and (8.51) we find from (8.53) ($r\Omega' = 1.5\Omega$) (Gor'kavyĭ 1985b)

$$x^5 + x^4 - \alpha x^3 + \beta \tau^2 (x + 1) = 0, \tag{8.54}$$

where

$$\alpha = \frac{3s\psi}{4(1 - q^2)b^2\xi}, \qquad \beta = \frac{\psi^2}{b^2}.$$

If in (8.54) we let the optical depth go to zero we obtain a balance equation close to the one used by Stewart and Kaula (1980). From (8.53) we find the ratio t_g/t_{cont} in an optically thin disc, $\Omega t_g \gg 1$, for different expressions of the shear velocity coefficient (Fridman et al. 1990):

$$\frac{t_g}{t_{\text{cont}}} = \frac{s(r\Omega')^2}{3(b\Omega)^2(1 - q^2)} \approx \begin{cases} 0.172 & \text{(Goldreich and Tremaine 1978a)}, \\ 0.163 & \text{(Stewart and Kaula 1980)}, \\ 0.225 & \text{(Gor'kavyĭ 1986b)}. \end{cases}$$

It is clear that the gravitational collisions occur about five times as often as the contact collisions in a disc with a small optical depth τ. In a disc with a large optical depth this ratio is even larger for the following reason.

It is clear from Fig. 8.6a that x decreases with increasing τ and according to Table 8.1 becomes less than 1 already when $\tau \gtrsim 1$. The ratio t_g/t_{cont}, which does not involve τ explicitly, turns out to be proportional to x and therefore also decreases with increasing τ. This proves the statement made earlier that the ratio t_g/t_{cont} is small for all values of τ, and it justifies taking into account in zeroth approximation in the collision integral only the elastic gravitational collisions (see Chap. 7).

The condition for the stability of the energy balance (8.53) takes the form

$$x^8 + \alpha x^6 + 2\beta\tau^2 x^4 - 3\alpha\beta\tau^2 x^2 + \beta^2\tau^4 > 0. \tag{8.55}$$

The balance equation (8.54) has in the range $0 < \tau < \tau_{\text{max}}$ two real roots but not a single one for $\tau > \tau_{\text{max}}$ (Fig. 8.6a). Substituting the solutions into (8.55) shows that only one root is energetically stable; this has a maximum value at $\tau = 0$ and decreases until τ_{max} (Gor'kavyĭ 1985b): $dx/d\tau = \infty$ at $\tau = \tau_{\text{max}}$ and the disc becomes energetically unstable, cooling rapidly and,

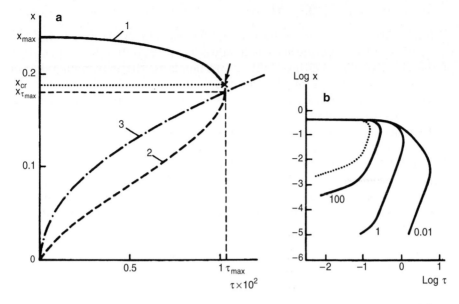

Fig. 8.6. Solution of the energy balance equation. (a) Energetically stable (*1*) and unstable (*2*) solutions of the energy balance equation; *3* is the boundary of energy stability. The arrow indicates the point where the diffusion instability starts; (b) similar solution obtained by Hämeen-Antilla et al. (1988). The numbers near the curves indicate the particle radius.

clearly, going into a lower energy state connected with the non-local particle viscosity. This state is not described by (8.54) but it was shown by Shukhman (1984) that even absolutely inelastic particles, deprived of all other energy sources, will have a velocity dispersion $c \simeq \Omega a$. It is interesting that Hämeen-Antilla et al. (1988) have recently used a more complicated transport theory and obtained a similar form of the energy balance (Fig. 8.6b). We note that the energy instability starts for $(b\Omega t_{\mathrm{g}})^2 \approx 2.5$.

If we use the fact that $E^+ = \nu(r\Omega')^2$ we get from (8.40) a simpler condition for the negative diffusion instability:

$$\frac{\partial \nu \sigma_0}{\partial \sigma_0} \frac{\partial E^-}{\partial T_0} - \frac{\partial \nu}{\partial T_0} \frac{\partial E^- \sigma_0}{\partial \sigma_0} < 0. \tag{8.56}$$

Using (8.50) to (8.52) we get from (8.56) the instability condition in the form

$$2x^5 + x^4 - 3\beta\tau^2(x+1) < 0. \tag{8.57}$$

We can determine the point $x_{\mathrm{cr}}, \tau_{\mathrm{cr}}$ where the instability starts by eliminating τ from (8.53) and (8.57) which gives

$$5x_{\mathrm{cr}}^3 + 4x_{\mathrm{cr}} - 3\alpha = 0, \tag{8.58}$$

whence

$$x_{cr} = 0.4 \left[\sqrt{1 + 3.75\alpha} - 1 \right]. \tag{8.59}$$

Substituting the value we have found for x_{cr} into (8.57) we get τ_{cr}. For $\alpha = 0.08$ to 0.16 we get rather small values $\tau_{cr} \approx (2$ to $9) \times 10^{-3}$. The start of the negative diffusion instability corresponds to the maximum value of the viscosity as function of Ωt_c – for $\Omega t_c \approx 0.8$. When the disc reaches the point of maximum viscosity it changes its structure and stratifies into ringlets which leads to an effective lowering of the viscosity. After the formation of the rings they show at once an energy instability and the gaps between the rings shift now to the region of a stable energy balance (Fig. 8.7). We give in Table 8.1 a few numerical values of the parameters of the disc for which the energy and the diffusion instabilities start. It is clear that for the parameters corresponding to a proto-planetary cloud of planetesimals the velocity dispersion corresponds to the calculations of Safronov (1969) and other authors.

Table 8.1. Values of the temperature and optical depth at which the energy and diffusion instabilities set in.

α	x_{max}^{a}	$\tau_{max} \times 10^2$ energy instability	$x_{\tau max}$	$\tau_{cr} \times 10^2$ diffusion instability	$x_{\tau cr}$
0.08	0.074	0.24837	0.0558	0.24835	0.0561
0.12	0.108	0.5345	0.0812	0.5324	0.0817
0.16	0.140	0.9050	0.1051	0.9048	0.1060
0.735[b]	0.492	1.361	0.3671	1.357	0.3752

[a] The maximum temperature as $\tau \to 0$.
[b] For a proto-planetary cloud we get, putting the Coulomb logarithm equal to 9, $\alpha = 0.735$, $\beta = 45.6$.

Let us estimate the velocity dispersion of the large particles of the planetary rings. From the balance equation (8.54) for $\alpha = 0.12$ and small τ we get $V = 0.26V_g \approx 1$ mm/s for a particle with $a = 10$ m and $\varrho = 0.15$ g/cm^3. This means that the centres of the large particles are distributed in a layer with a thickness of around 10 m.

Let us consider the dynamics of the small particles which always have a stable energy balance, increasing their random velocity on collisions when scattered in the gravitational field of the large particles and decreasing it when they collide with one another – we assume here that the optical depth of the small particles is much larger than that of the large particles. The criterion for the diffusion instability of a layer of small particles can be written in the form

$$\tau_2^2 - 1 + 2\frac{T_2}{E_2^+}\frac{\partial E_2^+}{\partial T_2} > 0, \tag{8.60}$$

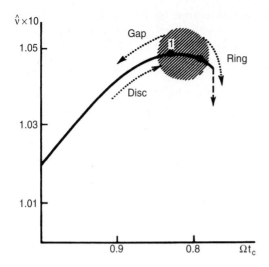

Fig. 8.7. The dimensionless viscosity as function of Ωt_c. The termination of the curve corresponds to the energy instability. The shaded region corresponds to the start of the diffusion instability. The disc is unstable when the viscosity is a maximum and stratifies into a structure with a lower viscosity.

where the index 2 indicates the small particles.

For different mechanisms of energy transfer from the large to the small particles we get critical optical depths τ_{cr} from 1 to $\sqrt{3}$ (Fridman and Poly-achenko 1984, Ward and Harris 1984, Gor'kavyĭ 1986b). The random veloci-ties of the small particles may exceed the velocities of the large particles by a large factor as a result of which the small particles form a thicker layer of several tens of meters (Cuzzi et al. 1979).

We note that the results of Sects. 8.2.1 to 8.2.4 are general in nature and are independent of the actual shape of the particles or the kind of energy balance. At the same time the models considered in Sect. 8.2.6 – not to mention Sect. 8.2.5 – are very incomplete and neglect such important particle properties as rotation around their axes and the finite size of the particles – which is neglected in the expression for the viscosity – which causes the viscosity to be non-local. The assumption that the range of the gravitational effect of the particles is independent of the presence of the planet is very simplifying. It is therefore difficult to make a reliable estimate of the critical optical depths at which the energy and diffusion instabilities occur. One can speak more confidently about the characteristic scales of the instabilities: for instance, the Jeans and the diffusion instabilities cause a stratification of the disc into ringlets with a width equal to a few times their thickness, the quasi-secular instability breaks the disc up into ringlets with a width $\lambda_0 \approx c^2/G\sigma_0 \simeq 0.1$ to 1 km (for $c \simeq 0.1$ cm/s). For the current rings $\sigma_0 \simeq 1$ to 10 g/cm^2 (if we use a particle density of 0.1 to 0.15 g/cm^3) and the quasi-secular instability can thus cause stratification on kilometer scales. In the proto-stage, at the time when the rings were formed, the magnitude of σ_0 was possibly lower and the stratification could be $\simeq 10$ to 100 km. Nonetheless, these instabilities cannot explain the large-scale stratification of the rings of $\simeq 1000$ km.

For an explanation of the condition for the start of the diffusion and other instabilities numerical methods offer great promise as they are able to take into account most of the basic particle characteristics. Such studies are actively pursued (see, for instance, Salo et al. 1988 and Brophy and Esposito 1989) but the results do not always agree with the analytical conclusions (Brophy and Esposito 1989). One must therefore consider the results of Sect. 8.2.6 as preliminary and it is as yet too early to take a position.

8.3 Analysis of the Axisymmetric Oscillations of a Disc with Non-diffusion Fluxes;
Accretion Instability – the Cause of the Large-Scale Structure of the Rings

The instabilities considered above lead to the growth of short-wavelength waves whereas Voyager photographs of the Saturnian rings distinctly showed a regular structure up to 1000 km (see Chap. 2). The search for an instability which may be responsible for the observed stratification pattern can start with a qualitative analysis of the dispersion equation and the system of linearised equations considered in the preceding section.

Indeed, the dissipative roots depend on the wavenumber to various powers and the lowest power occurring in the diffusion instability is k^2. One understands easily that one can obtain a large-scale instability if in that solution there occur terms which are linear in k. What new physical factor can lead to the occurrence of such terms in the dispersion equation? If we look at the system (8.32) of linearised equations – for the moment neglecting the terms with N^+ and N^- – it is clear that the equation of motion and the energy equation depend already on all macroscopic variables: σ, T, V_r, V_φ. The introduction of a new factor in those equations can therefore not add to the complication of the final dispersion equation and the appearance of new instabilities. The only promising way is to make the equation of continuity, which so far depended only on σ and V_r, more complicated.

The kind of generalisation we need exists in chemically reacting media where the number of particles can change due to chemical reactions. We note that taking chemical reactions into account in diffusive media led to the discovery of the Turing instability (Turing 1952) which is responsible for a whole series of morpho-energetic processes in biology such as the colouration of animal skins. Murray (1977) has described the impressive mathematical modeling of these processes. If we attempt to treat as chemical reactions the break-up of particles, which leads to a transformation of particles into one another, when the small particles stick together to form large ones and those again break up to form small particles, we achieve nothing – the mass of all particles in the disc remains constant and when summing the equations of continuity for all components we get again the classical equation of continuity without "chemical reactions". However, this is not a surprising result – it is well known that there do not appear structures in a closed "chemical reactor". The well known structure-forming Turing instability or the Belousov-Zhabotinskii reaction take place in systems where there is a supply of chemical reagents from outside. The role of "external" matter fluxes can for the rings be played by the particles which are not subject to the general diffusive dynamics. They arrive at a point in the disc and leave it again without satisfying the equations which we have written down for the bulk of the particles. This is the behaviour of the fine dust flowing through the ring system due to aerodynamic braking or the interaction with the solar radiation – the Poynting–Robertson effect – and the ejecta formed through meteoritic bombardment of the ring satisfy the same non-diffusive dynamics and transfer mass and angular momentum rather efficiently (Cuzzi and Durisen 1990) – there is even a name for it: ballistic transfer. Finally, such external fluxes exist in the proto-stage when the rings have already been formed, but around them there is a gas-dust halo from which condensing particles "rain down".

An interesting analogy with the required large-scale instability is the mechanism for the formation of sand-hills in the desert. Such hills are necessarily formed when there are at least two components present: large particles and smaller ones. The wind carries the smaller fraction and "drives" the large sand along the earth. As a result there occurs an instability and bands are formed where the sand precipitates more strongly – the hill grows.

A similar mechanism may occur also in our case of the planetary rings: a particle flux moving – in the plane of the rings – towards the planet "jams" in the fluctuations in the rings which have a higher density and hence a greater stopping power. Such an instability – which we shall call an accretion instability – must generate a large-scale stratification of the rings since small-scale fluctuations are unable to produce "hills" due to their fast diffusive spreading during a time $t \simeq (\nu k^2)^{-1}$.

Let us consider the problem of the existence of an accretion instability in more detail, starting from the set (8.32) of linearised equations. We have already mentioned at the start of this section that we are interested in the largest length scales of the regular structure, up to the width of the rings of about 1000 km; this corresponds to wavelengths of the perturbations satisfying the inequalities $h \ll \lambda \ll r$, or, if we introduce the radial wavenumber $k = 2\pi/\lambda$, the inequalities $r^{-1} \ll k \ll h^{-1}$.

We noted in Sect. 8.2.1 that Saturn's B ring, the large-scale structure of which can be seen on Plate 5, is at the limit of gravitational stability. The maximum of the instability growth rate for the gravitational-acoustic "branch" of the oscillations described by the dispersion relation (8.14) lies in the region $k \sim h^{-1}$. This instability generates scales several times larger than the thickness of the disc. All three terms on the right-hand side of (8.14) turn in that case out to be of the same order of magnitude and, in particular, we have $2\pi G \sigma_0 \sim \frac{5}{3} k c^2$.

However, we are now interested in a low-frequency instability on the other, dissipative branch of the oscillations, with a growth rate $\gamma \sim \nu k^2 \ll \Omega$ which generates the largest scale perturbations with $k \ll h^{-1}$. In that case $2\pi G \sigma_0 \gg \frac{5}{3} k c^2$ and the first term, $2\pi i G \hat{\sigma}$, on the right-hand side of the second equation of the set (8.32) turns out to dominate. For radial oscillations, with $m = 0$, we find from that equation that $\hat{V}_\varphi = \pi i G \hat{\sigma}/\Omega$, as we can neglect the first term on the left-hand side of that equation when $\gamma \ll \Omega$. We have also used here the second inequality, $k \gg r^{-1}$, which makes it possible for us to neglect terms containing gradients of unperturbed quantities in comparison with terms proportional to gradients of perturbed quantities.

Putting $\gamma \sim \nu k^2$ we can neglect the first terms on the left- and the right-hand sides of the third equation of the set (8.32) when $k \ll h^{-1}$.

Using inequality (8.29) and the conditions stipulated above the last three terms turn out to dominate in the last equation of the set (8.32). We can thus finally rewrite the set (8.32) in the following form:

$$
\begin{aligned}
\gamma \hat{\sigma} &= -i k \sigma_0 \hat{V}_r - \left[\frac{\partial N^-}{\partial \sigma_0} - \frac{\partial N^+}{\partial \sigma_0}\right] \hat{\sigma} - \left[\frac{\partial N^-}{\partial T_0} - \frac{\partial N^+}{\partial T_0}\right] \hat{T}, \\
\hat{V}_\varphi &= -i \frac{\pi G}{\Omega} \hat{\sigma}, \\
\frac{\kappa^2}{2\Omega} \hat{V}_r &= -i k \left[\frac{\partial \nu \sigma_0 (-r \Omega')}{\partial \sigma_0}\right] \frac{\hat{\sigma}}{\sigma_0} - i k \left[\frac{\partial \nu (-r \Omega')}{\partial T_0}\right] \hat{T}, \\
\left[\frac{\partial E^-}{\partial T_0} - \frac{\partial E^+}{\partial T_0}\right] \hat{T} &= - \left[\frac{\partial E^- \sigma_0}{\partial \sigma_0} - \frac{\partial E^+ \sigma_0}{\partial \sigma_0}\right] \frac{\hat{\sigma}}{\sigma_0} - 2 i k \nu (-r \Omega') \hat{V}_\varphi.
\end{aligned}
\tag{8.61}
$$

From (8.61) we get the following dispersion equation (retaining only terms with k^n, $n \leqslant 2$) (Gor'kavyĭ 1986a, Fridman and Gor'kavyĭ 1987, 1988, 1989):

$$
\gamma = -D k^2 + A k + B, \tag{8.62}
$$

where

$$D = \left[\frac{\partial \nu \sigma_0}{\partial \sigma_0} - \frac{\partial \nu}{\partial T_0}\left[\frac{\partial E^- \sigma_0}{\partial \sigma_0} - \frac{\partial E^+ \sigma_0}{\partial \sigma_0}\right]\left[\frac{\partial E^-}{\partial T_0} - \frac{\partial E^+}{\partial T_0}\right]^{-1}\right] \frac{2\Omega(-r\Omega')}{\kappa^2},$$

$$A = \left[\frac{\partial N^-}{\partial T_0} - \frac{\partial N^+}{\partial T_0}\right]\left[\frac{\partial E^-}{\partial T_0} - \frac{\partial E^+}{\partial T_0}\right]^{-1} \nu \left[-\frac{r\Omega'}{\Omega}\right] 2\pi G,$$

$$B = \frac{1}{\sigma_0}\left[\frac{\partial N^-}{\partial T_0} - \frac{\partial N^+}{\partial T_0}\right]\left[\frac{\partial E^- \sigma_0}{\partial \sigma_0} - \frac{\partial E^+ \sigma_0}{\partial \sigma_0}\right]\left[\frac{\partial E^-}{\partial T_0} - \frac{\partial E^+}{\partial T_0}\right]^{-1}$$
$$- \left[\frac{\partial N^-}{\partial \sigma_0} - \frac{\partial N^+}{\partial \sigma_0}\right].$$

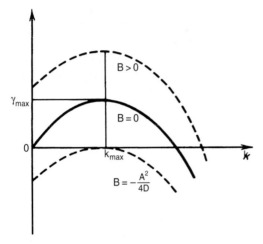

Fig. 8.8. The growth rate γ as function of k for the accretion instability.

If $D > 0$ so that the disc is "diffusion stable", but $A > 0$ and $B > -A^2/4D$, an instability develops which is connected with the external particle fluxes (Fig. 8.8).

Let us study the instability criterion in more detail. From (8.29) we have: $E^+ = \nu(r\Omega')^2$. We substitute here for ν its value from (8.52). If the collisions between the large particles are sufficiently rare so that the gravitational interactions dominate (see Sect. 8.2.6), we have $\Omega t_g \gg 1$. Using the relation $r\Omega' = \frac{3}{2}\Omega$ we then get from the left-hand side of (8.53) $E^+ = 9T_0/16t_g$. Using (8.51) for t_g we find

$$E^+ = \frac{9T_0\psi\Omega\tau}{16x^2} \propto \frac{\sigma_0}{T_0}, \qquad \frac{\partial E^+}{\partial \sigma_0} = \frac{E^+}{\sigma_0}, \qquad \frac{\partial E^+}{\partial T_0} = -\frac{E^+}{T_0}.$$

From the right-hand side of the balance equation (8.53) we have:

$$E^- \simeq \frac{3(1-q^2)T_0}{t_{\text{cont}}} \simeq 9(1-q^2)T_0\Omega\tau(1+x^{-1}).$$

Let us consider two limiting cases.

1) $x \ll 1$, that is, $x^{-1} \gg 1$.

$$E^- \propto \frac{T_0 \sigma_0}{v^2} \propto \sigma_0, \qquad \frac{\partial E^-}{\partial T_0} = 0, \qquad \frac{\partial E^-}{\partial \sigma_0} = \frac{E^-}{\sigma_0},$$

$$\frac{\partial E^- \sigma_0}{\partial \sigma_0} - \frac{\partial E^+ \sigma_0}{\partial \sigma_0} = 0, \quad \Delta E_T \equiv \frac{\partial E^-}{\partial T_0} - \frac{\partial E^+}{\partial T_0} = \frac{E^+}{T_0} = \frac{v(r\Omega')^2}{T_0} > 0.$$

2) $x \gg 1$, that is, $x^{-1} \ll 1$. $\hspace{5cm}$ (8.63)

$$E^- \propto T_0 \sigma_0, \qquad \frac{\partial E^-}{\partial T_0} = \frac{E^-}{T_0}, \qquad \frac{\partial E^-}{\partial \sigma_0} = \frac{E^-}{\sigma_0},$$

$$\frac{\partial E^- \sigma_0}{\partial \sigma_0} - \frac{\partial E^+ \sigma_0}{\partial \sigma_0} = 0, \quad \Delta E_T \equiv \frac{\partial E^-}{\partial T_0} - \frac{\partial E^+}{\partial T_0} = 2\frac{E^+}{T_0} = \frac{2v(r\Omega')^2}{T_0} > 0.$$

We note also that from the relations $E^+ = v(r\Omega')^2$ and $E^+ \propto \sigma_0$ it follows that $v \propto \sigma_0$, that is, $\partial(v\sigma_0)/\partial\sigma_0 = 2v$. Using this relation and (8.63) we find for D, A, and B:

$$D = 6v,$$

$$A = \frac{4}{3\alpha} \frac{\pi G T_0}{\Omega^2} \left[\frac{\partial N^-}{\partial T_0} - \frac{\partial N^+}{\partial T_0} \right], \quad \alpha = \begin{cases} 1 & \text{for } x \ll 1, \\ 2 & \text{for } x \gg 1, \end{cases} \qquad (8.64)$$

$$B = - \left[\frac{\partial N^-}{\partial \sigma_0} - \frac{\partial N^+}{\partial \sigma_0} \right].$$

As the criterion for the development of the accretion instability, far from the boundary of the diffusion instability, we may assume the following inequalities to be satisfied:

$$\frac{\partial N^-}{\partial T_0} > \frac{\partial N^+}{\partial T_0}, \tag{8.65}$$

$$\frac{\partial N^-}{\partial \sigma_0} < \frac{\partial N^+}{\partial \sigma_0}. \tag{8.66}$$

We have in inequality (8.66) neglected the existence of the region $A^2/4D < B < 0$ where the instability is also possible. Let us consider the realisability of inequalities (8.65) and (8.66) starting from the most general considerations.

Let the mechanism by which the mass of the disc is increased be the accretion of "external" particles forming a layer with a volume density $\varrho_{\text{ext}} \approx \sigma_{\text{ext}}/h_{\text{ext}}$, where h_{ext} is the thickness of the layer of such particles. We shall now assume that the number of particles in the disc is fixed: we get then for N^+ the following rate of increase in disc density through accretion (Safronov 1969):

$$N^+ = \frac{d\sigma}{dt} = n_0 \frac{dm_{\text{ext}}}{dt} = n_0 \pi R_{\text{ext}}^2 \varrho_{\text{ext}} V = \tau \varrho_{\text{ext}} V \left[1 + \frac{V_g^2}{V^2} \right], \tag{8.67}$$

where n_0 is the surface density of the particles in the disc, πR_{ext}^2 the cross-section of the disc particles for collisions with particles from the external flux, τ the optical depth of the rings, $\tau = \pi a^2 n_0$, $R_{\text{ext}}^2 = a^2[1 + (V_{\text{g}}^2/V^2)]$, $V_{\text{g}} \equiv 2Gm/a$, and $V \propto \sqrt{T_0}$ the random velocity of the disc particles. For $V_{\text{g}}^2/V^2 \simeq 4 > 1$ (see Sect. 8.2.6) we get

$$\frac{\partial N^+}{\partial \sigma_0} \simeq \frac{N^+}{\sigma_0} > 0, \qquad \frac{\partial N^+}{\partial T_0} \simeq -\frac{N^2}{2T_0} < 0. \tag{8.68}$$

The rate N^- at which the disc density decreases is determined in a much more complicated way. We shall assume that this rate is independent of the parameters of the disc itself or that there is no decrease in the density at all so that $\partial N^-/\partial \sigma_0 \simeq \partial N^-/\partial T_0 \simeq 0$ and inequalities (8.65) and (8.66) are satisfied. Of course, these estimates are simple-minded and the problem about the existence of the accretion instability remains to be further studied. For its solution one must construct a more detailed model of the interaction between a small and a large ring particle. However, it is very important that there is in principle a possibility of a large-scale instability of the rings.

Let us estimate the wavelengths at which the accretion instability may occur. The maximum growth rate is reached for $k_{\text{max}} = A/2D$ which gives

$$k_{\text{max}} \simeq \frac{N^+}{3\Omega^2} \frac{2\pi G}{12\nu}. \tag{8.69}$$

Using the expression for the shear viscosity in the rare-collision limit, $\nu \sim V_{\text{g}}^2 \tau/\Omega$, which follows from (8.52) and (8.50), we get an estimate for the resulting stratification scales:

$$\lambda_{\text{max}} \simeq \frac{\Omega V_{\text{g}}^2}{G\varrho_{\text{ext}} V} \simeq \frac{\Omega V_{\text{g}}}{G\varrho_{\text{ext}}}. \tag{8.70}$$

We have put here $V_{\text{g}} \simeq V$. When the flux of "external" particles diminishes we have $\varrho_{\text{ext}} \to 0$ and it then follows from (8.70) that $\lambda_{\text{max}} \to \infty$ which means that the system is stable. We note that, if by "external" and "independent" we understand dust particles drifting due to the action of aerodynamic or radiation friction forces, the drift velocity will determine indirectly the density of the layer of "external" particles which means that we can assume that only those particles which after the characteristic time for the development of the accretion instability have travelled a distance longer than a single wavelength can be considered to be "external" particles as otherwise they cannot be used to transfer mass.

Let us estimate the density of the layer of "non-diffusive" particles in the proto-stage – that is, of dust particles contained in the gas-dust cloud surrounding the rings whence the accretion onto the rings takes place. The characteristic density of the proto-disc round Jupiter or Saturn is 10^6 g/cm^2 (Safronov 1969); assuming that the fraction of dust suspended in the disc is between 10^{-3} and 10^{-2} and that $h_{ext} \simeq 10^9$ cm, we find $\varrho_{ext} \simeq 10^{-6}$ to 10^{-5} g/cm^3. From (8.70) we find that

$$\lambda \simeq \frac{100}{\varrho_{ext}} \simeq \begin{cases} 10^3 \text{ km} & \text{for } \varrho_{ext} \simeq 10^{-6} \text{ g/cm}^3, \\ 50 \text{ km} & \text{for } \varrho_{ext} \simeq 2 \times 10^{-5} \text{ g/cm}^3. \end{cases} \tag{8.71}$$

The characteristic growth time for such rings is $t \sim \gamma^{-1} \sim (\nu k^2)^{-1} \sim \lambda^2/40\pi^2 \sim 10^6$ and 2.5×10^3 year, respectively. We have used here $\nu \simeq 10$ cm^2/s for $\tau \simeq 0.1$, $c \simeq 0.1$ cm/s. The accretion instability may thus be the main cause for the large-scale (50 to 1000 km) stratification of the Saturnian rings.

8.4 Analysis of Non-axisymmetric Oscillations of a Disc – Ellipse Instability

All the instabilities discussed so far produce a ring structure with circular symmetry. However, all the Uranian rings and some of the narrow Saturnian rings are eccentric. Could there be a mechanism which transforms an initially circular ring into an elliptical one? If there is such a mechanism, it must be a collective one, that is, connected with an instability, since the existence of it means that a circular ring is unstable under a perturbation with an azimuthal $m = 1$ mode. Does one know of such a ring instability?

The first study of the stability of the Saturnian rings by Maxwell (1859) revealed just such an instability. Maxwell assumed that the ring was absolutely rigid and showed that any shift of the ring in its plane leads to an exponential growth with time of the amplitude of the displacement and the subsequent falling of the ring onto the planet. Almost 130 years later this instability was studied by considering a more general system – an elastic, absolutely flexible filament in the field of a central mass (Morozov and Fridman 1986). Let the ring ABCD as the result of a $m = 1$ type perturbation be changed into the ellipse A′BC′D (Fig. 8.9). A particle at A′ which approaches the centre will then be decelerated by particles at B and D which, in turn, will accelerate a particle at C′ which has moved away from the centre. As a result the latter tends to move further away from the centre, while the particle at A′ will move towards the centre. The elastic filament which had been stretched out as the result of the initial perturbation will be stretched out even further – this is the picture of an instability.

Using the dispersion relation found by Morozov and Fridman (1986), Fridman, Morozov, and Polyachenko (1984) found in the limit of an absolutely solid body Maxwell's dispersion relation, the solution of which leads, as we just mentioned, to the falling of a rigid ring onto the planet. Of course, Maxwell concluded by combining this result with the observation that Saturn's ring existed that the ring must have a meteoritic structure. This conclusion, even though it was not new, followed from the first rigorous proof

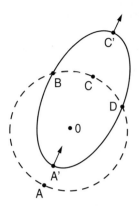

Fig. 8.9. An $m = 1$ perturbation changing the flexible ring ABCD into the ellipse
A'BC'D. The element at A' which is closest to the centre and which therefore has
the largest rotational velocity is decelerated by the other elements of the ring and
continues to "fall" towards the centre. The element at C' which is furthest from the
centre and has the lowest rotational velocity is accelerated by the other elements
of the ring and continues to move away from the centre.

of the instability of a continuous Saturnian ring in the approximation of an
absolutely rigid body. By making that assumption Maxwell used the con-
dition $c/v_{\mathrm{rot}} \to \infty$ where c is the sound velocity in the ring and v_{rot} the
rotational velocity of the ring relative to the planet. However, if one takes
into account realistic values of these parameters: the velocity of sound propagation in ice, $c \simeq 3.3$ km/s, and the rotational velocity of the Saturnian ring,
$v_{\mathrm{rot}} \simeq 16$ to 20 km/s, the ratio c^2/v_{rot}^2 which occurs in Fridman, Morozov,
and Polyachenko's theory turns out to be a small, rather than a large para-
meter. Under those conditions the dominating instability of a hypothetical
continuous Saturnian ice ring is not the instability with the $m = 1$ mode, but
a small-scale instability which leads not to the ring falling onto the planet,
but to short-scale bending of the ring, breaking it down into small lumps
(Fridman, Morozov, and Polyachenko 1984). The result of this instability is
the same meteoritic structure of the ring as was mentioned by Maxwell who
used for the study of the stability of a continuous ice ring rotating around a
planet a model which was wrong in principle.

The cosmogony of rings excludes the existence of a primary continuous
ring. There arises thus the problem of the presence of a similar instability of
a ring consisting of a flux of particles. Let us attempt to solve that problem
analytically.

Let us consider non-axisymmetric modes, $m \neq 0$. If we neglect in (8.34)
all dissipative processes and also the pressure and the self-gravitation, the
roots of the resulting equation will depend on m as follows (Table 8.2). The
only zero root for $m > 0$ occurs for $m = 1$. It is natural to expect that
when the dissipative processes are "switched on" it is just there that a low-

Table 8.2. Roots of the dissipationless dispersion equation for different values of m.

Solution	$m = 0$	$m = 1$	$m = 2$	$m = 3$
ω_1	$-\Omega$	0	Ω	2Ω
ω_2	0	Ω	2Ω	3Ω
ω_3	0	Ω	2Ω	3Ω
ω_4	Ω	2Ω	3Ω	4Ω

frequency branch of oscillations which may become unstable appears. Putting $\omega \ll \Omega$, $kh \ll 1$ we find for $m = 1$ (Fridman and Gor'kavyĭ 1987, 1988, 1989, Gor'kavyĭ and Fridman 1990)

$$\omega = -\frac{1}{2\Omega}\left(sk^2c^2 - 2\pi G\sigma_0 k - 2\Omega\omega_{\mathrm{p}}\right)$$
$$+ \mathrm{i}\frac{\Delta E_T}{3\Omega^2}\left(s'k^2c^2 - 2\pi G\sigma_0 k - 2\Omega\omega_{\mathrm{p}}\right), \tag{8.72}$$

where

$$s = \frac{5}{3} + \frac{14}{9}\frac{\nu\Delta E_T}{c^2} + \frac{2}{3}\frac{\alpha\mu}{c^2} - \frac{4}{3}\frac{\sigma_0}{\Omega c^2}\left(\beta\Delta E_T - \alpha\Delta E_\sigma\right),$$
$$s' = 1 + \frac{\Omega(2\alpha c^2 + 3\beta\sigma_0) - \sigma_0\Delta E_\sigma - 5\Omega^2\nu}{\Delta E_T c^2},$$

and ω_{p} is the rate of precession due to the asphericity of the gravitational field of the planet (see Sect. 3.3):

$$\omega_{\mathrm{p}} = \frac{\Omega^2 - \kappa^2}{2\Omega} \approx \Omega - \kappa = \frac{3}{2}J_2\left[\frac{R_{\mathrm{p}}}{r}\right]^2\Omega. \tag{8.73}$$

If the disc is energetically stable, $\Delta E_T > 0$ – for instance, in the case of rare collisions (see (8.63)) – the condition for the ellipse instability can be written in the form

$$s'k^2c^2 - 2\pi G\sigma_0 k - 2\Omega\omega_{\mathrm{p}} > 0. \tag{8.74}$$

In the case of a small precession the unstable wavelengths are $\lambda \lesssim c^2/G\sigma_0 \approx 1$ km (for $c \approx 0.1$ cm/s and $\sigma_0 \approx 1$ g/cm^2) with a growth rate $\gamma \simeq \nu k^2$. The characteristic growth time is $\approx \gamma^{-1} \lesssim 0.1$ year. One may assume that it is just this instability of non-axisymmetric perturbations which is responsible for the appearance of a non-zero eccentricity of the Uranian and some of the Saturnian rings. Borderies et al. (1985) proposed the idea that the eccentricity of narrow rings may arise in a similar fashion. The modes with $m > 1$ differ qualitatively from the one considered by the presence of a real part of the frequency with $\omega \propto \Omega$. This, however, does not exclude the possibility of a slow dissipative pumping of fast oscillations – for instance, of spiral waves. We note that the ellipse instability arising in a symmetric disc is an example of spontaneous symmetry breaking.

As we now know the characteristic scales of the perturbations and the magnitude of the growth rates γ of the dissipative instabilities described above we can show the validity of the conditions for the hydrodynamic approximation: $\gamma \ll \omega_{\mathrm{c}}$, $l \simeq V/\omega_{\mathrm{c}}$, where ω_{c} is the particle collision frequency, V a characteristic particle velocity (in the rotating frame), and l the particle mean free path. Bearing in mind that $\gamma \simeq \nu k^2$ and the frequency $\omega_{\mathrm{c}} \simeq \Omega$ we find $\nu k^2 \sim k^2 c^2/\omega_{\mathrm{c}} \sim k^2 c^2/\Omega \ll \Omega$, or $(kh)^2 \ll 1$.

A similar condition $(kh \ll 1)$ follows from the inequality $l \ll \lambda$, since $V \simeq c$. The condition for the applicability of the hydrodynamic approximation is thus valid for wavelengths longer than the thickness of the disc, that is, $kh \ll 1$. We find a similar condition in the case of rare collisions when l cannot be larger than the size of an epicycle, which is of the order of h. This condition is satisfied for all above mentioned instabilities – only for the diffusion instability have we used the condition $kh \simeq 1$ to get estimates.

9 Resonance Effects in Planetary Rings
I Spiral Waves

> Prospero: "Ye elves of hills, brooks, standing lakes, and groves; And ye, that on the sands with printless foot do chase the ebbing Neptune, and do fly him when he comes back; you demi-puppets that by moonshine do the green-sour ringlets make ... ... by whose aid – weak masters though ye be – I have bedimm'd the noontide sun, call'd forth the mutinous winds, and 'twixt the green sea and the azured vault set soaring war: to the dread rattling thunder ..."
>
> *W. Shakespeare, The Tempest, Act V, Scene I.*

9.1 Density Waves

We have already met the formation of a spiral wave in Sect. 6.1 where we calculated the trajectories of particles near a separate satellite. When the distance of the particle from the satellite increases the wavelength of the excited wave increases and its amplitude decreases – as can be seen in Fig. 6.2. Nonetheless there are special, resonance regions in the rings where the effect of the satellite is again very strong. The physics of this effect is well known: an oscillator – such as a molecule, a pendulum, or a swing – is excited if an eigenfrequency of its oscillations is the same as the frequency of an external driving force or stands in simple ratio, such as 1:2, 2:3, 3:4, ..., to that frequency. A particle in the ring, moving along a circle, can undoubtedly be considered as an oscillator with eigenfrequency Ω. A satellite with its gravitational field is a driving force with frequency Ω_s. If, for instance, $\Omega = 2\Omega_s$ for each two orbits of the particle the satellite has completed a single one and if the particle approaches the satellite at the apocentre of its orbit all subsequent approaches will take place at the same point: the perturbations add up and the particle trajectory changes significantly. However, if this resonance is slightly disturbed the encounters between the particle and the satellite will take place at different points of the particle orbit and the effect of the satellite will be negligibly small.

The simpler the resonance condition $\Omega_s/\Omega = n/m$ – or, the lower the order of the resonance – the more often the particle and the satellite approach one another. The wavelength – in the azimuthal direction – excited by the satellite becomes in the case of a simple resonance comparable with the length of the orbit. The number of waves which can be fitted along the perimeter is equal to m, that is, in the case of the $\Omega_s/\Omega = \frac{1}{2}$ resonance just two wavelengths can occupy the whole length of the orbit. If the resonance occurs at the edge of a ring this edge will start to oscillate in the m mode. For instance, the outer edge of the B ring oscillates in the $m = 2$ mode – due to the 2:1 Mimas resonance and the outer edge of the A ring oscillates in the $m = 7$ mode – due to the 7:6 Janus resonance. If the resonance orbit is positioned in a continuous disc it causes a spiral wave with the same azimuthal number. Let us study this process in more detail following the simple exposition of this problem given by Meyer-Vernet and Sicardy (1987) and by Rosen (1989).

9.1.1 Frequency Multiplication in an Aspherical Field

Let us write down the set of equations in an inertial frame of reference in cylindrical coordinates:

$$\ddot{r} - r\dot{\theta}^2 = -\frac{\partial \Phi_p}{\partial r}, \tag{9.1}$$

$$\frac{d}{dt}\left(r^2\dot{\theta}\right) = 0, \tag{9.2}$$

$$\ddot{z} = -\frac{\partial \Phi_p}{\partial z}. \tag{9.3}$$

The solution of (9.2) is the conservation law for the z-component of the angular momentum:

$$r^2\dot{\theta} = l_z = \text{const.} \tag{9.4}$$

We rewrite (9.1) in the form

$$\ddot{r} - \frac{l_z^2}{r^3} = -\frac{\partial \Phi_p}{\partial r}. \tag{9.5}$$

If we define an effective potential

$$\Phi_{\text{eff}} = \Phi_p + \frac{l_z^2}{2r^2}, \tag{9.6}$$

we can write the equations of motion in the simple form

$$\ddot{r} = -\frac{\partial \Phi_{\text{eff}}}{\partial r}, \qquad \ddot{z} = -\frac{\partial \Phi_{\text{eff}}}{\partial z}. \tag{9.7}$$

The potential has a minimum when the conditions

$$\frac{\partial \Phi_{\text{eff}}}{\partial z} = 0, \qquad \frac{\partial \Phi_{\text{eff}}}{\partial r} = 0 \tag{9.8}$$

are simultaneously satisfied.

The first equation is satisfied in the $z = 0$ plane. The second is satisfied in the case of a circular orbit ($\ddot{r} = 0$); we then get from (9.1) the following relation:

$$\frac{\partial \Phi_{\rm p}}{\partial r}\bigg|_{r=r_0} = \frac{l_z^2}{r_0^3} = r_0 \dot{\theta}^2. \tag{9.9}$$

We define $\delta r = r - r_0$, where r_0 is the radius of the circular orbit, and expand the effective potential in a Taylor series in the point $r = r_0$, $z = 0$:

$$\Phi_{\rm eff}(r,z) = \Phi(r_0,0) + \underbrace{\frac{\partial \Phi_{\rm eff}}{\partial r}\bigg|_{(r_0,0)}}_{=0} \cdot \delta r + \underbrace{\frac{\partial \Phi_{\rm eff}}{\partial z}\bigg|_{(r_0,0)}}_{=0} \cdot z$$

$$+ \frac{1}{2}\frac{\partial^2 \Phi_{\rm eff}}{\partial r^2}\bigg|_{(r_0,0)} \cdot \delta r^2 + \underbrace{\frac{\partial^2 \Phi_{\rm eff}}{\partial z \partial r}\bigg|_{(r_0,0)}}_{=0} \cdot z\delta r$$

$$+ \frac{1}{2}\frac{\partial^2 \Phi_{\rm eff}}{\partial z^2}\bigg|_{(r_0,0)} \cdot z^2 + \cdots. \tag{9.10}$$

The first derivatives vanish as we expand the potential in the point $r_0, 0$ which is a minimum of the potential. The mixed second derivative vanishes as the effective potential is symmetric relative to $z = 0$. We neglect the terms with the higher derivatives; this is the so-called epicyclic approximation. Substituting (9.10) into (9.7) we get the equations of motion in the epicyclic approximation:

$$\ddot{\delta r} = -\kappa_{\rm c}^2 \, \delta r, \tag{9.11}$$

$$\ddot{z} = -\mu_{\rm c}^2 \, z, \tag{9.12}$$

where $\kappa_{\rm c}$ is the epicyclic frequency and $\mu_{\rm c}$ the frequency of the vertical oscillations of the particles:

$$\kappa_{\rm c}^2 = \kappa^2(r_0) = \frac{\partial^2 \Phi_{\rm eff}}{\partial r^2}\bigg|_{(r_0,0)} , \quad \mu_{\rm c}^2 = \mu^2(r_0) = \frac{\partial^2 \Phi_{\rm eff}}{\partial z^2}\bigg|_{(r_0,0)} . \tag{9.13}$$

The circular frequency is given by (9.9):

$$\Omega_{\rm c}^2 = \Omega^2(r_0) = \frac{1}{r}\frac{\partial \Phi_{\rm p}}{\partial r}\bigg|_{(r_0,0)} = \frac{l_z^2}{r_0^4}. \tag{9.14}$$

Using the potential of an aspherical body we get the following expression for the squares of these three frequencies (see (3.92)):

$$\left\{ \begin{array}{c} \Omega^2(r) \\ \kappa^2(r) \\ \mu^2(r) \end{array} \right\} = \frac{GM_{\rm s}}{r^3}\left[1 + \left\{ \begin{array}{c} A_2 \\ B_2 \\ C_2 \end{array} \right\} J_2 \left(\frac{R_{\rm s}}{r}\right)^2\right], \tag{9.15}$$

with $A_2 = \frac{3}{2}$, $B_2 = -\frac{3}{2}$, $C_2 = \frac{9}{2}$. The presence of aspherical harmonics thus leads to having instead of a single characteristic frequency three slightly different ones: $\kappa < \Omega < \mu$. We can call this a multiplication of frequencies. If we bear in mind that earlier we spoke about resonances there arises the obvious question whether in the aspherical field of a planet there will be different types of resonance in accordance with the existence of different frequencies.

9.1.2 Resonance Interaction of a Satellite with Ring Particles (Two-Dimensional Case)

We write down a simple set of equations

$$
\left[\frac{\partial}{\partial t} + (\boldsymbol{V} \cdot \nabla)\right] \boldsymbol{V} = -\nabla \left(\Phi_{\mathrm{p}} + \Phi_{\mathrm{s}}\right),
\tag{9.16}
$$

$$
\frac{\partial \sigma}{\partial t} + \mathrm{div}(\sigma \boldsymbol{V}) = 0,
\tag{9.17}
$$

where Φ_{p} and Φ_{s} are the potentials of the planet and the satellite:

$$
\Phi_{\mathrm{p}} = -\frac{GM_{\mathrm{p}}}{r}, \quad \Phi_{\mathrm{s}} = -\frac{GM_{\mathrm{s}}}{|\boldsymbol{r} - \boldsymbol{r}_{\mathrm{s}}|} = -\frac{GM_{\mathrm{s}}}{\sqrt{r^2 + r_{\mathrm{s}}^2 - 2rr_{\mathrm{s}}\cos(\varphi - \Omega_{\mathrm{s}}t)}}.
\tag{9.18}
$$

Let us consider a perturbation $\boldsymbol{V} = \boldsymbol{V}_0 + \boldsymbol{V}_1 + \cdots$, $|V_1| \ll |V_0|$; $\sigma = \sigma_0 + \sigma_1 + \cdots$, $|\sigma_1| \ll |\sigma_0|$. We linearise (9.16) and (9.17). Dropping the index 1 of the perturbations we get

$$
\left[\frac{\partial}{\partial t} + \Omega\frac{\partial}{\partial \varphi}\right] V_r - 2\Omega V_\varphi = -\frac{\partial \Phi_{\mathrm{s}}}{\partial r},
\tag{9.19}
$$

$$
\left[\frac{\partial}{\partial t} + \Omega\frac{\partial}{\partial \varphi}\right] V_\varphi + \tfrac{1}{2}\Omega V_r = -\frac{1}{r}\frac{\partial \Phi_{\mathrm{s}}}{\partial \varphi},
\tag{9.20}
$$

$$
\left[\frac{\partial}{\partial t} + \Omega\frac{\partial}{\partial \varphi}\right] \sigma = -\frac{\sigma_0}{r}\left[\frac{\partial}{\partial r}(rV_r) + \frac{\partial V_\varphi}{\partial \varphi}\right].
\tag{9.21}
$$

We expand the satellite potential in a Fourier series:

$$
\Phi_{\mathrm{s}}(r, \varphi, t) = \sum_{m=-\infty}^{\infty} \Phi_{sm}(r)\, \mathrm{e}^{\mathrm{i}m(\varphi - \Omega_{\mathrm{s}}t)}.
\tag{9.22}
$$

By virtue of the linearity of the set (9.19) to (9.21) we can choose for our investigation one of the harmonics, $\simeq \Phi_{sm}(r)\, \mathrm{e}^{\mathrm{i}m(\varphi - \Omega_{\mathrm{s}}t)}$. We must look for the perturbed functions V_r, V_φ, and σ in the same form. As a result we find that (Meyer-Vernet and Sicardy 1987)

$$
V_{rm}(r) = -\frac{\mathrm{i}m}{rD}\left[(\Omega - \Omega_{\mathrm{s}})r\frac{d}{dr} + 2\Omega\right]\Phi_{sm},
\tag{9.23}
$$

$$
V_{\varphi m}(r) = \frac{1}{2rD}\left[\Omega r\frac{d}{dr} + 2m^2(\Omega - \Omega_{\mathrm{s}})\right]\Phi_{sm},
\tag{9.24}
$$

$$\sigma_m(r) = -\frac{\sigma_0}{imr(\Omega - \Omega_s)}\left[\frac{\partial}{\partial r}(rV_{rm}) + imV_{\varphi m}\right]. \qquad (9.25)$$

These expressions describe the spiral wave caused by a resonance with a satellite.

Putting the denominators of these expressions equal to zero, we find $\Omega - \Omega_s = 0$, $D(r) \equiv \Omega^2(r) - m^2[\Omega(r) - \Omega_s]^2 = 0$ which gives us three kinds of resonance between the disc and the satellite:

1) $\Omega(r_{\text{cor}}) = \Omega_s$ – the corotational resonance;
2) $\Omega(r_{\text{in}}) = \frac{m}{m-1}\Omega_s$ – inner Lindblad resonance;
3) $\Omega(r_{\text{out}}) = \frac{m}{m+1}\Omega_s$ – outer Lindblad resonance.

Let us consider the angular momentum L transferred from the satellite to the whole of the disc. We write $L = L^{(0)} + L^{(1)} + \cdots$ where

$$L^{(0)} = -\int_{r_1}^{r_2}\sigma_0(r)r\,dr\int_0^{2\pi}\frac{\partial\Phi_s}{\partial\varphi'}\,d\varphi'$$

$$= -\int_{r_1}^{r_2}[\Phi_s(2\pi) - \Phi_s(0)]\,\sigma_0(r)r\,dr. \qquad (9.26)$$

We now write down the perturbed angular momentum transferred from the satellite to the whole of the disc:

$$L^{(1)} = -\int_{r_1}^{r_2}r\,dr\int_0^{2\pi}\sigma_1(r,\varphi')\frac{\partial\Phi_s(r,\varphi')}{\partial\varphi'}\,d\varphi'. \qquad (9.27)$$

For the Fourier component $L_m^{(1)}$ we get in the vicinity of the inner Lindblad resonance $(r = r_m)$

$$L_m^{(1)} = \frac{2\pi m\sigma_0 A^2}{r_m D_*(r_m)}\,\text{Im}\int_{-\varepsilon}^{\varepsilon}\frac{dx}{x}, \qquad (9.28)$$

where

$$A(r_m) \equiv -\frac{GM_s}{2r_s}\left[2mb + \beta\frac{db}{d\beta}\right]_{\beta=r_m/r_s}, \qquad x = \frac{r - r_m}{r_m},$$

$$b \equiv \frac{2}{\pi}\int_0^{\pi}\frac{\cos m\varphi'\,d\varphi'}{\sqrt{1+\beta^2 - 2\beta\cos\varphi'}}, \qquad D_* \equiv \frac{\partial D}{\partial r}\bigg|_{r=r_m}.$$

We redefine x as follows:

$$x = \lim_{\alpha\to 0}(x + i\alpha).$$

We then have (Fig. 9.1)

$$\lim_{\alpha\to 0}\text{Im}\int_{-\varepsilon}^{\varepsilon}\frac{dx}{x + i\alpha} = -\lim\alpha\int_{-\varepsilon}^{\varepsilon}\frac{dx}{x^2 + \alpha^2} = -\pi\,\text{sgn}\alpha, \qquad (9.29)$$

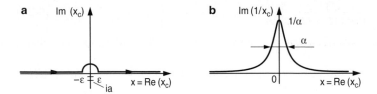

Fig. 9.1. Integration contour (**a**) and $1/\alpha$ as function of x in the resonance region (**b**).

and hence, finally,

$$L_m^{(1)} = -\frac{4\pi^2\sigma_0 A^2(r_m)}{3\Omega(r_m)\Omega_{\rm s}}\,\mathrm{sgn}\alpha. \tag{9.30}$$

From physical considerations it follows (Meyer-Vernet and Sicardy 1987) that $\alpha > 0$. Hence, $L_m^{(1)} < 0$: in the case of the inner Lindblad resonance the satellite removes angular momentum from the disc and in the case of the outer resonance it gives it to the disc. This takes place in the region of the resonance itself where spiral waves are excited, as we shall show in the next subsection. If there is viscosity, a non-linear spiral wave causes a radial flow of mass which leads to the formation of extended gaps, such as the Cassini Division, and of narrow ringlets (see Appendix V). If there is no viscosity, the spiral wave produces three-dimensional convection flows in the form of four giant vortices (see Appendix V).

9.1.3 Spiral Waves Taking into Account the Self-gravitation and Pressure of the Disc (Two-Dimensional Case)

The characteristics of a spiral wave depend not only on the satellite, but also on the disc parameters. We shall follow Meyer-Vernet and Sicardy (1987) and illustrate this dependence using the following set of equations:

$$\left[\frac{\partial}{\partial t} + (\boldsymbol{V}\cdot\nabla)\right]\boldsymbol{V} = -\nabla\left(\Phi_{\rm p} + \Phi_{\rm s} + \Phi_{\rm d}\right) - \frac{1}{\sigma}\nabla p, \tag{9.31}$$

$$\frac{\partial\sigma}{\partial t} + \mathrm{div}(\sigma\boldsymbol{V}) = 0, \qquad \frac{dp}{d\sigma} = c^2, \tag{9.32}$$

$$\left[\frac{\partial^2}{\partial z^2} + \frac{\partial^2}{\partial r^2} + \frac{1}{r^2}\frac{\partial^2}{\partial\varphi^2}\right]\tilde{\Phi}_{\rm d} = 4\pi G\delta(z)\sigma, \qquad \tilde{\Phi}_{\rm d} = {\rm e}^{-k|z|}\Phi_{\rm d}. \tag{9.33}$$

Linearising this set of equations we reduce it to a single equation for the Fourier component V_{rm}; near the inner Lindblad resonance it has the form

$$-\alpha_\omega^3\frac{d^2V_{rm}}{dx^2} + \alpha_G^2\frac{dV_{rm}}{dx} - {\rm i}xV_{rm} = c_m, \tag{9.34}$$

where

$$\alpha_\omega^3 \equiv i\alpha_p^3 + \alpha_\nu^3, \qquad c_m = -\frac{\Omega(r_m)A}{r_m^2 D_*}, \qquad \alpha_\nu^3 \equiv \frac{7\nu}{9mr_m^2 \Omega_s},$$

$$\alpha_p^3 \equiv -\frac{c^2}{3mr_m^2 \Omega(r_m)\Omega_s}, \qquad \alpha_G^3 \equiv \frac{2\pi\sigma_0 G}{3mr_m \Omega_s \Omega(r_m)}. \qquad (9.35)$$

Let us consider some exact solutions of (9.34).

1) Let $|\alpha_G| \ll |\alpha_p|$ so that

$$V_{rm} = c_m \int_0^\infty e^{i(kx - \alpha_p^3 k^3/3)} \, dk.$$

From the definition (9.35) of α_p^3 we see that $k < 0$ so that this acoustic wave propagates from the inner Lindblad resonance point to the planet. We can represent V_{rm} as a combination of Airy type functions (Fig. 9.2a):

$$V_{rm} = \frac{\pi c_m}{|\alpha_p|} \left[\text{Ai}\left(\frac{x}{|\alpha_p|}\right) + i\,\text{Gi}\left(\frac{x}{|\alpha_p|}\right) \right]. \qquad (9.36)$$

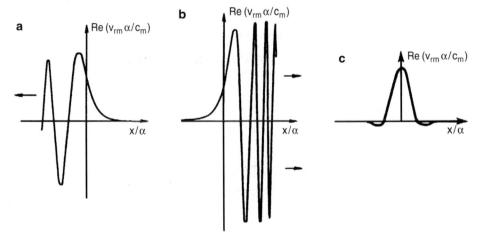

Fig. 9.2. Plots of exact solutions of spiral waves in the limiting cases when the pressure (**a**), the viscosity (**b**), or the self-gravitation (**c**) of the disc dominates.

2) If $|\alpha_G|, |\alpha_p| \ll |\alpha_\nu|$ we have (Fig. 9.2c)

$$V_{rm} = c_m \int_0^\infty e^{i(kx - \alpha_\nu^3 k^3/3)} \, dk = \frac{\pi G}{\alpha_\nu} \text{Hi}\left(\frac{ix}{\alpha_\nu}\right). \qquad (9.37)$$

3) If $|\alpha_p| \ll |\alpha_G|$ we find (Fig.9.2b)

$$
\begin{aligned}
V_{rm} &= c_m \int_0^\infty e^{i(kx - \alpha_G^2 k^2/2)} \, dk \\
&= \frac{\sqrt{\pi} c_m}{|\alpha_G|} \left[g\left(\frac{-x}{\sqrt{\pi}|\alpha_G|} \right) + i f\left(\frac{-x}{\sqrt{\pi}|\alpha_G|} \right) \right],
\end{aligned}
\tag{9.38}
$$

where g and f are Fresnel integrals. The acoustic wave therefore propagates from the resonance point towards the planet and the gravitational wave propagates away from the planet. In the Saturnian rings Voyager observations have identified spiral density waves propagating only "outwards" from the inner resonance point. We note that these spiral waves are caused by resonances between the satellite frequencies and the radial oscillations of the disc particles. Viscous processes damp these waves – that effect is used to estimate the viscosity and the density in the disc. In spite of recent progress in analytical studies of the non-linear stage of the spiral waves, taking various factors into account (see Appendix V), great use is made of numerical methods to study this problem (Shu, Yuan, and Lissauer 1985, Shu, Dones, and Lissauer 1985).

9.2 Bending Waves

Let us follow Rosen (1989) and consider small displacements of disc particles from the $z = 0$ plane. We write the equation for vertical displacements of the disc, $z(r, \theta, t)$, in the form

$$
\left[\frac{\partial}{\partial t} + \Omega(r) \frac{\partial}{\partial \theta} \right]^2 z(r, \theta, t) = g,
\tag{9.39}
$$

where g is the local acceleration in the vertical direction connected with the gravitation of the planet (g_p), the self-gravitation of the disc (g_d), and perturbations from a satellite (g_s). For small displacements of the disc we can write for the acceleration due to the planet (compare (9.12))

$$
g_p = -\mu^2 z.
\tag{9.40}
$$

Let us consider the Fourier components of all the quantities which are of interest to us:

$$
z(r, \theta, t) = \mathrm{Re}\left\{ h(r) e^{i(\omega t - m\theta)} \right\},
\tag{9.41}
$$

$$
g_s(r, \theta, t) = \mathrm{Re}\left\{ f_s(r) e^{i(\omega t - m\theta)} \right\},
\tag{9.42}
$$

$$
g_d(r, \theta, t) = \mathrm{Re}\left\{ f_d(r) e^{i(\omega t - m\theta)} \right\}.
\tag{9.43}
$$

If we neglect the self-gravitation of the disc we get the following equation:

$$
h(r) = \frac{f_s(r)}{\mu^2 - (\omega - m\Omega)^2}.
\tag{9.44}
$$

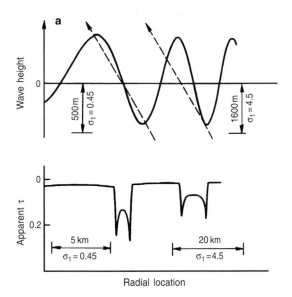

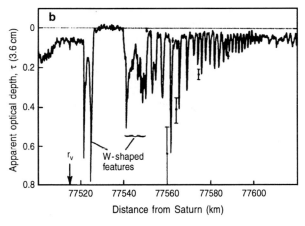

Fig. 9.3. Connection between optical depth and geometry in a bending wave (Rosen 1989). (**a**) the model. Oblique incidence of a radio signal (dashed lines) on the warped disc in the region where a bending wave propagates. We show $Z \to \tau$ mapping for the first few cycles of a model wave, propagating with amplitude $|A_W|$ = 500 m in a region with surface mass density $\sigma_1 = 0.45$ g/cm^2. This value of σ_1 was chosen specially to bring the model and the observed morphologies of the apparent τ near the wave onset into the best agreement. We obtain the same morphology of apparent τ and the proper radial scale, if we take σ_1 and the wave amplitude $|A_W|$ to be 4.5 g/cm^2 and 1600 m, respectively, which corresponds to the relation $|A_W| \propto \sigma_1^{-1/2}$ (Rosen 1989). (**b**) observed optical depths in a bending wave due to the 1:0 Titan resonance.

Hence we get the condition for resonance between the orbital frequency of the satellite and the frequency of the vertical oscillations of the disc particles:

$$D = \mu^2 - (\omega - m\Omega)^2 = 0. \tag{9.45}$$

For the acceleration due to the self-gravitation of the disc we find (Rosen 1989)

$$f_{\mathrm{d}} = 2\pi \mathrm{i} G\sigma s_k \frac{dh}{dr}. \tag{9.46}$$

Hence we find the equation for the vertical motions:

$$\left[-2\pi \mathrm{i} G\sigma s_k \frac{d}{dr} + D \right] h = f(r). \tag{9.47}$$

A bending density wave propagates from the inner Lindblad resonance point in the direction of the planet – in contrast to the nodal resonance. Such a wave is produced by a satellite with an orbit which is considerably inclined with respect to the plane of the rings. Voyager-2 photographed bending waves with a displacement amplitude up to several hundred meters. A low Sun illuminates the crests of such a fantastic wave and the troughs can be hidden in the darkness. Due to the aspherical harmonics of Saturn's gravitational field the Lindblad resonance point and the point of the resonance with the vertical oscillations can differ by several tens, even hundreds of kilometers – the "vertical" resonance point lies closer to the planet than the Lindblad resonance point. We show in Fig. 9.3 the optical depth profile of one of the bending waves and its connection with the geometry of the wave.

10 Resonance Effects in Planetary Rings
II Narrow Ringlets and Satellites

"... You are always badgering me about my ring; but you have never bothered me about the other things that I got on my journey."

"No, but I had to badger you" said Gandalf. "I wanted the truth. It was important. Magic rings are – well, magical; and they are rare and curious. I was professionally interested in your ring, you may say; and I still am."

J.R.R. Tolkien, The Lord of the Rings.

10.1 Hypotheses About the Origin of the Uranian Rings

10.1.1 The Remarkable Properties of the Uranian Rings

The discovery on 1977 March 10 of the Uranian rings caused enormous interest amongst researchers. Indeed, the origin and stability of narrow elliptical rings turned out to be very problematic. First of all, a narrow ring must spread out because of the exchange of angular momentum between colliding particles – after a number of centuries, with a characteristic "viscous" time $t_\nu \sim (\nu k^2)^{-1}$ – increasing its width and diminishing the steepness of its edges. Secondly, the asphericity of the Uranian field causes a precession of elliptical orbits at a rate which depends on the major semi-axis. The difference in the precession of particles at the outer and the inner edges of the ring – differential precession – must after a few hundred years separate the lines of apsides of the particles, destroy an elliptical narrow ring and form a circular, wider ring. Nonetheless, the rings have not spread out and have clearly defined boundaries, and precess as a whole.

The basic problems of the origin and the dynamics of the Uranian rings are (Fridman and Gor'kavyĭ 1989):

1) How were the Uranian rings formed? What collected the circumplanetary matter into narrow rings, well away from one another?
2) How did the eccentricity of the rings arise?
3) Why are the rings not destroyed?

Voyager-2 added yet another problem:

4) Why did the rings not fall on the planet due to the strong aerodynamic friction by the atmosphere?

The enigmas of the Uranian rings produced a multitude of the most divergent hypotheses.

10.1.2 Hypotheses About the Connection Between the Rings and the Five Known Uranian Satellites

Elliot et al. (1977) when discussing the discovery of the Uranian rings expressed the idea that the distances between the rings can be explained by resonance with the five known Uranian satellites. Dermott and Gold (1977) associated the positions of the five rings observed in 1977 and designated α, β, γ, δ, and ε with three-frequency resonances with Ariel and Titania and with Ariel and Oberon – when the orbital frequency of the ring Ω_1 satisfies the relation $q\Omega_1 - (q + p)\Omega_2 + p\Omega_3$, where Ω_2 and Ω_3 are the orbital frequencies of the two satellites and where p and q are integers; this kind of resonance has been observed between the three Jovian satellites Io, Europa, and Ganymede with $q = 1$, $p = 2$.

It was assumed that the particles "stick" to the resonance orbits when moving towards the planet – Gold (1975) has considered this capturing model. Later Aksnes (1977) and Goldreich and Nicholson (1977) showed that in the ring zone three-frequency resonances with Miranda rather than Ariel are more significant;[1] the strongest three-frequency resonances can here control the particle motion in a very narrow zone – a few tens of meters wide – which is much smaller than the width of the narrowest rings. This complicated the position of the "resonance" hypotheses but even the critics themselves did not abandon the idea of a resonance nature of the Uranian rings: Aksnes (1977) expressed the opinion that only some well defined kinds of resonances would capture matter and also mentioned Colombo's observation of the approximate resonance relations between the rings themselves. Goldreich and Tremaine (1978b) expressed the hypothesis that the Uranian rings are strongly non-linear waves which were excited through resonances in an optically thin disc. Steigman (1978) modified the Dermott-Gold hypothesis connecting the position of the rings with three-frequency resonances of Miranda-Ariel and Miranda with undiscovered satellites in an orbit with a radius of 105 221 km. However, in 1978 another four Uranian rings (η, 4, 5, and 6) were detected and it became difficult to associate all nine rings with three-frequency resonances with the outer satellites. Together with the critical remarks from Aksness (1977) and Goldreich and Nicholson (1977) this seriously undermined the position of the Dermott-Gold model and of other resonance hypotheses.

---

[1] Goldreich and Nicholson's paper is called "Revenge of the tiny Miranda".

10.1.3 Hypotheses About Unknown Satellites in the Rings and "Shepherd" Satellites

In 1979 a number of hypotheses were published suggesting the existence inside the ring zone of several undiscovered sateliites. Van Flandern (1979) and Dermott at al. (1979) assumed each ring corresponded to a satellite and that in each ring there were either (Van Flandern 1979) constantly renewed gaseous "trails" of invisible satellites or (Dermott et al. 1979) swarms of particles on complicated banana orbits near the satellite. Goldreich and Tremaine (1979a) assumed that each ring was positioned between two "shepherd" satellites and if not giving out ring particles would disperse. The influence of the shepherds might also produce eccentric rings (Goldreich and Tremaine 1981). The stability of the rings against a differential precession could be adequately explained by the self-gravitation forces (Goldreich and Tremaine 1979b). At the same time the sharp edges of the rings were connected with the action of shepherds (Borderies et al. 1982, 1989). In November 1980 Voyager-1 detected near the narrow elliptical Saturnian ring two shepherds – Pandora and Prometheus – after which the idea of shepherd satellites became wide-spread.

10.1.4 Hypothesis About the Resonance Nature of the Uranian Rings and the Existence of a Series of Undiscovered Satellites Beyond the Boundary of the Rings

As a rule, the formation of satellites in the rings is prohibited due to the strong collisional destruction of particles (Gor'kavyĭ and Fridman 1985a, Gor'kavyĭ and Taĭdakova 1989, Fridman and Gor'kavyĭ 1992; see Chap. 5). The "alternate" coexistence of rings and satellites is characteristic only for sufficiently narrow zones between ring and satellite regions. Models which assume the existence in each ring zone of from 9 to 18 satellites which are closely "mixed" with the nine rings clearly contradict this idea of the formation of rings.

At the start of the eighties several narrow, sometimes very eccentric, rings were discovered which were connected not with shepherd satellites but with resonances with outer satellites (see Figs. 2.3, 2.8, and 2.9). This brought into doubt that the shepherd model is necessary also for the Uranian rings.

Rings and satellites are formed through the condensation of the same proto-satellite disc. The matter in the proto-disc is distributed continuously. From this point of view the huge ($\approx 80\ 000$ km) empty space between the rings and Miranda poses a problem. Recently series of small satellites have been discovered near the outer boundary of the Saturnian and Jovian rings. It was natural to assume that beyond the outer boundary of the Uranian rings there also exist unknown satellites. Has, perhaps, the resonance interaction of these satellites produced the remarkable system of narrow elliptical Uranian rings?

The discovery (Gor'kavyĭ and Fridman 1985b,c) of a remarkable regularity in the positions of the Uranian rings gave the deciding argument used to advance this hypothesis. It was found that near the outer edge of the rings there exist several orbits each of which is at once in resonance with a pair of rings.

10.1.5 Calculation of the Orbital Radii of Hypothetical Satellites

Gor'kavyĭ and Fridman (1985b,c) advanced the hypothesis that the positions of the Uranian rings correspond to lower-order (1:2, 2:3, 3:4) Lindblad resonances with a series of not yet discovered satellites beyond the outer boundary of the rings. Inside the ring zone no satellites could exist. From the positions of the rings they calculated five "most satellite probable" orbits. Resonances with each satellite in any of these orbits determined at once the position of two – or more – rings. When we prepared our paper we thought that one of these orbits was superfluous – the details of that orbit were published later (Gor'kavyĭ and Fridman 1986) – as a satellite in that orbit partially duplicated the action of another satellite. Moreover, for the explanation of the features of the outer, broadest and most eccentric ring ε we introduced near the external edge of that ring a shepherd satellite which through a 3:4 resonance determined simultaneously the position of ring 4. The resultant picture of the hypothetical satellites and their resonances is shown in Fig. 10.1. It is clear from the figure that the satellite Z_0 which was published later duplicates the action of the satellite Z – this follows because the orbits of Z_0 and Z are in a 9:10 resonance with one another.

Table 10.1. Comparison of the orbits of the predicted and the detected Uranian satellites

Orbital radius of satellites [km]		Accuracy of the coincidence [km]	Number of 1:2, 2:3, and 3:4 resonances in the ring zone between 41 500 and 52 000 km	Satellite diameter [km]
predicted	detected			
	86 000		0	155
	75 260		1 (1:2)	60
	69 940		1 (1:2)	60
66 450	66 090	+360	2 (1:2, 2:3)	110
	64 350		1 (2:3)	80
62 470	62 680	−210	2 (2:3, 3:4)	60
61 860	61 780	+80	2 (2:3, 3:4)	60
58 600	59 170	−570	2 (2:3, 3:4)	50
55 380	53 800	(+1580)	1 (3:4)	30
51 580	49 770	(+1810)	0	25

In Fig. 10.2 we show the algorithm for selecting the narrow "satellite probable" zones starting from the structure of the Uranian ring system. Afanas'ev

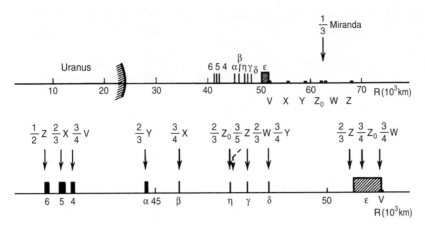

Fig. 10.1. Hypothetical system of Uranian satellites according to Gor'kavyĭ and Fridman (1985b,c)

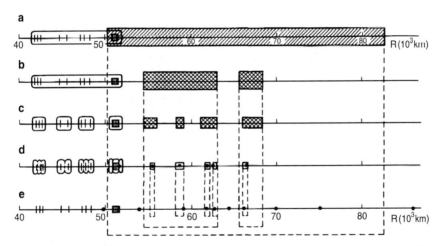

Fig. 10.2. Algorithm for splitting up the zones of the positions of unknown Uranian satellites according to the position of the rings.
(**a**) zone of satellites leading to at least one lower-order $-1:2$, $2:3$, or $3:4$ $-$ Lindblad resonance in a ring zone; (**b**) zone of satellites giving two resonances in a ring zone; (**c**) zone of satellites giving resonances in two ring zones; (**d**) zones of the positions of the individual satellites $-$ the dots indicate the orbital radii which were chosen; (**e**) position of the satellites detected by Voyager-2.

et al. (1985) estimated the average diameter of the unknown Uranian satellites as 100 km and discussed the possibility of observing these satellites by ground-based telescopes taking into account the predicted orbital radii and periods. The presumably known orbital periods of the unknown satellites and the position of the plane of the Uranian satellite system almost at right angles to the Earth-Uranus direction made it possible to track the satellites by superimposing photographs taken with the orbital frequency of the satellites – this appreciably enhanced the signal to noise ratio. An estimate of the stellar magnitude of the unknown satellites made it possible to hope that it was possible to detect them using modern radiation detectors.

Table 10.2. Comparison between the Gor'kavyĭ-Fridman hypotheses and the Voyager-2 observations

Nr	Hypothesis	Observation
1	Beyond the outer boundary of the Uranian rings there is a series of small satellites.	Nine out of the ten Uranian satellites lie beyond the outer boundary of the rings.
2	No satellites are formed inside the ring zone.	In the intermediate zone (near the outer edge of the rings) only a single, the smallest, satellite is found.
3	The positions of the rings are determined by 1:2, 2:3, 3:4 type resonances with undiscovered satellites (which are positioned in the 50 to 82×10^3 km zone).	Eight of the ten new satellites are found in that zone and they have such resonances in the ring zone. There is a very high correlation between the positions of the rings and resonances.
4	Each of the five predicted satellites determines at once the position of two rings.	Four out of the ten satellites determine simultaneously the position of two (or more) rings; their orbits coincide closely with the orbits of the predicted satellites.
5	The peculiar features of the ε ring are explained by the presence of a "shepherd" satellite.	The ε ring is the only one near the discovered "shepherd" satellite.[1]
6	The satellite diameters are roughly 100 km.	The average satellite diameter is roughly 70 km (ranging from 25 to 155 km).

[1] The inner shepherd of the rings satisfies all predicted functions and determines the position of ring 4 – except that it is a 4:5 rather than a 3:4 resonance – see Sect. 10.2.3 and Chap. 11.

10.1.6 Detection of New Uranian Satellites

The American spacecraft Voyager-2 discovered in January 1986 10 new satellites and thereby concluded the discussion of the nature of the rings–though it would be better to say that the Voyager data gave a new twist to this discussion. Only one ring, the ε ring, was surrounded by shepherds. In this way Voyager completely confirmed the hypothesis of the resonance nature of the Uranian rings (Gor'kavyĭ and Fridman 1986, 1987b). We compare in Table 10.1 the orbital radii of the predicted and the discovered satellites. We note that the basic points of our hypothesis were correctly predicted. In Table 10.2 we compare the main points of our hypothesis with the data from Voyager-2.

10.2 Correlation Between the Uranian Rings and Satellite Resonances

The coincidence of the orbits of the predicted and discovered satellites is the main proof of the resonance origin of the Uranian rings. Nonetheless a more detailed analysis of the relative position of the rings and the resonances is necessary since the positions of the rings are shifted relative to the resonance orbits. We shall show in what follows that there is a deep physical reason for this fact. We follow a paper by Gor'kavyĭ, Taĭdakova, and Fridman (1988) in this analysis.

10.2.1 Distribution of the Distances Between the Rings and the Resonances

The nine main Uranian rings–the most noticeable ones, discovered in 1977– lie in a 40 to 53 000 km zone from the centre of the planet (see Table 2.4). Voyager discovered also a series of less noticeable narrow ring structures; the most reliable are the narrow 1986U1R ring and the broad 1986U2R dust ring (Broadfoot et al. 1986, Lane et al. 1986, Smith et al. 1986; see Table 2.4). Let us associate the Uranian rings with lower-order resonances (1:2, 2:3, 3:4, 4:5, 1:3, 3:5) of which there are 25 in the 40 to 53 000 km zone and 31 in the 36 to 53 000 km zone. One can easily find the radii of these resonance orbits from the data given in Table 2.4, if we neglect the insignificant effect of the aspherical harmonics of the gravitational field of Uranus on the resonance relation: $n\Omega = m\Omega_{\mathrm{s}}$ where Ω is the orbital frequency in the resonance orbit and Ω_{s} the orbital frequency of the satellite. From this it follows that the radius of the resonance orbit, R_{res}, is connected with the orbital radius of the satellite R_{s}: $R_{\mathrm{res}} = (n/m)^{2/3}R_{\mathrm{s}}$.

Let us calculate the distances Δ_{r} from each resonance orbit to all nearest rings–not further away than 1000 km. We consider the magnitude distribution of the Δ_{r}, splitting the 1000 km into a few intervals. Let $\overline{N}$ be the number of distances Δ_r in a separate interval, divided by the number of rings. If $\overline{N} = 1$ in some interval this means that on average there is in each ring one distance from this interval. We have constructed in Figs. 10.3a, b histograms of the distribution according to intervals of 125 and 100 km for the case of 13 rings in the 40 to 53 000 km zone. One can see very clearly two features of the $\overline{N}$ distribution: there is a dip in the first and a peak in the second interval, that is, there are practically no rings near the resonances, but, on the other hand, practically each ring is positioned at a distance between 100 and 250 km from a lower-order resonance.

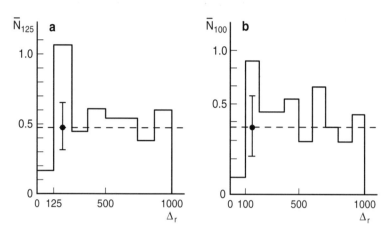

Fig. 10.3. Distribution of radial distances between rings and resonances in the Uranian system. $\overline{N}$ is the number of ring-resonance distances between Δ_r and $\Delta_r + \delta$, divided by the number of rings in the system; (**a**) $\delta = 125$ km, (**b**) $\delta = 100$ km. The bar indicates the mean-square error in the distribution for random positions of the rings.

Let us examine the statistical significance of these features of the distribution. To do this we use a random number databank, which would give a uniform distribution, to "throw" a fictitious system of 13 rings on the real set of resonance orbits in the 40 to 53 000 km zone. Altogether we generated 5000 of such systems. We then calculated for each interval the average magnitude of $\overline{N}$ and the magnitude of the error σ. In Figs. 10.3a, b the random $\overline{N}$ distribution is shown by a dashed line which is almost a straight line with a slight drop (of about 2%) in the 1000 km range. It is clear from Figs. 10.3a, b that the peaks and dips significantly go beyond the indicated limits of the error. For various cases we show in Table 10.3 by how much the dip and the peak exceed the error σ and also the corresponding probability that the given feature is not random. We note that the peak of $\overline{N}$ in the 125 to 250 km range is not random according to the most rigorous statistical criteria.

These characteristic features of the $\overline{N}$ distribution are a rigorously observed proof of the resonance origin of the Uranian rings and at the same

Table 10.3. Gap and peak magnitude (in units of σ) for different length of the intervals ($\delta = 100$ and 125 km) and the statistical significance of the features of the distribution of the ring-resonance distances (in brackets; in %)

Variants		Gap in first interval, σ		Peak in second interval, σ	
Number of rings	Size of the region $[10^3$ km$]$	$\delta =$ 100 km	$\delta =$ 125 km	$\delta =$ 100 km	$\delta =$ 125 km
9	40-53	1.43 (84.7)	1.26 (79.2)	1.99 (95.3)	3.05 (99.8)
13	40-53	1.95 (94.9)	1.94 (94.8)	2.31 (97.9)	3.45 (99.9)
15	36-53	1.65 (90.1)	1.70 (91.1)	2.53 (98.9)	3.52 (99.95)

time make the correspondence between the positions of the rings and the resonances less obvious – because of the shift of the rings from the resonances.

10.2.2 Correlation Between the Positions of the Rings and Resonances

Let us analyse the positions of the rings and resonances by a different, independent method. We show in Fig. 10.4b the total disposition of the narrow Uranian rings and of the resonances of the known satellites. The full-drawn vertical lines indicate the main Uranian rings and also η' – the broad component of the η ring – as well as the densest of the rings discovered by Voyager-2, 1986U1R. The dashed vertical lines indicate the less well studied narrow details of the ring system and also the edge of the broad dust ring 1986U2R. The 1:2, 2:3, 3:4, and 4:5 Lindblad resonances of each satellite are indicated by dots and connected by dash-dot lines, while in the names of the satellites the year of discovery (1986) has been omittted. Non-Lindblad resonances of the 1:3 kind are indicated by upright crosses and those of the 3:5 kind by lying crosses. The height of the dots and crosses for the different kinds of resonances is equal to $3/(n+m)$ and corresponds to the characteristic of these resonances.

If we divide the whole of the ring zone from 36 to 53 000 km into 1000 km intervals we notice that the average number of resonances in an interval containing a ring is more than twice the average number of resonances in an empty interval – and in the zone of the main rings, from 40 to 53 000 km, more than by a factor of 2.5. In Fig. 10.4a we show by arrows of a different height, $H_r = 3/(n+m)$, the general picture of the resonances in the ring

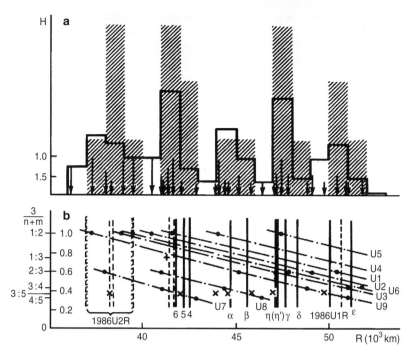

Fig. 10.4. Correlation of the positions of rings and resonances in the Uranian system. The points, crosses, and arrows indicate lower-order resonances; (**a**) histogram characterising the positions of the rings (hatched region) and histogram of the resonances (full-drawn lines); (**b**) main rings (discovered from the Earth, apart from 1986U1R; full drawn lines) and rings discovered by Voyager-2 (dashed lines). 1986U2R is a diffuse dust ring containing two denser details of the ring system. The numbers to the right of the coordinate axis give the characteristic of the resonance involved. The Lindblad resonances of each satellite are combined by the dash-dot lines (in the names of the satellites the year of discovery, 1986, has been omitted). The non-Lindblad resonances are indicated by upright crosses for the 1:3 kind or by lying crosses for the 3:5 kind.

zone and we have constructed a histogram (full-drawn line) which is the sum of the heights of the resonance arrows in each interval which characterises the spatial distribution of the resonance orbits. The hatched regions form the histogram of the ring distribution with each detail of the ring system contributing the same amount to the histogram. Using a standard method (Shchigolev 1969) we calculate the correlation coefficient between the heights of the two histograms: $Q = 0.7 \pm 0.1$ in the 36 to 53 000 km zone and $Q = 0.8 \pm 0.1$ in the 40 to 53 000 km zone. Taking higher-order resonances of the kind 5:6, ..., 10:11 and 5:7, ..., 9:11 ($n+m < 22$, $m-n < 3$) into account has practically no effect on the values of Q which remain equal to 0.7 ± 0.1 and 0.8 ± 0.1. We note that if we consider only low-order resonances there is no need to introduce different weights for resonances with the same value of $3/(n + m)$. The correlation coefficient between the number of rings and the number of low-order resonances in each interval then reaches the value 0.84 ± 0.08. Such a high value of the correlation coefficient is yet another proof of the resonance origin of the Uranian rings.

10.2.3 A Study of the Resonance System of Uranian Rings Using the Correlation Coefficient

Let us study the following problems. Which satellites turned out to have the maximum effect in the formation of the ring system? Resonances of which satellites made the decisive contributions to the formation of the ring systems? Let us determine for each satellite and for each kind of resonance the specific contribution to the correlation coefficient: $\Delta Q = Q - Q_x$, where Q_x is the correlation coefficient when one eliminates from the total picture the given resonances – corresponding to one type or to one satellite. We show in Fig. 10.5 the quantity ΔQ for each satellite; the order of the satellites corresponds to the actual situation – with the planet to the left. The full-drawn line is obtained from an analysis of the 40 to 53 000 km zone ($Q = 0.782$) and the dashed line from an analysis of the 36 to 53 000 km zone ($Q = 0.727$). The year of the discovery of the satellites, 1986, has been dropped; U1* was discovered on 1985 December 31. The numbers by the named satellites indicate the number of lower-order resonances ($n + m < 10$, $n - m < 3$) of that satellite in the 41 to 53 000 km zone. It is interesting to compare these data – which, in fact, reveal the satellites that are most significant for the formation of rings by a satellite – with the predicted system of satellites (see Table 10.1). The two furthest satellites – 1986U5 and 1985U1 – have only a single resonance in the ring zone so that it was impossible to calculate their orbits sufficiently reliably from the position of the rings. Of the eight other satellites only the five predicted ones give a positive contribution to the correlation function – in the 40 to 53 000 km zone. Cordelia (1986U7) corresponds exactly to her predicted function as a satellite – she is a shepherd satellite for the ε ring, except from the outside, rather than from the inside, and she has a strong resonance near ring 4 (though it is a 4:5 rather than a 3:4 resonance). If we consider a 4:5 rather than a 3:4 resonance with ring 4, one can "predict" the orbital radius of the inside shepherd as 49 410 km which differs from the actual radius by just 360 km. If we use this we find that the average deviation of the actual orbits from the ones calculated from the position of the rings is 316 km. This deviation is mainly connected with the physically conditioned shift of the rings from the resonances by several hundred km (see Chap. 11). From Fig. 10.5a we see that two satellites – 1986U2 and 1986U4 – clearly did not take part in the formation of the rings. We discuss in Chap. 11 the reasons why these satellites are exceptions. If we exclude the resonances from these two satellites from the general picture, the correlation coefficient becomes much larger:

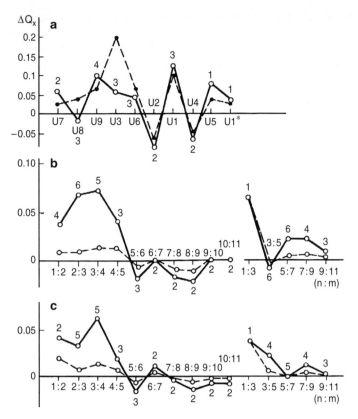

Fig. 10.5. Magnitude of the specific contribution to the correlation coefficient between the positions of the rings and the satellite resonances: of separate satellites (**a**), and of separate kinds of ten (**b**) or of eight (**c**; without 1986U2 and 1986U4) satellites. (**a**) the full-drawn line corresponds to the 40 to 53×10^3 km zone and the dashed line to the 36 to 53×10^3 km zone; the numbers *1, 2, 3, 4* determine the number of lower-order resonances ($m + n < 10$, $m - n < 3$) of a given satellite in the 40 to 53×10^3 km zone. The satellites are positioned along the abscissa axis in correspondence with their distance from Uranus. We have dropped the discovery year, 1986. U1* is the 1985U1 satellite, discovered 1985 December 31. (**b, c**) the dashed lines indicate the relative contribution of a single resonance of a given kind. The numbers along the full-drawn curve are the number of resonances of a given kind in the 40 to 53×10^3 km zone. The $n/(n + 1)$ resonances are positioned to the left and the $n/(n + 2)$ resonances to the right.

$$\left. \begin{aligned} Q_{36-53} &= 0.84 \pm 0.07; \\ Q_{40-53} &= 0.92 \pm 0.04; \end{aligned} \right\} \quad \text{for } m + n < 10, \quad m - n < 3,$$

$$\left. \begin{aligned} Q_{36-53} &= 0.82 \pm 0.08; \\ Q_{40-53} &= 0.90 \pm 0.05; \end{aligned} \right\} \quad \text{for } m + n < 22, \quad m - n < 3.$$

We show in Fig. 10.5b the contributions of the different kinds of resonances to the correlation coefficient (full-drawn line). The dashed line shows the relative contribution of a single resonance of each kind. We considered the 40 to 53 000 km zone ($Q = 0.780$); the numbers along the full-drawn line indicate the number of resonances of each kind. The regularity is particularly clear: the main positive contribution to Q comes from resonances of the kinds 1:2, 2:3, 3:4, 4:5, 1:3, and 3:5 ($m+n < 10$, $m-n < 3$). The higher-order resonances have a ΔQ which is close to zero and negative, that is, they are distributed randomly. This regularity is retained if resonances of all kinds contribute the same to the histogram ($H_r = \text{const}(m, n)$). In Fig. 10.5c we show ΔQ for the case when we omit the two satellites 1986U2 and 1986U4 ($Q = 0.899$). The tendency to a lower value of ΔQ when the order of the resonances is increased is no less clearly visible. This regularity is yet another proof of the resonance origin of the Uranian rings. All these conclusions enable us to consider the resonance origin of the Uranian rings to be a reliable fact established by two independent methods on the basis of an analysis of the observational data.

11 Formation and Stability
of the Uranian Rings

Portia: "What ring gave you, my lord?
Not that, I hope, which you receiv'd of me."
Bassanio: "If I could add a lie unto a fault,
I would deny it; but you see my finger
Hath not the ring upon it, – it is done."
Portia: "Even so void is your false heart of truth.
By heaven, I will ne'er come in your bed
Until I see the ring."

W. Shakespeare, The Merchant of Venice, Act V, Scene I.

In its proto-stage Uranus was surrounded by a thin, continuous gas-dust disk with a radius of about a million kilometers. We showed in Sect. 5.1.2 that one may think of the proto-disc around a planet as consisting of two parts: an inner part – the ring zone – and an outer part – the satellite zone. The qualitative difference between these two zones is connected with the different nature of the particle collisions in them. The inner zone is close to the planet and the azimuthal (rotational) velocity of the particles is large and, consequently, the relative velocity of the particles when they collide is large. In that region the processes in which the particles break up in a collision will dominate over those in which they stick together. In such a region satellites cannot be formed – it is a ring zone.

On the other hand, in the outer zone which is distant from the planet the azimuthal particle velocities are small. The relative velocities of the particles are consequently also small. In that zone of "slow" particles the adhesive processes dominate over the break-up processes. This is the satellite zone.

As we move radially away from the planet, to start with we escape the zone of future rings where the relative velocities of colliding particles are large and enter the zone of future close satellites. In the regions close to the zone of future rings the relative velocities of colliding particles are still large. The predominance of adhesive processes, when particles collide, over destructive processes is as yet small. The satellites grow slowly in this zone. As we move away from the planet and approach the satellites which are further away the domination of the adhesive process turns out to become ever larger. Apart from the fact that the azimuthal rotational velocity decreases with increasing radius, the radial distribution of the surface density also affects the rate of growth of the satellites. The maximum growth rate turns out to occur for the furthest satellites. They were therefore also formed first. However, since they are far removed from the zone of future rings, for the latter the formation of these far satellites passed unnoticed: none of them affected the ring zone through Lindblad resonances.[1] Therefore considering the proto-planetary disc of Uranus at that remote stage when only the far satellites had been formed (see Table 11.1) we do not envisage in it its present-day – narrow and dense – rings. These rings will appear later – the close satellites must first be formed; they are the cause of the generation of the famous Uranian rings. These near satellites started to be formed at once after the far ones (see Table 11.1). As they grow they begin to affect resonantly the zone of future rings which will be formed at the positions of such resonances. Since the first close satellites did not originate at all close to the boundary with the zone of future rings, their resonance interaction was restricted to the first Lindblad resonances – all their other resonances occurred in the satellite zone.

From what we have just said we can understand the meaning of the prediction of the orbits of unknown Uranian satellites which the present authors published in 1985 half a year before Voyager-2 discovered the first of these satellites. The manner of the prediction consisted in calculating the orbits of unknown close satellites from an observational knowledge of the localisation of the rings, and the theoretical assumption that they are positioned in the position of the lower Lindblad resonances with those unknown satellites makes it possible to calculate the orbits of the latter.

The disc around Uranus was non-uniform and the accumulation of particles in some regions and the rarefaction in other regions was connected with different competing forces and drift processes: viscous spread, aerodynamic forces, and ballistic drift caused by the bombardment of the disc by interplanetary particles. Often one and the same drift process may cause the motion of matter both towards the planet (negative drift) and away from it (positive drift). For instance, the motion of grains in a gaseous disc is traditionally associated with drag, but when a gaseous turbulent disc is rapidly spread outwards through viscosity the disc can drag behind it many slow grains.

The balance of the drift processes in the Uranian proto-disc is complicated but, fortunately, we have eyewitnesses for these processes: the satellites of the planets have reliably fixed the distribution of matter in the circumplanetary disc at the moment they were created (see Fig. 2.19). This distribution gives evidence for the fact that in the region of the Uranian rings there existed a powerful positive drift causing the lowering of the density in the region up to 60 000 km and its increase in the region from 60 000 to 68 000 km – near Neptune a similar positive drift emptied an inner zone up to 45 000 km and

---

[1] Here and henceforth we shall understand by the resonance action of a satellite on the rings the $n/(n+1)$ Lindblad resonances since they are the most efficient ones for rings with a small eccentricity. We shall call resonances with small values of n lower resonances and those with $n \gg 1$ higher resonances, in contrast to the definitions in a paper by Hamilton (1994) where all Lindblad resonances are called first-order resonances, resonances such as the 3:5 and 3:1 ones, second-order resonances, and so on.

the matter displaced by it formed a heavy region from 45 000 to 68 000 km. We shall discuss in Sect. 11.2 several positive drift mechanisms amongst which also those which are continuing to operate up to the present in the Uranian rings.

Each satellite of a planet causes a series of exterior and interior resonances. The Lindbald resonances cannot exist more than 37% inside the satellite orbit or more than 59% outside it – at those distances from the satellite the furthest Lindblad resonances (1/2 and 2/1) are situated; the higher the order of the Lindblad resonance, the closer to the satellite and the stronger the action of the satellite. Hence, in reality only satellites which are not further away from the ring zone than $0.59 \times 50\,000 = 30\,000$ km, that is, with an orbital radius less than 80 000 km, can interact with the Uranian ring zone. We shall call satellites within that region "close" satellites. The first close satellites were formed in the peak of the density at 60 000 to 68 000 km thanks to the considerable surface density in that region and its distance from the "millstones" of the ring zone. From those satellites which are at a distance of 10 000 to 15 000 km from the ring zone only the lower 1/2, 2/3, and 3/4 Lindblad resonances are "reached".

Particles in the ring zone, drifting outwards, hit the inner Lindblad resonances of the close satellites. A ring particle, being closer to the planet rotates faster than the outer planet and it will therefore, when interacting resonantly, lose some of its angular momentum and be decelerated. The resonance with the satellite will thus slow the particle down and attempt to dump it onto the planet, while the positive drift accelerates the particle and tends to move it away from the planet. If the positive drift cannot overcome the effect of the resonance the particle will be stopped – resonance trapping ensues. In front of the resonance a great number of trapped particles are stored – thus appears the nucleus of a future Uranian ring.

Sooner or later particle collisions become very frequent in the growing ring nucleus near the resonance barrier. Inelastic collisions tend to suppress all non-circular oscillations of the continuous medium leaving only the purely circular ($m = 0$) or elliptical ones, with $m = 1$ (see Sect. 8.4). Hence, the viscosity rearranges the initial set of almost independent resonance particles with an appreciable eccentricity – caused by the resonance interaction with a satellite – into a completely circular ringlet (with the $m = 0$ mode) or a weakly elliptical ring (with the $m = 1$ mode).

The decrease in the eccentricity of the resonance particles and the frequent changes of their orbits when they collide cause for many particles a disruption of the resonance coupling with the satellite (see Sect. 11.3). Thanks to the positive drift of such non-resonant particles which are dragged behind their resonant neighbours the ring breaks through the resonant barrier and starts to increase its orbital radius. However, the free drift epoch of the Uranian rings does not last long: a wave of satellite formation, moving inwards reaches the outer boundary of the rings where the satellites Cordelia and Ophelia appeared which are closest to the rings and therefore have most effect on them. They created in the rings tens of strong resonance barriers which turned out to be insurmountable for the Uranian rings. Part of the rings stands now in front of the resonance barriers, "pushed forward" by the positive drift. Another part of the rings is balanced in the interbarrier space, interchanging space debris with its neighbours.

In what follows we shall consider the following problems:

1) Where and when did satellites appear in the continuous disc around Uranus (Sect. 11.1)?
2) What produced the positive particle drift in the ring zone (Sect. 11.2)?
3) What is the mechanism for the resonance formation of the narrow rings near Uranus (Sect. 11.3)?
4) How did the Uranian rings evolve to their present state (Sect. 11.4)?

11.1 Sequence of the Formation of the Uranian Satellites

Where and in what order did satellites appear near Uranus? We have already mentioned above that satellites cannot be formed in the zone of the planetary rings which rotate in the region near the planet where destructive collisions dominate. We shall consider the growth of satellites beyond the outer boundary of the rings – in the zone outwards from 50 000 km.

The reconstruction of the surface density in the plane of the disc (Fig. 2.19) shows that at the time when the close satellites were formed in the 60 000 to 68 000 km zone there was a peak in the surface density of matter. In the satellite zone – there where the shear destruction is small – the higher the frequency of collisions between disc particles, the faster satellites are formed. One sees easily from expression (8.50) for the free flight time that the characteristic time of growth for satellites must be inversely proportional to the surface density in the disc and directly proportional to the orbital period. We shall understand by the characteristic satellite growth time the time it takes them to grow to the same size. The large, far satellites reached their final mass therefore later than the small, close ones. Taking into account the reconstruction of the surface density we can construct Table 11.1 of the characteristic growth times of the Uranian satellites, taking the growth time of Portia as the unit since, unfortunately, we do not know the proportionality factor between the characteristic satellite growth time and the collision frequency of the disc particles.

As the lower boundary of the feeding zone of Cordelia we have taken the radius of the δ ring: 48 300 km. We have put the density of the present satellites equal to 1 g/cm^3. Notwithstanding its arbitrariness, this value does not lead to large errors since the density of the largest satellites is around 1.5 g/cm^3 and that of Miranda 1.26 g/cm^3 (Burns 1986). We gave in Sect. 2.8

Table 11.1. Characteristic growth times of the Uranian satellites

Satellite	Orbital radius [km]	Satellite radius [km]	Rotational period [hours]	Density in the disc [g/cm^2]	Characteristic growth time [arbitrary units]
	Far satellites	($>$ 80 000 km)			
Oberon	582596	775	322.510	362.9	4.3
Titania	435844	805	208.684	510.8	2.0
Umbriel	265969	595	99.481	396.3	1.2
Ariel	191239	580	60.654	981.8	0.3
Miranda	129783	242	33.909	133.8	1.2
Puck	86006	77	18.293	11.8	7.5
	Close satellites	($<$ 80 000 km)			
Belinda	75256	34	14.973	4.3	17.0
Rosalind	69942	29	13.415	5.0	12.8
Portia	66090	55	12.322	59.5	1.0
Juliet	64350	42	11.839	44.9	1.3
Desdemona	62675	29	11.380	20.1	2.7
Cressida	61776	33	11.136	22.3	2.4
Bianca	59172	22	10.439	3.0	16.6
Ophelia	53794	16	9.049	1.1	40.7
Cordelia	49771	13	8.053	1.1	36.8

the shape of the surface density in part of the disc. We are in first instance interested in the order in which the satellites near Uranus appeared so that the lack of information about their absolute growth times is unimportant. We note merely that near the planet the growth of satellites is additionally slowed down because of the large shear velocities in the disc so that the values given in the last column of Table 11.1 for the close satellites must be understood to be lower limits.

The first of the group of satellites close to Uranus to appear are those which are 10 000 to 25 000 km from the ring zone – this fact immediately determined the nature of the resonance interaction between the satellites and the future rings through lower-order Lindblad resonances since higher-order resonances simply do not "reach" the ring zone.

Thus, near Uranus, at approximately 15 000 km from the outer boundary of the rings, Portia was formed first; this gives a 1/2 resonance at the inner edge and a 2/3 resonance at the outer edge of the ring zone. Juliet, Cressida, and Desdemona were formed practically at once after Portia. The growth of the other satellites was quite a distance from this group of leaders. Next followed the group of Rosalind, Bianca, and Belinda, while the last in the Uranian system to appear were Cordelia and Ophelia. It is exactly in that order that in the Uranian ring zone (from 41 000 to 52 000 km) the resonances from those satellites appeared.

The lower Lindblad resonances of the first six satellites caused the formation of the nine main Uranian rings which, in accordance with the resonance picture are split up into four separate, spatially spaced groups of rings. The satellites Portia, Cressida, Desdemona, and Bianca have in the ring zone up to two resonances of the 1/2, 2/3, or 3/4 kind and these resonances lie for each satellite inside different groups of rings. It is clear from Fig. 10.2 that taking into account the division of the Uranian ring system into four separate groups (Fig. 10.2c) at once, even without considering a detailed correspondence between the resonances and the separate rings inside the group, determines four narrow zones for the positions of the invisible satellites in which four predicted satellites were discovered which have up to two resonances in the ring zone (see Chap. 10).

Table 11.2. Formation of the resonance structure of the Uranian rings

Characteristic growth time of close satellites [arbitrary units]	Satellite	Resonances and		Groups	of	Rings
		6–5–4	α–β	η–γ–δ	Cordelia region	λ–ε^*
1.0	Portia	1/2				2/3
1.3	Juliet				2/3	
2.4	Cressida			2/3		3/4
2.7	Desdemona			2/3		3/4
12.8	Rosalind		1/2			
16.6	Bianca		2/3	3/4		4/5
17.0	Belinda			1/2		
36.8	Cordelia	4/5...**	6/7–7/8...	12/13–22/23...		129/128–24/23
40.7	Ophelia		3/4...	5/6...		13/14

* The 1986U1R ring is here indicated as the λ ring.
** The dots indicate the presence of intermediate resonances up to the ones written to the right.

We emphasise here the fact that all Lindblad resonances, which are situated in the ring zone, of the first six satellites played their role in the formation of the rings. Belinda, the seventh satellite did, apparently, not form a ring in connection with the weak 1/2 resonance (see Sect. 11.3) because in the region of its resonance three stronger resonances appeared earlier which formed three resonance ringlets.

Cordelia and Ophelia, originating after all this, caused the strongest resonances in the ring zone (see Sect. 11.3) but up to the moment they were formed practically all the matter in the zone was assembled in the ring and the new satellites could only capture part of the already existing rings: η, γ, δ, and ε.

The resonances from Cordelia and Ophelia do not generate rings; they only stabilise earlier formed ringlets – this logically explains why ringlets are found only in some of the higher-order resonances (5/6, 12/13, 22/23) of Ophelia and Cordelia. Any theory which wants to defend the idea of storing the η, γ, δ, and ε rings in those higher-order resonances must be able to explain why there are no rings in many tens of other, stronger resonances of those satellites.

11.2 Particle Drift in the Uranian Proto-disc

Particle drift in the Uranian proto-disc played a crucial role in the formation of the observed system of satellites and rings. The present section is devoted to a qualitative consideration of drift motions in the Uranian system in chronological order – starting with the proto-stage and ending with the present state.

11.2.1 Aerodynamic Drift in an Expanding Proto-disc

Let us consider the gas-dust Uranian proto-disc. In the central plane of such a disc micron-size grains condense and gradually grow. One can easily show that a grain rotates faster than the stationary gaseous disc, is decelerated, and moves towards the planet. We shall study a gaseous disc which is made turbulent by some external or internal factors. Turbulent viscosity causes the transfer of angular momentum from the inner, fast rotating regions of the disc to the outer regions. We shall assume that each element of matter of such a disc participates in some stationary motion.

Let us use (8.1) to write down the stationary equation in the zeroth approximation:

$$V_{r0}\frac{dV_{\varphi 0}}{dr} + \frac{V_{r0}V_{\varphi 0}}{r} = \frac{1}{\sigma_0 r^2}\frac{d}{dr}\left[r^3\nu\sigma_0\Omega_0'\right], \qquad \Omega' \equiv \frac{d\Omega_0}{dr},$$

whence we can determine V_{r0}:

$$V_{r0} = \frac{2}{\Omega_0\sigma_0 r^2}\frac{d}{dr}\left[r^3\nu\sigma_0\Omega_0'\right].$$

We have used here the fact that for a Kepler disc $\Omega_0 \propto r^{-3/2}$. Assuming that $\nu\sigma_0 \propto r^{-q}$ we find

$$V_{r0} = 3\left[q - \frac{1}{2}\right]\frac{\nu}{r}. \tag{11.1}$$

Many disc instabilities (see Chap. 8) lead to turbulent viscosity; the turbulent viscosity coefficient ν can in the α-accretion disc model (Shakura and Syunyaev 1973) be written in the following form:

$$\nu \approx \alpha h c_{\mathrm{s}},$$

where $\alpha \equiv v_T/c_{\mathrm{s}}$, v_T is the turbulent velocity in the disc, c_{s} the sound velocity in the gas, and h the half-thickness of the disc. Using the estimate $h \approx c_{\mathrm{s}}/\Omega_0$ we get finally

$$\nu \approx \alpha \frac{c_{\mathrm{s}}^2}{\Omega_0}. \tag{11.2}$$

Substituting (11.2) into (11.1) we find the radial drift velocity of the gas:

$$V_{r0} \approx 3 \left[q - \frac{1}{2} \right] \alpha \frac{c_{\mathrm{s}}^2}{r\Omega_0}. \tag{11.3}$$

This is a sufficiently general expression for determining the radial drift velocity. However, assuming that the surface density in a typical proto-satellite disc decreases, $\sigma_0 \propto r^{-2}$, and that the temperature is proportional to r^{-1} (see, for instance, Weidenschilling 1981), that is, $c_{\mathrm{s}}^2 \propto r^{-1}$, and taking into account that $\Omega_0 \propto r^{-3/2}$, we get from (11.2) $q \approx \frac{3}{2}$. With $c_{\mathrm{s}} \approx 10^5$ cm/s, $\Omega_0 \approx 2 \times 10^4$ s^{-1}, $r \approx 5 \times 10^9$ cm, we find $V_{r0} \approx 3 \times 10^4 \alpha$ cm/s.

Hence, for reasonable values of the parameter $\alpha \sim 0.1$ to 0.01 we find that the velocity of the radial expansion of the gas reaches 3 to 30 m/s – an altogether fair "wind". Up to what radius will this "wind" blow and how long will this "storm" last? Is there an efficient reverse process transferring matter in the disc from the outer edge of the disc to the planet? Alas, a consideration of these interesting problems requires a separate study which lies in the future. We note solely that possibly the radial gas expansion is just the mechanism which took away the main part of the gas and the smallest submicron-size grains from the inner part of the Uranian proto-planetary disc to a radius of 200 000 km where a peak in the density of 1000 g/cm^2 was formed. This peak exceeds the present density of the near zone by 2 to 3 orders of magnitude. The brightest of the Uranian satellites, Ariel, is now situated in the 200 000 km region; it is very unlike its outside dark neighbour, Umbriel. It is not impossible that the turbulence is damped with increasing distance from the planet. In that case the parameter α will decrease with increasing radius and the positive gas flow will be slowed down. However, it may occur even under the conditions when the density increases outwards, provided the condition $q > \frac{1}{2}$ is satisfied for the function $\nu\sigma \propto r^{-q}$ (see (11.1)).

Let us consider the behaviour of the dust layer in the expanding gaseous Uranian disc. We shall for the grains write down the condition for a balance between the centrifugal, the gravitational, and the aerodynamic forces, caused by the radial wind, which is directed outwards:

$$m\frac{V_{\varphi 0}^2}{r} = \frac{GMm}{r^2} + \pi a^2 \varrho V_{r0}^2, \tag{11.4}$$

where a is the radius of the grain and ϱ the gas density. The expression for the aerodynamic force is transparent: the radial dynamic pressure is multiplied by the cross-sectional area of the particle. The aerodynamic friction coefficient will be assumed to be of the order of magnitude of unity. Substituting (11.3) into (11.4) we get:

$$V_{\varphi 0}^2 - V_{\mathrm{g}}^2 + \beta c_{\mathrm{s}}^2 = 0, \tag{11.5}$$

where

$$\beta \equiv \tfrac{27}{4}\left[q - \tfrac{1}{2}\right]^2 \alpha^2 \frac{c_{\mathrm{s}}}{r\Omega_0} \frac{\sigma}{\varrho_{\mathrm{p}} a}, \qquad V_{\mathrm{g}}^2 \equiv \frac{GM}{r}; \tag{11.6}$$

we have here used the fact that $\varrho \approx \sigma/h \approx \sigma\Omega/c_{\mathrm{s}}$, where σ is the surface density of the gaseous disc, and $m = \tfrac{4}{3}\pi\varrho_{\mathrm{p}} a^3$. It follows from (11.5) that the radial wind partially balances the gravitational force and leads to a diminution of the orbital velocity of the particle. For $c_{\mathrm{s}} = 10^5$ cm/s, $\Omega = 2 \times 10^{-4}$ s^{-1}, $r = 5 \times 10^9$ cm, $\alpha = 0.1$, $\sigma = 1000$ g/cm^2, $\varrho_{\mathrm{p}} = 1$ g/cm^3 we get from (11.6) $\beta \approx \tfrac{9}{4}\left(q - \tfrac{1}{2}\right)^2$ for centimeter-size particles. Decreasing the density of the gaseous disc by three orders of magnitude leads to the same estimate for β for 10-micron-size particles ($a = 10^{-3}$ cm). Let us compare the equation for the dynamic balance of a grain (11.4) with the condition for hydrodynamic equilibrium of a rotating gas (see (A3.14)):

$$\frac{V_{\varphi}'^2}{r} = \frac{GM}{r^2} + \frac{1}{\sigma}\frac{dP}{dr}, \tag{11.7}$$

which can approximately be written as

$$V_{\varphi}'^2 - V_{\mathrm{g}}^2 + \zeta c_{\mathrm{s}}^2 \approx 0, \qquad \zeta \equiv r \ln' \sigma. \tag{11.8}$$

We have in (11.8) neglected the turbulent pressure in comparison with the normal gas pressure, since the condition $v_T^2 < c_{\mathrm{s}}^2$, or

$$\alpha^2 < 1, \tag{11.9}$$

is satisfied. From a comparison of expressions (11.8) and (11.5) for the velocities of the orbital motion of a grain and of the gas in an expanding disc it follows that if $\beta < \zeta$ the grain moves faster than the gas and will be aerodynamically decelerated while if $\beta > \zeta$ – which is realised for small particles – the grain will move more slowly in its orbit than the gaseous disc and, hence, will be aerodynamically accelerated. If $\beta \sim \zeta$ the velocities of the grains and of the gas will be equal. It is clear from (11.6) that the parameter β decreases rapidly with increasing distance from the planet because of the decrease in the parameters σ and c_{s}. Hence, for each size a of particle there exists a distance from the planet r_{cr} such that we have $\beta > \zeta$ for $r < r_{\mathrm{cr}}$ and $\beta < \zeta$ for $r > r_{\mathrm{cr}}$. This means that the inner grains will be accelerated and the outer ones decelerated and the two grain fluxes will move against one another – to an equilibrium radius r_{cr} at which the grain matter will rapidly accumulate. The peak in the density observed at 60 000 to 68 000 km in the Uranian circumplanetary disc might have been caused by such countermotions of grains with sizes which are most characteristic for small particles. Large particles for which only aerodynamic deceleration is characteristic always move only towards the planet so that in the Uranian proto-disc there existed counterflows of particles, interpenetrating each other.

The positive aerodynamic drift might thus cause the distribution of matter in the zone of the close Uranian satellites and determine the order in which the close satellites are formed. A confirmation of this statement requires an analysis which exceeds significantly our illustrative estimates, both in rigour and in volume. For the main purpose of the present chapter, which is devoted to a theory of the formation of the Uranian rings, it is only important that in the ring zone there could in the proto-stage have existed a positive particle drift which is a necessary condition for the capture of drifting particles in Lindblad resonances. Most probably, the main role in the formation and stability of the rings was played by another positive drift – the ballistic drift which we shall now consider.

11.2.2 Qualitative Discussion of the Ballistic Drift

The planetary rings are not isolated from the interplanetary medium and are all the time subject to bombardment by micrometeorites. When an interplanetary grain, moving with a velocity of 20 to 30 km/s, collides with a ring particle, a micro-blast occurs. The high-speed grain penetrates into the loose interior of the stoney-icy particle and causes an evaporation of matter, the mass of which exceeds the mass of the grain-pellet by two orders of magnitude. The expanding gas, which is the product of these micro-blasts, drags along with it part of the loose surface material of the particle with the velocity of its expansion which is comparable to the sound velocity in the gas, that is, with a velocity of several hundred meters per second. We shall call this material which has been thrown out: ejecta. Since the mean ejection velocity is a few percent of the orbital velocity, the mean eccentricity of such ejecta is also small. If a grain collides with a micro-particle of the ring, the latter is completely annihilated and after the collision high-velocity ejecta are formed with appreciable eccentricities. Let us consider a separate narrow and circular Uranian ring. Grains overtaking the ring, and grains with a reverse "counterstream" motion will, when colliding with ring particles, form ejecta, respectively, with a large angular momentum and an increased major semi-axis or with small angular momentum and smaller major semi-axes. The trajectory of the ejecta always intersects the mother ring so that it sooner or later recovers all matter lost as the result of the bombardment, and increasing in weight due to the absorption of the interplanetary grains – we neglect the small part of ejecta which may acquire a hyperbolic velocity after the collisions. In the laboratory frame of reference about the same amount of grains falls on each section of the Uranian rings in the direct direction, that is, along the rotation of the ring, as in the reverse direction, that is, countering the rotation. In the frame of reference rotating with the ring particles more often collide with it if they are flying in the direction opposite to the rotation. Therefore, as a result of the collisions of the ring with the grains the angular momentum of the ring must decrease. The *specific* angular momentum of the matter in the ring must decrease even more after absorption of the interplanetary grains. The decrease in the specific angular momentum of the ring means a decrease of its radius. This mechanism for decelerating the Uranian rings when they are bombarded by interplanetary particles was studied by Cuzzi and Durisen (1990) who noted its significance. However, if we apply it to the actual system of several Uranian rings, rather than to a single abstract ring, this mechanism may lead to an additional drift of the rings.

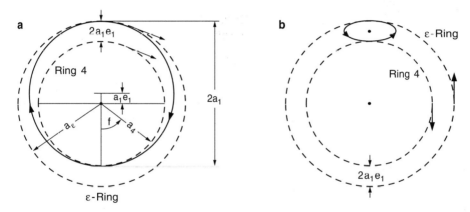

Fig. 11.1. Ejecta between the ε ring and ring 4. (**a**) In the laboratory frame of reference; the ejecta are the full-drawn elliptical curve. The continuous straight arrows indicate the azimuthal rotational velocities of the ε ring and ring 4. (**b**) In the frame of reference, rotating with the velocity of the centre of mass of the ejecta. The ejecta are shown by the full-drawn curve.

Let the ring being considered be the ε ring, the most massive ring of the Uranian system. Let us consider a second ring – for instance, ring 4, which lies closer to the planet (Fig. 11.1a). The inner ring, ring 4, has an appreciably lower optical depth so that we shall for the sake of simplicity assume that the interplanetary grains interact with it negligibly more rarely than with the main, outer ε ring. The pericentres of the orbits of the ejecta – the fragments of particles ejected from the main, outer ε ring – are situated in the zone of ring 4. The ejecta are for ring 4 an important source of angular momentum since near its pericentre the particles of the ejecta move faster than the particles of ring 4. Indeed, we can use (3.44) and (3.35) to write the expression for the azimuthal velocity of the particles of the ejecta, which is characterised by a major semi-axis a_1 and an eccentricity e_1 as follows:

$$V_\varphi = \frac{L}{r} = \frac{L}{p}[1 + e_1 \cos f] = \frac{L}{a_1(1 - e_1^2)}[1 + e_1 \cos f],$$

where f is the azimuthal angle. Using (3.39) to define L,

$$L = \sqrt{GMa_1(1 - e_1^2)},$$

we finally get the following expression for the azimuthal velocity of the particles of the ejecta:

$$V_\varphi = \sqrt{\frac{GM}{a_1(1 - e_1^2)}}[1 + e_1 \cos f]. \tag{11.10}$$

The excess of this velocity over the azimuthal velocity of ring 4 with a radius a_4 is

$$\Delta V_\varphi = \sqrt{\frac{GM}{a_1(1 - e_1^2)}}[1 + e_1 \cos f] - \sqrt{\frac{GM}{a_4}}. \tag{11.11}$$

It is clear from Fig. 11.1a that the ejecta touch ring 4 at the point $f = 0$, whence we get, using the relation $a_4 = a_1(1 - e_1)$ (see Fig. 11.1a)

$$\Delta V_\varphi = \sqrt{\frac{GM}{a_4}}\left[\sqrt{1 + e_1} - 1\right] > 0,$$

which we wanted to prove.

Let us now consider the interaction of the same ejecta with a ring of radius a in the range a_4 to a_ε. First of all, we note that a ring of radius $a = a_4 + a_1 e_1 = a_1$ intersect the injecta in the points $f = \pi/2$ and $f = 3\pi/2$ where $\Delta V_\varphi = \sqrt{GM/a_1(1 - e_1^2)} - \sqrt{GM/a_1} = 0$, if we assume that $e_1^2 \ll 1$. Hence, $\Delta V_\varphi > 0$ for $a \in (a_4, a_4 + a_1 e_1)$, that is, all rings between ring 4 and the 0 ring – the ring going through the centre of the ejecta (the point O) (see Fig. 11.1b) are accelerated when they absorb the ejecta. One can show similarly that $\Delta V_\varphi < 0$ for $a \in (a_4 + a_1 e_1, a_\varepsilon)$, that is, all rings between the 0 ring and the ε ring will be decelerated.

One can reach similar conclusions using simple considerations without using any formulae. Let us consider a hypothetical purely circular ring between ring 4 and the ε ring and passing through the point O (we have called this ring earlier an 0 ring). Let the radius of that ring be a_0 and the azimuthal velocity of its rotation be $V_{\varphi 0}$. The specific angular momentum $a_0 V_{\varphi 0}$ of this ring will remain constant for any radial displacement ξ_r up to terms proportional to this displacement. If we choose ξ_r equal to the distance from a_0 to the ε ring – or, what amounts to the same, to ring 4 – we find in the frame of reference rotating with the velocity $V_{\varphi 0}$ an epicycle in the form of the ellipse shown in Fig. 11.1b.

The ejecta shown in Fig. 11.1a have thus in the laboratory frame of reference a specific angular momentum $a_0 V_{\varphi 0}$ which is larger than the specific angular momentum of all rings with radii $a < a_0$ and which is smaller than the specific angular momentum of all rings with radii $a > a_0$. Hence follows the conclusion reached earlier.

The point O at a distance $a_1 e_1$ from the outer ε ring is thus an equilibrium point towards which particles or rings which enter the field of the drift effect of the ε ring will flow. In fact, there occurs an interchange interaction between the two Uranian rings which tends to lead to their approaching a well defined equilibrium distance. We note that in that exchange interaction the exterior ε ring also gains – in fact, it "fed" its interior ejecta (with a lower specific angular momentum) to ring 4, retaining only the exterior ejecta with an increased angular momentum. Hence, after absorbing these exterior ejecta the ε ring increases its specific angular momentum and moves away from the planet. We note that the massive outer ε ring and the light inner 4 ring, taken each separately, decelerate and fall onto the planet, but when they interact they both start to move away from the planet – ring 4 faster, and the ε ring more slowly. It is clear that here no laws of physics are broken: the total angular momentum of the two rings is slightly diminished and the mean specific angular momentum of the system decreases in the absorption of an interplanetary grain, but the outer ring becomes slightly lighter – due to the lost interior ejecta – and the inner ring becomes more massive by approximately the same amount. The total angular momentum of the ε ring decreases but its specific angular momentum increases; for ring 4 both the total and the specific angular momentum increase. The closer to one another the rings are, the more efficient their exchange interaction and the faster these rings move away from the planet – until the mass of the outer ring becomes comparable with the mass of the inner, light ring 4 which, in turn, cannot acquire more angular momentum than the ε ring had originally. This amazing construction of a pair of rings produces, in fact, positive angular momentum from the flow of the interplanetary dust for the acceleration of the outer ring, using the negative angular momentum as a means to transfer positive angular momentum from the outer ring to the inner one. The whole stability of the pair is guaranteed only at the cost of small losses of mass and total angular momentum by the outer ring.

A similar mechanism must operate in any sufficiently inhomogeneous disc; for instance, the outer density peak at 60 000 to 68 000 km must in its erosion cause a positive drift of particles in the inner region of the disc ($< 60\,000$ km). We shall call the drift caused by the ballistic transfer of angular momentum the ballistic drift. Below, in Sect. 11.2.4 we shall give the results of calculating the ballistic drift in the Uranian rings and now we shall describe a qualitative picture of the evolution of the rings, at the level of estimates.

11.2.3 Estimates of the Ballistic Drift and of the Aerodynamic Friction

After Voyager-2's fly-past near Uranus in 1986 there arose the problem of the aerodynamic friction which must dump the inner rings onto the planet after a few million years (Broadfoot et al. 1986). This problem practically closed the flow of theoretical papers about the Uranian rings because it has no sense to sweep the deck of a ship which has a gaping hole in its bottom.

It was shown by Porco and Goldreich (1987a, b) that only the ε ring is stable; it is the one which is furthest of all from the planet and has two shepherds as neighbours: Cordelia and Ophelia. The stability of the other rings is incomprehensible. Moreover, even if one assumes the existence of shepherd satellites near the inner rings with radii of about 10 km (larger satellites would have been detected by Voyager-2) it is all the same impossible to "save" the rings – for the neutralisation of the aerodynamic friction the flow of angular momentum from the satellites needs to be two orders of magnitude larger (Porco and Goldreich 1987b), and even the satellites themselves have to be saved.

Gor'kavyĭ (1989a) considered a fundamentally different stabilisation mechanism, based upon the transfer of angular momentum between rings through dust flows – basically from the ε ring which generates a peculiar dust trail. The origin of this trail is connected with such factors as the aerodynamic friction, the deceleration on magnetic field lines (for charged grains), the breaking up of the matter in the ring by micrometeorites, and fluxes of magnetosphere particles. When moving towards the planet the dust flow from the ε ring is captured and again emitted by the other Uranian rings. We shall show that the stability of the Uranian rings is connected with their ability to extract from the planetocentric dust flow the angular momentum necessary for their own stability.

To calculate the relative velocity between a dust grain on a circular orbit of radius a_d and the orbital velocity of particles near the apocentre of an elliptical ring of eccentricity e and major semi-axis a we use Fig. 11.2a. We have then according to (11.11)

$$\Delta V = \sqrt{\frac{GM}{a_d}} - \sqrt{\frac{GM}{a(1-e^2)}}(1 + e \cos f). \tag{11.12}$$

The intersection of the rings occurs at the point $f = \pi$ and it then follows from Fig.11.2a that $a_d = a(1+e)$. Using the fact that the eccentricity is small as compared to unity we find $\Delta V = \frac{1}{2} V_0 e = \frac{1}{2} \Omega Re$, with $V_0 = \sqrt{GM/a_d}$, $R =$

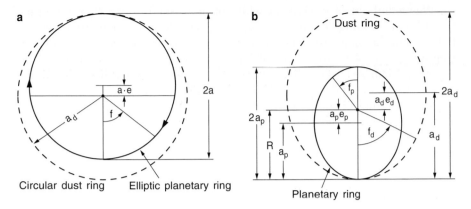

Fig. 11.2. Calculating the relative velocity between (**a**) a grain on a circular orbit and particles near the apocentre of an elliptical ring and (**b**) a grain at the pericentre and ring particles in the apocentre.

a_d. If the grains themselves move along an elliptical orbit with major semi-axis a_d and eccentricity e_d and in the pericentre collide with ring particles – in their apocentre; see Fig. 11.2b – the relative velocity will be given by the formula

$$\Delta V = \sqrt{\frac{GM}{a_d(1 - e_d^2)}}(1 + e_d \cos f_d) - \sqrt{\frac{GM}{a_p(1 - e_p^2)}}(1 + e_p \cos f_p)$$

If we use the relation $a_p(1 + e_p) = a_d(1 - e_d) = R$, which follows from Fig. 11.2b, and also the fact that the point where the two rings touch corresponds to $f_d = 0$ and $f_p = \pi$ and assume that the eccentricities are small, $e_d, e_p \ll 1$, we get $\Delta V = \frac{1}{2}\Omega R(e_d + e_p)$.

The stability condition for the rings, in the case when the acceleration W_{acc} imparted to it by the dust flow is compensated by the aerodynamic deceleration W_{dec}, is

$$W_{acc} = W_{dec},$$

with $W_{acc} = (\Delta V / M_r)(dm/dt)$, $W_{dec} = -\frac{1}{2}V_r\Omega$.

Here dm/dt is the rate at which grains are absorbed by a ring of mass M_r and angular velocity Ω, V_r is the velocity of the radial drift of the ring itself, $V_r = da_p/dt$, and a_p is the major semi-axis of the ring. We use here the notation W_{dec} for the aerodynamic deceleration instead of F used in (3.102).

Let us introduce the equilibrium coefficient of the rings which is equal to the modulus of the ratio of the accelerations:

$$S \equiv \left|\frac{W_{acc}}{W_{dec}}\right| = \frac{2\Delta V}{M_r V_r \Omega}\frac{dm}{dt}. \tag{11.13}$$

Equilibrium of the ring will correspond to $S = 1$.

We remind ourselves that the width of the ε ring is larger than the sum of the other dense rings by approximately a factor 2. Micrometeorites and fast particles in the magnetosphere produce when pulverising the matter in the ε ring fluxes of broken-up matter with a large eccentricity (Durisen et al. 1989, Cuzzi and Durisen 1990). The apocentres of these flows will be distributed outside the ε ring and the pericentres inside the orbit of the ε ring in the zone of the other rings. The absorption by the inner rings of the flow of broken-up matter from the ε ring is a considerable source of angular momentum.

We note that the mechanism of ballistic transfer of angular momentum and mass in the planetary rings has been studied closely by several authors (Durisen et al. 1989, Cuzzi and Durisen 1990) mainly for the Saturnian rings.

We can estimate the rate at which grains are absorbed by the inner rings of width ΔR_{ir}, radius R_{ir}, and optical depth τ_{ir}, as follows

$$\frac{dm}{dt} = \frac{2\pi\sigma_{\rm d}\sum_i \tau_{ir}R_{ir}\Delta R_{ir}}{t_{\rm typ}}, \tag{11.14}$$

where $\sigma_{\rm d}$ is the surface density of the dust cloud produced by the micrometeorite bombardment of the ε ring, $t_{\rm typ}$ is a typical time for the capturing process of a grain, and $2\pi\sum_i \tau_{ir}R_{ir}\Delta R_{ir}$ the grain capturing cross-section (of the inner rings) with $i = 1, 2, 3, \ldots$ the ordered numbering of the inner rings. We can calculate $\sigma_{\rm d}$ assuming that the source of the mass captured by the inner rings is the ε ring:

$$\sigma_{\rm d} = \frac{R_\varepsilon \Delta R_\varepsilon \sigma_\varepsilon}{\sum_i \tau_{ir}R_{ir}\Delta R_{ir}}, \tag{11.15}$$

where σ_ε is that part of the surface density of the ε ring which has been incident on the inner rings. Substituting (11.15) into (11.14) we get an expression for the rate of absorption of dust by the inner rings:

$$\frac{dm}{dt} = \frac{2\pi R_\varepsilon \Delta R_\varepsilon \sigma_\varepsilon}{t_{\rm typ}},$$

which could, of course, have been written down at once. The expression for the equilibrium of the whole system of the inner rings can be written in the form (see (11.13))

$$\overline{S} \approx \frac{\overline{Re_{\rm d}}}{\overline{V_r}t_{\rm typ}} \frac{M_\varepsilon}{\sum_i M_{ir}}, \tag{11.16}$$

where M_ε is that part of the mass of the matter lost by the ε ring which was captured by the inner rings, $\sum_i M_{ir}$ the total mass of the inner rings, and the bar indicates a value averaged over the whole of the ring system. One can write down a similar expression for any one of the rings without the averaging:

$$S_i \approx \frac{R_{\mathrm{d}i} e_{\mathrm{d}i}}{V_{ir} t_{\mathrm{typ}}} \frac{M_{\varepsilon i}}{M_{ir}}, \tag{11.17}$$

where the quantity $M_{\varepsilon i}$ is the mass of the matter from the ε ring (and also from the inner satellites) captured by the ith ring.

For $\overline{R} = 4.5 \times 10^9$ cm, $\overline{e_{\mathrm{d}}} = 0.1$ to 0.2, $t_{\mathrm{typ}} = 10^9$ year, $\overline{V_r} = 10^{-6}$ cm/s – which corresponds to an incidence time of 10^7 years for the rings – we find that $\overline{S} = 0.75$ to 1.5 if $M_\varepsilon / \sum M_r = 50$. If we use the fact that the present mass of the ε ring is 6.1×10^{18} g and the total mass of the two widest of the inner rings – the α and β rings – is 8×10^{16} g (according to Porco and Goldreich 1987b), such a relation is fully realistic. In fact, the massive ring only needs to lose around half of its mass during cosmological times in order to guarantee the stability of the inner rings and the neutralisation of the aerodynamic friction.

We note, however, that the action of the aerodynamic friction does not reduce to a simple deceleration – it can also facilitate, through interacting with other decelerating factors, the stability of the rings. Let us consider a separate ring on which two decelerating factors act simultaneously: the decrease in the specific angular momentum as the result of the bombardment by interplanetary particles and the aerodynamic deceleration. Let us study the behaviour of ejecta with small and with large angular momenta. The aerodynamic friction decelerates not only the ring itself, but even more efficiently the ejecta, consisting of small grains. This produces an obvious effect: part of the ejecta with small angular momentumn are slowed down by the atmosphere of the planet during their first revolution and do not return to the ring – for this to happen it is sufficient to decrease the apocentre of its orbit by the width of the narrow ring (see Fig. 11.1). The aerodynamic deceleration acts less strongly on the ejecta with larger major semi-axes – large angular momentum – and shifts them towards, rather than away from the ring. As a result we obtain, on the one hand, particles with small angular momentum which rapidly drift towards the planet and, on the other hand, the ring itself which compensates for its loss of angular momentum due to the aerodynamic deceleration and the deceleration due to the interplanetary dust by extracting positive angular momentum from ejecta with large angular momentum. The interaction of the deceleration due to the interplanetary dust with the aerodynamic friction can thus lead to a neutralisation of these two destabilising factors.

Let us now consider the system of two rings which we studied earlier (Fig. 11.1) when there is a simultaneous action of the ballistic drift and the aerodynamic friction. The friction causes, on the one hand, a negative drift of the particles of the ε ring and ring 4 and, on the other hand, it even more efficiently slows down small ejecta, transferring angular momentum and mass between the rings. The stronger the action of the aerodynamic friction, the larger the positive angular momentum transferred from the ε ring to ring 4. Indeed, thanks to the aerodynamic friction ring 4 "eats away" all the ejecta between it and the ε ring, even those the pericentre of which earlier did not reach ring 4 – the friction conveys them to ring 4 after some time. As was noted by Gor'kavyĭ (1989) the action of the aerodynamic friction to a large extent is neutralised by the ballistic drift mechanism which we considered.

The estimates given here show that the action of all above mentioned factors of the formation of dust flows guarantee the stability of the inner Uranian rings which turn out to be "pages" which are maintained by the dust of the most massive and stable ε ring. The stabilising action of the dust flow is appreciably larger than the action of hypothetical shepherd satellites the existence of which Porco and Goldreich (1987b) were forced to assume.

We note that on the background of the theoretical problems which remain unsolved in the framework of the shepherd satellite paradigm there appeared papers (Esposito and Colwell 1989) in which hypotheses were advanced that the Uranian rings were unstable formations arising due to catastrophic collisions amongst a multitude of invisible shepherd satellites. However, during the twelve years that the Uranian rings have been observed no indications whatever have been detected of the disruption or smearing out of the rings so that these hypothese are only attractive because in them the stability problems are removed from the arms of the theorists. Of course, the remarkable stability of the Uranian rings is the result of natural evolutionary selection: rings with an unstable shape or an insufficient supply of angular momentum could simply not have lasted until our times.

11.2.4 Numerical Calculation of the Ballistic Drift in the Present System of Rings

Let us calculate the contemporary field of the drift velocities caused by the ejecta from the ε ring. If a ring particle moves with a relative velocity V_0 in the medium of the ejected particles of density ϱ, it undergoes an acceleration

$$\frac{F}{m} = \tfrac{1}{2} C_D S \varrho V_0 |V_0| , \tag{11.18}$$

where m is the mass of a particle with cross-section S and C_D is the drag coefficient. If we are interested in the radial drift, to a first approximation we can take into account in V_0 only the azimuthal component – using the fact that perturbations of the radial velocities do not change the semi-axes of the particles (see Sect. 3.3.5). In the two-dimensional case one needs to know the density of the ejecta, their eccentricity e_1, and their major semi-axis a_1 in order to calculate the force (11.18) and the relative azimuthal velocity (11.11). We assume that the density of the ejecta formed decreases exponentially with increasing eccentricity, $\varrho \propto e^{-e_1/e_{cr}}$ – this assumption can be justified by the fact this is one of the strongest rates of decrease, while "weaker" dependences would lead to an even larger ballistic drift. We understand by e_{cr} a characteristic eccentricity of the ejecta which is determined by a typical

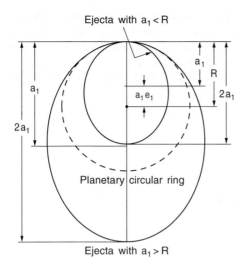

Ejecta with $a_1 < R$

a_1

a_1, e_1

a_1

R

$2a_1$

$2a_1$

Planetary circular ring

Ejecta with $a_1 > R$

Fig. 11.3. Illustration for (11.19); $a_1(1 + e_1) = R$ for $a_1 < R$, $a_1(1 - e_1) = R$ for $a_1 > R$.

ejection velocity for it of hundreds of meters per second. We shall assume that e_{cr} is equal to a few hundredths. We still must give the distribution law for the semi-axes of the ejecta. Here we consider two limiting cases: a) the ejecta are distributed uniformly over their semi-axes; b) all ejecta orbits are tangential to the orbit of their parent ring of radius R which enables us to write down the following equations for a_1 and e_1 (Fig. 11.3):

$$a_1(1 + e_1) \; = \; R \quad \text{for } a_1 \; < \; R, \quad a_1(1 - e_1) \; = \; R \quad \text{for } a_1 \; > \; R. \,(11.19)$$

The truth lies, as always, between the variants a) and b). The difference in the results of the calculations for the cases a) and b) consists basically in that in case a) the ejecta of the ε ring produce in the inner region only a positive drift–there is no negative drift in the range $R - a_1e_1 < r < R$. We assume that case b) corresponds better to reality since the ring is just for tangential trajectories of the micrometeorites the optimum target–the optical depth of the ring is maximum for a grazing collision. Moreover, the probability of finding a micrometeorite is a maximum near its pericentre (Kessler 1981), that is, exactly on grazing trajectories in contact with the ring. Hence, the majority of collisions of ring particles with micrometeorites will be grazing collisions. It is natural to expect that that part of the ejecta which is generated in grazing collisions will also have tangential orbits, that is, conditions (11.19) will be satisfied for them. Let us therefore consider case b) in detail. Calculating the distribution of the quantity $\varrho V_\varphi |V_\varphi|$ for ejecta in the region around the ε ring, we find from (11.18) the azimuthal acceleration of the interior test particles in the rings and we can use (3.102) to determine the radial drift velocity for the ring particles.

We show in Fig. 11.4 the change with distance of the radial drift velocity of a particle with a 1 cm radius for the present matter density distribution in the Uranian rings and $e_{cr} = 0.04$ – for a uniform mass distribution of the inner rings in the 41 000 to 48 500 km zone and a clear dominance by the mass of the ε ring. The drift velocity is inversely proportional to the particle size so that for a 1 m-size particle the drift will be two orders of magnitude smaller – so that, as we shall see in the next section, the probability for resonance capture of the particles is increased. We note the following features of the ballistic drift around the ε ring (curve *1*):

> on both sides there exist zones with a strong drift directed away from the ring;
> at distances larger than $a_1 e_{cr}$ the drift changes sign and is directed towards the ring;
> this drift passes through a maximum and then decreases with increasing distance from the ring.

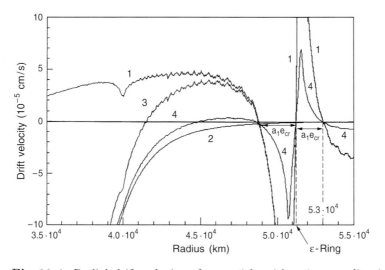

Fig. 11.4. Radial drift velocity of a particle with a 1 cm radius in the present Uranian rings as function of the planetocentric distance. Curve *1*: ballistic drift caused mainly by ejecta from the ε ring. Around the ε ring there exist zones with a strong drift directed away from the ring; at a distance of $a_1 e_{cr}$ the drift changes sign. Curve *2* shows the negative drift velocity of a particle with a 1 cm radius under the action of aerodynamic drift. Curves *3* and *4* demonstrate the resulting drift caused by the simultaneous action of ballistic and aerodynamic drifts. The ballistic drift is 5 times smaller (or the aerodynamic drift 5 times larger) for curve *4* than for curve *3*.

Any particle – or light ring – will thus drift to a massive ring until it reaches an equilibrium point at a distance equal to $a_1 e_{cr}$. Curve *2* in Fig. 11.4 corresponds to negative drift velocities of particles with a 1 cm radius under the action of the aerodynamic friction. Curves *3* and *4* show the sum of the ballistic and the aerodynamic drift velocities – without taking into account that they may neutralise one another. Curve *4* corresponds to a five times smaller ballistic drift – or a five times larger aerodynamic drift – than curve *3*. We note that for a wide range of ratios of the magnitudes of the two drifts the outer boundary of the zone of the inner rings remains practically unchanged and is approximately equal to 48 500 km. The inner boundary of the rings changes in a significant range. A very long-lived, stable positive drift zone is the zone from 45 000 to 48 300 km where the δ, γ, η, and β rings are situated. These rings – or at least the three outer ones of them – must always be subject to a resulting positive drift. The 4, 5, 6, α rings, and possibly the β ring are in a metastable region where the sign of the drift may change due to fluctuations in the atmosphere and in the dust flow, or may be close to zero.

The positive particle drift in the zone of the Uranian rings is thus a real process which agrees well with the observed distribution of rings in the zone. We can therefore proceed to an analysis of the behaviour of positively drifting particles in inner Lindblad resonance points.

11.3 Formation of the Uranian Rings in the Inner Lindblad Resonances

In this section we shall discuss the fundamental problem of the effect of resonances of the outer satellites on the inner rarefied disc in which there are drift motions; in particular we shall find out whether disc matter will accumulate in inner Lindblad resonance regions. We shall see that this problem is crucial for the formation of the Uranian rings and can be restated as follows: what is the effect of close satellites on the ring zone of the proto-disc.

The capture of matter in the outer – and not the inner! – Lindblad resonances when matter moved inside the system has been widely studied recently (see the literature review by Gor'kavyǐ and Ozernoy (1999) who have analysed in detail various capturing variants in the solar system for different directions of the drift). An important step in the capturing problem was made by Jackson and Zook (1989) who were the first to demonstrate numerically the capture of particles in many outer resonances – from 2:3 to 13:14 – from the Earth. In the same paper they predicted the existence of a heliocentric dust ring in the outer resonances from the Earth consisting of particles captured resonantly from a dust flow drifting towards the Sun due to the Poynting–Robertson effect. This ring was later discovered using the IRAS (Dermott et al. 1994) and COBE (Reach et al. 1995) spacecraft. Gor'kavyǐ et al. (1998) obtained theoretically the density distribution in this resonance ring by considering the motion of dust from 5000 asteroids and 217 comets and calculating the capture of particles in the ten most important outer Lindblad resonances from the Earth.

We present below the dynamics of the capture of matter in the inner Lindblad resonances which has not been studied in detail before (a discussion of the possibility for the capture of a particle in a different resonance can be found in a paper by Hamilton (1994). We note that Contopoulos (1975, 1977, 1979, 1981) has given some details of a general theory of inner Lindblad resonances in stellar discs.). The discovery of this new kind of capture is important for an understanding of the origin of the Uranian rings and of other structures in the solar system.

11.3.1 Elementary Capture Dynamics

In the framework of the two-dimensional problem Weidenschilling and Jackson (1993) obtained an analytical expression for the necessary condition for the capture of particles in external resonances, using the simple relation

$$-\left(\frac{da}{dt}\right)_{\text{PR}} + \left(\frac{da}{dt}\right)_{\text{res}} = 0, \tag{11.20}$$

the meaning of which is clear: the rate of decrease in the major semi-axis due to the Poynting–Robertson effect is balanced by the rate of increase of the major semi-axis due to the resonance transfer of angular momentum from the planet to the exterior particle – Weidenschilling and Jackson (1993) have studied a system of three bodies: Sun, planet, and a particle in a heliocentric orbit. The more slowly the particle moves through the resonance, the larger the chance that it gets captured.

The problem of capturing particles in an inner resonance appears to be more complicated. In that case the particle decreases its semi-axis both due to deceleration and due to a loss of angular momentum to the resonant satellite. In order to understand the resonance formation of the Uranian rings one must note an essentially important fact: resonance capture in an inner Lindblad resonance must occur, that is, an equation like (11.20) must have a solution, if there is positive particle drift – that is, when particles drift outwards.

Let this drift be a ballistic drift. In that case (11.20) becomes

$$\left(\frac{da}{dt}\right)_{\text{ball}} - \left(\frac{da}{dt}\right)_{\text{res}} = 0, \tag{11.21}$$

We can find the ballistic drift rate of change, $(da/dt)_{\text{ball}}$, by calculating it (see Fig. 11.4). To get an estimate of the slowing down of a particle in an inner Lindblad resonance by a satellite and obtain a qualitative picture of how the conditions for capture depend on the parameters of the resonance system, such as the satellite mass, the eccentricity of the particle, and the order of the resonance, we can use the following simple considerations. We consider a particle in a circular orbit with radius R_{p} between a planet of mass M and a satellite with a mass m_{s} and a circular orbit of radius R_{s}.

Up to the closest approach to the satellite, which attracts it, the particle is accelerated and after this point it is decelerated. We can estimate the gravitational acceleration due to the satellite to be

$$f \approx \frac{Gm_s}{(R_s - R_p)^2},$$ (11.22)

and the time during which there is an efficient action of the satellite on the particle as

$$\Delta t \approx \frac{R_s - R_p}{V_p - V_s},$$ (11.23)

where V_p and V_s are, respectively, the particle and satellite velocities. The change in the particle velocity is

$$\Delta V_i \approx f_i \Delta t_i,$$ (11.24)

where $i = 1$ corresponds to the first phase of the encounter (acceleration) and $i = 2$ to the second phase (deceleration). If the particle orbit is circular, we have

$$\Delta t_1 = \Delta t_2, \qquad f_1 = -f_2, \qquad \Delta V_1 = -\Delta V_2.$$ (11.25)

However, as the result of the gravitational interaction between the particle and the satellite the symmetry in the region of their encounter is broken: the particle orbit becomes elliptical. It is clear from Fig. 11.5 that the apocentre of the particle trajectory is at the maximum distance from the centre: $R'_p = a(1 + e)$ (see (3.38b)) while at the same time in the point where the particle is closest to the satellite orbit it turns out to be in the region where it is decelerated. Since the quantities a and R_p differ by an amount proportional to e^2, we can neglect this difference as compared to unity and we then get $R'_p = R_p(1 + e)$. To the same approximation the azimuthal particle velocity in the apocentre is, according to (11.10), equal to $V'_p = V_p \sqrt{(1 - e)/(1 + e)}$ with $V_p = \sqrt{GM/R_p}$. Hence, we get from (11.22) and (11.23) in the second phase for such a particle trajectory

$$
\begin{aligned}
f'_2 &= -\frac{Gm_s}{(R_s - R'_p)^2} = -\frac{Gm_s}{[R_s - R_p(1 + e)]^2} \\
&\approx f_2 \left\{ 1 + \frac{2eR_p}{R_s - R_p} \right\},
\end{aligned}
$$ (11.26)

$$
\begin{aligned}
\Delta t'_2 &= \frac{R_s - R'_p}{V'_p - V_s} = \frac{R_s - R_p(1 + e)}{V_p \sqrt{(1 - e)/(1 + e)} - V_s} \\
&\approx \Delta t_2 \left\{ 1 + \frac{eV_p}{V_p - V_s} - \frac{eR_p}{R_s - R_p} \right\}.
\end{aligned}
$$ (11.27)

We have used here the fact that

$$f_2 = -\frac{Gm_s}{(R_s - R_p)^2},$$

The unbalanced change in the particle velocity then takes the form

$$
\begin{aligned}
\Delta V &= f_1 \Delta t_1 + f_2' \Delta t_2' \\
&= f_1 \Delta t_1 + f_2 \left\{ 1 + \frac{2eR_p}{R_s - R_p} \right\} \Delta t_2 \left\{ 1 + \frac{eV_p}{V_p - V_s} - \frac{eR_p}{R_s - R_p} \right\} \\
&= e f_2 \frac{R_s - R_p}{V_p - V_s} \left\{ \frac{R_p}{R_s - R_p} + \frac{V_p}{V_p - V_s} \right\} \\
&= e f_2 \left\{ \frac{R_p}{V_p - V_s} + \frac{V_p (R_s - R_p)}{(V_p - V_s)^2} \right\}.
\end{aligned}
\tag{11.28}
$$

We have used here the symmetry properties of the phases 1 and 2 when the particle approaches the satellite moving along an unperturbed circular orbit (see Fig. 11.5),

$$f_1 \Delta t_1 + f_2 \Delta t_2 = 0, \tag{11.29}$$

which follows from (11.25).

Since the ring region is situated closer to the planet than the satellite region we have $R_p < R_s$ and $V_p > V_s$. Hence, the quantity ΔV, given by (11.28), is negative since the expression within the braces is positive and f_2 is negative.

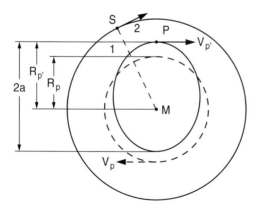

Fig. 11.5. Interaction between a ring particle and a satellite. The orbit of a particle P when there is no gravitational interaction between it and the satellite S is shown by the dashed curve, and the orbit of the same particle when the interaction is taken into account by the full-drawn curve. The numbers *1* and *2* indicate the regions of acceleration and deceleration, respectively, which are separated by the dashed straight line. The apocentre, where the particle at the given time is situated lies in the deceleration region. The point M indicates the central planet of mass M.

Starting from the above mentioned asymmetry the answer was obtained from the assumptions. Indeed, it follows from Fig. 11.5 that a particle in region 2 which is closer to the satellite orbit is decelerated more efficiently than it is accelerated in region 1. If the particle and the satellite are in a $n/(n+1)$ resonance the encounter passes through the same phase after a time interval $nT_s = 2\pi n/\Omega_s$, where T_s and Ω_s are the period and angular velocity of the satellite, the decrease in the particle velocity accumulates and the change in the major semi-axis, a_p, of the particle is given by (3.102):

$$\frac{da_p}{dt} = \frac{2f_p}{\Omega_p}, \tag{11.30}$$

where f_p is the azimuthal deceleration of the particle which causes the decrease in the major semi-axis – the radial acceleration leaves the semi-axis practically unchanged (see Sect. 3.3.5). The deceleration f_p which causes the decrease in the major semi-axis is an integral characteristic of the orbit. In other words, f_p determines the magnitude of the mean deceleration over the period between encounters with the satellite, rather than reflect the instantaneous value of the deceleration of the particle. According to this definition we have

$$f_p = \frac{dV_\varphi}{dt} \approx \frac{\Delta V \Omega_s}{4\pi n}. \tag{11.31}$$

We assume here that on average $V_\varphi \approx \frac{1}{2}\Delta V$. From (11.28), (11.30), and (11.31) we have

$$\frac{da_p}{dt} \approx \frac{e\Omega_s Gm[R_s V_p - R_p V_s]}{2\pi n\Omega_p(R_s - R_p)^2(V_p - V_s)^2}. \tag{11.32}$$

After some simple transformations we find:

$$\frac{da_p}{dt} \approx -e\frac{m}{M}V_s A_n, \tag{11.33}$$

where

$$A_n = \frac{\Omega_s[V_p/V_s - R_p/R_s]}{2\pi n\Omega_p(1 - R_p/R_s)^2(V_p/V_s - 1)^2}. \tag{11.34}$$

Using the resonance relations $\Omega_s/\Omega_p = n/(n+1)$, $R_p/R_s = [n/(n+1)]^{2/3}$, $V_p/V_s = [(n+1)/n]^{1/3}$, we get

$$A_n = \left(2\pi[n^5(n+1)]^{1/3}\left\{\left[\frac{n+1}{n}\right]^{1/3} - 1\right\}^2\left\{\left[\frac{n+1}{n}\right]^{2/3} - 1\right\}^2\right)^{-1}. \tag{11.35}$$

For large n we get $A_n = 3.2n^2$.

Both this last estimate and the general expression (11.35) demonstrate an interesting fact: the higher the order of the resonance, the stronger the action of the resonance satellite on the particle. This is qualitatively explained by the fact that as the order of the resonance increases, even though the particle rarely encounters the satellite,[2] the distance between the interacting objects becomes less and the characteristic time of interaction becomes larger. According to (11.35) the 2/3 resonance is 3.2 times stronger than the 1/2 resonance and the 3/4 resonance twice as strong as the 2/3 one. The 5/6 resonance is 17 times stronger than the 1/2 resonance, and the 12/13 one 90 times stronger. Our simple calculation agrees well with much more rigorous celestial mechanics calculations; it turns out that we have estimated the relative importance of the resonances correctly up to a factor 2. As far as the generality of (11.33) is concerned one must take into account that resonance is possible only for well defined ratios of the pericentre angles of the particle and satellite orbits. We can write down for the two-dimensional case the necessary condition for the capture of particles in an inner Lindblad resonance in a form which is analogous to the condition for capture in an outer resonance – the latter was obtained by Weidenschilling and Jackson (1993) who based their calculation of the second term on a paper by Greenberg (1973):

$$\left[\frac{da}{dt}\right]_{\text{drift}} - e\frac{m_{\text{s}}}{M}V_{\text{s}}A_n \sin\phi \;=\; 0, \tag{11.36}$$

where $(da/dt)_{\text{drift}}$ is the positive drift rate (the increase in the major semi-axis), A_n is a dimensionless coefficient which depends on the order n of the resonance and which, in the general case, can be expressed in terms of Laplace coefficients, e is the eccentricity of the particle orbit, m_{s} is the satellite mass, M is the planetary mass, V_{s} is the velocity of the resonance satellite, and ϕ is the angle from the particle pericentre to the longitude of its conjunction with the planet. We note that in the case of capture in an inner resonance we have $\sin\phi > 0$ and for the angle $0 < \phi < \pi$. For an outer resonance, which for a long time was assumed to be the only one adequate to capture particles, we have for the angle $\pi < \phi < 2\pi$ and the sign of the sine becomes negative.

11.3.2 Numerical Calculation of Particle Capture in Inner Lindblad Resonances

Let us look at the possibility for the capture of resonance particles in an inner Lindblad resonance through a direct numerical solution of the equations of celestial mechanics. Taidakova and Gor'kavyĭ (1998) have developed a new numerical scheme for the analysis of the dynamics of non-conservative

---

[2] We noted earlier that the time between encounters between the particle and the satellite under $n/(n + 1)$ resonance conditions is nT_{s} where T_{s} is the period of revolution of the satellite.

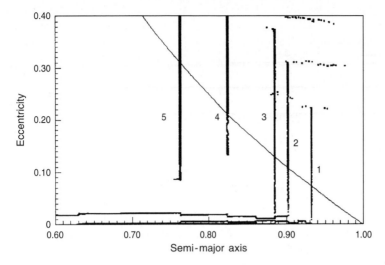

Fig. 11.6. Evolution of five drifting particles in a, e space (a is the major semi-axis and e the eccentricity; the major semi-axis of the satellite is equal to unity). Three particles starting with their initial eccentricity in the range from 0.01 to 0.001 and at a distance of 0.5 move across lower-order Lindblad resonances without being retarded but only suffering a change in their eccentricity. In curve *1* we note capture in the 9/10 resonance. Curve *2* shows capture in the 6/7 resonance and curve *3* in the 5/6 resonance. For particle 5 the initial eccentricity is 0.086, it starts close to the 2/3 resonance and is captured by it. Particle 4 is located at the 3/4 resonance and demonstrates the stability of the capture. All captured particles increase their eccentricity until they approach the satellite after which they suddenly change to non-resonance orbits.

systems and given a calculation of the evolution of a particle captured in the inner 5:6 resonance of the Earth. Using this scheme – which for the dissipationless case is the same as the scheme described in Chap. 5 – we have studied the motion of particles slowly drifting away from the planet and passing through inner Lindblad resonance points. We show in Fig. 11.6 the evolution of five particles in the (a, e) space, that is, the major semi-axis, eccentricity space; the major semi-axis of the satellite is taken to be equal to unity. Three particles (1, 2, and 3) starting with an initial eccentricity in the range from 0.01 to 0.001 pass through lower-order Lindblad resonances without being captured and suffer small jumps in eccentricity. Curve *1* shows capture in the 9/10 resonance, curve *2* in the 6/7 resonance, and curve *3* in the 5/6 resonance. To speed up the calculation we took the radial drift velocity of these particles two orders of magnitude larger than realistic – between 0.03 and 0.09 cm/s. To compensate for the increased drift velocity of these particles and to satisfy the capture conditions (11.36) we took the relative mass of the satellite to be 10^{-6} which is by the same two orders of magnitude larger than, say, Portia. The calculations showed that satellites with a mass

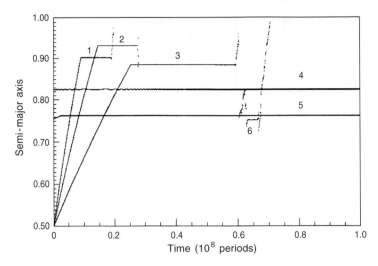

Fig. 11.7. Change with time of the major semi-axes of five drifting particles. The numbering of the curves corresponds to the cases described in Fig. 11.6. The time scale is up to one million revolutions of the satellite. The particle 3 abandoning its 5/6 resonance is briefly captured after a series of perturbations from the satellite in a non-Lindblad 17/26 resonance, indicated by the number 6.

of 10^{-8} (Portia) and 10^{-9} (Cordelia) are also successful in capturing drifting particles provided the drift velocity is decreased by two to three orders of magnitude as compared to the one used in the calculations presented in Fig. 11.6. The capture of particles in Lindblad resonances is simplified for sufficiently eccentric particles. For particle 5 the initial eccentricity was chosen to be equal to 0.086. The drift velocity was then lowered to 0.005 cm/s. To save time in calculating the drift in the non-resonance region we put the particle close to the 2/3 resonance and afterwards it was captured by this resonance. Particle 4 was simply placed in the 3/4 resonance and showed the stability of the capture.

In all cases the captured particles started increasing their eccentricity until they approached the planet and changed to non-resonance orbits. For the lower (2/3, 3/4) Lindblad resonances the eccentricity increased up to the given upper limit of the time for the calculation – up to one million revolutions of the satellite. One can easily see in Fig. 11.7, where we show how the major semi-axis changes with time, how long the particles stay in each resonance. We note an interesting fact: particle 3, leaving its 5/6 resonance, after a series of jumps – caused by changes in the orbital elements due to approaching the planet – was captured in the non-Lindblad 17:26 resonance (indicated by the number *6*) where it stayed a short time. We show in Fig. 11.8 the growth of the eccentricity of resonantly captured particles as function of time. We

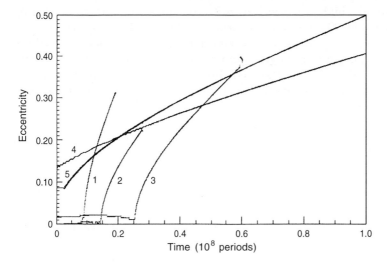

Fig. 11.8. Growth of the eccentricity of resonantly captured particles. The numbering of the curves corresponds to the cases described in Figs. 11.6 and 11.7. In an actual system inelastic collisions between the particles must fix the eccentricity at an equilibrium level.

note that in reality inelastic collisions between the particles rapidly limit the growth in the eccentricity of resonance particles.

The numerical calculations are in accordance with the necessary condition (11.36) for resonance capture and confirm the increase in the probability for capture with increasing n, e, and m_s, in accordance with (11.36). The probability for capture decreases with increasing drift velocity of the particle.

In the case of resonance capture the particle encounters the satellite always in the same place in its orbit, completing between encounters with the satellite $n+1$ revolutions around the planet, or $n+1$ epicyclic oscillations – in the comoving frame of reference. In the case of the $1/2$ resonance the particle completes two revolutions between encounters, in the case of the $2/3$ resonance three, and so on. Correspondingly each rank of resonance particles gives two humps on the average orbit (the $m = 2$ oscillation mode for the $1/2$ resonance), or three humps ($m = 3$ for the $2/3$ resonance), or $n + 1$ humps ($m = n + 1$ for the $n/(n + 1)$ resonance). This kink in the arrangement is shown by the particles at the outer edge of the B ring ($m = 2$ from the $1/2$ resonance of Mimas) and at the outer edge of the A ring ($m = 7$ from the $6/7$ resonance of Janus). If in the nucleus of a ring so many particles are accumulating that they form a collective, collisional medium the ring is reorganised: inelastic collisions can abruptly change the eccentricity of the colliding particles and the inelastic collisions can suppress the higher oscillation modes of the ring, except the very lowest: the circular ($m = 0$) or elliptic ($m = 1$) ones (see Sect. 8.4). Hence, viscosity rearranges the growing resonance ring from its initial state in a high mode with $m > 1$ into a quasi-circular or elliptic ring in the $m = 1$ mode. The resonance capture condition (11.36) is then violated due to the decrease in eccentricity e and due to the fact that some of the particles have changed their angle ϕ and moved away from the resonance influence of the satellite. Under those conditions a weak lower-order resonance cannot maintain a viscous ring which abandons the "parent" resonance and proceeds to a slow journey along the ring zone. If new drift matter enters the liberated resonance it may form yet another ring. Of course, if the ring meets with a much stronger resonance it will be captured in it as a single object. The condition for such a capture is given by an expression similar to (11.36):

$$\left[\frac{da}{dt}\right]_{\text{drift}} - K e \frac{m}{M} V_s A_n \sin \phi = 0, \tag{11.37}$$

where K is the ratio of the mass of the resonantly captured particles in the ring to the total mass of the ring.

The results given here of independent analytical and numerical calculations show a fact which is basic for the formation of the resonance structure of the Uranian rings: ringlets accumulate in the points of the inner Lindblad resonances when there is a positive particle drift.

We note once again that the preference for lower-order Lindblad resonances for the formation of the Uranian rings is determined by the history of the system, rather than by the dynamics: one can see from Table 11.2 that the close satellites which were first formed turned out to be rather far from the ring zone. Therefore there were simply no other resonances apart from the lower-order Lindblad ones in the ring zone for a long time.

Whereas the formation of an individual narrow ringlet is the result of the external resonance action of satellites, the further evolution of such a ringlet and its acquiring of an elliptical shape takes, apparently, place practically independently.

It is clear from the results of Sect. 8.4 that in a disc of inelastic particles one can observe a dissipative ellipse-instability of a non-axisymmetric ($m = 1$) perturbation of a circular ringlet – a typical spontaneous symmetry breaking of the system.

It is just this dissipative instability which made the narrow Uranian and Saturnian rings eccentric. We note that the resonance perturbations from the satellites are too insignificant and are unable to account for the observed eccentricity of the rings. It is clear from Sect. 8.4 that the narrowest ringlets have the largest chance of becoming unstable. The densest and narrowest rings, in which the self-gravitation stabilises the differential precession, can also retain their eccentricity. These stability problems are common to both the Uranian and the Saturnian elliptical rings.

11.4 The Present-Day Uranian Ring System

After the calculations we have performed we can now draw a rather detailed picture of the formation of the Uranian rings. The reliability of the main details of this picture is based upon numerical calculations and an analytical analysis, but also in a few cases on simple estimates and qualitative physical assumptions. We assume that this picture is rather logical and completely explains the observed features of the Uranian ring system and can serve as a basis for a further more penetrating study of the phenomenon of narrow eccentric planetary rings. We note that in the theory presented here the origin of the Uranian rings no longer needs a large number of inner, unseen shepherd satellites for its stability.

The history of the Uranian rings splits into two physically different temporal stages. The first one is the epoch of the formation and the free drift of the rings. It concludes by the formation of the latter from close satellites – Cordelia and Ophelia which are lying closer than the others to the ring zone and therefore cause in that zone the strongest Lindblad resonances. The second is the present stage of a stable state of the rings. A few hundred million years ago an exciting historical action unfolded, the melodramatic aspects of which were caused by the active participation of satellites with the names of well known Shakespearian heroines.

11.4.1 Epoch of Free Drift of the Rings and Its Finale with the Participation of Cordelia and Ophelia

In the Uranian ring zone there existed thus in the proto-stage a continuous disc with a positive ballistic, or aerodynamic, particle drift. Portia, the first to appear, starts to fill ring 4 through the 1/2 resonance and the ε ring through the 2/3 resonance. The 2/3 resonance is a few times stronger than the 1/2 resonance so that the ε proto-ring grows faster and becomes very massive.[3] The resonantly captured particles start to collide and form a ring which leaves the resonance and, dragged along by the positive drift, starts to move away from the planet. During the free drift epoch ring 4 travelled 950 km and the ε ring 720 km. Juliet, comparable in mass with Portia, had formed its own ring using the 2/3 resonance near the ε ring. This "baby of Juliet" intercepting part of the drifting "feeder" matter near "Portia's child" – the ε ring – grew to a considerable mass and also proceeded to become a drifting traveller. As it slowly drifted it managed to create the satellites Cressida and Desdemona which through 2/3 resonances formed the group of the η and γ rings and through 3/4 resonances filled two rings outside the ε proto-ring. By this time

---

[3] An additional favourable condition for the fast growth of the ε ring is its favourable localisation at the outer boundary of the ring zone where the destruction process which dominates in the rings when particles collide is stabilised by the sticking process which is not smaller in strength than the former.

the longer-lived of the rings – the 1/2 resonance of Portia – had long ago left ring 4 and managed to form ring 5 and also started it in its positive drift. Ring 5 managed only to move 610 km from the point where it originated. By that time a new generation of satellites – Rosalind and Bianca – formed yet another group of rings, the α and β rings, and added the δ ring to the already existing group of rings (η and γ). All these rings leave the resonances and start to move away from the planet.

The second act of the cosmogonical spectacle starts with the formation of the satellite Cordelia from the massive proto-ring of Juliet. Why did Cordelia originate from it rather than from the ε ring which lies further out? It is possible that the reason is that Juliet's proto-ring was somewhat more massive – due to its favourable position as compared to the ε ring or thanks to the significant eccentricity of Juliet itself. We see that a similar situation is observed also near the Saturnian rings where some of the satellites (Pan, Atlas, Prometheus, and others) were formed closer to the planet than the outer F ring and, apparently, earlier than it was possible for a satellite to grow from the F ring. Now the F ring – like the ε ring – perturbed as it is by the effect of its shepherds can no longer coagulate into a satellite.

It now becomes clear why the 2/3 resonance from Juliet does not contribute to the correlation between resonances and rings (see Chap. 10): it formed a satellite rather than a ring! It is very probable that from the ring formed earlier by the 3/4 resonance of Desdemona at a radius of 51 740 km and departing further from its point of origin than all other rings Ophelia was formed – the outer shepherd satellite of the ε ring.

The birth of Cordelia and Ophelia greatly changed the arrangement of the forces in the Uranian proto-disc. Thanks to their great mass and small size these satellites turned out not to be subject to the drift. Ophelia and Cordelia through their differently directed resonances "shoveled up" all ringlets in the zone between them – the ε proto-ring from the 2/3 Portia resonance, the ring from the 3/4 Cressida resonance, and, possibly, a "young" ringlet from the 4/5 Bianca resonance – into a single ring, the present ε super-ring. Cordelia and Ophelia halted the drift of this ring, capturing it in the 24/23 and 24/25 Cordelia resonances and the 13/14 Ophelia resonance (Goldreich and Porco (1987) have studied the angular momentum balance in such a system). A not less dramatic change was produced by these two young "heroines" in the fate also of the inner rings, inhibiting their further positive drift. The strongest resonances in the Uranian system are at present according to (11.35) and (11.36) the higher-order Lindblad resonances of Cordelia and Ophelia which are by orders of magnitude stronger than the lower-order Lindblad resonances from the larger, further satellites. In fact, after the formation of Cordelia and Ophelia a dense lattice of many tens of strong resonance barriers was imposed on the ring zone which prohibited a further positive drift of the inner rings. For the innermost group of rings, the 4-5-6 rings – where the 1/2 Portia resonance already managed to form a third ring (6) in that group which moved 210 km from its point of origin – the 4/5 resonance from Cordelia became such a barrier. The 6/7 resonance became the outer limit of the positive drift for the α ring and the 8/9 resonance for the β ring, from the same "stern" Cordelia. We note that even if some of the resonances from Cordelia or Ophelia turned out to be surmountable for a ring, the next stronger barrier from a resonance of higher order would certainly stop the ring. The history of the drift motions of the five inner Uranian rings was finished. However, the history of the drift of the η, γ, and δ rings continued because they were close to the massive ε ring.

11.4.2 Contemporary Picture of the Drift Equilibrium in the Rings and the Formation of the 1986U1R or λ Ring

The ε ring, capturing in it more than 90% of the mass of the ring system, produced around it, according to Fig. 11.4, a zone of drift motions directed away from it. All matter remaining between the ε ring and the satellite Cordelia started to be pushed away from the ε ring and stopped in its escape only near Cordelia itself in front of the very strong outer 129/128 resonance (the order of this resonance depends strongly on the determination of the radii of Cordelia and of the λ ring). The ringlet 1986U1R or λ ring was thus formed which is transparent and rich in dust due to strong resonance perturbations. However, the zone of the drift motion around the ε ring repelling matter beyond the equilibrium point $a_1 e_{\mathrm{cr}}$ extended much further than Cordelia and caused the opposite, negative drift of the η, γ, and δ rings. They moved towards the planet, easily surmounting the resonance barriers from the "back" until they returned to the positive drift zone with a radius of less than 48 300 km. Here there is such a dense grid of resonances of Cordelia and Ophelia that the rings no longer remained free but were captured in the closest resonance: the η ring in the 12/13 Cordelia resonance, the γ ring in the 5/6 Ophelia resonance, and the δ ring in the 22/23 Cordelia resonance. This comprises the present-day picture of the rings and their relations with resonances; it is closely connected with the angular momentum balance.

Let us write down the drift balance for rings in the region of the inner Lindblad resonances in the form

$$V_{\mathrm{ball}} - V_{\mathrm{drag}} - V_{\mathrm{res}} = 0, \tag{11.38}$$

where V_{ball} is the radial positive ballistic drift velocity, V_{drag} the negative aerodynamic deceleration, and V_{res} the resonance drift localised in the resonance regions. For the η, γ, and δ rings we can neglect the aerodynamic friction (Goldreich and Porco 1987) and write down for them the stable balance

$$V_{\mathrm{ball}} - V_{\mathrm{res}} = 0. \tag{11.39}$$

Small displacements of the ring – without shifting it through a resonance barrier – lead to a return of the ring to a stable point of balance. The 4-5-6 and α-β rings are situated in a zone of strong aerodynamic friction and a weakened ballistic positive drift from the ε ring (see Fig. 11.4) so that they cannot manage to "reside" in resonances which take away from them the last angular momentum. These rings leave the resonance regions thus getting rid of the negative term V_{res} in their balance relation:

$$V_{\mathrm{ball}} - V_{\mathrm{drag}} = 0. \tag{11.40}$$

This is just the reason why one cannot notice even a single, be it ever so weak, resonance in the immediate vicinity of the inner five Uranian rings – they carefully avoid them! We note that the balance of their angular momentum even under such an economical "form" of life all the same cannot be recognised as brilliant. One shows easily that the balance relation (11.40) is unstable: a small radial shift of the ring can become fatal for it and it can fall to the planet – the reverse variant of an unlimited moving away from the planet will be suppressed by the closest Cordelia resonance. It is difficult to say how unstable the balance (11.40) is especially since we have not studied quantitatively the amplification of the positive ballistic drift connected with the presence of aerodynamic friction. This ballistic-aerodynamic drift, like the aerodynamic friction, is amplified as one approaches the planet and that may make the balance (11.40), if not completely stable, at least close to such a situation.

Survival under conditions of marginal stability could be through mutual assistance of the inner rings. It is clear from Fig. 11.4 that each ring has around it an attractive zone and can capture a neighbouring ring in a stable point at a distance $a_1 e_{\mathrm{cr}}$. Hence, a ring can form coupled groups of rings which are able, as we showed earlier, to guarantee their stability more successfully than a single one, absorbing ejecta from neighbouring rings. Apparently, the α and β rings form a doublet – a pair of coupled rings. The 6-5-4 rings form a stable drift triplet in which the inter-ring distances are smaller than in the α-β doublet since each outer ring of the triplet simultaneously attracts two rings.

Other variants of angular momentum balance are realised for the ε ring:

$$V_{\mathrm{res}} - V_{\mathrm{drag}} = 0, \tag{11.41}$$

where the total (and positive) resonance action of the shepherd satellites neutralises the deceleration in the atmosphere and other negative drifts. The λ ring shows a balance between the positive action of an outer Cordelia resonance and a negative ballistic drift from the ε ring:

$$V_{\mathrm{res}} - V_{\mathrm{ball}} = 0, \tag{11.42}$$

Essentially the Uranian rings are a remarkably balanced and stable system which very efficiently uses the supply of its angular momentum, 99% of which is contained in the outer massive ε ring.

11.4.3 Dust Structures in the Rings

The slower drift motions may turn out to be essential for the evolution of the rings when the main drifts equilibrate each other in the Uranian rings; they cause a pulverisation of the matter of thin rings – especially this is true for the inner group of Uranian rings which are situated in a metastable region.

We show in Fig. 11.9 the ballistic drift velocity of particles with a 1 cm radius on the present Uranian rings, assuming that only the nine narrow Uranian rings – except the ε ring – are the sources for the ejecta. We take e_{cr} to be 0.03. We draw attention to the fact that the ballistic drift velocity must be rather high for dust particles with a size of a fraction of a millimeter. It is clear that as a result there appear in the rings a number of equilibrium drift flows – there where the slope of the drift curve is negative and the curve intersects the line of zero drift. Wandering particles in the Uranian rings – which are not part of stable rings – must drift to this point, absorbing fragments from the main rings. A comparison of this curve with the distribution of dust in the Uranian rings (see Fig. 2.14) shows that the dust structure at a radius of 41 200 km – which we shall call the 6a ring – lies close to the inner equilibrium drift point of the 6-5-4 triplet. Close to the outer equilibrium point of that triplet there is situated a symmetric dust structure – which we shall call the 4a ring – at a distance of 43 200 km. The dust structure at a distance of 46 500 km corresponds to the equilibrium drift point between the β and η rings. Between the δ ring and the satellite Cordelia – at 48 500 km to 49 600 km – there is a region filled with dust – to be called the δ_a ring – corresponding to the zone near the drift point of the ε ring – where the negative drift from this super-ring changes sign and becomes positive. Apparently, small dust particles drifting in the zone of the Uranian rings can also be captured in both lower-order and higher-order inner Lindblad resonances. A sufficient condition for such a capture must be the positive drift at the resonance circumference. Such a drift, caused by the ε ring, certainly exists in the inner part of the Uranian rings – from 46 000 to 48 500 km. It is just here, between the main η, γ, and δ rings, that a large number of narrow dust ringlets are situated which must correspond exactly to resonances in that region. In the remaining ring zone – from 41 000 to 46 000 km – the sign of the ballistic drift is determined by the arrangement of the inner rings. In the 41 000 to 41 300 km and 42 750 to 43 000 km regions where there must exist a positive ballistic drift from the 6-5-4 triplet one should note dust structures corresponding to the 2/3 resonance of Ophelia and the 3/4 and 4/5 resonances of Cordelia. The δ_a dust region is mainly situated in the negative drift region of the ε ring so that it is relatively uniform. Resonances can only locally lower the density in such regions, accelerating the negative drift of particles in them.

It is possible that the broad and transparent dust ring 1986U2R situated at between 37 000 and 39 500 km corresponds to an equilibrium point of the whole Uranian ring system. Its features can be noticed in Fig. 11.4 – there where in the 39 000 km region the zone is situated of the second maximum

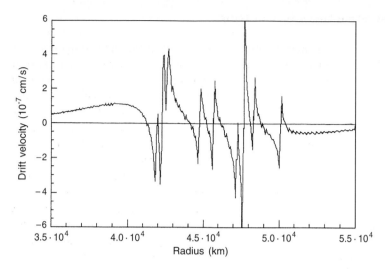

Fig. 11.9. Small-scale behaviour of the radial ballistic drift velocity of a particle with a 1 cm radius in the present Uranian rings. We assume that the ballistic drift from the ε ring is completely balanced by aerodynamic friction, that is, that only the nine other light narrow Uranian rings provide the sources for the ejecta.

of the ballistic drift curve *1*. When all drift factors are taken into account more rigorously this maximum must cause the appearance of a new stable drift point in the 38 000 to 39 000 km region.

The existence of stable drift points and local drift motions can clarify also many features of the structure of the Saturnian rings. It is not impossible that the existence of a system of small satellites and the F ring itself near Saturn are also connected with a drift point formed by the pulverisation of matter of the A ring– with $e_{cr} = 0.03$ to 0.04. An estimate of the resonance exchange of angular momentum between the A ring and the closest satellites indicates a rapid increase in the distance between the ring and the satellites (Goldreich and Tremaine 1982). A ballistic exchange of angular momentum between the A ring and satellites can significantly change the drift time of small satellites which is due to the viscous exchange of angular momentum with the rings. Drift motions in the Cassini Division caused by the pulverisation of the edges of the B and A rings may in the equilibrium drift point cause the formation of the ring structure observed at two Saturn radii (see Fig. 2.3). In the zone where there is a strong positive drift near the B ring the $1/2$ resonance of Mimas may cause the formation of the narrow eccentric Huygens ringlet proceeding after its formation to drift positively inside the Cassini Division. We draw attention to the similarity of the structure of the regions near the inner edge of the A (from 1.5 to 1.525) and of the B rings (from 2 to 2.025; see Fig. 2.3). Apparently, aligning the diminishing region near the steep edge, this is a negative ballistic drift zone which as one goes away from the steep edge is changed into a positive drift zone. While the drift effects of the A and B rings cancel one another in the Cassini Division, the positive drift zone in the C ring extends rather far – quite possibly up to the 0:1 Titan resonance. This also caused the abundant formation of narrow eccentric rings which are maximally densely scattered over the outer region of the C ring – for details see App. VI. We must not forget the fact that it is normal for the rings to leave their "parent" resonances and move away from the planet. One of such ring-wanderers is possibly the Maxwell ringlet which at this moment is not connected with a resonance. Apparently, it was formed in the 1/2 Atlas resonance, left it and shifted towards the B ring.

11.4.4 On the Stability of the Sharp Edge of Non-resonance Elliptical Rings

We have shown (Fridman and Gor'kavyĭ 1989a, Gor'kavyĭ and Fridman 1990) that the main reason for the diffusive stability of the edge of rings is their non-circular shape. Indeed, the spreading out of a circular ring is the transition of colliding – or drifting – particles into neighbouring quasi-circular orbits which do not intersect with the main ring. When an elliptical ring spreads out such neighbouring orbits will also be elliptical with eccentricities which are almost similar. In an aspherical field the ring precesses at a rate which differs from the precession rate of the orbit of a separate particle – due to the difference in major semi-axes. Because of this the orbit of the particle inevitably intersects the ring and the particle, using up the energy of its relative motion in inelastic collisions, sticks to the main bulk of the particles. We can estimate how far a ring "releases" its particles before it captures them again. This distance will give us an estimate of how sharp the edge of the ring is.

Let us consider two embedded non-intersecting ellipses – with the same foci and almost the same lines of apsides – with major semi-axes a and $a + \delta a$ and eccentricities e and $e + \delta e$ ($\delta a \ll a$, $\delta e \ll e$) Let us find the angle $\Delta\varphi_{\min}$ which is the minimum angle between the lines of apsides for which the ellipses come into contact:

$$\Delta\varphi_{\min} \approx \arccos(-A) - \arccos A, \tag{11.43}$$

where $A = \delta a/2ae + a\delta e/\delta a$.

Taking into account that δe is connected with collisions and comparable with the thermal eccentricities ($\delta e \simeq h/a$) we get the condition for "overtaking": $h < \delta a < 2ae$. If the strong inequalities $h \ll \delta a \ll 2ae$ hold we have: $\Delta\varphi_{\min} \approx \delta a/2ae + a\delta e/\delta a$.

The edge of the ring overtakes the particle due to the differential precession $\Delta\omega_{\mathrm{p}}$ in a characteristic time

$$t_{\mathrm{p}} \approx \frac{\Delta\varphi_{\min}}{\Delta\omega_{\mathrm{p}}} \approx \frac{4Aa^3}{21J_2\Omega R_{\mathrm{p}}^2\delta a}. \tag{11.44}$$

Comparing (11.44) with the time for the diffusive smearing out of the edge, $t_d \simeq (\delta a)^2/\nu$, we find the equilibrium value of δa. For instance, for the ε ring the sharpness of the edge is $\delta a = 400$ to 750 m for free flight times of the particles equal to $t_c = (1 \text{ to } 10)/\Omega$. Such a sharp edge agrees well with the observational data. We note that even a very small eccentricity ($e = 10^{-6}$ to 10^{-5}) must steeply enhance the diffusive stability of the edge of the ring. Of course, the balance of angular momentum inside a narrow elliptical ring has not yet been studied adequately and here much remains unexplained. For instance, the variable width of the ring may play an important role in the diffusive stability of elliptical rings, that is, the fact that the eccentricity of the outer edge is larger than that of the inner edge. Such an eccentricity gradient leads to the outer layers of the ring near the pericentre moving faster than the inner ones and it transfers angular momentum to them. Not less important is the ability of the particles to accrete onto one another in inelastic collisions and this is one of the most important stabilising factors for the Neptunian arcs.

On the whole we can say that the main holes in the bottom of the vessel – the model of the Uranian rings – have been closed up and all remaining problems can be solved without appealing to additional undiscovered satellites and without predicting a short life of the observed rings.

11.4.5 Biographical Information About the Uranian Rings

We collect the results obtained in Table 11.3, giving the data about the formation, total drift, and causes for the stability of the Uranian rings.

After ascertaining the fundamental differences in the balance of angular momentum of the different groups of rings many features of the Uranian rings become clear, for instance:

The steepest edges in the whole ring system are the outer edges of the γ and δ rings and the inner edge of the ε ring, that is, just those edges of rings on which they "pile up" at the resonance barriers of Cordelia and Ophelia (Gresh 1989);

the η, γ, δ, and λ rings are the most circular of all the Uranian rings with the γ and δ rings poorly approximating a Kepler ellipse due to large resonance perturbations (French et al. 1986);

the five inner rings which have a non-resonance balance and are free of resonance perturbations have a shape close to an ellipse with a significant eccentricity.

Table 11.3. Curricula vitae of the Uranian rings

Ring	Cause of formation	Shift from point of formation [km]	Cause of present stability	Kind of balance
6	1/2 Portia	+ 211	part of 6-5-4 triplet	$V_\mathrm{ball} = V_\mathrm{drag}$
5	1/2 Portia	+ 609	part of 6-5-4 triplet	$V_\mathrm{ball} = V_\mathrm{drag}$
4	1/2 Portia	+ 949	part of 6-5-4 triplet	$V_\mathrm{ball} = V_\mathrm{drag}$
α	1/2 Rosalind	+ 666	part of α-β doublet	$V_\mathrm{ball} = V_\mathrm{drag}$
β	2/3 Bianca	+ 512	part of α-β doublet	$V_\mathrm{ball} = V_\mathrm{drag}$
η	2/3 Cressida	+ 40	r.c.[a] (12:13 Cordelia)	$V_\mathrm{ball} = V_\mathrm{res}$
γ	2/3 Desdemona	− 197	r.c. (5:6 Ophelia)	$V_\mathrm{ball} = V_\mathrm{res}$
δ	3/4 Bianca	− 539	r.c. (22:23 Cordelia)	$V_\mathrm{ball} = V_\mathrm{res}$
Cordelia	2/3 Juliet	+ 663	its own supply of angular momentum	
λ	129/128 Cordelia	$\{- 700\}^\mathrm{b}$	r.c. (129:128 Cordelia)	$V_\mathrm{res} = V_\mathrm{ball}$
ε	2/3 Portia	+ 720	r.c. (24:23 Cordelia,	$V_\mathrm{res} = V_\mathrm{drag}$
	3/4 Cressida	− 161	13:14 Ophelia)	

[a] r.c. ≡ resonance capture.
[b] Drift value in braces gives the maximum possible shift from the ε to the λ ring.

11.5 Conclusions

Let us compare the conclusions of the theory constructed here for the origin, evolution, and stability of the Uranian rings with the statements in the Gor'kavyĭ–Fridman model, which were made more than thirteen years ago.

Of course, there follow, in addition to the points mentioned in Table 11.4, a number of more precise and more closely investigated conclusions from the complete theory of the formation of the Uranian rings:

4. The resonance formation and the present stability of the rings is determined by the positive drift in the Uranian rings.

5. After their formation all the rings left the resonances which produced them and moved away over distances of up to a few hundred kilometers. The outer rings (η, γ, δ, λ, and ε) turned out to be captured in strong higher-order resonances – from 5/6 to 129/128 – of the satellites Cordelia and Ophelia which were formed last of all. The inner rings escaped the resonances and formed the stable 6-5-4 triplet and α-β doublet.

6. The resonance capture and drift motion of the particles explains the origin of not only the main Uranian rings, but also of the finer dust structures between them.

Table 11.4. Comparison between the Gor'kavyĭ–Fridman model (1985) and the present theory of the origin of the Uranian rings

Gor'kavyĭ–Fridman model (1985)	Present theory of resonance formation of the Uranian rings
1. The position of the rings is determined by 1/2, 2/3, 3/4 type resonances of undiscovered satellites beyond the outer boundary of the rings, situated in the 50 000 to 82 000 km zone.	All nine main* Uranian rings originated as the result of resonance accumulation of matter in lower-order inner Lindblad resonances (1/2, 2/3, 3/4) of a series of satellites situated in the 59 000 to 76 000 km zone.
2. Each of the five predicted satellites determines simultaneously the position of two rings.	Each of the four satellites predicted in 1985 (Portia, Desdemona, Cressida, and Bianca) caused simultaneously the formation of two rings in different spatial groups through just those resonances which were indicated by Fridman and Gor'kavyĭ. (The action of the fifth predicted satellite is replaced by the 1/2 resonances of Rosalind and Portia.)
3. No satellites are formed inside the ring zone. The features of the outer ε ring is explained by the presence of a shepherd satellite.	The existing theory of the stability of the rings confirms this cosmological conclusion: the rings do not need shepherd satellites inside the ring zone. The only Uranian ring which is surrounded by shepherds is the peculiar ε ring.

* The most noticeable ones, discovered in 1977.

7. The theory for the formation of the Uranian rings when applied to the Saturnian rings explains the origin of the set of narrow ringlets in the C ring and several structures in the Cassini Division (for a mechanism for the formation of the division itself, see Appendix V).

We can state that the principal features of the origin and evolution, structure and stability of the remarkable Uranian rings have been excellently explained by a model for the formation of the Uranian rings in inner Lindblad resonances of close satellites which was published by the present authors in 1985, half a year before these Uranian satellites were observed by the Voyager-2 spacecraft.

12 Origin, Dynamics, and Stability of the Neptunian Rings

You are old, Father William, ... Yet you balanced an eel on the end of your nose – What made you so awfully clever?

Lewis Carroll, Alice's Adventures in Wonderland.

12.1 Hypotheses About the Dynamics of the Incomplete Neptunian Rings (Arcs)

After reports had been published about the Neptunian arcs in 1984-1985 theoreticians became very active and between that time and the Voyager-2 fly-past of Neptune on 1989 August 24 a number of hypotheses were advanced about the dynamics of such pathological formations.

12.1.1 Dynamical Models of the Neptunian Arcs in the Framework of the "Shepherd" Concept

The main problem in the dynamics of the arcs is: Why do they not spread out along the orbit? One can easily show that it takes only a few months for differential rotation to pull a bunch of particles of similar sizes apart. Lissauer (1985) put forward a very obvious model according to which the Neptunian arcs were the analogue of the asteroid swarms – the "Greeks" and "Trojans" – along the orbit of Jupiter in the Lagrangian L_4 and L_5 points – 60° along the orbit from Jupiter. A satellite preventing the spreading out of the Neptunian arcs along the orbit must be rather massive – with a diameter of about 200 km (Lissauer 1985). Frequent collisions cause an additional destabilisation of the cluster which may be prevented by the introduction of yet another shepherd satellite on the neighbouring inner orbit (Lissauer 1985). Lissauer's model is shown in Fig. 12.1a.

Goldreich, Tremaine, and Borderies (1986; to be quoted as GTB) published a paper in which they proposed a satellite near Neptune with a radius of 100 km (or more) in an inclined orbit. This satellite should stabilise a whole set of "arcs" on orbits in an $n/(n+1)$ resonance with its orbital frequency (see Fig. 12.1b). These two models were discussed at length in later papers by several authors (Borderies 1987, Brahic and Sicardy 1987, Lin et al. 1987, Sicardy 1987, Porco 1991). Lin et al. (1987) also studied the stabilising effect of the non-axisymmetric harmonics of the gravitational field of the planet

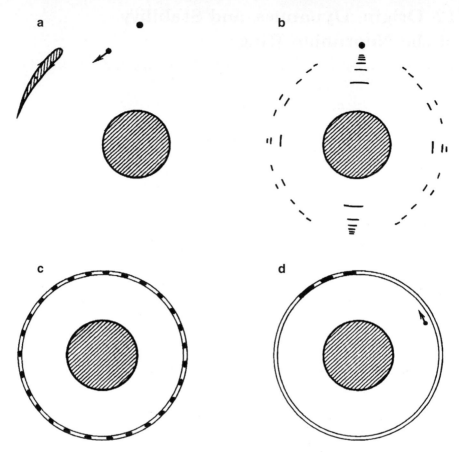

Fig. 12.1. Theoretical models of the Neptunian arcs. (**a**) Lissauer's model; (**b**) the Goldreich-Tremaine-Borderies model; (**c**) Gor'kavyĭ's model; (**d**) outer Neptunian ring according to the Voyager data.

in the case when there are non-axisymmetric oscillations of the planetary surface. One must note that a number of papers (Dobrovolskis et al. 1989) considered the possibility of the existence of a polar ring around Neptune. The massive satellite Triton on its inclined orbit would be the shepherd for such a ring.

12.1.2 Model of Intrinsically Stable Neptunian Arcs on a Continuous Ring

The present authors (Gor'kavyĭ and Fridman 1988a,b) presented the following important summary of the facts:

1) "The fact that the Neptunian rings exist indicates that the massive satellite Triton does not prohibit the formation near the planet of a regular system of rings and planets very similar to that of Uranus."
2) "As in the Uranian system the position – and even the stability – of the Neptunian rings may be determined by resonances from a series of as yet undetected satellites near the outer boundary of the rings."
3) "The observed arcs may be part of continuous rings with a considerable azimuthal variability in width and optical depth. ... Such an assumption makes it considerably easier to solve the stability problem of the Neptunian rings."

In a short note in the July 1989 Astronomical Circular – a month before the Voyager-2 fly-past of Neptune – Gor'kavyĭ (1989c) proposed a more detailed model for the origin and dynamics of the Neptunian rings:

1) "It is possible that the Neptunian rings in their proto-stage were narrow concentric rings similar to the Uranian rings and formed through the action of resonances with inner satellites. Afterwards the Neptunian rings – more massive than the Uranian ones – were subject to a gravitational instability and each ringlet split up into a series of elliptical vortices rotating in the opposite direction."
2) "Particles in such an elliptical cloud rotate around its centre of mass along an epicycle. All the particles in the vortex have the same major semi-axis but different eccentricities – from 0, at the centre of the cloud, to $e_{max} \simeq 10^{-4}$ to 10^{-3}, at its edge. The orbit of the centre of mass may also be eccentric. The strict equality of the major semi-axes guarantees the same orbital periods and precessional motion so that the vortex is stable and does not spread out along the orbit. The shear velocities in the vortex are small."
3) "If there is no differential rotation in the vortex the magnitude of the random particle velocities will be nearly zero. The maximum particle size of the particles is not determined by collisional break-up, as in the usual planetary rings, but by a break-up during the motion in an elliptical vortex with an axis ratio equal to $a/b = 2$. Near Neptune the particle radius may reach several hundred meters."
4) "In each orbit there may be several tens of vortices. Why are these particle accumulations not destroyed through mutual collisions? Resonance interactions with outer satellites may diminish the probability for collisions. Transparent dust rings, which cannot be seen from the Earth, may play an important role in the stabilisation of the rings. Such a circular and

rather narrow ring will lead to equal major semi-axes of the orbits of the nearest vortices moving along weakly elliptical orbits which intersect the dust ring. This guarantees equal velocities of the orbital and precessional motion of the clouds. The vortices will therefore move along an epicycle, each in its own part of the orbit and will not approach neighbouring clusters. The absorption of the dust component may play an important role in the angular momentum balance inside the vortex itself."

5) "In conclusion we note that when studying the outer orbits which have 1:2, 2:3, 3:4, 4:5, 1:3, and 3:5 resonances with the observed Neptunian rings we find orbits at 66 200, 73 500, 76 700, 82 900, 86 700, 89 000, and 95 500 km in which there may be small Neptunian satellites. Some of these orbits are in resonance with one another and they may turn out to be empty. The radii of the rings in resonance with these orbits are 41 700, 54 500, 56 500, 60 500, 63 300, and 68 000 km."

We show in Fig. 12.1c such proposed rings with vortices.

Von Weizsäcker (1943) was the first to consider in his cosmogonical model vortices with a similar organised particle motion. Sicardy (1987, 1991) noted in his numerical calculations that particles may carry out consistent epicyclic motions around the Lagrangian L_4 and L_5 points. In an unpublished note (Brophy 1989), which Brophy privately communicated in 1991 to one of the present authors, a similar model of the Neptunian arcs, which would consist of a set of epicyclic clumps, was proposed after the Voyager-2 fly-past.

12.1.3 The Voyager-2 Fly-Past near Neptune in August 1989

The results of this fly-past during which six new Neptunian satellites and several rings were discovered are given in Table 2.6 and Fig. 12.1d.

Smith et al. (Wilford 1989) noted that the standard shepherd model was unable to explain the stability of the observed arcs. Afterwards Porco (1991) produced evidence to show that there is a correlation between the positions of the arcs and the minima of the potential due to Galathea – an outer satellite 900 km from the ring with the arcs. Porco arrived at this conclusion by comparing the lengths of the arcs and the order in which they were arranged along the orbit with the angular pattern of the minima of the gravitational potential of Galathea which is in a 43:42 resonance. The period of the distribution of the minima along the orbit is 4.186°. This was considered to be an argument in favour of the GTB shepherd satellite model. However, the main problem – the stability of the Neptunian arcs and the fact that they do not spread out along the orbit – remains unsolved in Porco's paper. According to the GTB theory the shepherd satellite must have an appreciable inclination – in their paper they consider the hypothetical case of a satellite with $i = 0.1°$. Galathea's inclination – relative to the plane of the ring with the arcs – is statistically uncertain: $i = 0.02 \pm 0.01°$. Porco considered the maximum possible inclination, $i = 0.03°$, for Galathea and obtained a resonance width – within which the satellite can control the position of the particles – of 0.6 km. Using the GTB theory and the observed resonance perturbation of the arcs of 30 km she determined Galathea's mass to be $(2.12 \pm 0.08) \times 10^{21}$ g. With a value of 79 ± 12 km for the radius of Galathea this gives a density of the satellite of 1.0 ± 0.5 g/cm^3.

Such a small resonance width cannot explain the fact that the arcs do not spread out – with an observed width of 15 km – along the orbit. A comparison of the terrestrial and the Voyager data shows that the arcs are conserved during several years (Sicardy 1991). The existence in the arcs of fifteen separate clumps, discovered when the Voyager-2 data were analysed (Ferrari and Brahic 1994; see Plate 13), is a not less complicated problem for the GTB theory. One can easily show that such bunching with a size of between 10 and 15 km, which is "carved out" from a normal differentially rotating disc, will be destroyed within one or two hours. Finally, we note that for various models of the Neptunian arcs particle collisions are a destabilising factor (Horanyi and Porco 1993, Foryta and Sicardy 1996).

Let us consider the agreement between the model of the Neptunian rings proposed by the present authors (Gor'kavyĭ and Fridman 1988a,b, Gor'kavyĭ 1989c) and the Voyager-2 data. Indeed, it turned out that near Neptune there is a regular system of rings and satellites, in many ways similar to that near Uranus with the arcs simply being condensations in a continuous ring. The idea that the Neptunian rings were more massive than the Uranian ones was clearly confirmed. A comparative analysis (see Sect. 2.8) showed that the Neptunian rings were formed from a disc which was approximately two orders of magnitude more massive than the disc near proto-Uranus, which might lead to a Jeans instability (see Fig. 2.19). This fact also explains the existence inside the ring zone of a region with three satellites: clearly the mechanism of the collisional break-up of particles in a differentially rotating disc ceases to operate if the self-gravitation of the cluster is larger than the gravitational effect of the planet and the differential rotation is suppressed. The rings are therefore, as before, unable to penetrate further than their well defined limit, but satellites may be formed inside the gravitationally unstable region.

We note that even in a gravitationally unstable disc one cannot expect situations when near each ring there is a pair of satellites – as assumed by the shepherd model for the Uranian rings; in fact, one needs for this a fantastic tenfold alternation of gravitationally stable zones – where the rings are formed – and gravitationally unstable zones – where the satellites are formed – and on top of this one must satisfy the requirement that the zones where we have accretional feeding to the satellites do not overlap. Between the three inner Neptunian satellites, which clearly emerged from a single gravitationally unstable region, no rings have been observed which indicates the overlap of the feeding zones of the whole of the intersatellite space.

The hypothesis (Gor'kavyĭ 1989c) that there exist narrow dust rings stabilising the Neptunian arcs turned out to be viable. We shall in what follows

consider in detail the dynamics of vortices in a dust ringlet and we shall show
that such a configuration is stable and can explain all features of the arcs observed by Voyager. However, before that we shall consider the problem of the
connection between the observed Neptunian rings and satellite resonances.

12.1.4 Connection Between Satellite Resonances and the Neptunian Rings

The present authors (Gor'kavyĭ and Fridman 1988b, Gor'kavyĭ 1989c) attempted to calculate the radii of unknown Neptunian satellites using available data of terrestrial observations of the Neptunian arcs – while noting that
these calculations depended completely on the reliability of the observational
data. In the first paper the calculations were carried out on the basis of earlier published data (see left-hand side of Table 2.6) most of which had not
been confirmed. The second paper was based on later, improved data (see
right-hand side of Table 2.6). The data about the radii of six observed arcs
were reduced to the following values: 41 700, 54 500, 56 500, 60 500, 63 300,
and 68 000 km (see Sect. 12.1.2). For each proposed arc 6 outer resonance
orbits – for the simplest resonance ratios – were considered. From the 36 orbits found this way we chose seven pairs of orbits which were positioned very
close to one another. If we now replace each pair of these close orbits by
a single "smeared out" orbit and "set up" a satellite on it, it will interact
resonantly simultaneously with two rings. According to the principle formu-
lated in Sect. 10.1.5 about the maximum probability of finding a satellite on
an orbit where it can facilitate the simultaneous formation of two rings, the
number of orbits on which one could possibly find satellites was limited to
seven. The Voyager observations detected only 4 narrow ring details and only
two of the radii coincided closely with terrestrial observations – the 1989N3R
ring at 41 900 km and the 1989N1R ring at 62 900 km. The number of rings
turned out to be less than the number of satellites and the principle of double resonances – one satellite determines the position of two rings – scarcely
operates. Only the 1989N2 satellite at 73 600 km was predicted correctly: the
assumed value of the radius was 73 500 km. This "success" is connected with
the fact that the 1989N1R ring, used to "locate" this satellite, turned out to
be reliable. One "coincidence" from the seven above mentioned orbits is an
accuracy to be expected in the case of seven orbits if there is confirmation
for only two of the six rings which had been "observed" earlier.

We show in Fig. 12.2 the Neptunian ring and satellite system proposed
on the basis of terrestrial observations and Gor'kavyĭ's calculations (1989c)
as well as the actual system. We indicate the most noteworthy resonances
of the new Neptunian satellites. The orbital radii of several satellites occur
in the list of 36 orbits, but they could not have been picked out reliably
from the total number. On the whole one may think that the mechanism
for the formation of the Neptunian rings is close to the mechanism for the

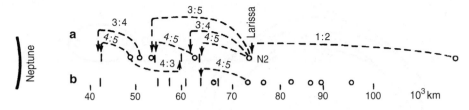

Fig. 12.2. Ring and satellite system discovered near Neptune by Voyager-2 (**a**) and Gor'kavyĭ's hypothetical system (**b**). (**a**) The most noteworthy resonances have been indicated; (**b**) The assumed resonance connections between satellites and rings. The agreement between the orbit of one of the new satellites with a predicted orbit is clear.

origin of the Uranian rings. It is true that it is complicated to prove this because there are so few rings and resonances – which makes it impossible to apply statistical methods. We note that the coincidence of even a single orbital radius for a predicted and detected satellite with an error of 100 km can – with a 99% probability – not be random.

12.2 Stability of a Separate Epiton

> ... she made out the proper way ... which was to twist it up into a sort of knot, and then keep tight hold of its right ear and left foot, so as to prevent its undoing itself...
>
> *Lewis Carroll, Alice's Adventures in Wonderland.*

This section reflects the studies of the dynamics of the Neptunian arcs following the Voyager fly-past (Gor'kavyĭ 1989b, Gor'kavyĭ and Taĭdakova 1991a,b, 1993, 1994, Gor'kavyĭ et al. 1991).

12.2.1 Particle Motion in an Epiton

Let us consider four elliptical orbits with the same eccentricity and major semi-axes lying in the same plane and differing only in their pericentre length (Fig. 12.3a). If we choose on each of these orbits the initial position of a single particle, for instance, in the way shown in Fig. 12.3a, the epicyclic motion of all four particles will be the same in the rotating frame of reference (Fig.12.3b).

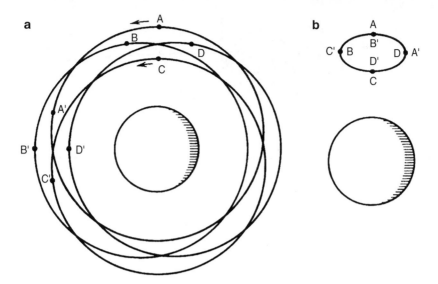

Fig. 12.3. Four elliptical orbits in the same plane with the same major semi-axes and eccentricities but different periapse arguments $(0°, 90°, 180°, 270°)$. (**a**) in the laboratory frame of reference; (**b**) in the frame of reference which rotates with the orbital frequency along the same epicyclical trajectory (Gor'kavyĭ and Taĭdakova 1994). A, B, C, and D indicate the initial positions of the four particles and the primed letters their positions after a quarter of an orbit.

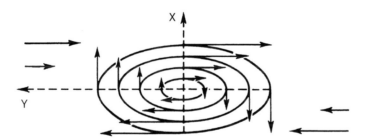

Fig. 12.4. Trajectories of particles in an epicyclic vortex (epiton) in the rotating frame of reference (Gor'kavyĭ and Taĭdakova 1993). The arrows with their appropriate length show the velocities of the particles in the vortex and – for comparison – those of particles moving along circular orbits with different radii in the same frame of reference.

We can similarly construct on a single epicyclic trajectory a whole chain of N particles which in the non-rotating frame of reference form a rosette of N non-coinciding elliptical orbits. Reducing the eccentricity of the particle orbits in the non-rotating frame to zero we get in the rotating frame a set of imbedded epicyclic trajectories densely occupied by particles. In the non-rotating frame of reference this will be a bunch of particles with their centre of

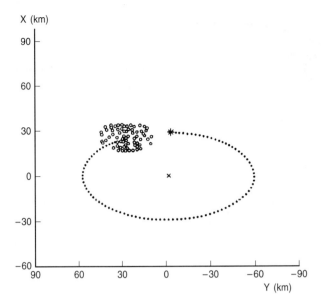

Fig. 12.5. Motion of an epiton of 50 particles in the rotating frame of reference along an epicyclic trajectory with eccentricity $e = 4.7 \times 10^{-4}$, $R = 62\,900$ km (Gor'kavyĭ and Taĭdakova 1993). The large asterisk indicates the initial position of the epiton and we show the trajectory of its centre of mass for an almost complete orbit.

mass moving around the planet and a very flat cloud, which without changing the orientation of the particles in the cloud – the line between two particular particles is always directed in the same direction (see Fig. 12.3a) – undergoes regular oscillations of its shape. In the rotating frame the shape and the position of the bunch of particles will be constant; in return the particles will perform in the cloud a reverse epicyclic rotation forming a distinctive vortex which rotates in the reverse direction (Fig. 12.4). All particles in such an epicyclic vortex, or clump, have the same orbital major semi-axis and, hence, also the same orbital period around the planet. For simplicity we shall call such a configuration an "epiton". It is clear that such a structure is a stationary one. We shall consider the problem of its stability in what follows.

If we separate in the rotating frame some part of the epiton and discard the other particles, such a part will also be an epiton. If the selected part has the shape of an ellipse with a 1:2 semi-axes ratio the shape of the new epiton will remain unchanged when it moves. If the centre of the new epiton is not the same as that of the old one, this epiton will move along an epicyclic orbit as a single entity (Fig. 12.5).

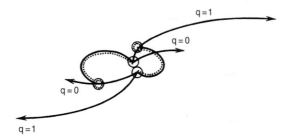

Fig. 12.6. Collision in the field of the planet of two spheres in the rotating frame for both the elastic and the inelastic case; the planet is at the bottom of the figure. The double line shows the motion of the particles before the impact. After an elastic collision the trajectories diverge strongly (shown by the long arrows) – as a result the particles move into more eccentic orbits. After an inelastic collision the trajectories diverge considerably less – see the short arrows with a non-vanishing length – since even for an absolutely inelastic collision the tangential components of the velocity are conserved.

12.2.2 Stability of an Epiton of Inelastic Particles

Let us consider an epiton consisting of a fixed number of particles which collide with one another with a constant inelasticity coefficient. How stable is such a structure against inevitable perturbations?

Gor'kavyĭ et al. (1991) have carried out a numerical simulation of the evolution of such a structure. Each particle was an inelastic sphere which moved only in the x, y-plane. Collisions between such spheres have been considered by many authors (see, for instance, Trulsen 1971). If the two spheres before the collision had the velocities V_{x10}, V_{y10} and V_{x20}, V_{y20} while the coordinates of their centres at the moment of impact were x_1, y_1 and x_2, y_2, their velocities after the collision can be found from the following equations (Trulsen 1971):

$$V_{x1} = V_{x10} - \tfrac{1}{2}B(1+q)(x_1 - x_2), \quad V_{y1} = V_{y10} - \tfrac{1}{2}B(1+q)(y_1 - y_2),$$
$$V_{x2} = V_{x20} + \tfrac{1}{2}B(1+q)(x_1 - x_2), \quad V_{y2} = V_{y20} + \tfrac{1}{2}B(1+q)(y_1 - y_2),$$

(12.1)

where

$$B = \frac{(V_{x10} - V_{x20})(x_1 - x_2) + (V_{y10} - V_{y20})(y_1 - y_2)}{(x_1 - x_2)^2 + (y_1 - y_2)^2}.$$

(12.2)

Here q is the restitution coefficient. We show in Fig. 12.6 the particle trajectories before and after the collision for different values of the inelasticity coefficient.

The calculation was carried out for Neptune with the orbital radius of the vortex equal to 63 000 km and the maximum eccentricity of the particles equal to 1.5×10^{-4} which corresponds to a 18.9×37.8 km size vortex. The initial arrangement of the particles was given by means of a databank of random

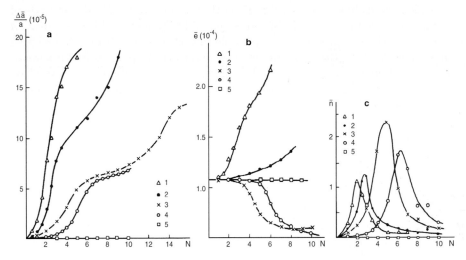

Fig. 12.7. Evolution of an elliptical vortex of size 18.9×37.8 km; the maximum eccentricity of the particles at the edge of the vortex is $e = 0.00015$ (Gor'kavyĭ et al. 1991). The calculation was carried out for 100 particles of radius $R = 0.25$ km and orbital radius 63 000 km; q is the restitution coefficient; *1*: $q = 1$; *2*: $q = 0.5$; *3*: $q = 0$ (at $t = 0$ the minimum distance between the centres of the particles GN $= 2R$); *4*: $q = 0$ (GN $= 3.5R$); *5*: $q = 0$ (GN $= 4R$; for GN $= 4R$ there are no collisions); (**a**) growth of the root-mean-square deviation of the major semi-axes of the orbits of the individual particles of the vortex from the initial major semi-axis – which was the same for all particles – as function of the number N of revolutions around the planet; (**b**) change in the mean eccentricity of the vortex particles during the evolution; (**c**) change in the collision frequency per particle and per revolution. It is clear that a collisional epiton is destroyed after a few revolutions – between 20 and 60 hours – while a collisionless vortex is stable.

numbers excluding cases where the particles would overlap. A typical epiton of 100 particles produced in this way is shown in Fig. 12.9 of Sect. 12.2.3. The initial velocities are determined in such a way that the particles perform an elliptical motion around the centre of the epiton (in the coordinate system shown in Fig. 12.4). There is to first approximation in the eccentricity a simple expression for these initial velocities in terms of the initial particle coordinates: $V_x = \frac{1}{2}y$, $V_y = -2x$. These formulæ are in a system of units in which the angular velocity is equal to unity (see Sect. 5.2) and in the rotating frame of reference. For circular orbits of the dust ring the velocities depend as follows on the coordinates: $V_x = 0$, $V_y = -1.5x$.

Calculations have shown (Fig. 12.7) that a vortex of tens or hundreds of rather large particles rapidly becomes unstable: frequent collisions cause an increase in the deviations of the particle major semi-axes from the original ones which leads to the appearance of different orbital periods and a spreading of the bunch along the orbit. For elastic particles this process would be faster and with a general increase in eccentricity, but for absolutely inelastic particles slower and with a reduction in eccentricity.

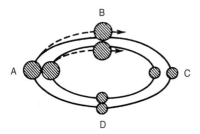

Fig. 12.8. The physics of the establishment of random velocities in an epiton of finite-size particles. The epiton consists of particles moving in a rotating frame of reference along similar ellipses (see Fig. 12.4). This means that the distance between two neighbouring trajectories along the major semi-axis is larger than along the minor semi-axis. Small particles moving along neighbouring orbits from the point C to the point D will therefore approach each other. Large particles which are in contact in the point A cannot continue to move towards the point B along their "own" trajectories. They displace each other in opposite directions, forcing a motion along the dashed trajectories. The dashed orbits have different rotational periods and the epiton "heats up" rapidly.

It is clear from our calculations that the larger the size of the particles the less stable the vortex. The physical meaning of this effect is rather simple and is illustrated by Fig. 12.8. If the initial particle coordinates are chosen randomly one always finds in the points A and C closely bunched particles – at distances less than two diameters. Such particles will, when moving to the points B and D along orbits which are getting close to one another, start to displace each other in their trajectories. The maximum rate of such a displacement is half a particle radius in a quarter orbit. This means that our particles in collisions acquire a velocity $\cong \Omega R$, where R is the radius of the sphere. This velocity arises even for absolutely inelastic particles. Later the "hot" particles start to collide with colder neighbours shifting them from their correct orbit and the particles acquire drift velocities. A shift in the centre of the epicyclic by Δy at once gives a relative collision velocity of $\cong \Omega \Delta y$ and this leads to a rapid scattering of the system. One shows easily that if we initially distribute particles in the vortex such that the interparticle distances in those parts of the epiton which in Fig. 12.8 are indicated by A and C are more than two particle diameters our vortex becomes collisionless

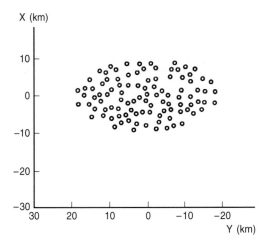

Fig. 12.9. Initial state of a typical epiton of 100 particles each with a radius of 0.4 km (Gor'kavyĭ and Taĭdakova 1993).

and practically eternal – of course, provided there are no perturbations. On the other hand, any perturbation – from a satellite or from an external particle on a circular orbit – will break up the epiton in a very short time.

Gor'kavyĭ et al. (1991) showed that an epiton can be stable if there exists a mechanism reducing the spreading of the major semi-axes before the epiton manages to disintegrate. In the next subsection we shall follow Gor'kavyĭ and Taĭdakova (1993) and consider the stability of an epiton of particles which stick together after inelastic collisions and break up when they reach a certain size under the action of tidal forces or collisions with fast particles from the continuous ring. The amalgamation – and subsequent break-up – of the particles turns out to be the natural factor which averages out the major semi-axes of the particle orbits and prevents the break-up of an epiton.

12.2.3 Evolution of an Epiton in Resonance with a Satellite

Let us consider a system of particles which stick together when they collide with one another – such a situation is typical for systems of gravitating particles with collision velocities less than the escape velocity from the particle surface. To more or less maintain a constant number of particles in the system we introduce a mechanism for the break-up of particles which reach a certain size. As a perturbing external factor we consider an inner satellite (Galathea) in a 42:43 resonance orbit. The initial arrangement of the particles in the epiton is shown in Fig. 12.9.

The passage of a massive satellite – with a mass which is 3.5×10^{-8} that of the planet – in resonance and near the epiton causes a strong perturbation in the major semi-axes of the particles (Fig. 12.10). Such a spread in the semi-

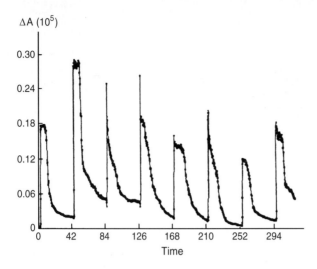

ΔA (10⁵)

Fig. 12.10. Root-mean-square deviation ΔA of the major semi-axes of particles in an epiton when passing through inner resonances of a satellite with a mass which is 3.5×10^{-8} times that of the planet M_p (Gor'kavyĭ and Taĭdakova 1993). The size of the epiton is 18.9×37.8 km, it consists of 100 particles with $R = 0.2$ km, $R_\mathrm{orb} = 62\,900$ km, and the initial eccentricity of the centre of mass of the epiton is 4.7×10^{-4}. The time is measured in orbital revolutions. The peaks correspond to the passing through of the satellite – every 42 revolutions.

axes corresponds to a dispersion in the velocities of the particles and thus also in the collision velocities by up to 2 cm/s. For an effective sticking together of the particles at such velocities the gravitational field of the particles must be sufficiently strong and this is possible for a diameter of $\cong 1$ km. An efficient energy loss in collisions may lower the requirements on the gravitational field and the size of the particles. It is clear from Fig. 12.10 that before the next passage of the satellite the spread in the orbital semi-axes is reduced by a factor of between 5 and 10. We note that an epiton of particles with a 1 km diameter is stable if it is a collection of 15 to 20 particles. The collision frequency of the particles and the number of particles in the epiton changes with a period of 42 revolutions (Figs. 12.11 and 12.12). The number of particles and the collision frequency also show faster oscillations with a period of a single revolution (Figs. 12.13 and 12.14).

The vortex can thus be considered to be a stable system which reacts on various actions as a single entity. In fact, it behaves like a satellite of the planet which itself may move along an elliptical orbit. In Fig. 12.15 we show the trajectory of the centre of mass of the epiton when it is perturbed by a satellite. After the passage of the satellite the eccentricity of the epiton increases. However, if the satellite passes close to the epiton slightly earlier, the eccentricity of the epiton may decrease (Fig. 12.16). We show in Fig. 12.17 the evolution of the eccentricity of the epiton during 320 revolutions.

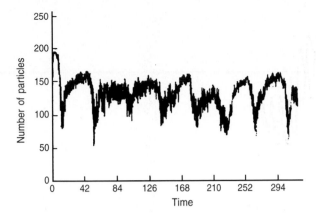

Fig. 12.11. Change with time in the number of particles in an epiton (Gor'kavyĭ and Taĭdakova 1993). One sees the 42 revolution period.

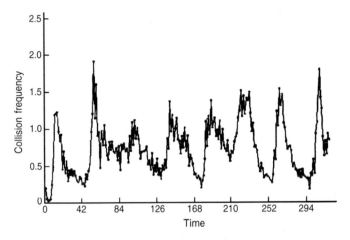

Fig. 12.12. Collision frequency (per particle and per revolution) of the particles in an epiton (Gor'kavyĭ and Taĭdakova 1993). One notes clearly the 42 revolution period.

Longer calculations confirm the periodicity of the changes in the eccentricity of the epiton with a period of about 350 revolutions.

Let us consider the changes in the major semi-axis of the centre of mass of an epiton which are connected with the effect of a close satellite. We show in Fig. 12.18 the change in the semi-major axis of an epiton during 90 revolutions. The times when the satellite passes by are noted by sudden changes in the semi-major axis by more than 10 km. However, more important than the decrease of the semi-axis at the moment the satellite passes by is the residual change in the major semi-axis of the epiton. One can note in

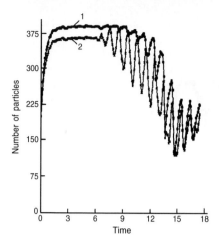

Fig. 12.13. Fast oscillations in the number of particles in an epiton after the passage of a satellite (*1*; Gor'kavyĭ and Taĭdakova 1993). *2*: Number of particles at the initial boundaries of the epiton. The amplitude of the oscillations indicates the large changes in shape and/or size of the epiton.

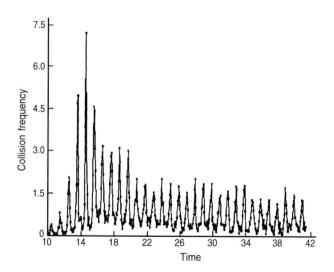

Fig. 12.14. Fast oscillations in the collision frequency in an epiton after the passage of a satellite (Gor'kavyĭ and Taĭdakova 1993).

Fig. 12.18 that as the result of the action of the satellite the epiton moves away from the satellite and its semi-axis is increased – we bear in mind that the satellite lies between the planet and the epiton.

Let us study the problem of the change in the major semi-axis of an epiton over long time intervals through the interaction with a satellite in resonance

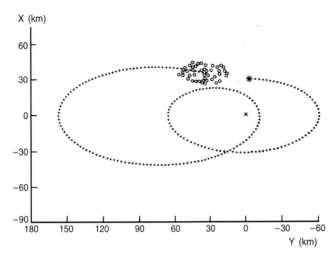

Fig. 12.15. Increase in the eccentricity of an epiton after the passage of a satellite (Gor'kavyĭ and Taĭdakova 1993).

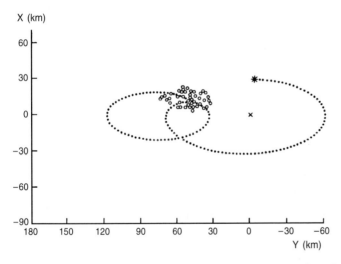

Fig. 12.16. Decrease in the eccentricity of an epiton after the passage of a satellite (Gor'kavyĭ and Taĭdakova 1993).

for different eccentricities of the satellite. We show in Fig. 12.19 the initial position of the epiton relative to the orbit of a satellite in resonance in an eccentric orbit. Altogether there are 43 such zones of length $360°/43 = 8.372°$ along the satellite trajectory. These zones are twice the length of the zones considered by Porco (1991). This is connected with the fact that we are considering an eccentric satellite in the plane of the rings which is producing

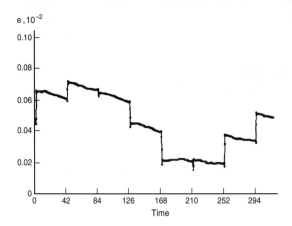

Fig. 12.17. Change in the eccentricity of an epiton during 320 revolutions (Gor'ka-vyĭ and Taĭdakova 1993). The characteristics of the epiton are given in the legend to Fig. 12.10.

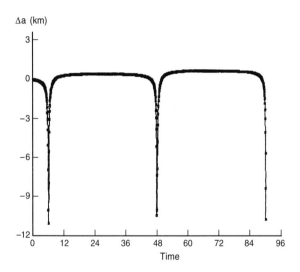

Fig. 12.18. Change in the major semi-axis a of the centre of mass of an epiton under the action of an inner satellite with mass $M = 3.5 \times 10^{-8} M_{\mathrm{p}}$ (Gor'kavyĭ and Taĭdakova 1993). The sharp changes in the semi-axis indicate the passage of a satellite.

an outer Lindblad resonance. Porco considered a corotational resonance connected only with the inclination of the satellite. We show in Fig.12.20 the evolution of the major semi-axis of the epiton during a period of 1350 revolutions under the action of a satellite with a 10^{-4} eccentricity. It is clear that during that time – which corresponds to 1.5 terrestrial years – the major

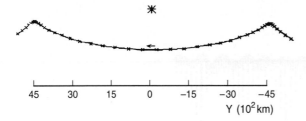

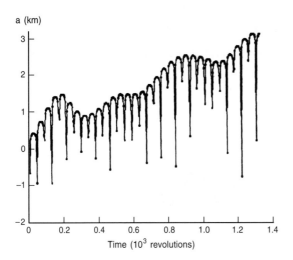

Fig. 12.19. Initial position of an epiton relative to the trajectory of a satellite, which is in resonance, with an eccentricity of 0.01 (Gor'kavyĭ and Taĭdakova 1993). The planet is below the picture, we are in the rotating frame of reference, and the satellite moves from right to left.

Fig. 12.20. Growth of the major semi-axis a of an epiton under the action of a satellite with mass $M = 3.5 \times 10^{-8} M_{\mathrm{p}}$ and $e = 10^{-4}$ (Gor'kavyĭ and Taĭdakova 1993). The depth of the peaks is changed due to a lack of resolution.

semi-axis has increased by more than 3 km. This means that the epiton has acquired a considerable drift velocity to the right and has been shifted over several thousand kilometers relative to our frame of reference (see Fig. 12.19). Hence, the epiton has left the stability zone in which it was initially.

The situation changes fundamentally when the eccentricity of the satellite is increased to $e_{\mathrm{cr}} \equiv 5 \times 10^{-4}$. A satellite with an eccentricity e larger than e_{cr} captures the epiton in its resonance and restricts the changes in its major semi-axis and its azimuthal coordinate. We show in Fig. 12.21 the evolution of the major semi-axis of an epiton of 20 to 30 particles with a particle diameter of about a kilometer under the action of a satellite with a mass equal to $2 \times 10^{-8} M_{\mathrm{p}}$ and an eccentricity equal to 2×10^{-3}. The calculation stretched over 13 500 revolutions (15 years) but in Fig. 12.21 we show only the

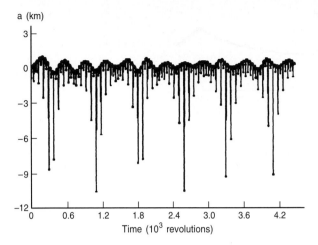

Fig. 12.21. Evolution of the major semi-axis of an epiton which is in resonance with a satellite of mass $M = 2 \times 10^{-8} M_p$ and $e = 2 \times 10^{-3}$ (Gor'kavyĭ and Taĭdakova 1993).

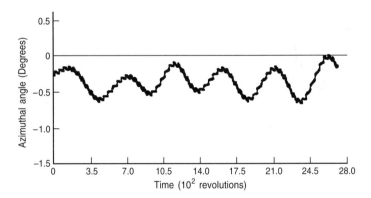

Fig. 12.22. Evolution of the azimuthal angle of an epiton which is in resonance with a satellite of mass $M = 2 \times 10^{-8} M_p$ and $e = 5 \times 10^{-4}$ (Gor'kavyĭ and Taĭdakova 1993). The straight line corresponds to the satellite perigee position – the minimum of the wave (see Fig. 12.19). The amplitude of the librations is $0.3°$, the period 350 revolutions, and the serrations correspond to a 42 revolution period.

data over the first five years (4500 revolutions). There is clearly a distinctive period of 350 revolutions – or 140 days.

Similar calculations were also carried out for a satellite mass equal to $3.5 \times 10^{-8} M_p$. The maximum length of one of the calculations was 20 000 revolutions, corresponding to 22 years in real time. We detected no signs whatever of epiton instability. We show in Fig. 12.22 the change during three years in the azimuthal angle of the epiton as a result of resonance capture.

The satellite mass was $2.1 \times 10^{-8} M_p$ and its eccentricity 5×10^{-4}. We note the following interesting features: as noted by Porco (1991), the arcs are 1.5 km further from the planet than the Lindblad resonance from Galathea. One can see in Fig. 12.21 that the major semi-axis is, indeed, for most of the time shifted from the resonance radius which corresponds to a zero shift of the major semi-axis. For a satellite mass of $3.5 \times 10^{-8} M_p$, corresponding to a satellite density of 1.7 g/cm^3 – close to the density which we assumed for the disc material and for the small Neptunian satelites (see Table 6.1) and to the observed density of Triton – this shift of the major semi-axis can reach 1.5 km. It is clear from Fig. 12.22 that the equilibrium position of the epiton in the "wave hole" is shifted to the right-hand edge (see Fig. 12.19). The shift sometimes reached 1.5° to 2°. This, together with the considerable librations of the epiton can cause considerable discrepancies between the pattern of the epitons and the positions of the potential maxima. We note that the passage of a satellite in the case of resonance capture always takes place when the epiton is in the apocentre, that is, during the encounter the satellite is close to its pericentre and the epiton close to its apocentre; the distance between them is the maximum possible during opposition. We note also that the eccentricity of an epiton which is not captured in resonance with a satellite cannot reach the observed values of 4.7×10^{-4} – which value corresponds to the measured wave-shaped perturbation of the arc (Porco 1991).

Gor'kavyĭ and Taĭdakova (1993) have thus shown that there exist multi-particle azimuthally limited objects which are stable against external perturbations and which may be in resonance with close satellites. Inelastic collisions between the particles in such objects are the main stabilising factor. It is natural to consider such objects – the epitons – as the main structural elements of the Neptunian arcs.

In the next section we shall follow Gor'kavyĭ and Taĭdakova (1991a, 1993, 1994) and consider the problems of the formation of the arcs from a set of epitons and their interaction with a continuous dust ring and with one another.

12.3 Formation of Arcs on a Continuous Ring

> First it marked out a race-course, in a sort of circle, ("the
> exact shape doesn't matter," it said), and then all the party
> were placed along the course, here and there. There was no
> "One, two, three, and away," but they began running when
> they liked, and left off when they liked, so that it was not
> easy to know when the race was over.

> *Lewis Carroll, Alice's Adventures in Wonderland.*

12.3.1 Break-Up of a Ring Under the Action of a Satellite Resonance

It is clear that in its proto-stage the 1989N1R ring was an "ordinary" continuous ring. How did it disintegrate into a set of epitons? Without pretending to give a final answer we shall present the result of one calculation by Gor'kavyĭ and Taĭdakova (1993).

Let us consider a ring with a length of 17° perturbed by an inner satellite with a mass equal to $3.5 \times 10^{-8} M_{\mathrm{p}}$ (Fig. 12.23a). A special method makes it possible to operate without simulating the rest of the ring and without interfering with the dynamics of the chosen section. After the satellite has produced a spiral wave in the ring the perturbed upper boundary of the ring starts to shift relative to the perturbed lower edge. This means that after some time there appear in the ring 42 regions where the particle orbits intersect. Taking the sticking together of the particles into account we may expect that the ring disintegrates into 42 particle agglomerations. We show in Fig. 12.23b two bunches which have appeared as the result of the collisional evolution of the continuous ring during 600 revolutions under the action of a satellite in resonance. Unfortunately Gor'kavyĭ and Taĭdakova (1993, 1994) did not follow the evolution of the ring up to the formation of epitons because at the moment an epiton is formed particles undergo triple, quadruple, ..., collisions which are difficult to describe numerically. These calculations can therefore only be considered as an illustration of the decay of a ring into epitons – the process we are discussing here.

There remain a whole series of unexplained problems. The possible role of the self-gravitation of the ring is unclear. Can two epitons merge to form a single larger one? Will there be 42 epitons which are formed from the proto-ring or are there factors which will change that number? Since there is no model of the decay of the ring it is impossible to give an answer: why did epitons appear only in the 1989N1R ring of the many narrow rings with shepherd satellites, others being Saturn's F ring, the ring in the Encke Division, the Uranian ε ring, or Neptune's 1989N2R ring. In the first case one may assume that in order to form an epiton it is necessary that some conditions on the particle density in the ring, its width, and the strength of the action of

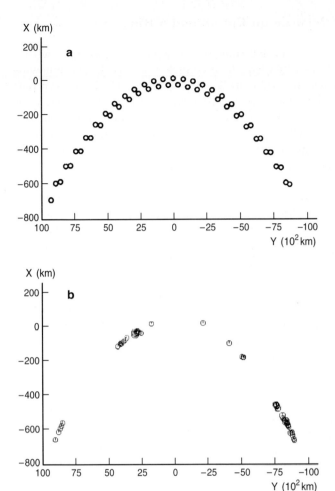

Fig. 12.23. Perturbation of a $17°$ section of the ring with a 40 km width of 50 particles with a satellite with $R = 15$ km, a mass $M = 3.5 \times 10^{-8} M_{\mathrm{p}}$, and $e = 10^{-3}$ (Gor'kavyĭ and Taĭdakova 1994). The scale along the x-axis is extended. (**a**) initial position; (**b**) after 600 revolutions, showing a tendency to pile up into 2 bunches.

the satellite are satisfied. It is clear that a perturbed transparent ring is not able to form epitons since at the point where different flows intersect there is not a sufficiently strong collisional interaction between the particles.

We shall leave these unexplained problems for further investigations and shall consider the dynamics of epitons which have already been formed in a continuous ring. The existence of a continuous ring is clearly connected with the fact that part of the material of the proto-ring did not get into epitons.

12.3.2 Interaction Between an Epiton and a Ring

The dynamics of an epiton is determined not solely by resonance with a satellite. The effect of Galathea can, for instance, not explain the following fact: Why do the epitons after they are formed not each stay in its own part of the orbit but accumulate in chains, forming the arcs? Why do epitons when approaching one another remain at distances of several hundred kilometers without sticking together? Clearly there must be yet another important factor in the dynamics of the epitons which acts two-fold: globally by collecting the epitons over an extension of hundreds of thousands of kilometers in the orbit and locally by not allowing them to get closer than a few hundred kilometers.

We shall follow Gor'kavyĭ and Taĭdakova (1991a) and consider an epiton with zero eccentricity imbedded in a circular dust flow – Fig. 12.4, where we show the velocities and the directions of motion of particles in a circular ring, may serve as an illustration. We shall call this ring a "dust ring" although we do not know the size of the particles in it which may reach appreciable magnitudes although they will nevertheless clearly be smaller than the particles in the epiton which have radii of several hundred meters. When the epiton is surrounded by the dust flow the small particles in the ring collide with the rather large particles in the vortex. We shall study the effect of this flow on the vortex analytically, neglecting the self-gravitation of the epicyclic cluster and considering the ring as a dissipative continuous medium flowing through the epiton without changing. Of course, the particles of the ring are absorbed by the particles of the epiton but, in turn, they must – because of the requirement that the system is stationary – dislodge some of the matter of the epiton into the continuous ring: in fact, the velocities of the small particles in the ring reach relative to the epiton values of a few meters per second and the gravitation of the large particles of the epiton cannot protect them from breaking up in collisions.

We write down from Eqs. (3.90) those which connect the change in the orbit of a body with the increases in the radial (ΔV_r) and transverse (ΔV_t) velocities in the inertial frame of reference:

$$\frac{da}{dt} = \frac{2}{\Omega\sqrt{1-e^2}} \left(\Delta V_r e \sin f + \frac{p}{R}\Delta V_t \right),$$

$$\frac{de}{dt} = \frac{\sqrt{1-e^2}}{\Omega a} \left[\Delta V_r \sin f + \Delta V_t (\cos E + \cos f) \right], \qquad (12.3)$$

$$\frac{d\varpi}{dt} = \frac{\sqrt{1-e^2}}{\Omega ae} \left[-\Delta V_r \cos f + \Delta V_t \left(1 + \frac{R}{p} \right) \sin f \right],$$

where f and E are the true and the eccentric anomalies. If the eccentricity is small ($e \ll 1$) we get from (12.3) the following approximate formulae:

$$\frac{da}{dt} = \frac{2}{\Omega} \left(\Delta V_r e \sin f + 2\Delta V_t \right), \qquad (12.4)$$

$$\frac{de}{dt} = \frac{1}{\Omega a}\left(\Delta V_{\mathrm{r}}\sin f + 2\Delta V_{\mathrm{t}}\cos f\right), \tag{12.5}$$

$$\frac{d\varpi}{dt} = \frac{1}{\Omega ae}\left(-\Delta V_{\mathrm{r}}\cos f + 2\Delta V_{\mathrm{t}}\sin f\right). \tag{12.6}$$

Let us consider the signs of the quantities occurring in Eqs. (12.4) to (12.6) as functions of the true anomaly f which is measured clockwise from the pericentre (see Fig. 12.4).

We have:

in the first quarter of the orbit $(y > 0, x < 0)$:
$$\sin f > 0, \quad \cos f > 0, \quad \Delta V_{\mathrm{t}} < 0, \quad \Delta V_{\mathrm{r}} < 0;$$
in the second quarter of the orbit $(y > 0, x > 0)$:
$$\sin f > 0, \quad \cos f < 0, \quad \Delta V_{\mathrm{t}} > 0, \quad \Delta V_{\mathrm{r}} < 0;$$
in the third quarter of the orbit $(y < 0, x > 0)$:
$$\sin f < 0, \quad \cos f < 0, \quad \Delta V_{\mathrm{t}} > 0, \quad \Delta V_{\mathrm{r}} > 0;$$
in the fourth quarter of the orbit $(y < 0, x < 0)$:
$$\sin f < 0, \quad \cos f > 0, \quad \Delta V_{\mathrm{t}} < 0, \quad \Delta V_{\mathrm{r}} > 0.$$

(12.7)

The physical meaning of the signs of the increases in the velocities is simple: in the upper half-plane the transverse component of the velocity of the particles in the epiton is smaller than the velocity of the grains in the ring on circular orbits so that collisions of the grains with the particles accelerate the latter and $\Delta V_{\mathrm{t}} > 0$ in the second and third quarter of the orbit. Correspondingly, in the lower half-plane the transverse component of the velocity decreases when moving in the slower ring medium: $\Delta V_{\mathrm{t}} < 0$ in the first and fourth quarters. When the particles in the vortex move away from the planet the radial component of the velocity is positive and decreases through braking in the dusty ring medium: $\Delta V_{\mathrm{r}} < 0$ in the first and second quarters of the orbit. When the particles move towards the planet their radial velocity component is negative and decreases in absolute magnitude so that $\Delta V_{r} > 0$ in the third and fourth quarters of the orbit.

From (12.4) and (12.7) it follows that there are two forms of drifting non-self-gravitating epitons in a dusty ring medium: a slow planetocentric drift connected with the first term in (12.4) and a gradient drift caused by the second term in (12.4).

If the density of the dust flow accumulating on the epiton in the upper half-plane is larger than the density of the opposite flow in the lower half-plane the resulting drift will be in the direction of the higher density. The drift in the zone of the highest density means that the position of the vortex in the dust ring is stable: any attempt to expel the vortex from the ring causes an opposite drift. The epiton will always "look for" the position of highest density in the flow. The first term in (12.4) is proportional to the eccentricity and becomes important only in the case of a uniform medium of a dusty ring when the radial drift connected with the density gradient vanishes. The slow planetocentric drift leads clearly to a shift of the vortex to the inner boundary of the ring until the drift directed towards the planet is cancelled by the gradient drift in the opposite direction.

It follows from (12.5) and (12.7) that the eccentricity of particles moving along an epicyclic in the dust flow must decrease.

Let us consider Eq. (12.6). It follows from (12.7) that if we neglect the self-gravitation of the epiton the azimuthal drift in the ring vanishes whatever the density distribution along the radius. We emphasise that we are dealing with a drift along the orbit without a change in the major semi-axis.

It was shown by Gor'kavyĭ and Taĭdakova (1991a) that these analytical conclusions are confirmed by numerical calculations which take the self-gravitation of the epiton into account. The gravitational field of the epiton was extrapolated by the gravitational field of a three-dimensional ellipsoid, strongly flattened along the z-axis. The corresponding expressions for the potential outside and inside this configurations have been given by Moulton (1914; see Appendix IV). The equations describing the dynamics of the grains are similar to those given in Sect.5.2 with the appropriate changes in the expression for the gravitational potential. In the numerical calculation of the motion and the compression of the epiton, it was no longer considered as a collection of particles but modelled by a set of imbedded epicyclic test trajectories the changes in which due to collisions with the particles in the continuous ring were calculated using the equations for the osculating orbital elements (see Chap. 3 and Eqs. (3.1) to (3.3)). The trajectories of the dust grains of the ring in the field of the planet and the gravitational potential of the epiton were calculated using the difference scheme of Sect. 5.2. The dust ring was usually simulated by 60 trajectories.

The numerical calculations showed: in a dust flow the epiton is compressed and drifts along the radius in the direction of the highest density in the ring. We note that a solid body would behave in the opposite way – it would be expelled from the dust flow.

In the case of a dust ring which is uniform along the radius the calculations revealed two kinds of drifts – the above-mentioned insignificant drift towards the planet and an azimuthal drift connected with the self-gravitation of the epiton and directly proportional to its surface density divided by the density of the continuous ring. The drift is directed in the direction opposite to that of the rotation of the ring. For an epiton with dimensions 18.9×37.8 km, a thickness of 100 m, and a surface density of 20 g/cm^2 the velocity of the azimuthal drift in a dust ring with a density of 1 g/cm^2 is equal to 5 cm/s. The radial planetocentric drift is independent of the self-gravitation and for such an epiton is equal to 2×10^{-3} cm/s.

It is clear that a narrow continuous ring which equalises the major semi-axes of all vortices facilitates the stability of a set of vortices such as were detected by Voyager-2 in the outer Neptunian ring.

12.3.3 Formation of a Stable Chain of Epitons (Arcs)

Let us consider a very important and delicate effect: how do self-gravitating epitons react on the presence in the ring of other epitons? It is clear that this is just the clue to the existence of a dense chain of epitons – we remind ourselves that they, 11 of them, have been observed only in one arc and that the regularity of their arrangement is incontrovertible (see Plate 13). Calculations (Gor'kavyĭ and Taĭdakova 1991a) have shown that the formation of a chain of epitons is connected with the interaction of the vortices through a perturbation of the ring. We note that we have already met in Sect. 6.2 with the mutual perturbation of the feeding zones of two satellites in the disc.

Let us consider the perturbation caused in a ring by an epiton which is modelled by a gravitating three-axial ellipsoid. The epiton absorbs part of the grains of a continuous ring and the main mass of the ring passes through itself acting on the grains through its gravitational field. The next epiton also absorbs part of the dust ring further passing through the material of the ring. The gravitational field of each epiton produces a spiral density wave in the continuous ring. Of course, an epiton sensitively reacting on all changes in the density and the particle velocities in the continuous ring does not remain indifferent to the fact that this ring is perturbed by a neighbouring epiton. It will shift in radius and azimuth: one can say that it moves in the field of an effective potential produced by a neighbouring epiton. Each vortex acts here in two forms – as a self-gravitating ellipsoid with a given surface density acting on the trajectories of the dust grains of the continuous ring and as an epicyclic orbit, osculating under the action of collisions with the dust grains of the continuous ring. In this problem we neglect the direct gravitational interaction between the vortices.

Calculations show that a pair of vortices with the same mass are in equilibrium – that is, at any distance from one another the azimuthal drift velocity of these two identical vortices is the same: they neither move away from nor approach each other. Let us now consider the case when the masses of the vortices are unequal. We show in Fig. 12.24 the results of calculations about the interaction between two vortices with different masses, which are in the same ring. It is clear that a spiral wave produces near the heavy vortex an alternating field of drift velocities and that there are then a number of stable points in which a light vortex can be captured. The distance along the y-axis of these points from the heavy vortex is very simply connected with the size of the vortex: $y_s \simeq N \cdot 3\pi L$ where L is half the height of the vortex in the x-direction while $N = 1, 2, \ldots$. One notes easily that this distance is a multiple of the maximum wavelength of a spiral wave produced by an epiton in the continuous ring. This is clearly connected with the fact that the strongest action on the light vortex comes from those particles in the ring which pass through the upper edge of the epiton and have the maximum velocity.

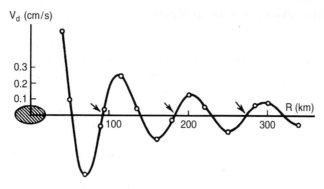

Fig. 12.24. Interaction between two gravitating epitons in a ring. The field of the relative drift velocities between the heavy (20 g/cm^2) and the light (10 g/cm^2) epitons (Gor'kavyĭ and Taĭdakova 1991a). R is the distance between the vortices and V_d the relative drift velocity. If $V_d > 0$ the epitons approach one another and if $V_d < 0$ they move away from one another. The arrows indicate stable points in which a light epiton can form a stable pair with the heavy one.

The formation of a stable pair of vortices can occur as follows: two epitons with different masses, positioned far from one another drift in the ring with different azimuthal velocities and gradually approach one another. When getting close they arrive at the same equilibrium point and form a stable pair. We note that direct gravitational action causes a mutual repulsion of the vortices. For the vortex parameters used – surface densities of 20 and 10 g/cm^2, epiton size 18.9×37.8 km, thickness 100 m, 60 analysed trajectories in a continuous ring of 0.1 g/cm^2 density, time step 0.1, or 63 steps per revolution, time of the calculation 1 revolution – the gravitational repulsion is unimportant, but for rather massive epitons with surface densities of 10^2 to 10^3 g/cm^2 it is possible that part of the curve in Fig. 12.24 in the negative value regions for small distances between the vortices may become "buried". In that case as the closest equilibrium point there may emerge a first point between the repulsive and attractive zones. Increasing the density of the continuous ring to 1 g/cm^2 increases the drift velocity by an order of magnitude. The density at which the alternating zone of the effective action of the epiton can propagate depends on its mass and may be many hundreds of kilometers.

It is interesting that vortices of the same mass do not form stable pairs, but stable triplets. Let us consider the interaction of three self-gravitating epitons in a continuous ring – with the same parameters and the same surface density of 20 g/cm^2 (Gor'kavyĭ and Taĭdakova 1991a). The epiton triplet turns out to be stable if the distance between the epitons is close to the equilibrium value of $N \cdot 3\pi L \simeq 100N$ km. It is clear that a pair of epitons acts on the third epiton like a single heavy epiton. We show in Fig. 12.25 the evolution of an epiton triplet for which initially the distance between the first epiton (the one on the left) and the second one was 98 km and that between the second and third ones 90 km which is less than the equilibrium distance.

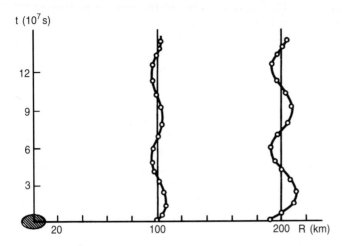

Fig. 12.25. Libration oscillations of epitons of the same mass forming a stable triplet (Gor'kavyĭ and Taĭdakova 1991a). R is the distance from the first epiton (the one on the left) and t is the time. The initial distances between the epitons are 98 and 90 km from left to right. The equilibrium distance at which the epitons are at rest is 99 km.

It is clear that there appear libration oscillations in the epiton triplet around the equilibrium values with a period of about two years. This period must decrease when the interaction between the epitons increases – when the self-gravitation of the epiton and the density of the continuous ring which give rise to the interaction increase.

It is clear that multiplets of several vortices with the same mass must also be stable. Epitons can thus form a stable chain also when they have the same mass.

The existence of stable epiton multiplets is a very important fact explaining the regular chain of bunches observed by Voyager-2 in the Neptunian arcs (see Plate 13). The presence of such a structure is a serious argument in favour of the epiton model; none of the other models provides for the possibility of the existence of an internal structure of the arcs. We note that the epitons are close to the kindred vortical formations in the atmospheres of the giant planets such as Jupiter's Large Red Spot and Neptune's Large Dark Spot.

12.3.4 General Scenario for the Origin of the System of Neptunian Arcs

To conclude this chapter we shall give a sketch of the general picture of the formation and the dynamics of the Neptunian arcs (Gor'kavyĭ and Taĭdakova 1994) combining the results described above into a complete picture – in which there are blank patches and speculative moments.

Let us then start with the moment the satellite Galathea was formed with a narrow ring near her. At some moment during the accretion when the ring – and perhaps also the satellite – reaches some well defined mass there appeared in the ring N regions where the particle trajectories intersected. The number N corresponds to the assumption that in the proto-stage the ring was in a $N/(N-1)$ resonance with Galathea. Perhaps in the proto-stage this resonance was not 43:42, but of a higher order – 44:43 or so. Sufficiently small particles of the disintegrating ring started to stick together to form larger ones and if this happened a bit farther from the planet the ring might turn into a new satellite, forming with Galathea a pair of co-orbital satellites like Janus and Prometheus. But the events occurred in the Roche zone – for Neptune and particles with a density of 1 g/cm^3 the radius of the Roche zone is 70 000 km while the radius of the ring with the arcs is 62 900 km – and, moreover, the particles of the ring are very loose bodies so that the growth of the particles ceased at a size of about 1 km – corresponding to a tensile strength of the particles of about 10^2 dyne/cm^2, the tensile strength of material such as lunar regolith. As a result the ring disintegrated into N epitons, each of which consisted of several tens of rather large particles – with a diameter of between a few hundred meters and a kilometer. The surface density of the epiton could be rather larger – several thousands of grams per square centimeter. The total mass of the ring with the epitons cannot be larger than the mass of a small satellite with a radius of 15 to 20 km. Epitons formed in different sections of the ring would have different masses. The azimuthal drift velocities which are connected with the self-gravitation of the epiton would therefore also be different. For massive epitons this drift velocity might reach tens of centimeters per second.

Let us consider the heaviest epiton in the ring. Its velocity is the largest: it moves rapidly along the ring overtaking lighter and slower epitons. When approaching a light epiton, it captures it in one of the equilibrium points of its effective potential – in that point where the repulsive potential is able to accelerate the slow epiton to the speed of the large epiton. The larger the epiton, the further the distance it will push the light neighbour which it has caught. This will continue until the "train" of captured light epitons has reduced the speed of the "engine" – the large epiton – which pushes them. The fastest then may turn out to be average epitons in other parts of the ring. They, collecting in turn shorter "trains" of light epitons form a complete circle, overtaking the "supertrain", and approach them from behind. This is apparently the situation which we observe now in the Neptunian arcs: a long arc with a tail in which there are several slightly smaller arcs. The term "tail" means that in the rotating frame of reference it is just this arc which must be the "leader" if there is any azimuthal drift – a reverse revolution. However, since the whole system of arcs rotates around the planet in the other direction it is just this arc which is said to be "lag behind". One can understand this the most easily if one imagines a series of trains which move from East to West along the Earth's equator. From the point of view of an observer, rotating with the Earth, the most Westerly train is the leader. From the point of view of an observer from the Moon this train is lagging behind since the Earth rotates from West to East faster than the motion of the trains.

The scenario for the formation of the arcs which we have described here has – apart from the fact that the arcs exist – the following observational confirmation: in the longest, A5, arc 11 epitons have been detected of which 6 were in the "head" of the train and 2 in the "tail" of the structure – there where the "engine" should be – the heaviest epiton. From the Voyager-2 data we can only estimate the distance between the epitons. It is interesting that the distance between two epitons in the tail is 200 km whereas the average distance in the epiton sextet is 550 km. Starting from the numerical calculations about the interaction between epitons in a dust ring we may conclude that the mass of at least one of the epitons in the tail of the arc is much larger than the mass of the epitons in the head of the arc. Even in the sextet itself one can observe an appreciable diminution of the distances between the epitons to the head of the "train". Moreover, in the neighbouring, shorter arc one can see in the tail of the arc a discontinuity of 1000 km. Porco (1991) took this discontinuity to be grounds for introducing a separate name for the short arc. This discontinuity indicating a considerable distance between epitons in the tail of the A4 arc can be used as an indication of the presence of a rather heavy epiton. The process of accumulating the epitons in the arc continued around a hundred years – during such a long time an epiton with a drift velocity of 10 cm/s will have performed a complete revolution around the planet from the point of view of a rotating observer.

We now combine this scenario with the fact that at the present time the arcs are captured in resonance and, apparently, the epitons can no longer travel along the ring surmounting the resonance "wells" from Galathea. Why were such travels allowed in the proto-stage? First of all, we note that the libration velocity of an epiton in resonance with Galathea is about 10 cm/s. An epiton with a drift velocity of a few tens of cm/s can then surmount the "barrier" restricting its movement along the orbit. This means, however, the destruction of the resonance connection with the satellite and an increase of the major semi-axis of the epiton at a rate of about 2 km/yr (see Fig. 12.20). We may therefore postulate the following approximate sequence of events: intially the ring is in a 44:43 resonance with the satellite. At the time the ring splits up – into 43 epitons? – the emerging drift velocities of the epitons would destroy the resonance connection with the satellite and the ring would start to increase it major semi-axis. The distance to the next resonance orbit (43:42) is 220 km. It will have taken about 100 years to travel this distance. During that time the epitons, not being restricted by Galathea's resonance potential, managed to congregate in one part of the orbit where they then were captured by the 43:42 resonance, which fixed the given arrangement of the arcs. At the present time the A1 and A2 arcs occupy neighbouring minima of the potential like the A3+A4 and A5 arcs (see Table 2.7 in Sect. 2.4). Between the A2 and the A3+A4 arcs there is an unoccupied potential minimum. We have already mentioned that librations of the arcs shifting them to the right-hand edge of the stability zone as well as the unequal lengths of these arcs distorts the arrangement of the stability zones and the observed position of the arcs. Nonetheless, the model with a distance of 8.372° between the potential maxima is in better agreement with the observations than Porco's model (1991) with a distance 4.186° where the A5 arc at a distance of 9.6° must occupy 2 to 3 resonance zones.

After the resonance coupling and the stationary pattern of the arrangement of the arcs have been established the satellite caused the spiral wave in the epiton chain with a 30 km amplitude which was recorded by Voyager-2. Both the epitons and the continuous ring, particles from which move along the epiton chain in both directions – in the rotating frame of reference – take part in the perturbation. The epitons which exchange mass with the dust ring help the 15 km ring to maintain the matching of the phases of the oscillations of the outer and the inner boundaries of the ring. In some small part of the material there is nevertheless a mismatch of the phases and that part of the dust forms at 50 km a halo of particles with a maximum eccentricity of $\simeq 5 \times 10^{-4}$ and mismatched phases of their oscillations.

Collisions between the particles in the epiton and fast particles in the ring and the halo produce an enhanced formation of dust in the epiton, a small part of which transfers to the ring while the remaining part stays in the epiton giving the main contribution to the optical depth of the clump. This is the reason why although the main mass of the clump is contained in large particles, observations record a large amount of dust in the arcs and the continuous ring. This is not astonishing if we bear in mind that the collision velocities of particles in the Saturnian rings are measured in millimeters per second, but in the Neptunian arcs in meters per second – these are the relative velocities of particles in the ring and in the clump; the relative velocities between particles in the clump are much smaller.

On the whole the outer Neptunian ring, containing a set of epitons, which are peculiar, vortex-like "quasi-particles", is a unique example of a dynamic structure, finely balanced as regards many parameters, and it considerably broadens our ideas about the possibilities of collective celestial dynamics.

13 Self-organisation of the Solar System

"Those hours, that with gentle work did frame
The lovely gaze where every eye doth dwell,
Will play the tyrants to the very same,
And that unfair which fairly doth excel ..."

W. Shakespeare, Sonnet 5.

13.1 Conditions for the Development of Spatial Structures

In the past it was assumed that natural forces would lead to disruption and the growth of chaos and that only intelligent activity could resist this. The assumption of the inexorable validity of the Second Law of thermodynamics led even to the idea of a "heat death" of the Universe. Only in the last decennia has it been realised that Nature has a strong tendency to move towards self-organisation and to the formation of structures which are ordered in space and time.

According to Feigenbaum's theorem (1978) the "intrinsically random" appearance of some ordered structures in a system is the consequence of the development of some instability in the system. However, generally speaking when an instability develops there are different situations possible which lie between two extreme possibilities.

In the one limiting case many modes develop in the system which interact in a complex manner with one another and which therefore need for their description a statistical approach. This is usually called a turbulent state.

In the other limiting case one of the modes for some reason or other out-distances the others in growth – or only a narrow wavepacket grows. This last regime occurs when the system is weakly subcritical and when the instability is weak – the system is on the stability boundary which is typical for astrophysical discs (Fridman et al. 1991); the first regime may occur when the dispersion curve has a sharp extremum. Depending on the peculiarities of the actual excitation mechanisms the medium changes in the second limiting case to a new laminar motion and a regular, ordered, and often structured flow develops in the system. However, also possible is the case of a structured turbulent motion (see, for instance, Di Prima and Swinorey 1981).

In an unstable medium structures such as regular bands, spots, circles, spirals, vortices, solitons, or modons will spontaneously appear in it. The transparent patterns on a dragon-fly wing and the bright clear colours of the wings of a butterfly turn out the consequences of general self-organisation processes in biological systems. Rigorous mathematical models have been developed for a number of morphogenetic processes; they describe both the distribution of spots on the skin of a giraffe and the nature of the coloured bands of a tiger (Murray 1977). The ordered pattern of clouds, sand and sea waves, "slick" bands of ripples on calm water, the stepwise flow of a slow stream on asphalt, the periodic oscillations in the number of game in hunting forests, the effective circles and spirals in the Belousov–Zhabotinskii reaction, the convective Bénard cells and Taylor vortices are all some of the innumerable manifestations of self-organisation, the most remarkable of which is the existence itself of highly organised life on Earth. There are also many examples of self-organisation in the Universe: solar granulation, sand dunes on Mars, vortex formations such as the Large Red Spot on Jupiter, spirals in galaxies, and the structure of the planetary rings. Finally, the stars themselves, star clusters, galaxies, and their associations are examples of the organisation of matter which initially was quasi-uniformly distributed.

13.1.1 Self-organisation of Open Systems

Various terrestrial structures chiefly originate due to the external influx of solar energy which, in turn, produces wind energy, running water, and vegetable mass. When the exchange with the outside medium ceases the ordering of the system vanishes – unless there is a mechanism to fix the structure which has been produced. We note that the "outside medium" is not a spatial concept just as "openness" does not mean necessarily its interaction with the surrounding space. Moreover, under conditions of complete spatial isolation a system may be able to produce a rather persistent self-organisation if there are internal energy sources. For instance, the loss of solar energy certainly leads to a rapid deterioration of the biosphere, but simple forms of life can exist quite long in places where there is a large accumulation of organic matter close to geothermal sources.

The self-organisation processes of the planetary rings are also connected with internal energy and matter sources. The energy of the orbital motion of the Saturnian rings is very large; due to gravitational and contact interactions of the snow particles it is transformed into the energy of the random motion and inelastic collisions transfer this energy of the relative motion into the heating of ice – into the energy of molecular motion which subsequently is spent in infrared radiation. Although all these transformations of the energy flux take place inside the rings when one considers the separate links in this energy chain one can consider the other processes to be "external". For instance, if we write down the balance equations for the energy of the random motion the viscous dissipation of the energy of the orbital motion is for this balance an "external" energy source and the decrease of the random velocities in inelastic collisions is "external" cooling. On the other hand, when we study the temperature balance of the particulate matter the inelastic collisions are an "external" source and the infrared radiation an "external" sink.

13.1.2 Gravitational Self-organisation

Jeans drew in 1902 attention to the possibility of the contraction of perturbations of a size larger than some critical (Jeans) size in an infinite uniform self-gravitating medium (Jeans 1929). Although there is no non-trivial stationary solution in the medium considered by him (see, for instance, Binney and Tremaine 1987), nonetheless the existence of the Jeans instability was later correctly established both in non-stationary (Lifshitz 1946, Bonnor 1957) and stationary (Fridman and Polyachenko 1984) gravitating systems. If there is no rotation, only two forces act on each element of matter: the gravitational force and the pressure gradient. When the wavelength of the perturbation increases, in the balance between the forces an ever larger mass gets involved, that is, the self-gravitation force increases. The characteristic scale over which the density and the pressure change also increases and, hence the pressure gradient decreases. It is clear that for a wavelength larger than some (Jeans) length the gravitational force will start to dominate over the pressure gradient force which leads to a gravitational instability, producing the whole of the hierarchy of scales observed in the Metagalaxy: from stars up to the so-called large-scale structure of the Universe.

When the system rotates there appear additional non-inertial forces, proportional to the radius. As a result, it turns out that for a disc, for instance, intermediate size perturbations are unstable, since those with a short wavelength are stabilised by the pressure force and those with sizes comparable to the disc radius are stabilised by the non-inertial forces. This leads to the appearance of structures in the disc: rings, spiral arms, and vortices. Since the main force in the gravitational instability process is the self-gravitation force, the energy source for the gravitational instability is the gravitational (potential) energy of the matter involved in the instability. We shall see that in that case an internal energy source "does the work".

13.2 The Law of the Planetary Distances

13.2.1 Tendency of the Solar System Towards Self-organisation

Let us consider the role of synergetic processes in the formation and development of the solar system; we can distinguish here three periods:

The formation of the Sun surrounded by a gaseous dust cloud.

The condensation of the proto-planetary cloud and the formation of the separate planets, satellites, and other objects.

The evolution of the planetary and satellite bodies.

The formation of single stars with proto-planetary discs is a special case of star formation and is studied in the field of stellar cosmogony. The second period of the development of the solar system is a traditional subject of planetary cosmogony. According to modern ideas the principal evolutionary mainspring of that stage is the accretional growth of small particles to planetary sizes. This process is the clue to the popular Safronov model (Safronov 1969, Vityazev et al. 1990) of the origin of the planetary system which has been developed in great detail. The sticking together of particles in collisions is the "internal" evolutionary factor for the proto-planetary disc, while the solar radiation, which causes complicated thermal and photochemical effects in the disc, is the "external" factor for the development. The Sun also exerts a strong force on the disc – for example through its magnetic field – imparting angular momentum to the disc and removing matter from the disc to its periphery. These days this process is intensively studied. In particular, many anomalies in the structure of meteorites, in which both volatile and refractory substances are found, are connected with it – this effect can easily be explained by the drift of matter from hot regions, close to the Sun, to the cold outskirts.

A characteristic feature of the second stage in the evolution of the solar system is the formation of a whole range of subsystems, especially the planets with proto-satellite discs from which the satellites and rings are formed. The evolution of the proto-satellite discs is in many ways similar to the development of the solar system itself. There have been many attempts in the past to produce a satisfactory theory about the origin of the solar system. An extensive review of such theories before 1967 can be found in a couple of review articles by ter Haar and Cameron (ter Haar and Cameron 1963, ter Haar 1967). We shall in this chapter be mainly concerned with the work of Safronov, but we may mention earlier papers by ter Haar (1948, 1950) whose theory has many similarities with Safronov's and who, building on the classic paper by von Weizsäcker (1943), showed how a careful study of the properties of the proto-planetary cloud is able to explain a number of the features of the solar system, such as the differences between the inner and the outer planets. In recent decades Prentice (for a survey of the main features of his work see Prentice 1978, Prentice and ter Haar 1979) has further developed this neo-Laplacian theory and, invoking the controversial concept of supersonic turbulence, has been able to predict the physical properties of many of the smaller bodies in the solar system – which subsequently were confirmed by observations during Voyager fly-pasts.

The third stage starts when the collection of planetesimals has been transformed into large bodies – the planets and satellites. During their evolution the interiors of the planets and satellites are heated up due to natural radio-activity, there occurs a gravitational stratification of matter into shells of different densities, and geological processes form surface structures. In this stage of the development meteoritic bombardment in many respects determines the surface structure, solar radiation plays an important role in the formation of the present atmospheres, and tidal interactions determine the synchronous rotation of many satellites as well as, in particular, the vulcanic activity of Io.

These are in broad terms the present ideas about the formation of the solar system. Undoubtedly, this scheme is only the fundamental "skeleton" of a future detailed cosmogonical theory. Clearly, one of the main qualitative additions to the existing scheme will be to take the self-organisation of the solar system into account, that is, the gradual appearance of a number of spatial structures such as spiral waves, rings, or vortices. Self-organised structures of a similar kind may appear in all three stages of the evolution of the planetary system which itself is an open system and consists of open subsystems in which the self-organising factors operate. The astonishingly regular arrangement of the planets and of the satellites in the systems of the outer giant planets indicates the vigorous self-organisation of the solar system in its proto-stage. The planetary rings are also a clear example of a self-organising disc system of macroparticles. Whereas in earlier times only one collective instability of a medium of macroparticles was known in planetary cosmogony – gravitation – the study of the planetary rings has uncovered such structure-forming instabilities as the diffusive, quasi-secular, accretion, and ellipse instabilities. Moreover, there have been studies of spiral waves which interact through resonance with satellites and which lead under well defined conditions to the formation of narrow elliptical rings near resonances.

We see thus that the following features of the solar system show the richness of the self-organising processes:

1) The presence of long-term energy sources:
   solar radiation – thermonuclear reactions;
   radio-active heating of large bodies – nuclear decay reactions;
   gravitational compression of the planets – especially of the giant planets;
   the energy of the orbital motion and the intrinsic rotation;
   collisional heating – by meteorites and planetesimals;
   gravitational action of neighbouring bodies – tides and resonances.

2) A variety of distributed media in which collective self-organisation processes can occur:
   solid media of Earth type planets, satellites, comets, and so on;
   liquid matter in terrestrial oceans, satellite envelopes, or planetary envelopes;

various media with a varying density – planetary, satellite, and cometary atmospheres, and the interplanetary and solar plasmas;
the gas-dust media of the proto-satellite and proto-planetary clouds;
the macroparticle media of the planetary rings, the proto-satellite clusters, and the disc of planetesimals;

3) Two kinds of structuring:
the appearance of ordered space-time structures in proto-planetary
and proto-satellite discs;
the occurrence of separate subsystems – such as planets, satellites,
comets. Subsystems are the consequence of self-organised structures
and themselves produce structures at a new level.

Modern cosmogonical theory must therefore consider the solar system as
a complex hierarchy of self-organised subsystems which are full of energy
sources and possess a remarkable wealth of collective processes. One of the
important observational tests of the completeness of cosmogonical models is
their ability to explain the great structural variety of the planetary rings.
The planetary rings are thus a unique proving ground for the validity of
cosmogonical theories.

Altogether taking into account synergetic – self-organising – evolutionary
factors in classical cosmogonical theories can and must lead us significantly
closer to a single scenario for the origin and the evolution of the solar system.
Planetary cosmogony becomes the synergy of the solar system. An important
result of all cosmogonical studies can be a single detailed model for the formation of the solar system in which one is given solely the initial parameters of
the proto-planetary cloud after which the internal regularities of the spontaneous development must lead to the planetary system which is as close to the
real system as is possible in principle. The appearance of qualitative evolutionary factors and the growth of computing power makes it possible to hope
that in the near future such a global cosmogonical theory can be produced.
A detailed cosmogonical model can be the foundation for understanding the
observed pattern of planets and satellites and for planning space studies.

13.2.2 Dissipative Instability
and the Law of the Planetary Distances

Let us consider a possible example of the development of spatial structures in
the solar system as the result of an instability. The problem of why the distribution of the planets satisfies the law of the planetary distances has a long
history. Polyachenko and Fridman (1972) studied the possibility of explaining the law of planetary distances on the basis of the idea of a gravitational
instability in the proto-planetary cloud. According to present-day cosmogonical ideas a gravitational instability in the solar proto-planetary cloud is
unlikely. We shall follow Gor'kavyĭ, Polyachenko, and Fridman (1989, 1990)

and consider the dissipative instability of the proto-disc as the cause for the appearance of a regular distribution of the planets.

If the planets were formed from perturbations of the rings, which are growing because of some instability, the distance between the planets would be the wavelength of the most unstable perturbations at a given point in the disc (Polyachenko and Fridman 1972). Since the wavelength of the most unstable waves is connected with the characteristics of the disc one can easily calculate the characteristics of the disc as function of the radius from the change of wavelength with changing radius (Polyachenko and Fridman 1972). A comparison of these relations with those assumed for the proto-planetary disc according to modern cosmogonical models (Safronov and Vityazev 1983) makes it possible to estimate how realistic the suggested scenario for the formation of the regular structure of the solar system is.

In Chap. 8 we studied the stability of a differentially rotating viscous disc. Let us apply the results of that analysis to the proto-planetary disc. We note that the transport equations can describe both a laminar and a turbulent disc (Morozov et al. 1985, 1986). In the case of a turbulent gas the velocities V_r and V_φ will refer to the large-scale motion of the gas and T will characterise the turbulent rather than the thermal velocities. The meaning of the energy sources and sinks and the expressions for the transport coefficients must also be changed appropriately. However, the general form of the transport equations is retained and their analysis gives the same instabilities which are characteristic for a differentially rotating medium.

The dissipative instabilities considered in Chap. 8 lead to two characteristic scales for the stratification of the disc which can be compared with the observed distances between the planets: $\lambda_0 \approx 10h$ – the diffusive instability – and $\lambda = c^2/G\sigma_0$ – the quasi-secular instability (Gor'kavyĭ, Polyachenko, and Fridman 1989). The characteristic time for the development of the instabilities is $t_x \simeq (\nu k^2)^{-1}$. One shows easily that in the case of molecular viscosity the growth time of waves with even the shortest wavelengths (0.3 a.u.) is larger than the cosmological time. Therefore, only in the case of turbulent viscosity in the disc can ring-shaped perturbations with $\lambda_0 \simeq 1$ a.u. form sufficiently rapidly. Using estimates of the turbulent spreading out of the proto-disc (Safronov and Vityazev 1983) one can take for the turbulent viscosity $\nu_T \simeq 10^{12}$ cm^2/s which gives a characteristic growth time of $t_x \simeq 2 \times 10^4$ yr for 0.3 a.u. and 2×10^7 yr for 10 a.u.. During the main part of its existence the proto-disc may be in a stable state and be "waiting" for the start of conditions for instabilities which might be satisfied only during a well defined, rather short period of its evolution.

13.2.3 Proposed Characteristics of the Proto-disc

It is well known that the distances of the planets from the Sun are rather well described by the following empirical law, established in the eighteenth century by Bode and Titius:

$$R_n = 0.4 + 0.3 \cdot 2^n, \tag{13.1}$$

where $n = -\infty, 1, \ldots, 6$, and R_n are the distances – in astronomical units – from Mercury, Venus, ..., Uranus, respectively.

Table 13.1. Comparison of the Observed Distances of the Planets from the Sun with the Empirical Titius–Bode Law

Planet	Distance according to the Titius–Bode Law, R_n	Observed distance
Mercury	0.4	0.39
Venus	0.7	0.72
Earth	1.0	1.00
Mars	1.6	1.52
Asteroid belt	2.8	2.90
Jupiter	5.2	5.20
Saturn	10.0	9.55
Uranus	19.6	19.20
Neptune	38.8	30.10
Pluto	77.2	39.50

The planets Neptune and Pluto were not known in the eighteenth century – if that had not been the case the authors of the law for the planetary distances would have been obliged to add corrections to (13.1), as can be seen from Table 13.1 showing the degree of agreement between the Titius–Bode law and the observational data. In the table we have given the distances of the planets from the Sun – in astronomical units – according to (13.1) as well as according to observations. We see that the agreement is good up to Uranus ($n = 6$). The distances between the last three planets turn out to the same and equal to ≈ 10 a.u.. This is also reflected by the fact that Neptune and Pluto do not satisfy the rule (13.1). One can rewrite (13.1) in the form

$$\lambda = \tfrac{2}{3}(R - 0.4), \tag{13.2}$$

where $\lambda \equiv R_{n+1} - R_n$ and $R \equiv \tfrac{1}{2}(R_n + R_{n+1})$.

This formula correctly shows that the distance between the planets from Venus to Uranus increases like a geometric progression. The distances between the first three planets, Mercury, Venus, and the Earth, turn out to be the same and are equal to approximately 0.3 a.u., just as they turn out to be the same between the last three planets, Uranus, Neptune, and Pluto, and equal to approximately 10 a.u..

In the exposition which follows we shall use the logic of a paper by Poly-achenko and Fridman (1972) which gives the solution of the following inverse problem: we find the parameters of the disc for which the growth rate of some instability leading to the growth of ring perturbations turn out to be maximal for wavelengths of the perturbations, $\lambda(r)$ corresponding to the observed distance between the planets. The diffusive instability turns out to be the most suitable instability for this purpose for the following reasons.

First of all, it leads to the development of ring perturbations. Secondly, it follows from Sect. 8.2.4 that the growth rate of this instability turns out to be maximal for $\lambda \equiv \lambda_0 = \alpha h$, $\alpha \approx 10$. The half-thickness h of the disc depends only on a single disc parameter – the temperature which according to the calculations of Safronov (1969) in the central plane of a gaseous dusty disc can be described by the formula $T \cong 100/\sqrt{R}$ K, where R is measured in astronomical units. Using, as in Sect. 8.2.1, for the half-thickness of the disc the expression $h = \sqrt{2/\gamma}(v_z/\Omega_0)$ we have for an isothermal disc, $\gamma = 1$,

$$
T = \frac{mh^2\Omega^2}{2k_{\mathrm{B}}} = \frac{mGM}{2\alpha^2}\frac{\lambda_0^2}{R^3}. \tag{13.3}
$$

Here T is the temperature in degrees Kelvin, m the mass of a hydrogen molecule, – we assume that the main constituent of the proto-disc is hydrogen – M the mass of the Sun, k_{B} the Boltzmann constant, and the coefficient $\alpha = \lambda_0/h$ determines the correct scale of the stratification.

We assume that the thickness of the disc is determined by the gas temperature since the turbulent motion has, apparently, a subsonic velocity (Safronov and Vityazev 1983). We show in Fig. 13.1 the result of the solution of this inverse problem: when we calculate the temperature of the proto-disc at the moment rings are formed at a given region of the disc from the actual distances between the planets. It is clear from Fig. 13.1 that the values of the temperature obtained in this way agree well with the temperature profile calculated by Safronov (1969; see also Safronov and Vityazev 1983). Exceptions are the regions between Mercury and Venus and between Mars and the asteroids. The heating of the disc regions closest to the Sun to almost the maximum possible values of the temperature is to be expected whereas the regions of the gas disc which are further away are screened and heated only by grazing solar radiation. We note that the values of the temperature calculated using Eq. (13.3) do not give a picture of the situation at a single time since the rings were not formed simultaneously. For instance, if one just takes into account the increase of λ_0 the characteristic time for the formation of rings in the zone between Mars and the asteroids increases by a factor 7 as compared with the region between Mars and the Earth and by a factor of 21 as compared with the region between Venus and the Earth. One should

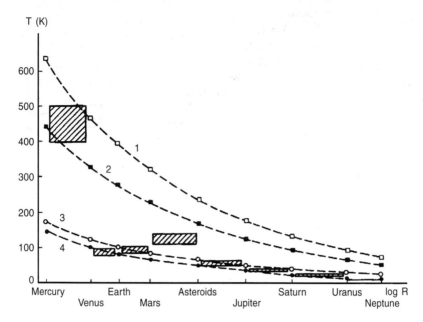

Fig. 13.1. Temperature of a proto-planetary disc as function of the distance from the Sun. The curves *1* and *2* show the decrease in temperature with increasing distance for a black slab at right angles to the line of sight (*open squares*) and for a black sphere (*black squares*). The curves *3* and *4* show the temperature of a gas-dust disc according to Safronov (1969) (open circles correspond to the central plane and black circles to the height of a uniform atmosphere). The hatched rectangles show the temperatures calculated at the positions of the planets using (13.2).

add that according to Safronov's estimates (1969, Safronov and Vityachev 1983) the time for the accretional growth of a planet steeply increases with increasing orbital radius – between Venus and Mars. Up to the moment that the Martian ring is formed the aggregation of dust and other factors might increase the transparency of the region near the Earth and that would lead to an increased heating of the region near Mars. In this way one may interpret the temperature rise from the region between Venus and the Earth to the region between Mars and the asteroids. The data of Fig. 13.1 may serve as an important source of information about the temperature and other conditions in the proto-cloud.

From the results of the present chapter one may conclude that the regularity in the planetary orbit distribution in the solar system may be connected with the smooth change in the thickness of the proto-disc with radius and the operation of the diffusive instability. At any rate the fact that the distances between the planets change in proportion to the thickness of the proto-cloud is noteworthy. We note that we were dealing with the structure of the gas component of the disc, but in Sect. 12.2 we showed that a body moving along an elliptical orbit and intersecting a circular dust – or gas – ring will be pulled into that ring. This mechanism may cause a redistribution of the planetesimals to correspond to the gaseous structure. In conclusion we should like to mention that the Titius–Bode law may also reflect a scaling similar to that observed according to Kolmogorov in the case of homogeneous isotropic turbulence. Chandrasekhar and ter Haar (1950) have shown that in the case of axi-symmetrical turbulence the scaling law corresponds, indeed, to the Titius–Bode law.

14 Space Studies of the Outer Planets

"We are given to boasting of our age being an age of science.
... Yet though we may exalt research and derive enormous
benefits from it, with what pettiness of spirit, poverty of
means and general haphazardness do we pursue truth in the
world today! ... we leave it to grow as best it can, hardly
tending it, like those wild plants whose fruits are plucked by
primitive peoples in their forests."

*Pierre Teilhard de Chardin, The Phenomenon of Man, Book
Four, Chap. III, Sect. 2A.*

14.1 Space Successes in the Period 1959–1989

Up to the time of Galileo six planets and one satellite–the Moon–were
known; during the time of telescopic observations–from the first observations
by Galileo in 1610 up to our times, that is, for nearly four centuries–three
planets, 32 satellites, the asteroids, comets, the Saturnian rings, the Uranian
rings, and the Neptunian arcs were discovered. By means of spacecraft during
about ten years–between 1979 and 1989–the Jovian rings and 28 satellites in
the systems of the giant planets were discovered. As a result of this "productivity" of the space projects the "epoch of the great geographical discoveries"
within the known limits of the solar system had been swiftly concluded. After
Voyager-2 in 1989 had flown past Neptune only Pluto remained to be visited
in the solar system and one can hardly expect the existence of a developed
satellite system near it. This means that the number of known satellites of
the main planets of a size of more than 100 km is now practically fixed.

After the discovery of a satellite of the asteroid Ida and of transplutonic
asteroids it is difficult to expect such fundamental discoveries as the observation of a new class of objects in the solar system. Galileo made this kind
of discovery when he found out that the Moon was not exceptional and that
there were also satellites near Jupiter. Such events also occurred at the end
of the seventies and the beginning of the eighties of the present century when
the Saturnian rings ceased to be an anomaly and ring systems were found
around all the giant planets. One can conclude that towards the middle of
the nineties all basic features of the composition of the solar system–within
its known spatial limits–had been established.

Starting from the pioneering spacecraft studies of recent years the time
has now arrived for a detailed study of the planets and their satellite systems

which requires the creation of spacecraft which can orbit, land, and return to the Earth. The transition to planned studies of the solar system requires preliminary estimates of the efficiency of space projects and the carrying out of observations and experiments which provide us with data which give us key information, rather than routine data, for solving the most fundamental planetological problems. In the present chapter we shall follow Gor'kavyĭ, Minin, and Fridman (1989a,b) and consider some aspects of modern planetary physics.

Table 14.1. Spacecraft Studies of Planets and Satellites

Planet	Number of spacecraft studying the planet		Name of spacecraft and year of encounter with planet
	USSR	USA	
Mercury	–	1	Mariner-10 (1974–1975)
Venus	15	6	Venus-4 to 16, Vega-1, 2 (1967–1985), Mariner-2, 5, 10, Pioneer-Venus-1, 2 (1962–1978), Magellaen (1990–1994)
Mars	6	6	Mars-2 to 6 (1971–1974), Phobos-2 (1989), Mariner-4, 6, 7, 9, Viking-1, 2 (1965–1976)
Jupiter	–	4[a]	Pioneer-10, 11, Voyager-1, 2 (1973–1979)
Saturn	–	3	Pioneer-11, Voyager-1, 2 (1979–1982)
Uranus	–	1	Voyager-2 (1986)
Neptune	–	1	Voyager-2 (1989)
Pluto	–	–	

[a] Galileo was launched in 1989.

The use of spacecraft for the study of planets and satellites is the main experimental basis of planetology and in many respects determines how successful its development will be. In Tables 14.1 and 14.2 we compare the space programmes of the USSR and the USA[1] in the field of the study of the planets and their satellites – excluding the Earth, the Moon, and interplanetary space. We add that for lunar studies 22 spacecraft were successfully launched in the USSR and 14 unmanned and 9 manned spacecraft, of which 6 landed on the Moon, in the USA; 5 American Pioneer spacecraft studied interplan-

---

[1] All data from the USSR refer to the period up to 1989.

Table 14.2. Results of Spacecraft Studies

Results	USSR	USA
Number of launches	30	23
Number of spacecraft providing information	21	17
Number of resulting planetary encounters	21	23
Number of planets studied	2	7
Number of satellites studied	1	49

etary space – and there were 9 failed Pioneer missions while the two Soviet Vega spacecraft – Vega-1 and Vega-2 studied Halley's comet.

It is clear from Tables 14.1 and 14.2 that in the period between 1959 and 1989 the efforts of the USSR and the USA to accomplish space missions for studying the planets were comparable, if we look at the number of spacecraft and the number of planetary encounters. However, the objects of the Soviet and American programmes were very different.

The Soviet programme was aiming at a detailed study of the planets closest to the Earth and the USSR had the lead in its studies of Venus; although the American Magellaen spacecraft which discovered the active volcanoes on Venus and obtained magnificent radio-location pictures of the surface led to a very rapid advantage of the American planetologists in the field of the cartography and geology of Venus. Later Soviet projects were oriented towards Mars and its satellites.

The American programme involved spacecraft studies of almost all the planets of the solar system displaying a heightened interest in the giant planets and their satellite systems. The next stage of the American projects is the production of artificial satellites of the outer planets: in 1989 the Galileo spacecraft was successfully launched; it had a complicated Earth-Venus-Earth-Jupiter orbit and in 1995 started an orbit around Jupiter where it probed Jupiter's atmosphere. The Cassini spacecraft was launched in 1997; it is a an orbiter for Saturn with the task of directly probing Titan's atmosphere. We note the complexity of the American projects: for instance, Voyager-2 studied the Jovian, Saturnian, Uranian, and Neptunian systems.

After the wave of experimental information of the last decennium there followed a wave of publications. The papers based on spacecraft data form the bulk of planetological publications and on the whole are determined by

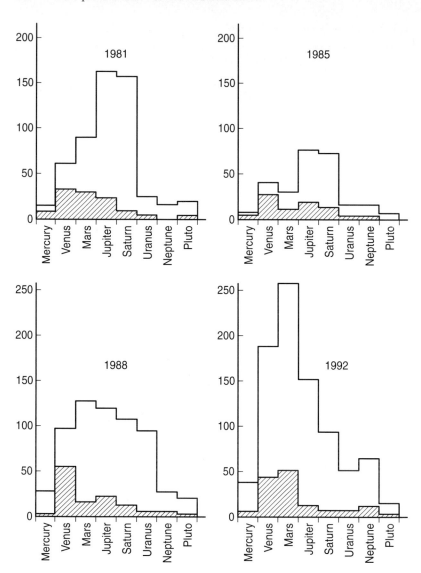

Fig. 14.1. Distribution of non-USSR and USSR (*shaded*) planetological publications as function of the various planetary systems. The data are taken from the 1981, 1985, 1988, and 1992 issues of the abstract journal "Astronomiya".

the space programmes. We show in Fig. 14.1 histograms of the distribution of Soviet and foreign papers over the various planetary systems.

It is crystal clear that Jupiter and Saturn are the centre of attention abroad while Venus occupies that position in our country. This is in complete accordance with the peculiarities of the Soviet and the American space programmes in the eighties. On the whole the growth of interest in the outer planets is an important trend of present-day planetology. From the beginning of the sixties up to the middle of the eighties the average annual number of papers increased: those dealing with the planets of the terrestrial group – including the Earth – by a factor 1.8, those dealing with the giant planets by a factor of between 4 and 5, those dealing with the satellites of the giant planets by a factor of between 40 and 60, and those dealing with planetary rings by a factor of between 14 and 16 – according to the data from "Astronomical and Astrophysical Abstracts". The outer planet systems give plentiful material for comparative planetology, volcanology, and atmospheric physics.

14.2 The Voyager Missions

In 1969 NASA started their considerations of the Grand Tour – a plan to send an interplanetary spacecraft to all the outer planets of the solar system. A favourable configuration for such a journey was expected in 1976 to 1977. The cost of this project was estimated to be 750 million dollars. One craft should study Jupiter, Saturn, and Pluto and another Jupiter, Saturn, Uranus, and Neptune. However, the programme of the lunar expeditions meant that there was less money for the other US space projects and the cost of the Grand Tour project was cut to 250 million dollars by omitting the furthest planets – Uranus, Neptune, and Pluto. The spacecraft were constructed on the basis of the Mariner spacecraft so that the project originally was called "Mariner–Jupiter–Saturn" and only in 1977 got the name Voyager.

Voyager-2 was launched, using the Titan–Centaur rocket, from Cape Canaveral on 1977 August 20. The whole craft originally weighed 2066 kg, and after the end of the operation of the additional stage it turned out to be equal to 815 kg. Its twin-craft Voyager-1 started later – on 1977 September 5 – but reached Jupiter earlier, whence its number 1.

Radio-isotopic thermoelectric generators with plutonium oxide guaranteed at the start of the flight about 470 W for the power of the internal electrical system of the craft. A 3.7 m diameter antenna was used for contact with the Earth, guaranteeing information transfer at the rate of 10^5 byte/s – one high-quality picture per minute.

The flight was not without its troubles. Voyager-2 ran into serious difficulties in April 1978 when, due to an error of the operator the on-board computer was irreversibly switched to the device for information from the Earth not from the main but from the secondary radio-receiver. However, this receiver was also damaged: one of the condensers was put out of order. As a result the transmission band – the frequency range over which the signal from the Earth is received – was narrowed by a factor 1000 and started to alter the frequency – particularly strongly because of oscillations in the temperature of the spacecraft. For a stable link it was necessary to calculate most accurately all effects which can shift the receiver frequency, including the thermal regime of the spacecraft. In 1981 the power drive which provided the rotation of the platform with the instruments on Voyager-2 jammed. For its further functioning the engineers found a special regime of slow motions of the platform and the spacecraft itself. Six days before the closest approach to Uranus strong distortions appeared on the photographs transmitted by Voyager-2 due to a disruption in the memory of the on-board computer. In two days the cause for this trouble was found and a programme was written bypassing the defective parts of the memory.

Because of the long exploitation of the Voyager-2 spacecraft the power of the on-board electrical system decreased to 400 W so that it was necessary severely to restrict the number of instruments and devices to be operated simultaneously. Notwithstanding the manifold problems, the state of the spacecraft was in the opinion of the American specialists in fact improved during the flight time. The functioning of most of the Voyager subsystems was reprogrammed and refined even during the voyage. This made it possible to surpass considerably the flight programme.

We show in Fig. 14.2 the flight trajectory of Voyager-2. Voyager-1 successfully studied Jupiter on 1979 March 5 (Voyager-2 did it on 1979 July 9), and Saturn on 1980 November 12 (Voyager-2 on 1981 August 25). After that Voyager-1 deviated considerably from the ecliptic plane while Voyager-2 was directed towards Uranus, which was lying above the plane, which it encountered on 1986 January 24. Finally Voyager-2 reached on 1989 August 24 the Neptunian system, thus practically completing the Grand Tour project as it had been originally conceived.

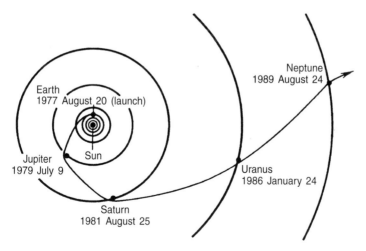

Fig. 14.2. Trajectory of Voyager-2.

Voyager-2 was the first spacecraft in history to encounter Uranus and Neptune. The volume of information obtained just by Voyager-2 is enormous: 20 000 single photographs were transmitted from the Jovian system, 18 500 from the Saturnian system, 6 000 from the Uranian system, and more than 9 000 from the Neptunian system. Using the data from the two Voyagers 26 new satellites of the giant planets were discovered and the total number of known satellites in the solar system was brought to 61. The two Voyagers overturned many of our ideas about the outer planets and also those about the solar system. One needs only mention the parade of planetary rings – from Jupiter to Neptune – and the observation of volcanoes on Io and Triton.

The Voyager results literally discovered again the solar system for us. The very rich set of geological stuctures, atmospheric effects, and celestial mechanical constructions gives us material to be analysed during many decades. The scientific significance of the Voyager project is unique; this project practically completed the epoch of geographical discoveries in the solar system where only tiny Pluto, with apparently a single satellite – Charon – remains to be studied from close up. We should like to congratulate all those who participated in this grandiose project for its brilliant completion.

14.3 The Cassini Mission

The Cassini project was conceived simultaneously by NASA and the European Space Agency ESA (Cassini Mission 1989, Coates 1997, Spehalski and Matson 1998). The interplanetary flight started 1997 October 15, using a Titan IV/Centaur rocket. The trajectory is shown in Fig. 14.3a. Cassini will first execute two gravity-assisted fly-pasts of Venus. After passing at a height of 300 km above the Earth's surface on 1999 August 16 and receiving extra acceleration through the gravitational field of our own planet the spacecraft will start on a trajectory to Jupiter; it will fly past Jupiter on 2000 December 30. It will encounter Saturn on 2004 July 1. The trajectory of the craft in Saturn's field and the place where it will encounter Titan on 2004 November 28 are shown in Fig. 14.3b. It is foreseen by the Cassini project that on 2004 November 6 the craft will split up into an orbital part and a probe for studying Titan – the probe was prepared by ESA. The probe must study the dense atmosphere of Titan and its surface, which up to now have been unaccessible for observations. The orbital stage must perform about sixty turns around Saturn with an average period of around one month (see Fig. 14.3c). Provisionally the period during which the orbital stage will function is four years – up to December 2008.

The mass of the scientific equipment of the orbiter is 197 kg and the power of the on-board electrical instruments is 213 W at the start of its operation near Saturn and 167 W at the end of the mission. The rate of information transfer is 249 kbyte per second. The orbiter must study Saturn itself, its

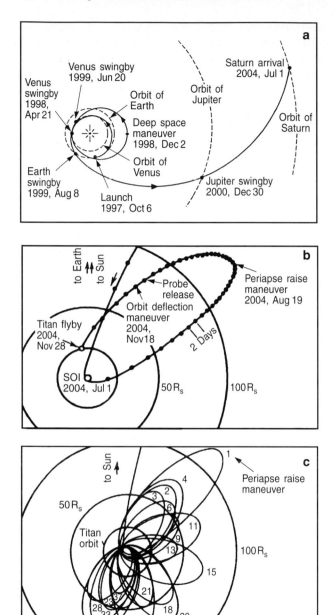

Fig. 14.3. (a) Flight trajectory of the Cassini probe; (b) entering its orbit; (c) its first thirty odd revolutions.

satellites and rings. Three or four equatorial and 18 oblique passages behind the Saturnian disk – as seen by a terrestrial observer – and five additional passages behind the rings are planned to be completed during the period that the orbiter is functioning. In those passages a radiosignal from the orbit will pass through the atmosphere of Saturn and the rings – as on the Voyagers there will be an antenna with a 3.66 m diameter on the craft. Seven such passages and radio-occultation of a duration of 10 to 18 minutes are also planned for Titan. Altogether 40 approaches to Titan at a relative velocity of about 6 km/s are intended to take place. Radar observations of the surface of Titan will give maps with a resolution of 300 to 600 m – in the case of narrow-band observations – and of 2 to 7 km – for broad-band observations.

The Cassini orbiter can obtain photographs of the rings with a resolving power of 2 km (private communication from L.W. Esposito). This makes it possible to solve many problems, especially problems about the existence of so far undetected satellites in the ring zone.

14.4 The Chronos Mission

The planetary rings are the only "nature reserve" in the whole of the solar system where not only the "proto-matter", out of which the solar system was formed, itself is conserved in its primordial form but also the original wealth of collective collisional processes which are the cause of the remarkable self-organisation of the planetary rings.

The studies in flight of the planetary rings carried out by American space-craft have produced a huge amount of fundamental problems to which various theoretical models give contradictory answers. Further study of the rings is therefore of great interest. A qualitatively new stage in the experimental study of the rings which would make it possible to solve many problems would be to obtain photographs of the rings with a resolution of 10 to 100 cm, that is, such that one could see the individual particles and the details of their structure; the resolution of the best photographs of the rings by the Voyager craft was several kilometers and even the one-dimensional profile of the optical depth obtained by observing a star through the rings have only a resolution of about 100 m – much larger than the radius of the largest particles.

Gor'kavyĭ, Minin, and Fridman (1989a,b) have proposed a project for direct probing of the Saturnian rings. We note that any future project for studying the outer planets must take into account the results which have already been obtained in that region of the solar system in order not to duplicate the known data. Of the four giant planet systems the Jovian system has been studied the most – four fly-past studies – and in 1995 the orbital probe Galileo operated near Jupiter. The Saturnian, Uranian, and Neptunian systems are therefore more attractive – in the sense that they have been less

intensively studied – and at the same time these are the ones which have rings
with a great variety of structures.

The most favourable way to reach the Saturnian system is by means of
a gravitational manoeuvre in Jupiter's field. This saves time and fuel and,
moreover, makes a fly-past study of the Jovian system possible. The optimum
arrangement of the planets for such a manoeuvre returns approximately every
twenty years. The realisation of such a programme may give a strong impulse
for the development of the physics of the rings, of cosmology, and of the whole
of planetary physics. This project of an artificial satellite of Saturn was called
Chronos by Gor'kavyĭ, Minin, and Fridman after the ancient Greek name
of Saturn which was also used in Russia, already at the beginning of this
millennium.

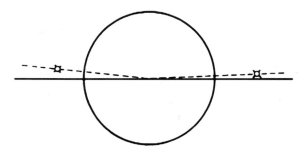

Fig. 14.4. Trajectory for probing Saturn's rings.

Let us consider the possible programme for the operation of the separate
probe. The entrance of the probe in the rings can naturally be accomplished
through a gravitational manoeuvre in the field of one of the natural Saturnian
satellites. It is necessary to achieve maximum coincidence of the plane of the
probe orbit and the plane of the rings. The smaller the angle between these
two planes the smaller the vertical velocity with which the probe approaches
the rings (Fig. 14.4). If the probe is at a height of several kilometers above
the plane of the rings a small engine can bring the probe into an orbit with
an almost zero inclination to the plane of the rings. The probe then "hovers"
over the plane of the rings surmounting the remaining tens of kilometers in a
few hours. During its slow descent to a height of a few kilometers the probe is
able to obtain a series of photographs of the rings and its individual particles
with a resolution of a few centimeters. Moreover, it can then also carry out
a chemical analysis of the material of the particles. When the probe comes
closest to the rings one can for an extremely small expenditure of fuel force
the probe to stop descending and to start a slow departure from the plane of
the rings. The probe will after several hours revolving around Saturn again
slowly approach the rings but now from the other, "shadow", side of the
rings. A manoeuvre which "reflects" the probe from the plane of the rings

can be repeated several times until the fuel is exhausted or until the orbit of the probe changes through a random encounter with a satellite. We can thus obtain multiple probing of the rings with "rest pauses" between the closest approaches to the rings which makes it possible to store information during the approach and when the probe passes outside the rings to transmit it to Earth through the orbiter. We note that a variant is possible in which the probe after studying the rings will fire a small "microprobe" for a direct study of the atmosphere of Saturn itself.

Let us consider the selection of the position of the probing. The most attractive position in this respect is the Cassini Division. It is rather transparent and large particles do not occult one another when they are projected on the plane of the rings. Near the inner edge of the division there exists a narrow elliptical ringlet similar to the Uranian rings, the dynamics of which have caused everlasting disputes for more than ten years. Similar rings exist also in Saturn's C ring which is also optically rather transparent.

The proposed programme of investigations enables us to solve the following problems:

One of the main problems of the dynamics of the planetary rings: are the narrow elliptical rings independent stable formations or are they controlled by shepherd satellites?

Is the mechanism destroying the ring particles collisional or tidal? This problem is one hundred and fifty years old but so far has not received a generally accepted answer.

What are the properties of a typical ring particle: its density, its tensile strength, its albedo, and its chemical composition? Is it covered by a layer of regolith? This last property determines how inelastic the collisions between particles are – the most important characteristic on which many theoretical models of the dynamics of the rings are based. There are no direct data about the density of the particles so that it is impossible to calculate even the total mass of the rings relatively accurately. If the rings are probed with a slight deformation of the trajectory of the probe one could calculate the mass of the rings and the density of the particles could be determined from the shape of the clouds of the dispersing fragments. An analysis of the chemical composition of the particles is extremely important for cosmology and would clarify the process of the condensation of the rings from the primordial proto-satellite swarm.

What is the size distribution of the particles? Are there particles in the rings larger than tens of meters – theorists often talk about those.

Is there a very thin – of a few hundred meters – stratification of the rings? A detailed study of the rings may clear up a whole lot of problems connected with the causes for the stratification of the rings into structures with varying sizes.

With what is the effect of the azimuthal brightness asymmetry of Saturn's A ring connected?

The Uranian rings can be probed using a similar scheme or even more simply: thanks to the extreme vacuum of the space between the rings the probe can move along an orbit intersecting the plane of the rings.

The solution of the above mentioned problems of the dynamics of the planetary rings is the most important stage on the path of constructing a detailed theory of the collective collisional processes in thin discs of macroparticles and may be of fundamental value for our understanding of the evolution of the solar system.

14.5 The Infrastructure of Planetary Physics

An analysis of the development of planetary science in the world shows that it has developed into an independent science which studies the solar system with the aim of producing a picture of its formation, evolution, and present state. Planetary science has its own specialised methods for obtaining experimental data: investigations using interplanetary spacecraft, observations from ground-based and orbiting telescopes, and laboratory studies of meteorites and lunar soil. An specialised organisational communication system has been developed in planetary physics which involves:

National and academical scientific centres.

Teaching departments in universities and technical colleges which prepare scientific and engineering teams.

National and international communication systems – journals, symposia, joint projects, and so on.

Let us consider "interplanetological" processes. One often divides the planetary sciences into three fields of activity:

1. Experimental Planetary Physics

Its main aim is the collection with all available means of empirical information about the solar system. This is the most expensive and highly technical part of planetology. Not only specialists in space travel and instrument building, and observers, but also scientists who process the primary information play an important role in experimental planetology.

2. General Planetary Physics and the Modelling
of Present-Day Phenomena

The aim of general planetology is the theoretical study of the *existing* solar system and the construction of theoretical models of objects, effects, and processes observed in the present state of the solar system, such as Jupiter's Large Red Spot, the hyperrotation of the Venerian atmosphere, or the volcanic activity on Io.

3. Planetary Cosmogony and the Modelling of Evolutionary Processes

The main aim of planetary cosmogony is the construction of a picture of the origin of the solar system and the evolution to its present state. It is possible to reconstruct the whole vast array of evolutionary processes, which, in the main, stretch milliards of years back, only by synthesising cosmogonical theories with models of present-day phenomena in the solar system and with contemporary achievements in mathematics and physics while at the same time actively carrying out experimental tests of the conclusions of cosmogonical models.

Let us compare the level of development of the infrastructure of planetary physics in the two leading members of the space club – the Soviet Union and the USA. This can be done by analysing published papers – the main results of scientific studies. According to the data from the abstract journal "Astronomiya" more than 4400 papers were published in the world in 1981 to 1987 dealing with planets, satellites – excluding the Moon – and solar system cosmology of which 800, that is, 18%, were published in the Soviet Union.

According to the data of the "Space Studies" abstract journal for 1986 the fraction of Soviet papers on comparative planetology and the study of bodies in the solar system by space agents was also 18%. If we take into account that the former Soviet Union had about half of all facilities in the world for the realisation of planetological space projects one must admit that the productivity of Soviet planetology was unsatisfactory. An analysis of the papers published in the international journal for solar system topics, "Icarus", shows that the authors – or coauthors – of 80% of non-Soviet papers are Americans. For one Soviet paper about the planets and their satellites there were therefore recently about four American papers.

An analysis of the publications shows the level of scientific communications between planetological groups. We give in Table 14.3 the result of an analysis of 882 papers published in "Icarus", the Russian "Astronomical Bulletin", and the Japanese planetological yearbook. It follows from Table 14.3 that scientists from the USA working in different cities interact much more actively than scientists from different Moscow institutes. It looks as if a colleague from another city is further away from a Russian scientist than an overseas colleague from France or Japan for an American. Clearly this picture of scientific communications is characteristic not only for planetary sciences. Such an isolation of scientific groups has an extremely negative effect on investigations.

We show in Table 14.4 an analysis of world-wide journals which publish planetological papers. The journals read by American planetary scientists –

Table 14.3. Level of scientific communications in the whole world, Japan, and the Soviet Union (analysis from 3 to 4 year periods)

Fraction of papers (in %) written	Whole world 1987/8	1990/1	Japan 1988/9	1991/2	Soviet Union 1986/9	1990/2
with coauthors	67	77	72	79	58	55
with coauthors from different organisations	44	52	41	46	26	14
with coauthors from different cities	41	50	35	42	13	10
with coauthors from different countries	13	19	2	7	0.6	1.4
with coauthors from different kinds of organisations, such as universities, scientific institutes, or observatories	33	33	18	31	11	7

Icarus (the indisputable leader), Nature, Science, the Journal of Geophysical Research, and the Astrophysical Journal – have been specially chosen.

Let us turn to such scientific characteristics as the existence of specialised scientific centres and the training of specialists. We give in Table 14.5 some data about the most important centres for planetary studies in the USA. The study of planets is carried out in such national scientific centres as the Goddard Space Flight Center in Greenbelt, the US Geological Service in Flagstaff and also industrial organisations such as the Rand Corporation. We add that specialised planetological scientific centres, or departments, exist also – according to authors' addresses – at Cornell, Pittsburgh, Stanford, and Washington Universities and at the Massachusetts Institute of Technology – this is a feature which occurs not only in the USA; for instance, in England there is a specialised department at the University of Lancaster. The study of the solar system is also carried out in the usual astronomy departments of such academic institutions as the Universities of Massachusetts, Colorado, New York, Florida, Kansas, Maryland, and so on.

We also note the well developed scientific connections between the planetological centres in the USA: each year there is a Lunar and Planetary conference in Houston and the University of Arizona publishes annually a large collective monograph – up to 1000 pages – about the planets and their satellites with the participation of a large circle of American scientists. The joint work on space project plays an important role in scientific communications. The JPL (Jet Propulsion Laboratory) in Pasadena controls the operation of the planetological spacecraft but in the scientific groups of these projects

Table 14.4. Distribution of planetological papers in the world over various journals during 1985 and references to these journals

Place of publication	Name of publication	Fraction of 543 papers published in the world on: cosmology, planets, and satellites*		Percentage of references to journal in reviews on the topics mentioned		
		Number	%	Planetary satellites (USSR)	Jovian satellites (USA)	Planetary rings (USA)
Internat.	Icarus	107	20	38	30	26
USA**	Review articles	52	10	9	21	13
USA	J. Geophys. Res.	28	5	6	7	10
USSR	Review articles	27	5	5	–	–
England	Nature	22	4	7	13	10
Europe	Astron. Astrophys.	18	3	2	1	3
Europe	Earth, Moon, and Planets	17	3	1	–	1
USSR	Astron. Lett.	14	3	1	–	–
USSR	Astron. Bull.	13	2	3	–	–
Internat.	IAU Circ.	13	2	–	–	–
USA	Meteorites	12	2	–	–	–
USSR	Kosm. Issled.	11	2	–	–	–
USSR	Astron. Tsirkulyar	11	2	–	–	–
USA	Astrophys. J.	11	2	3	6	7
Europe	Astron. Astrophys. Suppl.	11	2	–	–	–
USA	Science	10	2	7	14	8

* Analysis from the review journal "Astronomiya" for 1985.
** Predominantly.

scientists from all corners of the country are involved. For instance, about 100 scientists from 40 organisations from the USA – and also from France, Germany, England, and Canada – among which there were 20 universities, participated in the Voyager project. As an indication of the level of scientific communications we may mention the fact that about half of all planetological publications from the USA have coauthors working in different cities (see Table 14.3).

From this it is clear that a decisive role in the study of the planets in the USA is played by specialised planetological centres which have close connections with one another and with universities and observatories. Moreover, in the universities of the USA specialist planetologists are carefully trained. This means also that world-wide planetology has become an independent science, split off from astronomy.

Table 14.5. Largest centres of planetary science in the USA

City (State)	Organisation
Tucson (Arizona)	Planetary Science Institute
	University of Arizona with Department of
	Astronomy and Lunar and Planetary
	Laboratory with a 1.55 m telescope
Pasadena (California)	Jet Propulsion Laboratory (at CalTech)
	California Institute of Technology (CalTech) with
	Division of Geology and Planetary Science and
	Palomar Mountain Observatory with 1.2, 1.5,
	and 5 m telescopes
	Hale Observatory with 1.5 and 2.6 m telescopes
	at Mount Wilson
	University of California at Los Angeles (UCLA)
	with Department of Earth and Space Sciences
Moffett Field	NASA/Ames Research Center
(California)	Kuiper Flying Observatory with 91 cm telescope
Houston (Texas)	NASA/Johnson Space Center
Honolulu (Hawaii)	University of Hawaii with Department of
	Geology and Geophysics and Department
	of Astronomy
	Mauna Kea Observatory with 2.2, 3.6, 3.8,
	and 10 m telescopes

In Russia there are no specialised planetological centres and no universities or educational institutions with appropriate professorships or faculties. The study of the planets is carried out mainly in astronomy or physics departments and laboratories or similar institutions.

The study of the planets is retarded if planetology is not independent, especially through the absence of specialised planetological centres, The rate of the present development and the future outlook requires the urgent creation in Russia of the appropriate level of an organisation and communication structure. The creation of a coordination centre was discussed already at the First All-Union Congress on Cosmogonical Problems in 1951 (Cosmogony 1951).

The physics of the planets is the scientific basis for mankind to penetrate into space and also the foundation for understanding planetary processes on Earth. Future success in space projects and, possibly, in the control of the terrestrial climate must be guaranteed by the development of planetary physics to-day.

Conclusion

This book is an instantaneous photograph of the fast developing physics of planetary rings. There are still many interesting and complicated problems waiting to be solved, but the level of understanding of the nature of the rings reached at this moment enables us to expect that the picture of the origin and the dynamics of the planetary rings corresponds on the whole to the correct one.

The rings have produced a revolution in "discology": they have forced us to consider the dynamics of viscous discs from a completely new point of view. The variety of processes occurring in collective celestial dynamics have exceeded all expectations and further studies of collective effects in proto-solar and proto-satellite discs must lead to much greater detail – and to changes – in the existing models of the origin of the various bodies in the solar system. The study of the physics of the planetary rings shows that disc systems in the stage when the solar system was formed were saturated with rings, spiral waves, and all kind of vortices. Internal energy sources make such discs a very active medium likely to present us with hardly expected dynamical surprises and tropically intense self-organisation.

The Saturnian rings which have attracted the attention of researchers for more than three and a half centuries have completely justified this attention. This huge snow field, belonging to Saturn, is not at all dead, but alive, not in the least as it is as enigmatic as life. This book is the first biography of the rings of Saturn and of the other giant planets.

> "Hey, merchant, listen!
> Why not buy a hat from me,
> This hat, covered in snow?"
>
> *Base.*[1]

---

[1] Translation by Yulya Fridman.

Appendix I. The Possibility of Studying the Dynamics of Astrophysical Discs in a Two-Dimensional Approach[1]

1 Introduction

One might expect that it should be possible to describe the dynamics of thin discs by using solely two-dimensional functions, which depend on the coordinates in the plane of the disc, such as the surface density, the two-dimensional velocity field, or the gravitational potential in the plane of the disc. In fact, dynamical equations for such functions are widely used in the astrophysical literature.

It is just as natural to expect that such a description is valid only when a number of conditions are satisfied and that one should derive such equations, in the framework of these approximations, from the general three-dimensional equations of gas dynamics. However, as far as we know, up to the present there has been no such derivation of these equations or a consideration of the conditions for their applicability.

In the present appendix we derive a closed set of two-dimensional equations for a thin disc for two limiting cases: discs where the external gravitational field dominates and purely self-gravitating discs. In both cases we assume that the matter in the disc satisfies the barotropic equation of state $P = P(\varrho)$. The question of whether one can obtain a closed set in the case of equations of state of the general form $P = P(\varrho, S)$ remains unanswered.

If the barotropic equation of state $P = P(\varrho)$ has a general form the equations turn out to be a set of integro-differential equations with dynamical differential equations and an integral equation of state, in the form of a quadrature. In the particular case of a polytropic equation of state, $P = A\varrho^\gamma$, the integral equation of state for the two-dimensional functions reduces to an algebraic equation and the set of equations turn out to have a form very similar to the ones which have been used before.

However, there are significant differences.

Firstly, the dynamic equations differ from the ones proposed before by the presence of additional terms or – in the case of a self-gravitating disc – by the values of the constants occurring in them.

---

[1] Written with Khoruzhii and Libin as coauthors (see Fridman and Khoruzhii 1996b).

Secondly, the condition for the applicability of (two-dimensional) equations of this type involves not only putting a lower limit on the characteristic length scales in the plane of the disc $-$ $L \gg h$, where $2h$ is the thickness of the disc $-$ which is obvious as it gives us the thin-disc approximation, but also on the characteristic time scales of the process $-$ $T^2 \gg h^2/c^2$, where c is the sound velocity $-$ and this had not been considered earlier. In the general case this leads to an upper limit equal to the rotational frequency Ω for the characteristic frequency of a process: $\omega^2 \ll \Omega^2$.

2 Original Equations for the "Volume" Functions

2.1 Initial Dynamic Equations

The initial dynamic equations have the form

$$\hat{L}_1 v_r - \frac{v_\varphi^2}{r} = -\frac{1}{\varrho} \frac{\partial P}{\partial r} - \frac{\partial \varPhi}{\partial r}, \tag{A1.1}$$

$$\hat{L}_1 v_\varphi + \frac{v_r v_\varphi}{r} = -\frac{1}{\varrho r} \frac{\partial P}{\partial \varphi} - \frac{1}{r} \frac{\partial \varPhi}{\partial \varphi}, \tag{A1.2}$$

$$\hat{L}_1 v_z = -\frac{1}{\varrho} \frac{\partial P}{\partial z} - \frac{\partial \varPhi}{\partial z}, \tag{A1.3}$$

$$\frac{\partial \varrho}{\partial t} + \frac{1}{r} \frac{\partial}{\partial r} (r \varrho v_r) + \frac{1}{r} \frac{\partial}{\partial \varphi} (\varrho v_\varphi) + \frac{\partial}{\partial z} (\varrho v_z) = 0. \tag{A1.4}$$

Our notation is the standard one (Landau and Lifshitz 1987): ϱ and P are the density and pressure, $\varPhi$ is the gravitational potential, and v_r, v_φ, v_z are the radial and azimuthal velocity components and the velocity component parallel to the rotational axis. The $\hat{L}_1$ operator is given by the equation

$$\hat{L}_1 \equiv \frac{\partial}{\partial t} + v_r \frac{\partial}{\partial r} + \frac{v_\varphi}{r} \frac{\partial}{\partial \varphi} + v_z \frac{\partial}{\partial z}. \tag{A1.5}$$

The potential $\varPhi$ is a given function, if the disc is in an external gravitational field and its own gravitational field is negligibly small compared to the external field. In the opposite limiting case the gravitational potential is determined by the density of the disc through the Poisson equation:

$$\frac{1}{r} \frac{\partial}{\partial r} \left(r \frac{\partial \varPhi}{\partial r} \right) + \frac{1}{r^2} \frac{\partial^2 \varPhi}{\partial \varphi^2} + \frac{\partial^2 \varPhi}{\partial z^2} = 4\pi G \varrho. \tag{A1.6}$$

2.2 Equation of State

For the five – or six – unknown functions

$$v_r, v_\varphi, v_z, \varrho, P, (\varPhi) \qquad (A1.7)$$

we have so far written down four – or five – equations. The fifth – or sixth – equation which is needed to close the set of equations is the equation of state which has a different form depending on which thermodynamic functions are directly measured. If one measures the density and the temperature we can write the equation of state in the form

$$P = P(\varrho, T). \qquad (A1.8a)$$

If, however, apart from measurements of the density we have well defined ideas about the entropy function in the disc we can determine the pressure in the form

$$P = P(\varrho, S); \qquad (A1.8b)$$

we can consider (A1.8a) and (A1.8b) as the two general forms of writing down the equation of state.

Let the process studied by us be "slow", that is, be characterised by a time much longer than the heat transfer time. This means that after a characteristic time of our process any possible perturbations of the temperature profile will have relaxed. One can therefore consider this process to be *isothermal*, that is, at any point in space we have $T = T_0$ =const. For such a process we have instead of (A1.8a) the relation

$$P = P(\varrho, T_0) = P(\varrho). \qquad (A1.9a)$$

In the opposite case of a process, such as a sound wave, which is "fast" – as compared to a heat transfer process – this can be considered to be an *adiabatic* process because there is not enough time to exchange heat between two neighbouring regions; hence we have $S = S_0$ =const. For such a process we have instead of (A1.8b) the relation

$$P = P(\varrho, S_0) = P(\varrho). \qquad (A1.9b)$$

In both limiting cases of "slow" or "fast" processes we are therefore dealing with a barotropic equation of state.

3 Derivation of the Basic Equations for the "Plane" Functions

The derivation in this section is similar to the one of the equations of the shallow wave theory (Pedlosky 1982).

3.1 Order-of-Magnitude Estimates of the Terms in the Initial Equations

We shall give some "order-of-magnitude" estimates of the functions (A1.7) and their derivatives:

$$|v_r| \sim U, \ |v_\varphi| \sim V, \ |v_z| \sim W, \ |\varrho| \sim \varrho, \ |P| \sim P, \ (|\varPhi| \sim \varPhi),$$

$$\left|\frac{\partial}{\partial r}\right| \sim L^{-1}, \ \left|\frac{1}{r}\frac{\partial}{\partial \varphi}\right| \sim \zeta^{-1}, \ \left|\frac{\partial}{\partial z}\right| \sim D^{-1}, \ \left|\frac{\partial}{\partial t}\right| \sim T^{-1}. \quad (A1.10)$$

We have here introduced three characteristic length scales: L along the radius r, the scale of the radial perturbations; ζ along the azimuth, the scale of the azimuthal perturbations; and D along the rotational z-axis, the characteristic thickness of the disc, as well as a time scale T, the characteristic time of a process. Substituting (A1.10) into the continuity equation (A1.4) we find as to order of magnitude

$$W \leqslant \left(DT^{-1}, \varepsilon_1 U, \varepsilon_2 V\right)_{\text{max}}, \quad (A1.11)$$

where we must take the largest of the terms within the rounded brackets – we note that the terms in (A1.4) may partially cancel one another – and where we have introduced the dimensionless parameters ε_1 and ε_2 through the relations

$$\varepsilon_1 \equiv \frac{D}{L}, \qquad \varepsilon_2 \equiv \frac{D}{\zeta}. \quad (A1.12)$$

Substituting (A1.10) into (A1.1) and (A1.2) and dropping terms containing derivatives with respect to z for reasons which will become clear in what follows we find as to order of magnitude

$$\frac{P}{\varrho} \sim \left(\frac{L}{T}U, U^2, \frac{L}{\zeta}UV, \frac{L}{R}V^2, \varPhi, \frac{L}{R}\varPhi, \frac{\zeta}{T}V, \frac{\zeta}{L}UV, V^2, \frac{\zeta}{R}UV\right)_{\text{max}}. \quad (A1.13a)$$

The dimensionless quantities L/R, ζ/R, and L/ζ which occur within the brackets appear due to the difference in scale over which the pressure and the other functions – such as the potential and the velocities – change. Using (A1.10) we determine the order of magnitude of the ratio P/ϱ from (A1.3):

$$\frac{P}{\varrho} \sim \left(\frac{D}{T}W, \varepsilon_1 UW, \varepsilon_2 VW, W^2, \varPhi\right)_{\text{max}}. \quad (A1.13b)$$

Let us now find the conditions under which all terms inside the brackets, except the last one, turn out to be much smaller than the magnitude of P/ϱ following from (A1.13a). To obtain those conditions we substitute in (A1.13b) instead of W successively its expressions from (A1.11).

Substituting $W \sim D/T$ leads to the following estimate:

$$\frac{P}{\varrho} \sim \left(\frac{D^2}{T^2}, \varepsilon_1 \frac{D}{T} U, \varepsilon_2 \frac{D}{T} V, \Phi \right)_{max} . \tag{A1.13c}$$

Taking the ratio of the second term in (A1.13c) to the first one in (A1.13a) and the first term in (A1.13c) to the seventh one in (A1.13a) we find:

$$\frac{\varepsilon_1 D T^{-1} U}{L T^{-1} U} = \varepsilon_1^2, \qquad \frac{\varepsilon_2 D T^{-1} V}{\zeta T^{-1} V} = \varepsilon_2^2. \tag{A1.14}$$

We see thus that if the conditions

$$\varepsilon_1^2 \ll 1, \qquad \varepsilon_2^2 \ll 1 \tag{A1.15}$$

are satisfied we can neglect the second and third terms on the left-hand side of the equation of motion (A1.3). The inequalities (A1.15) define the "thin" disc approximation.

The first and fourth terms are of the same order of magnitude and the condition under which we may neglect them is $D^2/T^2 \ll P/\varrho$. The quantity on the right-hand side of this inequality is of the order of the square of the sound velocity, $P/\varrho \sim c^2$. As a result we get a lower bound for the characteristic time of a process:

$$T^2 \gg \frac{D^2}{c^2}. \tag{A1.16}$$

Whereas condition (A1.15) indicates that the thickness of the disc is small as compared to the "horizontal" scale of the perturbations, condition (A1.16) indicates that the processes considered are slow as compared to the time for establishing the pressure along the thickness of the disc. We shall analyse this condition in Sect. 3.3.

There thus remain only two terms in (A1.3) if conditions (A1.15) and (A1.16) are satisfied:

$$\frac{1}{\varrho} \frac{\partial P}{\partial z} = -\frac{\partial \Phi}{\partial z}. \tag{A1.17}$$

This equation expresses just the condition for hydrostatic equilibrium along the z-direction. It is clear that when there are no fast motions along the z-axis the attractive force towards the $z = 0$-plane can be cancelled by the pressure gradient in the z-direction.

Let us now in the right-hand side of (A1.13b) substitute for W its value from (A1.11): $W \sim \varepsilon_1 U$. We then find

$$\frac{P}{\varrho} \sim \left(\frac{D}{T}\varepsilon_1 U, \varepsilon_1^2 U^2, \varepsilon_2 U \varepsilon_1 V, \Phi \right)_{\max} . \tag{A1.13d}$$

Taking the ratios of the first three terms within the brackets in (A1.13d) to the first three terms within the brackets in (A1.13a), respectively, we find

$$\frac{\varepsilon_1 D T^{-1} U}{L T^{-1} U} = \varepsilon_1^2, \qquad \frac{\varepsilon_1^2 U^2}{U^2} = \varepsilon_1^2, \qquad \frac{\varepsilon_1 U \varepsilon_2 V}{L \zeta^{-1} U V} = \varepsilon_1 \varepsilon_2 \frac{\zeta}{L}.$$

From this it follows that if conditions (A1.15) and the inequalities

$$\varepsilon_1 \varepsilon_2 \frac{\zeta}{L} \ll 1, \qquad \varepsilon_1 \varepsilon_2 \frac{L}{\zeta} \ll 1, \qquad \varepsilon_1 \varepsilon_2 \frac{R}{\zeta} \ll 1 \tag{A1.18}$$

are satisfied – the second and third inequalities in (A1.18) follow if we use instead of the third term within the brackets in (A1.13a) the eighth and last terms – we can again find (A1.17) from Eq.(A1.3).

Finally, if in the right-hand side of (A1.13b) we substitute for W its last value from (A1.11): $W \sim \varepsilon_2 V$, we get

$$\frac{P}{\varrho} \sim \left(\frac{D}{T}\varepsilon_2 V, \varepsilon_1 U \varepsilon_2 V, \varepsilon_2^2 V^2, \Phi \right)_{\max} . \tag{A1.13e}$$

Taking the ratios of the first three terms within the brackets in (A1.13e) to the seventh, third, and fourth terms within the brackets in (A1.13a), respectively, we find

$$\frac{\varepsilon_2 D T^{-1} V}{\zeta T^{-1} V} = \varepsilon_2^2, \qquad \frac{\varepsilon_1 U \varepsilon_2 V}{L \zeta^{-1} U V} = \varepsilon_1 \varepsilon_2 \frac{\zeta}{L}, \qquad \frac{\varepsilon_2^2 V^2}{L R^{-1} V^2} = \varepsilon_2^2 \frac{R}{L}.$$

From this it follows that if conditions (A1.15) and the inequalities

$$\varepsilon_1 \varepsilon_2 \frac{\zeta}{L} \ll 1, \qquad \varepsilon_1 \varepsilon_2 \frac{L}{\zeta} \ll 1, \qquad \varepsilon_2^2 \frac{R}{L} \ll 1, \qquad \varepsilon_1 \varepsilon_2 \frac{R}{\zeta} \ll 1 \tag{A1.19}$$

are satisfied – the third and fourth inequalities in (A1.19) follow if we take the eighth and last terms in the brackets in (A1.13a) into account – we can again find (A1.17) from (A1.3).

One easily checks that all the inequalities of (A1.18) and (A1.19) can be reduced to the two conditions (A1.15) and a third inequality $\varepsilon_2^2(R/L) \ll 1$. Since we always have $R/L > 1$, we shall in what follows instead of the inequalities (A1.15), (A1.18), and (A1.19) use the conditions

$$\varepsilon_1^2 \ll 1, \qquad \varepsilon_2^2 \frac{R}{L} \ll 1. \tag{A1.20}$$

Equation (A1.3) thus reduces to (A1.17) if conditions (A1.20) and the additional condition (A1.16) are satisfied.

3.2 The Two Limiting Cases of Astrophysical Discs

We shall in the present appendix study in detail the two limiting cases of thin astrophysical discs: a disc in a predominant external gravitational field and a "purely" self-gravitating disc. We shall assume that we are dealing with the first case if we can neglect the force along the axis of rotation due to the intrinsic gravitational field of the disc in comparison with the pressure. We shall assume that we are dealing with the second case if the vertical structure of the disc – along the axis of rotation – is determined by the balance between the pressure and the gravitational force of the disc itself.

The last case requires an additional elucidation. Let us assume that there is no external gravitational field, that is, that the rotation curve of the disc is determined by its own gravitational field. For the linearised equations condition (A1.20) means that $(kh)^2 \ll 1$ where $2h$ is the thickness of the disc and $k = 2\pi/\lambda$ with λ the wavelength of the perturbations. In order that this condition be satisfied for perturbations with even the longest possible wavelength, that is, a wavelength of the order of the radius of the disc, $\lambda \sim R$, the inequality $(R/h)^2 \gg (2\pi)^2$ must be satisfied. We shall show that in that case the disc is highly unstable so that the characteristic time for the evolution of perturbations will be much shorter than a single revolution period of the disc. This time is even shorter for perturbations with shorter wavelengths.

Indeed, let us for the sake of simplicity consider a self-gravitating disc rotating with a constant velocity $\Omega_0 = \text{const}$ and a characteristic thickness $2h$ and radius R. Using the equilibrium condition of such a disc one can easily calculate the surface density as a function of the radius: $\sigma_0(r) = \sigma_0(0)\sqrt{(1 - 2^2/R^2)}$ as well as the rotational velocity as a function of the disc radius and of the value of the surface density at its centre (Fridman 1975, Fridman and Polyachenko 1984):

$$\Omega_0^2 = \frac{\pi^2 G \sigma_0(0)}{2R}.$$

The dispersion relation describing the frequency ω of small vibrations of such a disc as function of the radial wavenumber k has the form ($\sigma_0 \equiv \sigma_0(0)$)

$$\omega^2 = 4\Omega_0^2 - 2\pi G \sigma_0 |k| + k^2 c^2.$$

At the minimum of the dispersion curve which is determined from the equation $\partial \omega^2/\partial k = 0$ where

$$k = k_0 = \frac{\pi G \sigma_0}{c^2}, \tag{A1.21}$$

the dispersion relation reduces to

$$\omega^2 = 4\Omega_0^2 - \left(\frac{\pi G \sigma_0}{c}\right)^2.$$

Let us assume that the disc under consideration is isothermal ($\gamma_V = 1$, where γ_V is the adiabatic index for a three-dimensional process). From the equilibrium condition along the rotational axis we then get a relation between the half-thickness h and the volume density $\varrho_0(z = 0)$ in the plane of symmetry of the disc (Fridman and Khoruzhii 1999):

$$h = \frac{c}{\sqrt{2\pi G \varrho_0(z = 0)}} \approx \frac{c^2}{\pi G \sigma_0}.$$

We have here used the relation between σ_0 and $\varrho_0(z = 0)$ which one easily finds once one knows the density profile along the rotational axis: $\varrho_0(z) = \varrho_0(z = 0) \cosh^{-2}(z/h)$. Using (A1.21) we obtain the equation $k_0 h = 1$ which determines the shortest possible wavelength λ_0 for perturbations; this corresponds to $\lambda_0 = 2\pi h$.

Using the expression for h we find for the value of the square of the frequency at the point k_0 the expression

$$\omega^2 = 4\Omega_0^2 \left(1 - \frac{R}{2\pi h}\right),$$

whence follows the instability condition

$$\frac{R}{h} > 2\pi.$$

According to the definition of the radial size L of a perturbation with radial wavenumber k and wavelength λ we have $L \simeq k^{-1} \simeq \lambda/2\pi$. For perturbations with the longest wavelength, $\lambda_{\max} \simeq R$, $L_{\max} \simeq R/2\pi$, we get, by using the first of conditions (A1.20), $h^2/L_{\max}^2 \ll 1$, $R^2/h^2 \gg (2\pi)^2$. In this case there will already during a fraction of a single revolution appear structures in a gaseous self-gravitating disc as a result of which the disc will "heat up" and its thickness will increase until the system no longer finds itself on the instability boundary. One may thus assume that there is no such situation as an initial stationary self-gravitating disc if the condition $(R/h)^2 \gg (2\pi)^2$ is satisfied.

One checks easily that the condition on R/h, which we obtained a moment ago, for marginal stability does not depend significantly on the equation of state of the disc, that is, on the quantity γ_V, that is, the three-dimensional (ordinary) polytropic index (see (A1.32)). Indeed, let us obtain the condition for instability for the limiting case $\gamma_V = 2$ which is the opposite one to the isothermal case.

To present an argument which has the nature of a methodical statement we must look forward slightly and use some formulae and definitions which we shall obtain in what follows.

It follows from (A1.55) and (A1.58) that for arbitrary γ_V we have

$$-A_V \gamma_V \varrho^{\gamma_V - 2} \frac{\partial \varrho}{\partial z} = \sqrt{8\pi G A_V \left[\varrho_c^{\gamma_V} - \varrho^{\gamma_V}(z)\right]},$$

whence we find for the half-thickness of the disc

$$h = \int_0^h dz = \int_0^{\varrho_c} \frac{A_V \gamma_V \varrho^{\gamma_V - 2} \, d\varrho}{\sqrt{8\pi G A_V \left[\varrho_c^{\gamma_V} - \varrho^{\gamma_V}\right]}}$$

$$= \frac{A_V \gamma_V \varrho_c^{\gamma_V - 1}}{\sqrt{8\pi G A_V \varrho_c^{\gamma_V}}} \int_0^1 \frac{x^{\gamma_V - 2} \, dx}{\sqrt{1 - x^{\gamma_V}}} = \frac{c_c^2 I(\gamma_V)}{\sqrt{8\pi G P_c}}$$

$$= \sqrt{\frac{\gamma_V}{8\pi G \varrho_c}} \, c_c I(\gamma_V).$$

The index "c" indicates quantities referring to the (central) disc plane $z = 0$; the constant A_V is defined in (A1.32). We have used here the definition (A1.32) of P and the definition $c_c^2 = (\partial P/\partial \varrho)_c = \gamma_V A_V \varrho_c^{\gamma_V - 1}$ as well as

$$I(\gamma_V) \equiv \int_0^1 \frac{x^{\gamma_V - 2} \, dx}{\sqrt{1 - x^{\gamma_V}}}, \qquad x \equiv \frac{\varrho}{\varrho_c}.$$

For the special case $\gamma_V = 2$ we have $I(2) = \frac{1}{2}\pi$ and hence

$$h = \frac{\pi}{2} c_c \sqrt{\frac{1}{4\pi G \varrho_c}} = \sqrt{\frac{\pi c_c^2}{16 G \varrho_c}} = \sqrt{\frac{\pi A_V}{8G}}.$$

To find the connection between ϱ_c and σ we use the density profile along the rotational axis of a rigidly rotating disc with $\gamma_V = 2$ found by Goldreich and Lynden-Bell (1965):

$$\varrho_0(z) = \varrho_c \cos\left(\sqrt{\frac{2\pi G}{A_V}} z\right),$$

whence follows that

$$\sigma_0 = \int_{-h}^h \varrho_0(z) \, dz = \frac{4}{\pi} \varrho_c h.$$

The connection between h and σ_0 for the case considered thus turns out to be

$$h = \frac{c_c^2}{4G\sigma_0},$$

whence we find

$$\omega^2 = 4\Omega_0^2 \left(1 - \frac{R}{8h}\right).$$

We then find for the final instability condition for a disc with $\gamma_V = 2$:

$$\frac{R}{h} > 8.$$

We thus see that this instability condition hardly differs from the similar one for an isothermal disc. Everything we have said about the absence of an initial stationary state for a self-gravitating disc thus holds independent of the magnitude of γ_V.

A "purely" self-gravitating disc is thus highly unstable and most unstable for wavelengths corresponding to $kh = 1$ which do not satisfy the definition (A1.15) of the "thin" disc approximation, or of the long-wavelength approximation.

It is thus in principle impossible to describe the dynamics of a self-gravitating disc in the framework of a two-dimensional approximation in spite of the fact that this is traditionally done in the literature.

An obvious question is now whether it is possible to find such an external field that, while stabilising the disc, the disc remains self-gravitating along the rotational axis, that is, the structure of the disc along that axis is solely determined by its own field. If we require from the external field that it stabilises the disc the presence of such a field must bring the disc, if not to the stable region, at least to the boundary of this region. The only way to attain this is to increase significantly the rotational velocity of the disc. In other words, the rotational velocity of a stable disc must be mainly determined by the external field.

Let us assume that the external gravitational field is spherically symmetric. This is true, for instance, in the case of the spherical stellar halo and in the case of a large central mass for an accretion, circumstellar, or circumplanetary disc. In that case one can easily show that the derivatives along the rotational axis of the external gravitational field are of the same order of magnitude as the derivatives of the eigenpotential of the disc.

Indeed, to fix the ideas let us take the external potential of a central mass M:

$$\Phi_{\text{ext}}(r, z) = \frac{GM}{\sqrt{r^2 + z^2}}$$

and calculate its derivatives with respect to z:

$$\frac{\partial \Phi_{\text{ext}}}{\partial z} = -\frac{GMz}{\left(r^2 + z^2\right)^{3/2}} \approx \Omega^2 z, \quad \frac{\partial^2 \Phi_{\text{ext}}}{\partial z^2} \approx -\frac{GM}{\left(r^2 + z^2\right)^{3/2}} \approx \Omega^2.$$

On the other hand, one can use the Poisson equation to estimate the derivatives with respect to z of the potential of a thin self-gravitating disc:

$$\frac{\partial^2 \Phi_{\text{d}}}{\partial z^2} \approx 4\pi G \varrho, \quad \frac{\partial \Phi_{\text{d}}}{\partial z} \approx 4\pi G \int_0^z \varrho \, dz.$$

Here ϱ is the density in the disc. As the density of the disc decreases towards its periphery – in the z-direction – we can take for an upper estimate of the ratio of – both the first and the second – derivatives of the disc potential and of the external potential its value at the central plane of the disc – for $z = 0$. For an isothermal disc – and as to order of magnitude for any disc – we have $\varrho(z = 0) = \sigma/2h$. Using this and the earlier relations we find:

$$\frac{\partial^2 \Phi_{\mathrm{d}}/\partial z^2}{\partial^2 \Phi_{\mathrm{ext}}/\partial z^2} \leqslant \frac{2\pi G \sigma}{h\Omega^2} \approx \frac{1.7}{Q^2},$$

where Q is Toomre's safety coefficient (Toomre 1964) and where we have used the expression for the thickness of an isothermal disc which we gave earlier. We see that the ratio of the derivatives is proportional to Q^{-2}. The condition for the stability of the disc means that $Q^{-2} \leqslant 1$, that is, $\partial^2 \Phi_{\mathrm{d}}/\partial z^2 \leqslant \partial^2 \Phi_{\mathrm{ext}}/\partial z^2$, or $\partial^2 \Phi_{\mathrm{d}}/\partial z^2 \leqslant \Omega^2$. Integrating over z we get a similar relation for the first derivatives: $\partial \Phi_{\mathrm{d}}/\partial z \leqslant \partial \Phi_{\mathrm{ext}}/\partial z$. Going to the boundary of the stability region must thus inevitably lead to the structure of the disc along the rotational axis being determined both by its own as well as by the external gravitational field: the disc ceases to be self-gravitating.

We note that as the rotational curve of a stable disc is determined by the external field we have always

$$\frac{\partial^2 \Phi_{\mathrm{ext}}}{\partial r^2} \approx \Omega^2.$$

The conclusion arrived at a moment ago is therefore also valid for a non-spherical external field, if its anisotropy is not too great:

$$\frac{\partial^2 \Phi_{\mathrm{ext}}}{\partial r^2} \simeq \frac{\partial^2 \Phi_{\mathrm{ext}}}{\partial z^2}.$$

An exception will be the case when the external field has a cylindrical symmetry with a generatrix much longer than the thickness of the disc along the rotational axis of the latter. In that case the rotational velocity of the disc will be determined by the radial component of the gravitational force of the cylinder and may be rather too large to guarantee the stability of the disc. At the same time the structure of the disc along the rotational axis will be exclusively determined by its self-gravitation. This will be just the model which we shall have in mind when in what follows we speak about a self-gravitating disc. This model is the opposite limiting case of that where the external potential is predominant. Of course, this case is artificial as in real systems such "cylindrical" fields when we meet with them – needle-shaped galaxies, bridges between galaxies – do not have discs attached to them. However, the consideration of such a self-gravitating disc attached to a cylinder is, in our opinion, of methodological interest. Firstly, we can in this case split off effects which are the consequence of self-gravitation, so to speak, in its purest form, and thus estimate the limiting differences compared with the case of a predominant external field. Secondly, this model is essentially the one which has been considered in the majority of classical papers about astrophysical discs (see, for instance, Toomre 1964, Goldreich and Lynden-Bell 1965, Hunter 1972).

From the estimates given here it follows that real discs present an intermediate case between the two limiting cases we have considered and as their stability increases they approach the situation where the external field is predominant.

3.3 Limitations of the Characteristic Times of Processes Studied in the Two-Dimensional Approximation

We now turn to an explanation of the restrictions imposed by condition (A1.16) on the characteristic times – frequencies – of the processes which one can study in the two-dimensional approximation.

We note that if the matter of the disc is incompressible, that is, as $c_c^2 \to \infty$, the right-hand side of inequality (A1.16) tends to zero and this condition is automatically satisfied for all processes. It is just that situation which is realised in the "shallow water" case and as a result the only condition for the applicability of the two-dimensional equations is the "spatial" restriction (A1.20).

At first sight it should be easy to satisfy condition (A1.16) because the thickness D of the disc is small. In order to understand that is not, in fact, the case one must bear in mind that the thickness of the disc is small because the pressure forces – which are essentially equal to c^2 – are small as compared to the gravitational and inertial forces. As a result the quantity occurring on the right-hand side of (A1.16) is the ratio of two small quantities and needs to be studied closely.

Let us consider separately each of the cases we have described: a self-gravitating disc – in the sense we have discussed above – and a disc where the external field is dominant.

In the latter case we have (Fridman and Khoruzhii 1999) $c^2/D^2 \simeq \Omega_0^2$ which leads to the characteristic frequencies of the various processes being limited by the rotational frequency of the disc:

$$T^{-2} \simeq \omega^2 \ll \Omega_0^2. \tag{A1.16a}$$

This condition needs additional comments. In fact, the characteristic frequency of a process will, generally speaking, change depending on the rotating frame of reference in which it is considered, that is, the problem arises what we understand by the characteristic magnitude of a frequency. Since the restriction considered here is caused by the fact that the time for establishing vertical equilibrium is finite, from a physical point of view it is clear that the time to establish equilibrium in any actual position in the disc must be compared with the characteristic time of change at the same position in the disc, that is, the characteristic time for a process must be estimated in the comoving coordinate system. A more accurate – but also a more cumbersome – derivation of condition (A1.16a) confirms what we have said earlier. This is shown in Appendix II for the special case of small amplitude perturbations.

In the case of a self-gravitating disc its thickness can be calculated from the formulae given above. For an isothermal disc we have

$$h \approx \frac{c^2}{\pi G \sigma_0}$$

(as we showed earlier this result does not depend strongly on the actual form of the equation of state). Hence condition (A1.16) can be rewritten as follow:

$$T^{-2} \simeq \omega^2 \ll \frac{c^2}{h^2} \approx \left(\frac{\pi G \sigma_0}{c} \right)^2 \leqslant \kappa^2. \qquad \text{(A1.16b)}$$

This inequality is a condition for the epicyclic frequency κ of a marginally stable self-gravitating disc (Fridman and Khoruzhii 1999):

$$\kappa \geqslant \frac{\pi G \sigma_0}{c}.$$

For a self-gravitating disc, as for a disc where the external field is dominant, condition (A1.16) thus leads to a restriction on the characteristic frequencies of processes which can be studied in the two-dimensional approximation to be less than the rotational frequency of the disc.

An exception is the case when the external field acting on a gaseous disc is produced by a thicker and more massive stellar disc. Let the half-thickness of the stellar disc be h_* and the dispersion of the velocities along the rotational axis be c_{z*}; the half-thickness of the gaseous disc will then, according to Fridman and Polyachenko (1984) be

$$h_g = H_* \frac{c_g}{c_{z*}} = \frac{c_g}{\omega_{z*}} = \frac{c_g}{\kappa_*} \sqrt{\frac{h_*}{R_*}},$$

where ω_{z*} is the frequency of the oscillations of the stars along the rotational axis and R_* and κ_* are the radius and epicyclic frequency of the stellar disc. From this expression we find

$$\frac{c_g}{h_g} = \kappa_* \sqrt{\frac{R_*}{h_*}}.$$

Since as to order of magnitude we have $\kappa_* \simeq \kappa_g \simeq \kappa$ it follows that

$$\frac{c}{D} \simeq \kappa \sqrt{\frac{R_*}{h_*}} \gg \kappa.$$

However, we must note here that the inequality $\omega_{z*} = \kappa_* \sqrt{R_*/h_*} \gg \kappa_*$ used here is valid only in the case of a purely self-gravitating stellar disc without an external halo. We have already mentioned earlier that such a disc is highly unstable. On the other hand, if there exists an external spherical halo stabilising the stellar disc we are again led to condition (A1.16a).

Condition (A1.16) therefore turns out to be very rigid for any disc and one must take this into account together with conditions (A1.20) when one analyses the applicability of the two-dimensional equations for actual problems.

3.4 Closed System of Integro-differential Equations for a Barotropic Disc

We shall here consider barotropic processes; we have already met with some special cases earlier on – see (A1.9a) and (A1.9b): $\varrho = \varrho(P)$.

We shall introduce the following function of the pressure:

$$\mathcal{P}(P) \ = \ \int \frac{dP}{\varrho(P)}.$$

The gradient of this function is equal to

$$\nabla \mathcal{P} \ = \ \frac{d\mathcal{P}}{dP} \nabla P \ = \ \frac{1}{\varrho} \nabla P. \tag{A1.22}$$

Hence we have $\varrho^{-1} \partial P / \partial z = \partial \mathcal{P} / \partial z$ and instead of (A1.17) we find

$$\frac{\partial}{\partial z}(\mathcal{P} + \varPhi) \ = \ 0. \tag{A1.23}$$

The general solution of this equation is an arbitrary function which is independent of z:

$$\mathcal{P}(r, \varphi, z, t) + \varPhi(r, \varphi, z, t) \ = \ \chi(r, \varphi, t). \tag{A1.24}$$

We can use (A1.22) and (A1.24) to write the right-hand sides of (A1.1) and (A1.2), respectively, in the form of functions $\partial\chi/\partial r$ and $r^{-1}\partial\chi/\partial\varphi$, which are also independent of z. The left-hand sides of (A1.1) and (A1.2) must therefore also be independent of z, that is, they must contain the functions

$$v_r \ = \ v_r(r, \varphi, t), \qquad v_\varphi \ = \ v_\varphi(r, \varphi, t). \tag{A1.25}$$

Since $\partial v_r / \partial z = \partial v_\varphi / \partial z = 0$ it is clear why in deriving (A1.13a) we neglected terms containing derivatives with respect to z of v_r in (A1.1) and of v_φ in (A1.2). Using (A1.22) and substituting (A1.24) and (A1.25) into (A1.1) and (A1.2) we get the following dynamic equations for the "plane" functions:

$$\hat{L}_2 v_r - \frac{v_\varphi^2}{r} \ = \ -\frac{\partial\chi}{\partial r}, \tag{A1.26}$$

$$\hat{L}_2 v_\varphi + \frac{v_r v_\varphi}{r} \ = \ -\frac{1}{r}\frac{\partial\chi}{\partial\varphi}, \tag{A1.27}$$

where

$$\hat{L}_2 \equiv \frac{\partial}{\partial t} + v_r \frac{\partial}{\partial r} + \frac{v_\varphi}{r} \frac{\partial}{\partial \varphi}.$$

Integrating the equation of continuity (A1.4) over z from $-\infty$ to $+\infty$ we get

$$\frac{\partial}{\partial t} \left(\int_{-\infty}^{+\infty} \varrho \, dz \right) + \frac{1}{r} \frac{\partial}{\partial r} \left[r v_r \left(\int_{-\infty}^{+\infty} \varrho \, dz \right) \right]$$

$$+ \frac{1}{r} \frac{\partial}{\partial \varphi} \left[v_\varphi \left(\int_{-\infty}^{+\infty} \varrho \, dz \right) \right] + \int_{-\infty}^{+\infty} \left[\frac{\partial}{\partial z} (\varrho v_z) \right] dz \; = \; 0.$$

Obtaining this equation turns out to be possible thanks to the fact that v_r and v_φ are independent of z. The last integral vanishes as there is no matter at infinity. Introducing the surface density through the definition

$$\sigma(r, \varphi, t) \equiv \int_{-\infty}^{+\infty} \varrho(r, \varphi, z, t) \, dz, \tag{A1.28}$$

we obtain the "two-dimensional" continuity equation of a thin disc:

$$\frac{\partial \sigma}{\partial t} + \frac{1}{r} \frac{\partial}{\partial r} (r \sigma v_r) + \frac{1}{r} \frac{\partial}{\partial \varphi} (\sigma v_\varphi) \; = \; 0. \tag{A1.29}$$

To close the set of equations we use the barotropic equation of state

$$\varrho \; = \; \varrho(P) \; = \; \varrho(\mathcal{P}) \; = \; \varrho(\chi - \Phi), \tag{A1.30}$$

which we substitute into (A1.28):

$$\sigma(r, \varphi, t) = \int_{z_1}^{z_2} \varrho(r, \varphi, z, t) \, dz + \int_{-\infty}^{z_1} \varrho(r, \varphi, z, t) \, dz + \int_{z_2}^{\infty} \varrho(r, \varphi, z, t) \, dz$$

$$= \int_{z_1}^{z_2} \varrho \big[\chi(r, \varphi, t) - \Phi(r, \varphi, z, t) \big] \, dz, \tag{A1.31}$$

where the expression within the square brackets is the argument of the function $\varrho(\mathcal{P})$; the integrals outside the disc vanish, as we assume that the density in those regions are zero: $\varrho(z, \varphi, z, t) - 0$ for $z < z_1$ and $z > z_2$.

For the five (six) unknown functions

$$\varrho, \quad v_r, \quad v_\varphi, \quad \sigma, \quad \chi, \quad (\Phi)$$

we have now the five (six) equations (A1.26), (A1.27), (A1.29) to (A1.31) (and the Poisson equation (A1.6)).

We see that due to the integral form of (A1.31) – the equation for the surface density – our equations form a closed set of integro-differential equations.

For actual calculations the barotropic equation of state is already a rather general one. We shall restrict our class of equations of state to the polytropic one.

4 Closed Set of Differential Equations for a Polytropic Disc in an External Gravitational Field

4.1 Derivation of the Two-Dimensional Equations

We shall see in the present – and the next – section that for the derivation of a closed set of differential equations for the "plane" functions the above-mentioned conditions (A1.20) and (A1.16a) – or (A1.16b) in the next section – are sufficient conditions jointly with an assumption about the polytropic connection between P and ϱ:

$$P_V = A_V \varrho^{\gamma_V}, \tag{A1.32}$$

where A_V and γ_V are constants and γ_V is the "volume" polytropic index. From (A1.21) it then follows that the pressure function will be equal to

$$\mathcal{P} = \int_0^P \frac{dP}{\varrho} = A_V \frac{\gamma_V}{\gamma_V - 1} \varrho^{\gamma_V - 1}. \tag{A1.33}$$

The lower integration limit is taken here at $z = -\infty$ where the pressure – and the density – vanish.

Substituting (A1.33) into (A1.24) we find ϱ:

$$\varrho = \left[\frac{\gamma_V - 1}{A_V \gamma_V} (\chi - \Phi) \right]^{1/(\gamma_V - 1)}. \tag{A1.34}$$

After that we use (A1.31) to obtain an expression for σ:

$$\sigma(r, \varphi, t) = B_V \int_{z_1}^{z_2} [\chi(r, \varphi, t) - \Phi(r, \varphi, z, t)]^{1/(\gamma_V - 1)} \, dz, \tag{A1.35}$$

where

$$B_V \equiv \left[\frac{\gamma_V - 1}{A_V \gamma_V} \right]^{1/(\gamma_V - 1)}. \tag{A1.36}$$

Whereas in (A1.31) the integrand means some function ϱ of the difference of two other functions χ and Φ, the integrand in (A1.35) describes a completely well defined function of z, as Φ is the given external gravitational field. After integration over z we obtain a definite connection between the two "plane" functions σ and χ which closes the set of equations (A1.26), (A1.27), and (A1.29) for the "plane" functions v_r, v_φ, σ, and χ. The condition that the thickness of the disc be small makes it possible to describe the connection between σ and χ for any function Φ by using its expansion in the vicinity of the $z = 0$-plane:

$$\Phi(r, \varphi, z, t) = \Phi_0(r, \varphi, t) + \Phi_0'(r, \varphi, t) z + \tfrac{1}{2} \Phi_0''(r, \varphi, t) z^2.$$

The index "0" indicates here that the corresponding function is evaluated at the point $z = 0$ and the prime indicates differentiation with respect to z.

Substituting the expansion of the function Φ into expression (A1.35) we get

$$
\sigma(r, \varphi, t) = B_V \left(\frac{\Phi_0''}{2} \right)^{1/(\gamma_V - 1)} \int_{z_1}^{z_2} \left\{ \frac{2(\chi - \Phi_0)}{\Phi_0''} + \left(\frac{\Phi_0'}{\Phi_0''} \right)^2 \right.
$$
$$
\left. - \left[z^2 + 2\frac{\Phi_0'}{\Phi_0''} z + \left(\frac{\Phi_0'}{\Phi_0''} \right)^2 \right] \right\}^{1/(\gamma_V - 1)} dz
$$
$$
= B_V \left(\frac{\Phi_0''}{2} \right)^{1/(\gamma_V - 1)} \left[\left(\frac{\Phi_0'}{\Phi_0''} \right)^2 + \frac{2(\chi - \Phi_0)}{\Phi_0''} \right]^{1/(\gamma_V - 1) + \frac{1}{2}}
$$
$$
\times \int_{-1}^{1} (1 - x^2)^{1/(\gamma_V - 1)} \, dx, \tag{A1.37}
$$

where

$$
x \equiv \left(z + \frac{\Phi_0'}{\Phi_0''} \right) \bigg/ \sqrt{ \left(\frac{\Phi_0'}{\Phi_0''} \right)^2 + \frac{2(\chi - \Phi_0)}{\Phi_0''} } . \tag{A1.38}
$$

The limits of integration are found from the condition $\varrho(z_1) = \varrho(z_2) = 0$, that is, from the condition $x_{1,2} = \mp 1$, which follows from (A1.37). As a result we have:

$$
z_{1,2} = \mp \sqrt{ \left(\frac{\Phi_0'}{\Phi_0''} \right)^2 + \frac{2(\chi - \Phi_0)}{\Phi_0''} } - \frac{\Phi_0'}{\Phi_0''}. \tag{A1.39}
$$

It is clear that when the external gravitational field is symmetric with respect to the $z = 0$-plane, that is, when $\Phi_0' = 0$, the limits of the integration over z are also symmetric.

We rewrite the integral in (A1.37) in the form

$$
\int_{-1}^{1} (1 - x^2)^{1/(\gamma_V - 1)} \, dx = 2 \int_{0}^{1} (1 - x^2)^{\mu - 1} \equiv 2I, \tag{A1.40}
$$

where

$$
\mu = \frac{1}{\gamma_V - 1} + 1 = \frac{\gamma_V}{\gamma_V - 1}.
$$

From the tables in Gradshtein and Ryzhik's book (1965) we find that

$$
I = \tfrac{1}{2} B(\tfrac{1}{2}, \mu), \qquad \mu > 0, \tag{A1.41}
$$

where B is a beta-function.

This formula is valid only for positive μ which in our case is always the case since $\gamma_V > 1$. Moreover, we have (Gradshtein and Ryzhik 1965)

$$B(\tfrac{1}{2},\mu) = \frac{\Gamma(\tfrac{1}{2})\Gamma(\mu)}{\Gamma(\mu + \tfrac{1}{2})}, \tag{A1.42}$$

where Γ is a gamma-function.

Substituting (A1.42) into (A1.41) and the result into (A1.40) we have

$$\int_{-1}^{1} (1 - x^2)^{1/(\gamma_V - 1)}\, dx = \frac{\Gamma(\tfrac{1}{2})\Gamma(\mu)}{\Gamma(\mu + \tfrac{1}{2})} = \frac{\sqrt{\pi}\,\Gamma\left(\frac{\gamma_V}{\gamma_V - 1}\right)}{\Gamma\left(\frac{\gamma_V}{\gamma_V - 1} + \tfrac{1}{2}\right)}. \tag{A1.43}$$

Substituting (A1.36) and (A1.43) into (A1.37) we find

$$\sigma(r,\varphi,t) = \sqrt{\pi}\left[\frac{\gamma_V - 1}{A_V \gamma_V}\right]^{1/(\gamma_V - 1)} \frac{\Gamma\left(\frac{\gamma_V}{\gamma_V - 1}\right)}{\Gamma\left(\frac{\gamma_V}{\gamma_V - 1} + \tfrac{1}{2}\right)}$$

$$\times \left(\frac{\Phi_0''}{2}\right)^{1/(\gamma_V - 1)} \left[\left(\frac{\Phi_0'}{\Phi_0''}\right)^2 + \frac{2(\chi - \Phi_0)}{\Phi_0''}\right]^{(\gamma_V + 1)/2(\gamma_V - 1)}, \tag{A1.44}$$

whence follows that

$$\chi = C\sigma^{2\lambda} + \Phi_0 - \frac{(\Phi_0')^2}{2\Phi_0''}, \tag{A1.45}$$

where

$$C \equiv \left[\frac{\Phi_0''}{2} \frac{\Gamma^2\left(\frac{\gamma_V}{\gamma_V - 1} + \tfrac{1}{2}\right)}{\Gamma^2\left(\frac{\gamma_V}{\gamma_V - 1}\right)}\right]^{\lambda} \left(\frac{A_V \gamma_V}{\gamma_V - 1}\right)^{2/(\gamma_V + 1)}, \tag{A1.46}$$

$$\lambda = \frac{\gamma_V - 1}{\gamma_V + 1}. \tag{A1.47}$$

We note that in the general case $\Phi_0 = \Phi_0(r, \varphi, t)$ so that, generally speaking, C is not a constant but a function of the coordinates and the time: $C = C(r, \varphi, t)$. As Φ_0 is a given function, the connection (A1.45) between σ and χ gives us, together with (A1.26), (A1.27), and (A1.29) a closed set of equations for the "plane" functions.

Substitution of (A1.45) into (A1.26) and (A1.27) gives us

$$\hat{L}_2 v_r - \frac{v_\varphi^2}{r} = -2\lambda C\sigma^{2\lambda - 1}\frac{\partial\sigma}{\partial r} - \frac{\partial\Phi_0}{\partial r} - \sigma^{2\lambda}\frac{\partial C}{\partial r} + \frac{\partial}{\partial r}\left[\frac{(\Phi_0')^2}{2\Phi_0''}\right], \tag{A1.26a}$$

$$\hat{L}_2 v_\varphi + \frac{v_r v_\varphi}{r} = -\frac{2\lambda C\sigma^{2\lambda - 1}}{r}\frac{\partial\sigma}{\partial\varphi} - \frac{1}{r}\frac{\partial\Phi_0}{\partial\varphi} - \frac{\sigma^{2\lambda}}{r}\frac{\partial C}{\partial\varphi} + \frac{\partial}{\partial\varphi}\left[\frac{(\Phi_0')^2}{2r\Phi_0''}\right]. \tag{A1.27a}$$

We see that the equations of motion which we have obtained are significantly different from those traditionally used for a gaseous disc in an external gravitational field.

4.2 Special Case of the Potential $\Phi_0 = \Phi_0(r)$, $\Phi'_0 = 0$

Let us consider the special case when the external potential $\Phi_0(r, \varphi, t)$ satisfies the following three additional conditions:

1) It is stationary.
2) It is axisymmetric.
3) It is symmetric with respect to the central plane of the disc.

Under those conditions Eq.(A1.45) becomes

$$\chi(r, \varphi, t) = C(r)\,\sigma^{2\lambda}(r, \varphi, t) + \Phi_0(r). \qquad (A1.45a)$$

Substitution of (A1.45a) into the equations of motion (A1.26) and (A1.27) gives

$$\hat{L}_2 v_r - \frac{v_\varphi^2}{r} = -2\lambda C(r)\sigma^{2\lambda-1}\frac{\partial\sigma}{\partial r} - \frac{\partial\Phi_0}{\partial r} - \sigma^{2\lambda}\frac{\partial C}{\partial r}, \qquad (A1.26b)$$

$$\hat{L}_2 v_\varphi + \frac{v_r v_\varphi}{r} = -\frac{2\lambda C(r)\sigma^{2\lambda-1}}{r}\frac{\partial\sigma}{\partial\varphi}. \qquad (A1.27b)$$

Equations (A1.26b), (A1.27b), and (A1.29) are a closed set of equations under the restrictions given above.

Let us now write down the equations of motion for the "plane" functions in the "traditional" way:

$$\hat{L}_2 v_r - \frac{v_\varphi^2}{r} = -\frac{1}{\sigma}\frac{\partial P_S}{\partial r} - \frac{\partial\Psi}{\partial r}, \qquad (A1.48)$$

$$\hat{L}_2 v_\varphi + \frac{v_r v_\varphi}{r} = -\frac{1}{\sigma r}\frac{\partial P_S}{\partial\varphi} - \frac{1}{r}\frac{\partial\Psi}{\partial\varphi}, \qquad (A1.49)$$

where P_S is the "plane" pressure and Ψ the value of the gravitational potential in the $z = 0$-plane. The latter is the same as Φ_0. As to the first terms on the right-hand side they are only the same in the completely uniform case when we have $C(r) = \text{const}$.

Indeed, following the traditional method (Hunter 1972, Churilov and Shukman 1981) we introduce a "plane" polytrope

$$P_S = A_S\,\sigma^{\gamma_S}, \qquad (A1.50)$$

where A_S and γ_S are constants; γ_S is the plane polytropic index.

Substituting (A1.50) into the first terms on the right-hand sides of (A1.48) and (A1.49) we get

$$\frac{\nabla P}{\sigma} = A_S\gamma_S\sigma^{\gamma_S-2}\nabla\sigma. \qquad (A1.51)$$

Comparing (A1.51) with the first terms on the right-hand sides of (A1.26b) and (A1.27b) for the case when $C = \text{const}$ we find

$$A_S = \frac{2\lambda}{\gamma_S} C, \qquad \gamma_S = 2\lambda + 1,$$

or, finally, if we use (A1.47):

$$A_S = \frac{2(\gamma_V - 1)}{3\gamma_V - 1} C, \tag{A1.52}$$

$$\gamma_S = 3 - \frac{4}{\gamma_V + 1}. \tag{A1.53}$$

Churilov and Shukman (1981) were the first to derive these relations.

We see thus that in the case when $C(r) -$ const and when (A1.52) and (A1.53) are satisfied the "traditional" equations (A1.48) and (A1.49) are the same as the correctly obtained equations (A1.26b) and (A1.27b). However, if $C = C(r) \neq$ const, (A1.26b) and (A1.27b) differ from (A1.48) and (A1.49) through additional terms which are proportional to $C'(r)$. The differences become even more significant in the case where the external gravitational field has a more general form $\Phi_0 = \Phi_0(r, \varphi, t)$ when the additional terms appear not only in the terms describing the "pressure" but also in the expressions describing the gravitational potential (see (A1.45)).

4.3 The Applicability of $C =$ constant

If the function C is independent of the angle φ, this means that the external gravitational field is axisymmetric. This is practically always the case for galactic discs – if we neglect the regions near bars – for accretion dics, proto-planetary clouds, and planetary rings.

As regards the radial dependence of C we do not know any example where, strictly speaking, one can assume the coefficient C to be independent of r. However, for us it is important that the function C does not change very much over distances of the order of L – and we must bear in mind that L is the characteristic radial scale of the various structures.

If we talk about the large-scale spiral structure – the grand design – of the gaseous disc of our Galaxy the radial distance L between the main spiral arms in the vicinity of the Sun is not less than 4 kpc (Avedisova 1985). When one approaches the centre this distance decreases but not more than by a factor two, that is, it remains larger than 2 kpc. These distances are comparable with the radial scale of the exponential change in the density of the stellar component of 4 to 5 kpc.

The size of the "mini-spirals" in the centre of our Galaxy is also the same as the characteristic scale of the radial change in the density of the central stellar cluster (Blitz et al 1993).

For an accretion disc it has been shown (Bisikalo et al 1997, 1998) that the scale of its spiral structures is comparable to its size.

In all these cases it is therefore necessary to consider additional terms accounting for the r-dependence of $C(r)$.

In the ring structures of a proto-planetary cloud and especially in planetary rings the characteric size L of such a structure is much smaller than the scale over which $C(r)$ changes, remaining at least several times larger than the thickness of the disc. In order that in those cases the "traditional" two-dimensional equations, used earlier, be correct it is necessary that condition (A1.16a) be satisfied which is possible only for the dissipative branch – when in the absence of dissipation there is indifferent equilibrium $\omega = 0$. The basic instabilities of rings which are described in the present book develop just in that branch.

5 Closed Set of Differential Equations for a Polytropic Self-gravitating Disc

5.1 Derivation of the Two-Dimensional Equations

If we take inequality (A1.20) into account we can write the Poisson equation in the form

$$\frac{\partial^2 \Phi}{\partial z^2} = 4\pi G \varrho, \tag{A1.54}$$

where $-\mathcal{O}(D) < z < \mathcal{O}(D)$ with D as before the characteristic half-thickness of the disc. From (A1.23) and (A1.33) we have

$$\frac{\partial \Phi}{\partial z} = -A_{\mathrm{V}} \frac{\gamma_{\mathrm{V}}}{\gamma_{\mathrm{V}} - 1} \frac{\partial}{\partial z} \varrho^{\gamma_{\mathrm{V}} - 1} = -A_{\mathrm{V}} \gamma_{\mathrm{V}} \varrho^{\gamma_{\mathrm{V}} - 2} \frac{\partial \varrho}{\partial z}. \tag{A1.55}$$

Multiplying the left- and right-hand sides of (A1.54) and (A1.55) with one another we get

$$\left[(\Phi')^2 \right]' = 2\Phi' \Phi'' = -8\pi G A_{\mathrm{V}} \gamma_{\mathrm{V}} \varrho^{\gamma_{\mathrm{V}} - 1} \varrho' = -8\pi G A_{\mathrm{V}} (\varrho^{\gamma_{\mathrm{V}}})'.$$

Here and in what follows the prime indicates differentiation with respect to z. Integrating this equation from 0 to z we find

$$(\Phi')^2 - (\Phi'_c)^2 = -8\pi G A_{\mathrm{V}} \left[\varrho^{\gamma_{\mathrm{V}}} - \varrho_c^{\gamma_{\mathrm{V}}} \right], \tag{A1.56}$$

where the index "c" indicates the value of the function for $z = 0$.

Here we shall not be interested in membrane oscillations of the density of the disc, for example, caused by the bending instability (see, for instance, Fridman and Polyachenko 1984) but in perturbations in the plane of the disc. It is in this case very convenient for us to use the assumption that the density at all times has a symmetrical distribution with respect to the $z = 0$-plane. Following Hunter (1972) we apply this assumption which is convenient in order for us to compare our results with those from that classical paper.

This assumption means that the z-component of the gravitational force vanishes in all points of the $z = 0$-plane:

$$\Phi'_c = 0. \tag{A1.57}$$

We then find from (A1.56):

$$\Phi'(z) = \sqrt{8\pi G A_V \left(\varrho_c^{\gamma_V} - \varrho^{\gamma_V}(z)\right)}. \tag{A1.58}$$

We can now write down an expression for the surface density of a self-gravitating disc:

$$\sigma = 2 \int_0^D \varrho \, dz = \frac{1}{2\pi G} \int_0^D \Phi'' \, dz = \frac{\Phi'(D)}{2\pi G}.$$

Using Eq.(A1.58) we finally get

$$\sigma = \sqrt{\frac{2A_V}{\pi G}} \, \varrho_c^{\gamma_V/2}. \tag{A1.59}$$

We have used here the condition $\varrho(D) = 0$. We note that Hunter (1972) was the first to obtain (A1.59) which gives the relation between the surface density of the disc and the volume density in the $z = 0$-plane.

Equation (A1.24) holds for any value of z and, in particular, for $z = 0$:

$$\chi = \Phi_c + \mathcal{P}_c = \Phi_c + A_V \frac{\gamma_V}{\gamma_V - 1} \varrho^{\gamma_V - 1}. \tag{A1.60}$$

From (A1.59) we have

$$\varrho_c = \left(\frac{\pi G}{2A_V}\right)^{1/\gamma_V} \sigma^{2/\gamma_V}.$$

From the last two equations we can find the value of the pressure function in the $z = 0$-plane:

$$\mathcal{P}_c = A_V \frac{\gamma_V}{\gamma_V - 1} \left(\frac{\pi G}{2A_V}\right)^{(\gamma_V - 1)/\gamma_V} \sigma^{2(\gamma_V - 1)/\gamma_V}. \tag{A1.61}$$

According to (A1.26) and (A1.27) the right-hand side of the equation of motion is

$$\cdots = -\nabla\chi = -\nabla\Phi_c - \nabla\mathcal{P}_c. \tag{A1.62}$$

According to Hunter (1972) the right-hand side of the equation of motion can be written in the form

$$\cdots = -\nabla\Phi_c - \frac{1}{\sigma} \nabla P_S, \tag{A1.63}$$

where

$$P_S = A_S \, \sigma^{\gamma_S}. \tag{A1.64}$$

Using (A1.61) we get

$$\nabla \mathcal{P}_c = 2A_V \left(\frac{\pi G}{2A_V} \right)^{(\gamma_V - 1)/\gamma_V} \sigma^{(\gamma_V - 2)/\gamma_V} \nabla \sigma. \tag{A1.65}$$

On the other hand, from (A1.64) we have

$$\frac{\nabla P_S}{\sigma} = A_S \gamma_S \sigma^{\gamma_S - 2} \nabla \sigma. \tag{A1.66}$$

Equating the left-hand sides of (A1.65) and (A1.66) we find A_S and γ_S as functions of A_V and γ_V:

$$\gamma_S - 1 = \frac{2(\gamma_V - 1)}{\gamma_V}, \tag{A1.67}$$

$$\gamma_S A_S = 2A_V \left(\frac{\pi G}{2A_V} \right)^{(\gamma_V - 1)/\gamma_V}. \tag{A1.68}$$

Finally, from (A1.67) we get (Hunter 1972)

$$\gamma_S = 3 - \frac{2}{\gamma_V}. \tag{A1.69}$$

Substituting this expression into (A1.68) we find

$$A_S = \frac{2}{3\gamma_V - 2} A_V \left(\frac{\pi G}{2A_V} \right)^{(\gamma_V - 1)/\gamma_V}. \tag{A1.70}$$

Comparing this expression with the analogous one in Hunter's paper (1972),

$$A_S = \frac{\pi^{(3\gamma_V - 2)/2\gamma_V} \, \Gamma\left(\frac{2\gamma_V - 1}{\gamma_V} \right)}{2^{(2\gamma_V - 1)/\gamma_V} \, \Gamma\left(\frac{5\gamma_V - 2}{2\gamma_V} \right)} A_V \left(\frac{\pi G}{2A_V} \right)^{(\gamma_V - 1)/\gamma_V}, \tag{A1.71}$$

we see that these two expressions are completely different. The reason is simple: the right-hand side of the equation of motion contains the gradient of the pressure function and certainly not the gradient of the "plane" pressure, that is, the volume pressure integrated over z. One should note that this last quantity, in contrast to the surface – "plane" – density, does not have a physical meaning.

5.2 Why Does the Gradient of the Plane Pressure Not Have the Physical Meaning of a Force?

This can be most simply and clearly be shown by considering the flow of a fluid in a tube with a conical cross-section. It is clear that if the three-dimensional (ordinary) pressure in the fluid is everywhere constant there is no flow. At the same time, an analysis from the point of view of the "plane" pressure gives a completely different result. In the present case the "plane" pressure is equal to

$$P_S = SP_V,$$

where S is the area of the cross-section. The plane pressure thus has a non-vanishing gradient, as $S \neq$ const along the flow, and this should cause a flow in all calculations which use P_S. On the other hand, if the volume pressure changes inversely proportional to the cross-section the fluid will flow in the direction in which the cross-section of the tube increases, notwithstanding the fact that the gradient of the plane pressure vanishes ($P_S = SP_V =$ const). A correct analysis of the presence or absence of a flow is possible only if we take into account not only the plane pressure but also the longitudinal component of the reaction force from the walls of the tube. The latter can be determined from the condition of transverse equilibrium of the fluid in the tube.

The situation regarding a self-gravitating disc is completely similar with the sole difference that the role of the walls is now played by the gravitational potential of the disc. One could use the plane pressure in the case of a potential which is uniform in the disc – corresponding to a tube with a constant cross-section. However, for a self-gravitating disc this is equivalent to an infinite disc with a constant surface density. In that case the gradient of the plane pressure also vanishes. If, however, the gravitational potential changes one can only calculate the resulting pressure force in the plane of the disc if one takes the corresponding component of the gravitational force – the contribution from the walls – accurately into account and this can be determined from the condition that there be equilibrium along the rotational axis. This was just the kind of analysis which was carried out above which led to expression (A1.70) for the constant A_S describing the plane polytrope. The use of the plane pressure and hence of expression (A1.71) is incorrect in this case.

The substitution of expression (A1.56) into (A1.55) gives an equation for the volume density which can be integrated. In turn, knowing ϱ one can integrate the Poisson equation and find the function Φ_0. The set of equations which we have obtained is thus a closed set although its solution is certainly not without mathematical difficulties.

6 Conclusion

In the preceding sections we have thus derived a closed set of two-dimensional equations which describe the dynamics of rotating gravitating gaseous discs. They are integro-differential equations in the case of barotropic discs and the set consists of five or six equations: (A1.26), (A1.27), and (A1.29) to (A.31) (and (A1.6)), depending on whether the external gravitational field is dominant or the disc is self-gravitating. The set obtained consists of three or four differential equations in the case of polytropic discs: (A1.26a), (A1.27a), and (A1.44) (and (A1.6)). In the case of both barytropic and polytropic discs these equations differ from the dynamic equations used in the literature for astrophysical discs. Moreover, the equations obtained differ from the traditional equations of two-dimensional hydrodynamics.

We have, moreover, obtained the sufficient conditions which must be satisfied in order that one can describe the dynamics of a disc in the two-dimensional approximation. The first condition reflects the thin-disc approximation. It is well known and consists of the two inequalities (A1.20). The second condition (A1.16) imposes a restriction on the characteristic time for the processes studied in the two-dimensional approximation and we have not seen it before in the literature. In most cases this condition leads to an upper limit of the characteristic frequency of a process equal to the rotational frequency of the disc (see (A1.16a) and (A1.16b)). We have discused how rigid the restrictions imposed by these conditions turn out to be for real systems.

For discs where the external gravitational field is dominant the first "spatial" condition can easily be satisfied. However, the second "temporal" condition is in the overwhelming majority of cases not satisfied, except for the branch of oscillations with indifferent equilibrium for which $\omega = 0$.

Condition (A1.16) is relatively mild for a "purely" self-gravitating disc. However, it has no sense to consider long-wavelength processes in such a disc. The fact is that a purely self-gravitating thin disc is highly unstable and the maximum growth rate occurs in the region of large wavenumbers, for which $k \simeq D^{-1}$ and, hence, it is incorrect to study the instability itself in the two-dimensional approximation. The instability develops over a time less than a single revolution and leads to large changes in the disc parameters. The idea itself of an initial disc in which a long-wavelength perturbation can develop loses its meaning.

The instability of a "purely" self-gravitating disc has been known for more than thirty years (Safronov 1960, Toomre 1964) and the way to stabilise it has been known for more than twenty years. This is by taking into account the external gravitational field either of a central point mass (Ginzburg et al. 1971) in models of circumplanetary or circumstellar discs or of a stellar halo (Kalnajs 1972) in models of galactic discs.

However, we have shown in Sect. 3 of the present appendix that if the above-mentioned external sources are successful in stabilising the disc, bringing it to the boundary of the gravitational instability the "temporal" condition (A1.16) becomes very rigid and, as in the case of a disc where the external field is dominant, it restricts the frequencies of the processes considered to be less than the rotational frequency of the disc. The latter is basically determined by the external potential whereas the structure of the disc along the rotational axis and its thickness are determined equally by the external field and the field of the disc itself.

As a result it is not possible to satisfy condition (A1.16b) when studying long-wavelength processes. On the other hand, when studying self-gravitating discs close to the instability boundary the situation is the opposite. The frequency of oscillations in the locally corotating coordinate system is nearly zero and condition (A1.16b) is thus satisfied. However, the boundary of the unstable region lies in the large-wavenumber region, where $k \simeq D^{-1}$, and hence condition (A1.20) is not satisfied.

In the case where the external system is a self-gravitating stellar disc the thin-disc condition $kh \ll 1$ is satisfied for a gaseous disc at the stability boundary. However, the stellar disc itself is highly unstable. Its stabilisation by a spherical halo again returns us to situations which we have already considered.

It follows from what we have said here that the use of the two-dimensional equations for a description of the dynamics of real discs is possible only in rare cases. It is then necessary to use the equations obtained in the present appendix. In the overwhelming majority of cases a correct analysis is possible only in the framework of three-dimensional equations.

We shall turn to such an analysis for the case of small-amplitude perturbations in the next appendix.

Appendix II. Small-Amplitude Waves in a Disc Which Are Symmetric with Respect to Its $z = 0$-Plane[1]

1 Derivation of a Closed Set of Integro-differential Equations

As our initial set of equations we take (A1.1) to (A1.6) and we take for the gravitational potential Φ the gravitational potential of the disc – to be denoted by Φ_d – which is connected with its density through the Poisson equation (A1.6). As we shall be interested in waves generated in the region of the inner Lindblad resonance of a satellite, we must add the gravitational field of the satellite to the right-hand side of the equations of motion.

One of the requirements obtained in Appendix I, necessary to describe the dynamics of a disc in the two-dimensional approximation, is that the square of the eigenfrequencies of the perturbations of a self-gravitating disc ω^2 is small as compared to the square of its rotational velocity Ω_0^2: $\omega^2 \ll \Omega_0^2$. However, gravitational-acoustic waves and acoustic waves which have high frequencies do not satisfy this condition. Hence, we must study those waves using the three-dimensional equations and we shall now do this.

Linearising the original set of non-linear equations (A1.1) to (A1.4) we get

$$-\mathrm{i}\hat{\omega} v_{r1} - 2\Omega_0 v_{\varphi 1} = -\left(\mathcal{P}_1 + \Phi_{\mathrm{d}1} + \Phi_\mathrm{s}\right)', \tag{A2.1}$$

$$-\mathrm{i}\hat{\omega} v_{\varphi 1} + \frac{\kappa^2}{2\Omega_0} v_{r1} + \frac{\partial v_{0\phi}}{\partial z} v_{z1} = -\frac{\mathrm{i}m}{r} \left(\mathcal{P}_1 + \Phi_{\mathrm{d}1} + \Phi_\mathrm{s}\right), \tag{A2.2}$$

$$-\mathrm{i}\hat{\omega} v_{z1} = -\frac{\partial}{\partial z} \left(\mathcal{P}_1 + \Phi_{\mathrm{d}1} + \Phi_\mathrm{s}\right), \tag{A2.3}$$

$$-\mathrm{i}\hat{\omega} \varrho_1 + \frac{1}{r} \left(r\varrho_0 v_{r1}\right)' + \frac{\mathrm{i}m}{r} \varrho_0 v_{\varphi 1} + \frac{\partial}{\partial z} \left(\varrho_0 v_{z1}\right) = 0. \tag{A2.4}$$

The prime indicates here differentiation with respect to r, κ is the epicyclic frequency, and $\hat{\omega} \equiv \omega - m\Omega_0$.

The stationary quantities are indicated by an index "0" and depend on r and z.[2] The perturbed quantities are indicated by an index "1",

---

[1] Written with Khoruzhii as coauthor (Fridman and Khoruzhii 1996b).
[2] An exception is Ω_0; Poincaré has shown (see below) that it depends only on r.

$$A_1(r, \varphi, z, t) = \frac{1}{\sqrt{2\pi}} \int_{-\infty}^{+\infty} \sum_{m=-\infty}^{+\infty} A_{1m}(r, z, \omega) \, e^{i(m\varphi - \omega t)} \, d\omega.$$

The gravitational potential Φ_s of the satellite is defined as follows (compare expression (9.18))

$$\Phi_s(r, \varphi, z, t) = -\frac{GM_s}{\sqrt{r^2 + z^2 + r_s^2 - 2rr_s \cos(\varphi - \Omega_s t)}}. \tag{A2.5}$$

Writing $\zeta \equiv \varphi - \Omega_s t$ we expand the function $\Phi_s(r, \zeta, z)$ in a Fourier series, bearing in mind that stationary quantities do not depend on ζ:

$$\Phi_s(r, \zeta, z) = \sum_{m=-\infty}^{+\infty} \Phi_{sm}(r, z) \, e^{im\zeta}, \tag{A2.6}$$

where the coefficients in the series are given by the expressions

$$\begin{aligned}
\Phi_{sm}(r, z) &= \frac{1}{2\pi} \int_{-\pi}^{\pi} \Phi_s(r, \zeta, z) \, e^{-im\zeta} \, d\zeta \\
&= -\frac{GM_s}{2\pi} \int_{-\pi}^{\pi} \frac{e^{-im\zeta} \, d\zeta}{\sqrt{r^2 + z^2 + r_s^2 - 2rr_s \cos \zeta}}.
\end{aligned}$$

Since an integral of an odd function with symmetric limits vanishes we finally have

$$\Phi_{sm}(r, z) = -\frac{GM_s}{\pi} \int_0^{\pi} \frac{\cos m\zeta \, d\zeta}{\sqrt{r^2 + z^2 + r_s^2 - 2rr_s \cos \zeta}}. \tag{A2.7}$$

As the set (A2.1) to (A2.4) is linear we have chosen the one harmonic which is proportional to $\Phi_{sm}(r, z) \exp[im(\varphi - \Omega_s t)]$ and now we can drop the index m and also the amplitudes $A_{sm}(r, z, \omega)$ of the other functions. Of course, when in what follows we consider the wave which is generated at the inner Lindblad resonance $m : m-1$ we shall choose the harmonic m corresponding to this resonance.

In contrast to Sect. 9.1.3 all potentials are now functions of three spatial coordinates and $\mathcal{P}$ is the pressure function given by (A1.21): $\mathcal{P} = \int dP/\varrho$. If, by analogy with (A1.32), we make the assumption of a polytropic connection between the three-dimensional pressure P_V and the density ϱ, $P_V = A_V \varrho^{\gamma_V}$ with A_V and γ_V being constants and γ_V the "volume" polytropic index, we have from (A1.33): $\mathcal{P} = A_V \left(\gamma_V/(\gamma_V - 1)\right)\varrho^{\gamma_V - 1}$.

Using the standard definition $c^2 \equiv (dP_V/d\varrho)_{\varrho=\varrho_0} = A_V \gamma_V \varrho_0^{\gamma_V - 1}$ to introduce the sound velocity c we can find the perturbed pressure function:

$$\mathcal{P}_1 = \mathcal{P} - \mathcal{P}_0 = \left(\frac{d\mathcal{P}}{d\varrho}\right)_{\varrho=\varrho_0} (\varrho - \varrho_0) = A_V \gamma_V \varrho_0^{\gamma_V - 2} \varrho_1,$$

or, finally,

$$\mathcal{P}_1 = \frac{c^2 \varrho_1}{\varrho_0}. \tag{A2.8}$$

In the case of stationary functions (A1.1) and (A1.3) have the form

$$\frac{v_{0\varphi}^2}{r} = \frac{1}{\varrho_0} \frac{\partial P_0}{\partial r} + \frac{\partial \Phi_0}{\partial r}, \tag{A2.9}$$

$$\frac{1}{\varrho_0} \frac{\partial P_0}{\partial z} + \frac{\partial \Phi_0}{\partial z} = 0. \tag{A2.10}$$

If E is the internal energy and V the specific volume we have from thermodynamics

$$W = E + PV,$$

or

$$dW = dE + P\,dV + V\,dP.$$

Using the First Law of thermodynamics,

$$dE = T\,dS - P\,dV,$$

we can simplify the last equation:

$$dW = T\,dS + V\,dP = T\,dS + \frac{1}{\varrho}\,dP.$$

For adiabatic processes we have $dS = 0$ and hence

$$dW = \frac{1}{\varrho}\,dP.$$

We can then rewrite (A2.9) and (A2.10) in the form

$$\frac{v_{0\varphi}^2}{r} = \frac{\partial \chi_0}{\partial r}, \tag{A2.9a}$$

$$\frac{\partial \chi_0}{\partial z} = 0, \tag{A2.10a}$$

where $\chi_0 \equiv W_0 + \Phi_0$.

It follows from (A2.10a) that χ_0 is independent of z and hence that $v_{0\varphi}$ is also independent of z.

A similar result can also be obtained for barotropic systems. In that case we introduce the pressure function:

$$\mathcal{P} = \int \frac{dP}{\varrho(P)},$$

the differential of which is formally the same as the differential dW for adiabatic processes:

$$d\mathcal{P} = \frac{dP}{\varrho(P)}.$$

It is clear that one can use $\mathcal{P}$ to describe not only adiabatic but also isothermal processes and, in general, always when we can restrict ourselves to the barotropic approximation, $\varrho = \varrho(P)$.

Poincaré was the first to prove that $v_{0\varphi}$ is independent of z for barotropic systems.

If, in accordance with what we said above, we drop in (A2.2) the term with v_{z1} we are left with two equations of motion for the two velocity components v_{r1} and $v_{\varphi1}$ which we can now express in terms of the potentials:

$$v_{r1} = \left[-i\hat{\omega}\chi' + i\frac{2m\Omega_0}{r}\chi \right] \Big/ D,$$

$$v_{\varphi1} = \left[-\frac{\kappa^2}{2\Omega_0}\chi' + \frac{m\hat{\omega}}{r}\chi \right] \Big/ D,$$

where we have used (A2.8) and introduced the notation

$$\chi \equiv c^2\frac{\varrho_1}{\varrho_0} + \Phi_{d1} + \Phi_s, \tag{A2.11}$$

$$D \equiv \hat{\omega}^2 - \kappa^2. \tag{A2.12}$$

Substituting the expressions for v_{r1} and $v_{\varphi1}$ into the equation of continuity (A2.4) we see that, after a few transformations, that equation can be written in the form

$$\hat{\omega}\left[\varrho_1 + \frac{1}{r}\left(\frac{r\varrho_0\chi'}{D}\right)' - \frac{m^2\varrho_0}{r^2 D}\chi \right] - \frac{2m}{r}\left(\frac{\Omega_0\varrho_0}{D}\right)'\chi + i\frac{\partial}{\partial z}(\varrho_0 v_{z1})$$

$$= 0. \tag{A2.13}$$

If we integrate this equation over z we get

$$I \equiv \int_0^z \left\{ \hat{\omega}\left[\varrho_1 + \frac{1}{r}\left(\frac{r\varrho_0\chi'}{D}\right)' - \frac{m^2\varrho_0}{r^2 D}\chi \right] - \frac{2m}{r}\left(\frac{\Omega_0\varrho_0}{D}\right)'\chi \right\} dz$$

$$= -i\varrho_0(r,z)v_{z1}(r,\varphi,z,t). \tag{A2.14}$$

We have restricted ourselves here to perturbations ϱ_1, $v_{\varphi1}$, and v_{r1} which are symmetric and perturbations v_{z1} which are antisymmetric with respect to the $z = 0$-plane which means, in particular, that $v_{z1}(r,\varphi,0,t) = 0$. In other words, we are not interested in bending oscillations for which $v_{z1}(r,\varphi,0,t) \neq 0$. Multiplying both sides of (A2.14) by $\hat{\omega}$ and using the equation of motion (A2.3) we find

$$-i\hat{\omega}v_{z1}(r,\varphi,z,t) = \frac{I}{\varrho_0}\hat{\omega} = -\frac{\partial}{\partial z}\chi. \tag{A2.15}$$

Integrating (A2.15) over z from 0 to z and using the definition (A2.11) of χ we find the magnitude of the perturbed volume density for any value of z:

$$\varrho_1(r, z) = \frac{\varrho_0(r, z)}{c^2(r, z)} \left[-\Phi_{d1}(r, z) + \Phi_{d1}(r, 0) \right.$$

$$\left. + c^2(r, 0) \frac{\varrho_1(r, 0)}{\varrho_0(r, 0)} - \int_0^z \frac{\hat{\omega} I}{\varrho_0(r, z)} \, dz \right]. \quad \text{(A2.16)}$$

It follows from (A2.5) that the difference $\Phi_s(r, \varphi, z, t) - \Phi_s(r, \varphi, 0, t)$ is small,

$$\Phi_s(r, \varphi, z; t) - \Phi_s(r, \varphi, 0; t) = -\frac{\Phi_s(r, \varphi, 0; t) z^2}{2|r - r_s|^2}, \quad \text{(A2.17)}$$

and we have therefore dropped it in (A2.16). The Poisson equation (A1.6) together with (A2.16) forms a closed set of integro-differential equations for the two functions $\varrho_1(r, \varphi, z, t)$ and $\Phi_1(r, \varphi, z, t)$.

2 Derivation of the Dispersion Equation Describing the Three-Dimensional Perturbations

When one considers gravitational-acoustic waves in stratified atmospheres (Lighthill 1978) a similar problem arises. The atmosphere can act like a one-sided waveguide in which changes in pressure above the Earth's surface are most effectively transferred through low-frequency perturbations. However, we cannot use here the simple dispersion relation describing the strictly one-dimensional propagation of sound as the sound velocity c varies with height.

Lighthill (1978) in his monograph solved this analogous problem by proposing an iteration method which we now shall use also.

We use as the first approximation the solution obtained neglecting motion in the z-direction, that is, we shall assume that $v_{z1} = 0$. According to (A2.15) this approximation corresponds both to putting $I = 0$ and assuming that χ is independent of z. In that case we get from (A2.16):

$$c^2(r, z) \frac{\varrho_1(r, z)}{\varrho_0(r, z)} + \Phi_{d1}(r, z) = c^2(r, 0) \frac{\varrho_1(r, 0)}{\varrho_0(r, 0)} + \Phi_{d1}(r, 0).$$

The solution of the Poisson equation taking into account the fact that χ is independent of z, which is also reflected in the last equality, was given in Sect. A1.5.

Determining in this way the functions $\varrho_1(r, z)$ and $\Phi_{d1}(r, z)$ of the first approximation we can substitute them into the integral in (A2.16) and thus obtain these functions in the second approximation, and so on.

If we are interested in the dispersion relation for free waves or for waves generated in the region where there is a resonance with a satellite, we can obtain it by integrating (A2.16) from one edge of the disc to another. We must then use the zeroth-order boundary condition for the fluxes in the z-direction (Lighthill 1978)

$$(\varrho_0 v_{z1})_{-h} = (\varrho_0 v_{z1})_h = 0, \tag{A2.18}$$

where h is the half-thickness of the disc, $\varrho_0(\pm h) = 0$.

One can easily understand the necessity of the requirement (A2.18) by a reductio ad absurdum. Assume that the vertical fluxes at the boundaries of the disc are finite. If the density of the disc is equal to zero, this means that the velocity v_{z1} must be infinite – an impossibility as this means an infinite energy of the vertical motion at the boundaries of the disc (Lighthill 1978): $\left(\frac{1}{2}\varrho_0 v_{z1}^2\right)_{\pm h} \to \infty$.

We integrate (A2.13) over z from $-h$ to $+h$. Using (A2.18) we find

$$\int_{-h}^{+h} \left\{ \hat{\omega} \left[D\varrho_1 + \frac{D}{r} \left(\frac{r\varrho_0\chi'}{D} \right)' - \frac{m^2}{r^2}\varrho_0\chi \right] \right.$$
$$\left. - \frac{2mD}{r} \left(\frac{\varrho_0\Omega_0}{D} \right)' \chi \right\} dz = 0. \tag{A2.19}$$

This equation together with the boundary conditions in r are the boundary value problem for the determination of the frequencies of the eigenvibrations of $\hat{\omega}$. We can reduce (A2.19) directly to the dispersion relation in the case of perturbations with wavelengths short as compared to the radius of the system. To prove that we can analyse in detail the nature of the z-dependence along the rotation axis of the various terms occurring in the integrand.

The quantity D is independent of z and we have already used this as we obtained (A2.19) by integrating (A2.13) multiplied by D. (We also note that as a result of this we have no right to apply the relations obtained in any of the Lindblad resonance points.)

We have already shown earlier that the gravitational potential Φ_s is characterised by its z-dependence being negligibly small.

The density, pressure, and sound velocity change, in general, on a scale of the order of the thickness of the disc. An exception is the case when the equation of state is specially chosen – for instance, the density is constant in an incompressible disc and the sound velocity is constant in the isothermal case.

To determine how strongly the perturbations of the gravitational potential Φ_{d1} depend on z we must consider in more detail the solution of the Poisson equation for this case.

3 Solution of the Poisson Equation for a Disc of Half-Thickness h

We can consider a disc with a finite half-thickness h to be a set of infinitesimally thin layers at distances z_i from the central $z = 0$-plane and perturbed surface densities $d\sigma_i = \varrho(z_i)\, dz$. The perturbed surface density of the whole disc is equal to $\sigma = \sum d\sigma_i = \int_{-h}^{h} d\sigma$. (In contrast to the rest of this appendix we shall in the present section omit the index "1" of the perturbed quantities, but we shall continue to distinguish the unperturbed quantites by an index "0".) By virtue of the linearity of the Poisson equation the problem of determining the potential of a disc with a finite thickness can thus be reduced to the problem of determining the potential of an infinitesimally thin layer.

Outside the i-th layer the potential generated by that layer is described by the Laplace equation:

$$\left[\frac{\partial^2}{\partial r^2} + \frac{1}{r}\frac{\partial}{\partial r} + \frac{1}{r^2}\frac{\partial^2}{\partial\varphi^2} + \frac{\partial^2}{\partial(z - z_i)^2} \right] \Phi_i = 0. \tag{A2.20}$$

We shall look for a particular solution of the Laplace equation (A2.20) outside the disc in the form of a product:

$$\Phi_i(r, \varphi, z) = R(r)F(\varphi)Z_i(z) = R(r)Z_i(z)e^{im\varphi}. \tag{A2.21}$$

Substituting (A2.21) into (A2.20) we get

$$Z_i\left[\frac{\partial^2 R}{\partial r^2} + \frac{1}{r}\frac{\partial R}{\partial r} - \frac{m^2}{r^2}R \right] + R\frac{\partial^2 Z_i}{\partial(z - z_i)^2} = 0.$$

This equation is solved using the separation of variables:

$$\frac{1}{R}\left[\frac{\partial^2 R}{\partial r^2} + \frac{1}{r}\frac{\partial R}{\partial r} - \frac{m^2}{r^2}R \right] = -\frac{1}{Z_i}\frac{\partial^2 Z_i}{\partial(z - z_i)^2} = -k^2, \quad k^2 > 0. \tag{A2.22}$$

The separation constant is chosen to be negative in order that we can satisfy the homogeneous boundary conditions as $z \to \pm\infty$.

The last equation,

$$\frac{\partial^2 Z_i}{\partial(z - z_i)^2} - k^2 Z_i = 0,$$

has the general solution

$$Z_i(z) = A_1\, e^{-|k|(z - z_i)} + A_2\, e^{|k|(z - z_i)},$$

which if we take into account the homogeneous boundary conditions we can write in the form

$$Z_i(z) = A_1\, e^{-|k||z - z_i|}, \qquad z \neq z_i. \tag{A2.23}$$

The other equation (A2.22) turns out to be the equation for cylindrical functions:

$$\frac{\partial^2 R}{\partial x^2} + \frac{1}{x}\frac{\partial R}{\partial x} + \left[1 - \frac{m^2}{x^2}\right] R = 0,$$

with the general solution

$$R = C_1 J_m(x) + C_2 Y_m(x),$$

where we have $x \equiv |k|r$ while J_m and Y_m are Bessel functions of the first and second kind – the latter are also called Weber functions. Since $Y_m(x) \to \infty$ as $x \to 0$ we must put $C_2 = 0$.

The particular solution of the Laplace equation outside the i-th layer is thus

$$\Phi_i(r, \varphi, z) = A_i J_m(|k|r) e^{im\varphi - |k||z - z_i|}, \qquad z \neq z_i. \tag{A2.24}$$

The constant A_i is determined by joining this potential to the potential in the layer itself which is determined by the Poisson equation

$$\left[(\nabla^2)_\perp + \frac{\partial^2}{\partial(z - z_i)^2}\right] \Phi_i^{\text{in}} = 4\pi G \, d\sigma_i \, \delta(z - z_i). \tag{A2.25}$$

The potential Φ_i must be a continuous function for all z and, in particular, when $z = z_i$, as otherwise the z-component of the force would be infinite. This means that $\Phi_i = \Phi_i^{\text{in}}$ for $z = z_i$. Substituting in (A2.25) expression (A2.24) instead of Φ_i^{in} we get

$$-2\delta(z - z_i)|k| A_i J_m(|k|r) e^{im\varphi - |k||z - z_i|} = 4\pi G \, d\sigma_i(r, \varphi) \, \delta(z - z_i),$$

whence follows the following relation between the constant A_i and the perturbed density in the layer:

$$A_i J_m(|k|r) e^{im\varphi} = -\frac{2\pi G}{|k|} \, d\sigma_i(r, \varphi). \tag{A2.26}$$

Substituting (A2.26) into (A2.24) we get the final expression for the potential Φ_i:

$$\Phi_i(r, \varphi, z) = -\frac{2\pi G \, d\sigma_i}{|k|} e^{-|k||z - z_i|}. \tag{A2.27}$$

By summing the contributions from all the infinitesimally thin disks we get the gravitational potential of the original layer of finite thickness:

$$\Phi(r, \varphi, z) = \sum_i \Phi_i(r, \varphi, z) = -\frac{2\pi G}{|k|} \int_{-h}^{h} e^{-|k||z - z_i|} \, d\sigma_i$$

$$= -\frac{2\pi G}{|k|} \int_{-h}^{h} e^{-|k||z - z_i|} \varrho(z_i) \, dz_i = -\frac{2\pi G}{|k|} \left(1 - \alpha(|z|)|k|h\right), \tag{A2.28}$$

where α is a quantity of order unity:

$$\alpha(|z|) \;=\; \frac{1}{h\sigma} \int_{-h}^{h} |z - z_i| \varrho(z_i)\, dz_i, \tag{A2.29}$$

with σ defined by the relation

$$\sigma \;=\; \int_{-h}^{h} \varrho(z_i)\, dz_i.$$

We bear in mind that we are studying only perturbations which are symmetric with respect to $z = 0$ so that we can use the fact that $\varrho(z)$ is an even function and we can then transform the integral in (A2.29) to a more convenient form:

$$\int_{-h}^{h} |z - z_i| \varrho(z_i)\, dz_i \;=\; \int_{z}^{h} (z_i - z)\varrho(z_i)\, dz_i + \int_{-h}^{z} (z - z_i)\varrho(z_i)\, dz_i$$

$$=\; \int_{0}^{h} (z_i - z)\varrho(z_i)\, dz_i + 2\int_{0}^{z} (z - z_i)\varrho(z_i)\, dz_i + \int_{-h}^{0} (z - z_i)\varrho(z_i)\, dz_i$$

$$=\; 2\int_{0}^{h} z_i \varrho(z_i)\, dz_i + \int_{0}^{z} \sigma_{z_i}\, dz_i.$$

Here we have put $\sigma_{z_i} = \int_{-z_i}^{z_i} \varrho(z_i)\, dz_i$ which is the surface density of that part of the disc bounded by the planes $z = \pm z_i$. In the last expression the first term gives the value of the integral for $z = 0$ and the second one is clearly an even function of z. The final expression for α takes the form

$$\alpha(\tau) \;=\; 1 - \int_{\tau}^{1} f(x)\, dx, \tag{A2.30}$$

where $\tau = |z|/h$ and where we have introduced the function $f(x) = \sigma_x/\sigma$ which varies from zero to unity.

From this expression we easily get the change in α over the half-thickness of the disc:

$$\Delta\alpha \;=\; \int_{0}^{1} f(x)\, dx \;<\; 1.$$

The change in the perturbed disc potential when z varies from 0 to h is similarly

$$|\Delta\Phi| \;<\; 2\pi G|\sigma|h \;\approx\; |k|h|\Phi(z=0)| \;\ll\; |\Phi(z=0)|.$$

The variation of the potential over the half-thickness of the disc is thus small.

By differentiating the expressions we obtained for the potential we find its derivatives along the rotation axis:

$$\frac{\partial\Phi}{\partial z} \;=\; 2\pi G\sigma_z \;=\; 4\pi G \int_{0}^{z} \varrho\, dz, \tag{A2.31}$$

$$\frac{\partial^2 \Phi}{\partial z^2} = 4\pi G\varrho. \tag{A2.32}$$

We have thus, as to order of magnitude,

$$\left|\frac{\partial \Phi}{\partial z}\right| \simeq |k||\Phi(z=0)| \simeq \left|\frac{\partial \Phi}{\partial r}\right|, \tag{A2.31a}$$

$$\left|\frac{\partial^2 \Phi}{\partial z^2}\right| \simeq \frac{|k|}{h}|\Phi(z=0)| \simeq \frac{1}{|k|h}\left|\frac{\partial^2 \Phi}{\partial r^2}\right|. \tag{A2.31b}$$

4 Dispersion Relation for Waves in the Plane of the Disc

We shall now use (A2.19) to derive a dispersion relation for waves in the plane of the disc, assuming for the sake of simplicity that the initial density in the disc is independent of the radius ($\varrho_0(r) = $ const). Neglecting terms of order $|k|h \ll 1$ we can instead of the exact values of the gravitational potential use its values in the central plane of the disc:

$$\Psi_1(r,\varphi,t) \equiv \Phi_{d1}(r,\varphi,0,t), \qquad \Psi_s(r,\varphi,t) \equiv \Phi_s(r,\varphi,0,t). \tag{A2.33}$$

In D there occur the values of the rotational velocity and of the epicyclic frequency which, as we have shown earlier, are independent of z in a barotropic disc.

Using what we have just said we find that (A2.19) takes the form

$$\int_{-h}^{+h} \left\{ \hat{\omega} \left[D\varrho_1 + \frac{D}{r} \left(\frac{r\left(c^2\varrho_1 + \varrho_0\Psi_1 + \varrho_0\Psi_s\right)'}{D} \right)' - \frac{m^2}{r^2} \left(c^2\varrho_1 + \varrho_0\Psi_1 + \varrho_0\Psi_s \right) \right] \right.$$
$$\left. - \frac{2mD}{r}\left(\frac{\Omega_0}{D}\right)'\left(c^2\varrho_1 + \varrho_0\Psi_1 + \varrho_0\Psi_s\right) \right\} dz = 0. \tag{A2.19a}$$

Integrating over z we get

$$\hat{\omega}\left[D\sigma_1 + \frac{D}{r}\left(\frac{r\left(\overline{c^2}\sigma_1 + \sigma_0\Psi_1 + \sigma_0\Psi_s\right)'}{D} \right)' - \frac{m^2}{r^2}\left(\overline{c^2}\sigma_1 + \sigma_0\Psi_1 + \sigma_0\Psi_s\right) \right]$$
$$- \frac{2mD}{r}\left(\frac{\Omega_0}{D}\right)'\left(\overline{c^2}\sigma_1 + \sigma_0\Psi_1 + \sigma_0\Psi_s\right) \right\} = 0, \tag{A2.34}$$

where we have introduced the notation

$$\sigma_1 \equiv \int_{-h}^{+h} \varrho_1\, dz, \qquad \overline{c^2} \equiv \frac{1}{\sigma}\int_{-h}^{+h} c^2\varrho_1\, dz. \tag{A2.35}$$

We can considerably simplify (A2.34) in the main order of the WKB approximation: we can neglect the derivatives of the unperturbed quantities and use (A2.28) for a simple local relation between Ψ_1 and σ_1:

$$\Psi_1 = -\frac{2\pi G\sigma}{|k|}. \tag{A2.36}$$

If there is no satellite we find the following dispersion relation:

$$\hat{\omega}\left[\hat{\omega}^2 - \kappa^2 + 2\pi G\sigma_0|k| - k^2\overline{c^2}\right] = 0. \tag{A2.37}$$

We see that it contains two branches of eigenperturbations. One with $\hat{\omega} = 0$ describes indifferent equilibrium. However, instabilities can develop on this branch in the presence of dissipation; these can lead to an exponential growth of perturbations with a negative energy (see Fridman and Polyachenko 1984, Morozov et al. 1986, Fridman and Khoruzhii 1999, and also Chap. 8).

The other branch – the gravitational-acoustic branch – is also well known except that here the square of the sound velocity is not determined by the planar pressure P_s: $c^2 = \partial P_s/\partial\sigma$, but is a weighted average of the square of the normal (three-dimensional) sound velocity over the thickness of the disc (see (A2.35)). The weight function reflects the profile of the perturbed volume density of the disc, that is, $\overline{c^2}$ will, in general, differ for different perturbations. An exception is the case of an isothermal disc when $\overline{c^2} = c^2 = $ const for all perturbations.

5 The Role of Perturbations Along the Rotation Axis

We have shown in Appendix I that one can certainly not always consider the dynamical equations of astrophysical discs in a two-dimensional formulation. In the present appendix we can study the role of perturbations along the rotational axis in more detail using the linearised equations. The analysis carried out in what follows confirms the conditions found in Appendix I for the spatial and temporal scales of the perturbations which can be considered in a two-dimensional approximation. For small-amplitude perturbations these conditions take the form: $(kh)^2 \ll 1$ and $\hat{\omega}^2 \ll \Omega_0^2$. Moreover, the analysis given here enables us to estimate whether it is possible to consider viscous effects in the two-dimensional approximation.

In the WKB approximation $(k^2r^2 \gg m^2)$ the equation of continuity (A2.13) and (A2.3) for the motion along the z-axis take the form

$$\hat{\omega}\left[\varrho_1 + \frac{k^2\varrho_0}{D}\left(c^2\frac{\varrho_1}{\varrho_0} + \Phi_1\right)\right] + i\frac{\partial(\varrho_0 v_{z1})}{\partial z} = 0, \tag{A2.38}$$

$$i\hat{\omega}v_{z1} = \frac{\partial}{\partial z}\left(c^2\frac{\varrho_1}{\varrho_0} + \Phi_1\right). \tag{A2.39}$$

The presence of a satellite is not essential for the problem which we are considering here so that we can put $\Psi_s = 0$. Together with the Poisson equation these equation form a closed set of equations for the three unknown functions ϱ_1, Φ_1, and v_{z1}. It is just the necessity to take the last of these variables, that is, the motion along the rotation axis, into account which makes these equations truly three-dimensional.

One can see from (A2.38) and (A2.39) that the role of the motion along the rotation axis is twofold. In the equation of motion (A2.39) the term with v_{z1} describes the action of the inertial effects connected with motions in the z-direction. It is clear that these effects turn out to be unimportant when the frequency of the perturbations is low. This is just the reason for the condition, found in Appendix I, to have low frequencies: $\hat{\omega}^2 \ll \Omega_0^2$.

In the continuity equation (A2.38) the term with v_{z1} describes the transfer of mass along the z-axis, that is, its redistribution connected with the motion along the rotation axis. As the other terms in the equation of continuity are proportional to the eigenfrequency, neglecting the contributions from the motions in the z-direction is possible only for high-frequency processes.

In one way or another it thus turns out that the effect produced by the motions along the rotation axis is always important. There are, strictly speaking, no perturbations for which the role played by v_{z1} is negligibly small, except possibly in some specially selected cases.

Let us obtain the conditions on the eigenfrequency for which one of the factors of the effect of the motions in the z-direction turns out to be unimportant.

5.1 Conditions for Neglecting Mass Transfer Along the Rotation Axis

5.1.1 General Case

Neglecting mass transfer along the rotation axis and using the dispersion relation (A2.37) we get for the density perturbation from (A2.38)

$$\varrho_1 = \varrho_0 \frac{k^2 \Phi_1}{k^2 \left(\overline{c^2} - c^2\right) - 2\pi G \sigma_0 |k|}, \tag{A2.40}$$

and for the perturbed potential χ

$$\chi = c^2 \frac{\varrho_1}{\varrho_0} + \Phi_1 = \frac{D}{k^2} \frac{\varrho_1}{\varrho_0}.$$

Substituting into the latter expression the value of the perturbed density from (A2.40) we find

$$\chi = c^2 \frac{\varrho_1}{\varrho_0} + \Phi_1 = \frac{D \Phi_1}{k^2 \left(\overline{c^2} - c^2\right) - 2\pi G \sigma_0 |k|}. \tag{A2.41}$$

When we substitute this expression into the equation of motion (A2.39) we must bear in mind that c^2 is the function in (A2.41) which varies most strongly with z. As a result we get for the velocity in the z-direction the following expression:

$$v_{z1} = -\frac{i}{\hat{\omega}} \frac{k^2 D\Phi_1}{\left[k^2\left(\overline{c^2} - c^2\right) - 2\pi G\sigma_0|k|\right]^2} \frac{\partial(c^2)}{\partial z}$$

$$\sim \frac{k^2 D\Phi_1 c^2}{\hat{\omega}h\left[k^2\left(\overline{c^2} - c^2\right) - 2\pi G\sigma_0|k|\right]^2} \sim \frac{Dc^2}{\hat{\omega}2\pi G\sigma_0|k|h} \frac{\varrho_1}{\varrho_0}.$$

Since $D \sim 2\pi G\sigma_0|k|$, we find finally that the quantity v_{z1} turns out to be of the order of

$$v_{z1} \sim \frac{c^2 \varrho_1}{\hat{\omega}h\varrho_0}. \qquad (A2.42)$$

Let us now estimate the order of magnitude of the neglected term describing the mass transfer along the z-axis:

$$\frac{\partial}{\partial z}(\varrho_0 v_{z1}) \sim \frac{\partial}{\partial z}\left[\frac{c^2}{\hat{\omega}h}\varrho_1\right] \sim \frac{c^2}{\hat{\omega}h^2}\varrho_1. \qquad (A2.43)$$

The condition that one can neglect mass transfer along the z-axis follows from the equation of continuity (A2.38) and is

$$\frac{\partial}{\partial z}(\varrho_0 v_{z1}) \ll \hat{\omega}\varrho_1. \qquad (A2.44)$$

Hence we get, using (A2.43)

$$\frac{c^2}{h^2} \ll \hat{\omega}^2. \qquad (A2.45)$$

We can transform this lower bound for the frequency by using the condition that the disc is in equilibrium. For a disc in an external field we have (Fridman and Khoruzhii 1999)

$$h = \frac{c}{\Omega_0}, \qquad (A2.46)$$

and we then find from (A2.45)

$$\hat{\omega}^2 \gg \Omega_0^2. \qquad (A2.47)$$

In the case of a self-gravitating disc we have (Fridman 1996)

$$\frac{c}{h} \sim \Omega_0\sqrt{\frac{R}{h}}, \qquad (A2.48)$$

and it follows from (A2.45) that

$$\hat{\omega}^2 \gg \Omega_0^2 \frac{R}{h}, \tag{A2.49}$$

We do not know of a single case when conditions (A2.47) and (A2.49) have been satisfied for long-wavelength perturbations.

During the derivation it was clear that an exception is the case of an isothermal disc with $\overline{c^2} = c^2 = $ const. Let us consider that case in some detail.

5.1.2 Isothermal Disc

Expression (A2.41) which was derived for the case when we neglect mass transfer along the rotation axis for an isothermal disc, $\overline{c^2} = c^2$, has the form:

$$\chi = c^2 \frac{\varrho_1}{\varrho_0} + \Phi_1 = -\frac{D\Phi_1}{2\pi G \sigma_0 |k|}. \tag{A2.50}$$

From the equation of motion (A2.39) it then follows that

$$v_{z1} = \frac{iD}{2\pi G \sigma_0 |k| \hat{\omega}} \frac{\partial \Phi_1}{\partial z}. \tag{A2.51}$$

Using (A2.31a) and (A2.28) we have, as to order of magnitude,

$$v_{z1} \sim \frac{D}{\hat{\omega}k} \frac{\sigma_1}{\sigma_0} \sim \frac{D}{\hat{\omega}k} \frac{\varrho_1}{\varrho_0}, \tag{A2.52}$$

where we are assuming in the remainder of this appendix that $k > 0$.

Instead of (A2.43) we get now for an isothermal disc

$$\frac{\partial}{\partial z}(\varrho_0 v_{z1}) \sim \frac{\partial}{\partial z}\left[\frac{D\varrho_1}{\hat{\omega}k}\right] \sim \frac{D\varrho_1}{\hat{\omega}hk}. \tag{A2.53}$$

In obtaining the last estimate we used the fact that ϱ_1 is the function which depends most strongly on z – it changes by its own magnitude over a half-thickness.

Substituting (A2.53) into the condition (A2.44) that we may neglect mass transfer in the z-direction we get

$$D \ll \hat{\omega}^2 kh. \tag{A2.54}$$

Using the definition (A2.12) of D and the dispersion relation (A2.37) we can reformulate condition (A2.54) as follows:

$$-2\pi G \sigma_0 |k| + k^2 c^2 \ll \hat{\omega}^2 kh. \tag{A2.55}$$

In the case when the contributions from the acoustic effects and of the self-gravitation are comparable, $2\pi G \sigma_0 |k| \sim k^2 c^2$, we get

$$k^2 c^2 \ll \hat{\omega}^2 kh. \tag{A2.56}$$

Let us consider two cases:

1. *Disc in an External Field.* Using the equilibrium condition (A2.46) we get from (A2.56) the condition for neglecting the mass flux in a disc in an external field:

$$\Omega^2 kh \ll \hat{\omega}^2. \tag{A2.57}$$

We see that this condition can be satisfied in the case of long-wavelength perturbations. A systematic comparison of two- and three-dimensional numerical models has shown that the solutions obtained are qualitatively similar only in near isothermal discs (Bisikalo et al. 1999).

2. *Self-gravitating Disc.* Using the equilibrium condition (A2.48) we get from (A2.56) the condition for neglecting the mass flux in a self-gravitating disc:

$$\Omega^2 kR \ll \hat{\omega}^2. \tag{A2.58}$$

As $kR \gg 1$ it is impossible to satisfy this condition in the case of long-wavelength perturbations with $kh \ll 1$.

Comparing the condition (A2.49) for neglecting the mass flow in the general case of a self-gravitating disc with the analogous condition for the particular case of an isothermal self-gravitating disc we see that in the latter case the condition for the neglect turns out to be much softer for long-wavelength perturbations. However, we noticed nonetheless earlier that it can still not be satisfied.

5.2 Condition for Neglecting the Inertial Term in the Equation of Motion in the z-Direction – Condition for Neglecting Oscillations Along the Rotation Axis

We shall now derive more rigorously than was done in Appendix I the condition for neglecting the inertial term in (A2.39) for motion in the z-direction. It is clear that this is a low-frequency approximation. Neglecting the left-hand side of (A2.39) in comparison with the right-hand side we have

$$c^2 \frac{\varrho_1}{\varrho_0} + \Phi_1 = \chi(r, \varphi, t), \tag{A2.59}$$

where the function χ is independent of z. We now rewrite Eq. (A2.38) in the form

$$\frac{\partial}{\partial z}(\varrho_0 v_{z1}) = i\hat{\omega} \left[\varrho_1 - \frac{k^2}{D} \varrho_0 \chi \right]. \tag{A2.60}$$

It is clear from (A2.59) that in the general case the z-dependence of ϱ_1 is not the same as that of ϱ_0. Indeed, it follows from (A2.28) and (A2.29) that the z-dependence of function Φ_1 is in the general case not the same as that of c^2: the latter is determined by the equation of state which, generally speaking, cannot have an arbitrary form. Both $\Phi_1(z) - \Phi_0(z)$ and c^2 change over the thickness of the disc by an amount of the order of their own value. Hence, the ratio ϱ_1/ϱ_0 can, generally speaking, also change over the thickness of the disc by an amount equal to this ratio.

It follows from the fact that the z-dependence of ϱ_1 and ϱ_0 is different that the difference in the square brackets on the right-hand side of (A2.60) cannot be not much less than the two terms themselves for all z. Hence, we can estimate the magnitude of the left-hand side of (A2.60) by using, for instance, the first term on the right-hand side:

$$v_{z1} \sim \hat{\omega} h \frac{\varrho_1}{\varrho_0}. \tag{A2.61}$$

Knowing v_{z1} we can now find the condition for neglecting the inertial force in (A2.39) as compared to the pressure force:

$$\hat{\omega}^2 \ll \frac{c^2}{h^2}. \tag{A2.62}$$

Condition (A2.62) corresponds to the restriction (A1.20) of Appendix I on the frequency – characteristic time – of processes which can be described in the framework of the two-dimensional equations.

The only exception is the case of a self-gravitating disc with a polytropic index $\gamma = 2$ in a cylindrical external field (see Appendix I for the role played by the latter).

For this case the density profile of the disc in the z-direction has the form (see Appendix I)

$$\varrho_0(z) = \varrho_c \cos\left(\sqrt{\frac{2\pi G}{A_V}}\, z\right).$$

Let the profile of the perturbed density have the same shape:

$$\varrho_1(z) = \varrho_{1c} \cos\left(\sqrt{\frac{2\pi G}{A_V}}\, z\right).$$

For such a ϱ_1 profile we have

$$\sigma_z = \int_{-z}^{z} \varrho_1(x)\,dx = 2\varrho_{1c}\sqrt{\frac{A_V}{2\pi G}}\,\sin\left(\sqrt{\frac{2\pi G}{A_V}}\, z\right).$$

From (A2.28) and (A2.29) we then get for the perturbed potential in the disc

$$\Phi_1(r, \varphi, z) = \text{const} - c^2 \frac{\varrho_1}{\varrho_0}.$$

For the case considered the conditions $\varrho_1/\varrho_0 = $ const and $\chi = $ const thus turn out to be compatible and the mass transfer along the rotation axis is negligibly small.

6 Conclusion

We have thus seen that one can completely neglect the contribution of the vertical motions only in two separate cases: an isothermal disc with the external field dominant and a self-gravitating disc with polytropic index $\gamma = 2$. In all other cases a perturbation in the disc leads to a perturbation of the motions along the z-axis. This may be important if one takes into account viscous effects for slow perturbations ($\hat{\omega}^2 < c^2/h^2$). It was shown in Appendix I that in that case the dynamical equations can be reduced to the two-dimensional ones, if one neglects viscosity. However, since even in that case the necessary redistribution of mass by motions in the z-direction leads to the appearance of velocities along the rotation axis, $v_{z1} \sim v_{r1}kh$, and it is impossible to neglect the corresponding contribution to the viscous stress tensor which is proportional to $\partial v_{z1}/\partial z \sim \partial v_{r1}/\partial r$. Thus, although the set of equations when viscosity is neglected turns out to be effectively two-dimensional, taking the viscosity into account turns out to be impossible unless one takes the motion along the rotation axis into account.

Appendix III.
Derivation of the Linearised Equations for Oscillations of a Viscous Disc

1 Derivation of the Linearised Equations for Oscillations of a Viscous Uniformly Rotating Disc

Let us consider the two-dimensional equation of continuity

$$\frac{\partial \sigma}{\partial t} + \frac{1}{r}\frac{\partial}{\partial r}(r\sigma v_r) = 0, \tag{A3.1}$$

or

$$\frac{\partial \sigma}{\partial t} + \frac{\sigma v_r}{r} + v_r \frac{\partial \sigma}{\partial r} + \sigma \frac{\partial v_r}{\partial r} - 0. \tag{A3.2}$$

We now split every quantity into a stationary and a perturbed part: $\sigma = \sigma_0 + \sigma'$, $v_r = v_{r0} + v_r'$. Putting for the moment $v_{r0} = 0$ – we shall in the next section take into account the radial flow of matter – we find

$$\frac{\partial \sigma_0}{\partial t} + \frac{\partial \sigma'}{\partial t} + \frac{\sigma_0 v_r'}{r} + \frac{\sigma' v_r'}{r} + v_r'\frac{\partial \sigma_0}{\partial r} + v_r'\frac{\partial \sigma'}{\partial r} + \sigma_0\frac{\partial v_r'}{\partial r} + \sigma'\frac{\partial v_r'}{\partial r} = 0. \tag{A3.3}$$

If we now write every quantity in the form $A' = \hat{A}e^{\gamma t + ikr}$ and use the short-wavelength approximation,

$$kr \gg 1, \tag{A3.4}$$

where $k \equiv 2\pi/\lambda$, we obtain as a result the linearised equation of continuity:

$$\gamma\hat{\sigma} + ik\sigma_0\hat{v}_r = 0. \tag{A3.5}$$

Using the relations between the pressure, the temperature, and the sound velocity,

$$p = \sigma T, \qquad T_0 = c_0^2, \tag{A3.6}$$

we can write down the equation for the radial component of the hydrodynamic velocity

$$\sigma\left[\frac{\partial v_r}{\partial t} + v_r\frac{\partial v_r}{\partial r} - \frac{v_\varphi^2}{r}\right] = -\frac{\partial \sigma}{\partial r}T - \sigma\frac{\partial T}{\partial r} - \sigma\frac{\partial \Psi}{\partial r}$$
$$+ \tfrac{4}{3}\nu\sigma\left[\frac{\partial}{\partial r}\frac{v_r}{r} + \frac{\partial^2 v_r}{\partial r^2}\right] + \tfrac{2}{3}\frac{\partial(\nu\sigma)}{\partial r}\left[2\frac{\partial v_r}{\partial r} - \frac{v_r}{r}\right]. \tag{A3.7}$$

Let us consider the non-dissipative – non-viscous – part of (A3.12), using the fact that $v_{r0} = 0$:

$$(\sigma_0 + \sigma') \left[\frac{\partial v_r'}{\partial t} + v_r' \frac{\partial v_r'}{\partial r} - \frac{(v_{\varphi 0} + v_\varphi')^2}{r} \right]$$

$$= - \frac{\partial \sigma_0}{\partial r} (T_0 + T') - \frac{\partial \sigma'}{\partial r} (T_0 + T') - (\sigma_0 + \sigma') \frac{\partial T_0}{\partial r}$$

$$- (\sigma_0 + \sigma') \frac{\partial T'}{\partial r} - (\sigma_0 + \sigma') \frac{\partial \Psi_0}{\partial r} - (\sigma_0 + \sigma') \frac{\partial \Psi'}{\partial r}. \qquad (A3.8)$$

The linearised form of (A3.8) has for short-wavelength perturbations the form

$$\sigma_0 \left[\frac{\partial v_r'}{\partial t} - \frac{2 v_{\varphi 0} v_\varphi'}{r} \right] = - \frac{\partial \sigma'}{\partial r} T_0 - \sigma_0 \frac{\partial T'}{\partial r} - \sigma_0 \frac{\partial \Psi'}{\partial r}, \qquad (A3.9)$$

or, writing again $A' = \hat{A} e^{\gamma t + ikr}$ and substituting into (A3.9) the relation

$$\frac{\partial \hat{\Psi}}{\partial r} = -2\pi i G \hat{\sigma}, \qquad (A3.10)$$

which follows from the Poisson equation, we get

$$\gamma \hat{v}_r - 2\Omega_0 \hat{v}_\varphi = - \frac{ik c_0^2 \hat{\sigma}}{\sigma_0} - ik\hat{T} + 2\pi i G \hat{\sigma}. \qquad (A3.11)$$

Let us consider the viscous terms of (A3.7):

$$\tfrac{4}{3} \nu \sigma \left[\frac{\partial}{\partial r} \frac{v_r}{r} + \frac{\partial^2 v_r}{\partial r^2} \right] + \frac{2}{3} \frac{\partial (\nu \sigma)}{\partial r} \left[2 \frac{\partial v_r}{\partial r} - \frac{v_r}{r} \right]. \qquad (A3.12)$$

Taking into account that in the $kr \gg 1$ approximation the higher derivative always dominates for perturbed quantities, and dropping non-linear terms, we see that there remains only one term from all the viscous terms in the radial equation of motion, namely,

$$\tfrac{4}{3} (\nu \sigma)_0 \frac{\partial^2 v_r'}{\partial r^2}, \qquad \text{or} \qquad - \tfrac{4}{3} \nu_0 \sigma_0 k^2 \hat{v}_r. \qquad (A3.13)$$

Equation (A3.11) together with the term (A3.13), divided by σ_0, leads to (8.7).

We can similarly write down the equation for the azimuthal velocity component v_φ:

$$\sigma_0 \left[\frac{\partial v_\varphi'}{\partial t} + v_r' \left\{ \frac{\partial v_{\varphi 0}}{\partial r} + \frac{v_{\varphi 0}}{r} \right\} \right] = \nu_0 \sigma_0 \frac{\partial^2 v_\varphi'}{\partial r^2}, \qquad (A3.14)$$

where in writing down the last term we have used the fact that in the case of a uniformly rotating disc we have

$$\frac{\partial}{\partial r}\frac{v_{\varphi 0}}{r} = \frac{\partial \Omega_0}{\partial r} = 0.$$

Linearising the energy equation,

$$\frac{3}{2}(\sigma_0 + \sigma')\left[\frac{\partial T_0}{\partial t} + \frac{\partial T'}{\partial t} + v_r'\frac{\partial T_0}{\partial r} + v_r'\frac{\partial T'}{\partial r}\right] + (p_0 + p')\left[\frac{\partial v_r'}{\partial r} + \frac{v_r'}{r}\right]$$

$$= \frac{\partial(\chi_0 + \chi')}{\partial r}\frac{\partial(T_0 + T')}{\partial r} + \frac{(\chi_0 + \chi')}{r}\frac{\partial(T_0 + T')}{\partial r}$$

$$+ (\chi_0 + \chi')\frac{\partial^2(T_0 + T')}{\partial r^2} + [(\nu\sigma)_0 + (\nu\sigma)']\left[r\frac{\partial}{\partial r}\left\{\frac{v_{\varphi 0}}{r} + \frac{v_\varphi'}{r}\right\}\right]^2$$

$$+ \frac{4}{3}[(\nu\sigma)_0 + (\nu\sigma)']\left[\left(\frac{\partial v_r'}{\partial r}\right)^2 + \left(\frac{v_r'}{r}\right)^2 - \frac{\partial v_r'}{\partial r}\frac{v_r'}{r}\right], \tag{A3.15}$$

and using the short-wavelength approximation, we find (8.9) from (A3.24):

$$\frac{3}{2}\gamma\hat{T} + \frac{ikp_0\hat{v}_r}{\sigma_0} = -\chi_0 k^2\hat{T}. \tag{A3.16}$$

2 Derivation of the Linearised Equations for Oscillations of a Viscous Differentially Rotating Disc of Inelastic Particles with Account of External Matter Fluxes

Let us consider the set of equations (8.22) to (8.25). We write the equation of continuity (8.22) in the form $(A = A_0 + A')$

$$\frac{\partial \sigma_0}{\partial t} + \frac{\partial \sigma'}{\partial t} + \frac{(\sigma_0 + \sigma')v_r'}{r} + v_r'\frac{\partial(\sigma_0 + \sigma')}{\partial r} + (\sigma_0 + \sigma')\frac{\partial v_r'}{\partial r}$$

$$+ \frac{(v_{\varphi 0} + v_\varphi')}{r}\frac{\partial(\sigma_0 + \sigma')}{\partial \varphi} + \frac{(\sigma_0 + \sigma')}{r}\frac{\partial(v_{\varphi 0} + v_\varphi')}{\partial \varphi}$$

$$= N_0^+ + N^{+'} - N_0^- - N^{-'}. \tag{A3.17}$$

We can split off from (A3.17) the equation for the stationary quantities:

$$\frac{\partial \sigma_0}{\partial t} = N_0^+(\sigma, T) - N_0^-(\sigma, T). \tag{A3.18}$$

The inflow N^+ of mass into the disc and its outflow N^- must thus cancel one another in order that the condition

$$\frac{\partial \sigma_0}{\partial t} \ll \frac{\partial \sigma'}{\partial t}, \tag{A3.19}$$

be satisfied – which expresses the fact that the time for the growth of the total surface density of the disc must be much longer than the characteristic times of the oscillations or instabilities which we are considering since otherwise the linearisation of the equations which we are carrying out would be incorrect. Using (A3.18) and (A3.19), writing the changes in $N^{+'}$ and $N^{-'}$ in the form

$$
N^{+(-)'}(\sigma, T) = \frac{\partial N_0^{+(-)}}{\partial \sigma_0} \sigma' + \frac{\partial N_0^{+(-)}}{\partial T_0} T', \tag{A3.20}
$$

and dropping non-linear terms, we get

$$
\frac{\partial \sigma'}{\partial t} + \sigma_0 \frac{\partial v_r'}{\partial r} + \Omega_0 \frac{\partial \sigma'}{\partial \varphi} + \frac{\sigma_0}{r} \frac{\partial v_\varphi'}{\partial \varphi}
$$
$$
= \left[\frac{\partial N_0^+}{\partial \sigma_0} - \frac{\partial N_0^-}{\partial \sigma_0} \right] \sigma' + \left[\frac{\partial N_0^+}{\partial T_0} - \frac{\partial N_0^-}{\partial T_0} \right] T'. \tag{A3.21}
$$

As a result of substituting $A' = \hat{A} e^{\gamma t + ikr + im\varphi}$ in the last equation we get the first equation of the set (8.32):

$$
(\gamma + im\Omega_0)\hat{\sigma} + ik\sigma_0 \hat{v}_r = \left[\frac{\partial N_0^+}{\partial \sigma_0} - \frac{\partial N_0^-}{\partial \sigma_0} \right] \hat{\sigma} + \left[\frac{\partial N_0^+}{\partial T_0} - \frac{\partial N_0^-}{\partial T_0} \right] \hat{T}. \tag{A3.22}
$$

Let us now consider (8.23) for v_r. We write its left-hand side in the form

$$
\frac{\partial v_{r0}}{\partial t} + \frac{\partial v_r'}{\partial t} + v_r' \frac{\partial v_r'}{\partial r} + \frac{(v_{\varphi 0} + v_\varphi')}{r} \frac{\partial v_r'}{\partial \varphi} + \frac{(v_{\varphi 0} - v_\varphi')^2}{r}. \tag{A3.23}
$$

We shall study the stationary terms $\partial v_{r0}/\partial t$ and $v_{\varphi 0}^2/r$ later on; dropping the non-linear terms we get from (A3.23)

$$
\frac{\partial v_r'}{\partial t} + \Omega_0 \frac{\partial v_r'}{\partial \varphi} - 2\Omega_0 v_\varphi'. \tag{A3.24}
$$

We have already studied the first four terms on the right-hand side of (8.23) – for the uniformly rotating case (see also (A3.8) and (A3.12)). This analysis remains the same for the case of a differentially rotating disc. We compare the new dissipative terms on the right-hand side of (8.23) with the terms with the main derivative which we have already studied:

$$
\frac{4}{3}\nu_0 \frac{\partial^2 v_r'}{\partial r^2} - \frac{2}{3\sigma_0 r} \frac{\partial (\nu \sigma)_0}{\partial r} \frac{\partial v_\varphi'}{\partial \varphi} - \frac{2}{3\sigma_0 r} \frac{\partial (\nu \sigma)'}{\partial r} \frac{\partial v_\varphi'}{\partial \varphi} - \frac{2(\nu_0 + \nu')}{3r} \frac{\partial^2 v_\varphi'}{\partial r \partial \varphi}
$$
$$
- \frac{4(\nu_0 + \nu')}{3r^2} \frac{\partial v_\varphi'}{\partial \varphi} + \frac{1}{r\sigma_0} \frac{\partial (\nu \sigma)'}{\partial \varphi} \left[\frac{1}{r} \frac{\partial v_r'}{\partial \varphi} + r \frac{\partial}{\partial r} \frac{(v_{\varphi 0} + v_\varphi')}{r} \right]
$$
$$
+ \frac{(\nu_0 + \nu')}{r} \left[\frac{1}{r} \frac{\partial^2 v_r'}{\partial \varphi^2} + r \frac{\partial^2}{\partial \varphi \partial r} \frac{(v_{\varphi 0} + v_\varphi')}{r} \right]. \tag{A3.25}
$$

We have used here the fact that all stationary quantities, $v_{\varphi 0}$, ν_0, σ_0, and v_{r0}, are independent of φ and, moreover, that $\sigma_0 \gg \sigma'$ and $v_{r0} = 0$. Using the relation between derivatives of stationary and of perturbed quantities and the fact that $\nu' \ll \nu_0$, and dropping non-linear terms we find

$$\frac{4}{3}\nu_0\frac{\partial^2 v_r'}{\partial r^2} - \frac{2\nu_0}{3r}\frac{\partial^2 v_\varphi'}{\partial r\partial\varphi} - \frac{4\nu_0}{3r^2}\frac{\partial v_\varphi'}{\partial\varphi}$$

$$+ \frac{1}{r\sigma_0}\frac{\partial(\nu\sigma)'}{\partial\varphi}r\frac{\partial}{\partial r}\frac{v_{\varphi 0}}{r} + \frac{\nu_0}{r^2}\frac{\partial^2 v_r'}{\partial\varphi^2} + \frac{\nu_0}{r}\frac{\partial^2 v_\varphi'}{\partial\varphi\partial r} - \frac{\nu_0}{r^2}\frac{\partial v_\varphi'}{\partial\varphi}. \tag{A3.26}$$

Let us consider all seven terms in expression (A3.26). The third term is small as compared to the second one, since $kr \gg 1$, as is the seventh as compared to the sixth or the second. We postpone the discussion of the fourth term. Let us rewrite the higher-order terms, again using the fact that we look at perturbations proportional to $e^{\gamma t + ikr + im\varphi}$:

$$-\tfrac{4}{3}\nu_0 k^2 \hat{v}_r + \frac{2\nu_0 km}{3r}\hat{v}_\varphi - \frac{1}{r\sigma_0}\frac{\partial(\nu\sigma)'}{\partial\varphi}(-r\Omega') - \frac{\nu_0 m^2}{r^2}\hat{v}_r - \frac{\nu_0 km}{r}\hat{v}_\varphi. \tag{A3.27}$$

We see that the first term in (A3.26) dominates since $kr \gg m$ and $\hat{v}_\varphi \sim \hat{v}_r$. Let us now consider the fourth term in (A3.26):

$$\frac{1}{r\sigma_0}\frac{\partial(\nu\sigma)'}{\partial\varphi}(r\Omega') = \frac{1}{r\sigma_0}\left[\frac{\partial(\nu\sigma)_0}{\partial\sigma_0}\frac{\partial\sigma'}{\partial\varphi} + \frac{\partial(\nu\sigma)_0}{\partial T_0}\frac{\partial T'}{\partial\varphi}\right](r\Omega')$$

$$\rightarrow \frac{\mathrm{i}}{\sigma_0}\frac{\partial(\nu\sigma)_0}{\partial\sigma_0}\frac{m}{r}(r\Omega')\hat{\sigma} + \mathrm{i}\frac{\partial\nu_0}{\partial T_0}\frac{m}{r}(r\Omega')\hat{T}. \tag{A3.28}$$

If we compare this with the terms arising when we linearise the pressure term $(\partial p/\partial r)/\sigma$ (see (A3.11)),

$$-\frac{\mathrm{i}kc^2}{\sigma_0}\hat{\sigma} - \mathrm{i}k\hat{T}, \tag{A3.29}$$

we see that the pressure terms (A3.29) dominate the viscosity terms (A3.28) provided

$$kc^2 \gg \frac{\partial(\nu\sigma)_0}{\partial\sigma_0}\frac{m}{r}(r\Omega'), \tag{A3.30}$$

$$k \gg \frac{\partial\nu_0}{\partial T_0}\frac{m}{r}(r\Omega'). \tag{A3.31}$$

Let us consider the validity of inequalities (A3.30) and (A3.31), bearing in mind that

for a molecular gas : $\quad \nu \sim c^2 t_c, \qquad \Omega t_c \ll 1,$ \hfill (A3.32)

for a gas of macroparticles : $\quad \nu \sim \dfrac{c^2}{\Omega^2 t_c}, \qquad \Omega t_c \gg 1,$ \hfill (A3.33)

Using the fact that

$$\frac{\partial(\nu\sigma)_0}{\partial\sigma_0} \sim \nu_0, \qquad \frac{\partial\nu_0}{\partial T_0} \sim \frac{\nu_0}{c^2}, \tag{A3.34}$$

we find for the case of a molecular gas from (A3.30) to (A3.32) the general inequality

$$k \gg \frac{\Omega t_c m}{r},$$ (A3.35)

which is satisfied as $kr \gg m$ and $\Omega t_c \ll 1$.

For the case of a rarefied gas we find from (A3.30) and (A3.31) the general inequality

$$k \gg \frac{m}{r \Omega t_c},$$ (A3.36)

which is satisfied as $kr \gg m$ and $\Omega t_c \gg 1$.

The "viscous" terms in (A3.28) are thus negligibly small as compared to the "pressure" terms in (A3.29). As a result this linearised equation for the radial velocity component finally takes the form

$$\frac{\partial v'_r}{\partial t} + \Omega_0 \frac{\partial v'_r}{\partial \varphi} - 2\Omega_0 v'_\varphi = -\frac{\partial \sigma'}{\partial r} \frac{T_0}{\sigma_0} - \frac{\partial T'}{\partial r} - \frac{\partial \Psi'}{\partial r} + \tfrac{4}{3}\nu_0 \frac{\partial^2 v'_r}{\partial r^2}.$$ (A3.37)

The balance of the stationary terms split off from (A3.23) and equated to the gravitational and pressure terms is the same as in the case of the rigidly rotating disc. Taking the perturbed terms to have a factor $e^{\gamma t + ikr + im\varphi}$ we get from (A3.37) the second equation of the set (8.32):

$$(\gamma + im\Omega_0)\hat{v}_r - 2\Omega_0 \hat{v}_\varphi = i\frac{2\pi G\sigma_0 - kc^2}{\sigma_0}\hat{\sigma} - ik\hat{T} - \tfrac{4}{3}\nu_0 k^2 \hat{v}_r.$$ (A3.38)

Let us now consider the linearisation of (8.24) for the azimuthal velocity component – assuming that the stationary quantities are independent of φ:

$$\frac{\partial v_{\varphi 0}}{\partial t} + \frac{\partial v'_\varphi}{\partial t} + (v_{r0} + v'_r)\frac{dv_{\varphi 0}}{dr} + (v_{r0} + v'_r)\frac{\partial v'_\varphi}{\partial t} + \frac{(v_{r0} + v'_r)v_{\varphi 0}}{r}$$
$$+ \frac{(v_{r0} + v'_r)v'_\varphi}{r} + \frac{(v_{\varphi 0} + v'_\varphi)}{r}\frac{\partial v'_\varphi}{\partial \varphi}$$
$$= -\frac{(T_0 + T')}{\sigma_0 r}\frac{\partial \sigma'}{\partial \varphi} - \frac{(\sigma_0 + \sigma')}{\sigma_0 r}\frac{\partial T'}{\partial \varphi} - \frac{1}{r}\frac{\partial \Psi_G}{\partial \varphi}$$
$$+ \left\{ \frac{3[(\nu\sigma)_0 + (\nu\sigma)']}{\sigma_0} + \frac{r}{\sigma_0}\left[\frac{\partial}{\partial r}[(\nu\sigma)_0 + (\nu\sigma)']\right] \right\} \frac{\partial}{\partial r}\frac{(v_{\varphi 0} + v'_\varphi)}{r}$$
$$+ \frac{r}{\sigma_0}[(\nu\sigma)_0 + (\nu\sigma)']\frac{\partial^2}{\partial r^2}\frac{(v_{\varphi 0} + v'_\varphi)}{r} + \frac{1}{\sigma_0 r}\left[\frac{\partial}{\partial r}[(\nu\sigma)_0 + (\nu\sigma)']\right]\frac{\partial v'_r}{\partial \varphi}$$
$$+ \frac{1}{r\sigma_0}[(\nu\sigma)_0 + (\nu\sigma)']\frac{\partial^2 v'_r}{\partial r\partial \varphi} + \frac{(\nu_0 + \nu')}{r^2}\frac{\partial v'_r}{\partial \varphi}$$
$$+ \tfrac{4}{3}\frac{1}{\sigma_0 r}\frac{\partial(\nu\sigma)'}{\partial \varphi}\left[\frac{1}{r}\frac{\partial v'_\varphi}{\partial \varphi} + \frac{v'_r}{r} - \frac{1}{2}\frac{\partial v'_r}{\partial r}\right]$$
$$+ \tfrac{4}{3}\frac{1}{\sigma_0 r}[(\nu\sigma)_0 + (\nu\sigma)']\left[\frac{1}{r}\frac{\partial^2 v'_\varphi}{\partial \varphi^2} + \frac{1}{r}\frac{\partial v'_r}{\partial \varphi} - \frac{1}{2}\frac{\partial^2 v'_r}{\partial r\partial \varphi}\right].$$ (A3.39)

Dropping the non-linear terms and taking into account that $v'_\varphi, v_{r0} \ll v_{\varphi0}$, $\nu' \ll \nu_0$, $\sigma' \ll \sigma_0$, we get

$$
\frac{\partial v_{\varphi0}}{\partial t} + \frac{\partial v'_\varphi}{\partial t} + (v_{r0} + v'_r)\left[\frac{dv_{\varphi0}}{dr} + \frac{v_{\varphi0}}{r}\right] + \frac{v_{\varphi0}}{r}\frac{\partial v'_\varphi}{\partial \varphi} = -\frac{T_0}{\sigma_0 r}\frac{\partial \sigma'}{\partial \varphi}
$$

$$
-\frac{1}{r}\frac{\partial T'}{\partial \varphi} - \frac{1}{r}\frac{\partial \Psi'}{\partial \varphi} + 3\nu_0 \frac{d}{dr}\frac{v_{\varphi0}}{r} + \frac{3}{\sigma_0}(\nu\sigma)'\frac{d}{dr}\frac{v_{\varphi0}}{r} + 3\nu_0 \frac{\partial}{\partial r}\frac{v'_\varphi}{r}
$$

$$
+\frac{r}{\sigma_0}\frac{\partial(\nu\sigma)_0}{\partial r}\frac{d}{dr}\frac{v_{\varphi0}}{r} + \frac{r}{\sigma_0}\frac{\partial(\nu\sigma)_0}{\partial r}\frac{\partial}{\partial r}\frac{v'_\varphi}{r} + \frac{r}{\sigma_0}\frac{\partial(\nu\sigma)'}{\partial r}\frac{d}{dr}\frac{v_{\varphi0}}{r}
$$

$$
+rv_0\frac{d^2}{dr^2}\frac{v_{\varphi0}}{r} + rv_0\frac{\partial^2}{\partial r^2}\frac{v'_\varphi}{r} + \frac{r}{\sigma_0}(\nu\sigma)'\frac{d^2}{dr^2}\frac{v_{\varphi0}}{r} + \frac{1}{\sigma_0 r}\frac{d(\nu\sigma)_0}{dr}\frac{\partial v'_r}{\partial \varphi}
$$

$$
+\frac{\nu_0}{r}\frac{\partial^2 v'_r}{\partial r \partial \varphi} + \frac{\nu_0}{r^2}\frac{\partial v'_r}{\partial \varphi} + \frac{4\nu_0}{3r}\left[\frac{1}{r}\frac{\partial^2 v'_\varphi}{\partial \varphi^2} + \frac{1}{r}\frac{\partial v'_r}{\partial r} - \frac{1}{2}\frac{\partial^2 v'_r}{\partial r \partial \varphi}\right]. \quad (A3.40)
$$

The unperturbed quantities are connected with one another through the following relations:

$$
v_{r0} = \frac{2\Omega_0}{\kappa^2}\left[3\nu_0 \frac{d\Omega_0}{dr} + \frac{1}{\sigma_0}\frac{d(\nu\sigma)_0}{dr}\left[r\frac{d\Omega_0}{dr}\right] + \nu_0 r\frac{d^2\Omega_0}{dr^2}\right], \quad (A3.41)
$$

or

$$
v_{r0} = \frac{2\Omega_0}{\kappa^2}\frac{1}{\sigma_0 r^2}\frac{d}{dr}r^3\nu_0\sigma_0\frac{d\Omega_0}{dr}, \quad (A3.42)
$$

where

$$
\kappa^2 \equiv 4\Omega_0^2\left(1 + \frac{r}{2\Omega_0}\frac{d\Omega_0}{dr}\right).
$$

This equation describes the slow spreading out of the disc due to viscous transfer of angular momentum from the interior regions of a differentially rotating disc to the outer ones (Lynden-Bell and Pringle 1974). The terms on the left-hand side of (A3.40) become the left-hand side of the third equation of the set (8.32).

Let us now consider the right-hand side of (A3.40). We group together the terms for the various perturbed quantities:

$$
\frac{3}{\sigma_0}(\nu\sigma)'\frac{d\Omega_0}{dr} + \frac{r}{\sigma_0}\frac{\partial(\nu\sigma)'}{\partial r}\frac{d\Omega_0}{dr} + \frac{r}{\sigma_0}(\nu\sigma)'\frac{d^2\Omega_0}{dr^2}, \quad (A3.43)
$$

$$
\frac{3\nu_0}{r}\frac{\partial v'_\varphi}{\partial r} - 3\nu_0\frac{v'_\varphi}{r^2} + \frac{r}{\sigma_0}\frac{\partial(\nu\sigma)_0}{\partial r}\frac{\partial}{\partial r}\frac{v'_\varphi}{r} + \nu_0\frac{\partial^2 v'_\varphi}{\partial r^2} + \frac{4\nu_0}{3r^2}\frac{\partial^2 v'_\varphi}{\partial \varphi^2}, \quad (A3.44)
$$

$$
\frac{1}{\sigma_0 r}\frac{d(\nu\sigma)_0}{dr}\frac{\partial v'_r}{\partial \varphi} + \frac{7\nu_0}{3r^2}\frac{\partial v'_r}{\partial \varphi} + \frac{\nu_0}{3r}\frac{\partial^2 v'_r}{\partial r \partial \varphi}. \quad (A3.45)
$$

In the $kr \gg m$ approximation only the term

$$\frac{r}{\sigma_0} \frac{\partial(\nu\sigma)'}{\partial r} \frac{d\Omega_0}{dr} \tag{A3.46}$$

remains in (A3.43). In (A3.44) the term

$$\nu_0 \frac{\partial^2 v'_\varphi}{\partial r^2} \tag{A3.47}$$

dominates and in (A3.45) the term

$$\frac{\nu_0}{3r} \frac{\partial^2 v'_r}{\partial r \partial \varphi}. \tag{A3.48}$$

If we bear in mind that $v'_r \sim v'_\varphi$ we see that in the $kr \gg 1$ approximation the term (A3.48) is much smaller than (A3.47). Let us compare (A3.46) with the first three pressure and gravitation terms on the right-hand side of (A3.40). Using the fact that $kr \gg 1$ and the expansion

$$(\nu\sigma)' = \frac{\partial \nu_0 \sigma_0}{\partial \sigma_0} \sigma' + \sigma_0 \frac{\partial \nu_0}{\partial T_0} T',$$

we find from (A3.46) for the perturbation of the temperature

$$\frac{r}{\sigma_0} \frac{\partial(\nu\sigma)'}{\partial r} \frac{d\Omega_0}{dr} \rightarrow ik \frac{\partial \nu_0}{\partial T_0} \frac{3}{2} \hat{T} \Omega_0. \tag{A3.49}$$

This term is much larger than the term

$$-\frac{1}{r} \frac{\partial T'}{\partial \varphi} \rightarrow -\frac{mi}{r} \hat{T} \tag{A3.50}$$

in the case in which we are interested:

$$\Omega_0 k \frac{\partial \nu_0}{\partial T_0} \gg \frac{m}{r}. \tag{A3.51}$$

We write

$$\frac{\partial \nu_0}{\partial T_0} \sim \frac{\nu_0}{c^2} \sim \begin{cases} t_c & \text{for a molecular gas,} \quad \Omega t_c \ll 1, \\ \dfrac{1}{\Omega^2 t_c} & \text{for a gas of macroparticles,} \quad \Omega t_c \gg 1. \end{cases} \tag{A3.52}$$

Inequality (A3.51) takes for a molecular gas the form

$$k\Omega t_c \gg \frac{m}{r}, \tag{A3.53}$$

where

$$\Omega t_c \ll 1,$$

and this needs to be checked by an estimate. We shall study small m and such stratification scales that $kr \sim 10^3$ to 10^6. Hence, the inequality is satisfied if

10^{-2} to $10^{-5} \lesssim \Omega t_c \ll 1$. However, the collision frequency is in a molecular gas, as a rule, several orders of magnitude larger than the orbital rotational frequency. This means that for a molecular gas the opposite inequality holds:

$$k\Omega t_c \ll \frac{m}{r}, \tag{A3.54}$$

and the term (A3.50) is larger than (A3.49).

Let us now consider the case which is of the most interest for us – a rarefied gas of macroparticles – when (A3.51) takes the form

$$\frac{k}{\Omega t_c} \gg \frac{m}{r}, \tag{A3.55}$$

where

$$\Omega t_c \gg 1.$$

Since $kr \sim 10^3$ to 10^6 inequality (A3.55) is satisfied when $1 \ll \Omega t_c \lesssim 10^2$ to 10^5. This range agrees well with the observed optical depths, $\tau \sim 1/\Omega t_c$, of planetary rings; hence we have

$$10^{-2} \text{ to } 10^{-5} \lesssim \tau \ll 1. \tag{A3.56}$$

Inequality (A3.55) is thus satisfied for the case in which we are interested and we can drop the term (A3.50) in comparison with (A3.49). It is clear that the same inequalities enable us to drop also the other terms from the first three on the right-hand side of (A3.40). Only for very rarefied discs of macroparticles for which $\tau \ll 10^{-2}$ to 10^{-5} is it necessary to take into account the angular dependence of the pressure and the self-gravitation. The general form of the dispersion relation is not changed in that case since in the equation for $\hat{v}_\varphi$ there are already terms with the density and temperature perturbations and only the coefficients of these perturbed quantities become more complicated. Taking all this into account we can write down the equation for v'_φ:

$$\frac{\partial v'_\varphi}{\partial t} + v'_r \left[\frac{d(\Omega_0 r)}{dr} + \Omega_0 \right] + \Omega_0 \frac{\partial v'_\varphi}{\partial \varphi} = \frac{r}{\sigma_0} \frac{\partial(\nu\sigma)'}{\partial r} \frac{d\Omega_0}{dr} + \nu_0 \frac{\partial^2 v'_\varphi}{\partial r^2}. \tag{A3.57}$$

Equation (8.32) follows from this equation through the standard procedure:

$$(\gamma + im\Omega_0)\hat{v}_\varphi + \frac{\kappa^2}{2\Omega_0}\hat{v}_r = -\nu_0 k^2 \hat{v}_\varphi - ik\alpha\hat{T} - ik\beta\hat{\sigma}, \tag{A3.58}$$

where α and β are coefficients (see (8.33)). They are the ones which change in the case of greatly rarefied discs when $\Omega t_c \gg kr/m$.

Let us now consider the linearisation of the energy equation (8.25). Neglecting non-linear terms and spatial derivatives of T_0 and χ_0 we find

$$\frac{3}{2}\left[\frac{\partial T_0}{\partial t} + \frac{\partial T'}{\partial t} + \frac{v_{\varphi 0}}{r}\frac{\partial T'}{\partial \varphi}\right] + T_0\left[\frac{\partial v_r'}{\partial r} + \frac{v_r'}{r} + \frac{1}{r}\frac{\partial v_\varphi'}{\partial \varphi}\right]$$

$$= \frac{\chi_0}{\sigma_0 r}\frac{\partial T'}{\partial r} + \frac{\chi_0}{\sigma_0}\frac{\partial^2 T'}{\partial r^2} + \frac{\chi_0}{\sigma_0 r^2}\frac{\partial^2 T'}{\partial \varphi^2} + v'\left(r\frac{d\Omega_0}{dr}\right)^2$$

$$v_0\left[r\frac{d\Omega_0}{dr} + \frac{\partial v_\varphi'}{\partial r} - \frac{v_\varphi'}{r}\right]^2 + 2v_0\frac{\partial v_r'}{\partial \varphi}\frac{d\Omega_0}{dr} - E_0^- - E'^-. \tag{A3.59}$$

Bearing in mind that $kr \gg m$ we find

$$\frac{3}{2}\left[\frac{\partial T_0}{\partial t} + \frac{\partial T'}{\partial t} + \frac{v_{\varphi 0}}{r}\frac{\partial T'}{\partial \varphi}\right] + T_0\frac{\partial v_r'}{\partial r} + \frac{T_0}{r}\frac{\partial v_\varphi'}{\partial \varphi}$$

$$= +\frac{\chi_0}{\sigma_0}\frac{\partial^2 T'}{\partial r^2} + v'\left(r\frac{d\Omega_0}{dr}\right)^2 + v_0\left(r\frac{d\Omega_0}{dr}\right)^2 + 2v_0\left(r\frac{d\Omega_0}{dr}\right)\frac{\partial v_\varphi'}{\partial r}$$

$$+ 2v_0\frac{\partial v_r'}{\partial \varphi}\frac{d\Omega_0}{dr} - E_0^- - E'^-. \tag{A3.60}$$

From (A3.60) we split off the equation for the unperturbed quantities:

$$\frac{3}{2}\frac{\partial T_0}{\partial t} = v_0\left(r\frac{d\Omega_0}{dr}\right)^2 - E_0^-, \tag{A3.61}$$

which, if the term E_0^- which is connected with the cooling of the disc through the inelastic collision is not there, describes the heating of a viscous differentially rotating disc. Hence, for a correct analysis of the linear oscillations it is necessary for such an energy balance that

$$\frac{\partial T_0}{\partial t} \ll \frac{\partial T'}{\partial t}.$$

Let us compare the terms with v_r' and v_φ':

$$T_0\frac{\partial v_r'}{\partial r} \quad \text{and} \quad 2v_0\left(r\frac{d\Omega_0}{dr}\right)\frac{1}{r}\frac{\partial v_r'}{\partial \varphi}, \tag{A3.62}$$

$$\frac{T_0}{r}\frac{\partial v_\varphi'}{\partial \varphi} \quad \text{and} \quad 2v_0\left(r\frac{d\Omega_0}{dr}\right)\frac{1}{r}\frac{\partial v_\varphi'}{\partial r}. \tag{A3.63}$$

For a rarefied gas of macroparticles for which $v_0 \sim c_0^2/\Omega_0 t_c$ and $\Omega_0 t_c \gg 1$ we find from (A3.62):

$$kc_0^2 \gg \frac{3m}{r}v_0\Omega_0 \sim \frac{3mc_0^2}{\Omega_0 t_c r}. \tag{A3.64}$$

This inequality is also satisfied for a molecular gas for which $v_0\Omega_0 \sim c_0^2/\Omega_0^2 t_c$ and $\Omega_0 t_c \ll 1$. Let us now consider (A3.63) for a gas of macroparticles:

$$\frac{mc_0^2}{r} \ll \frac{2kc_0^2}{\Omega_0 t_c}. \tag{A3.65}$$

This inequality is the same as (A3.55) and is therefore satisfied for the case in which we are interested, except for very rarefied discs. Correspondingly, this inequality is not satisfied for a molecular gas. Using this we can write (A3.60) in the form

$$\frac{3}{2}\left[\frac{\partial T'}{\partial t} + \Omega_0 \frac{\partial T'}{\partial \varphi}\right] + T_0 \frac{\partial v_r'}{\partial r} = \frac{\chi_0}{\sigma_0}\frac{\partial^2 T'}{\partial r^2}$$

$$+ v'\left(r\frac{d\Omega_0}{dr}\right)^2 + 2v_0\left(r\frac{d\Omega_0}{dr}\right)\frac{\partial v_\varphi'}{\partial r} - E'^{-}. \tag{A3.66}$$

From this we obtain the required linearised fourth equation of the set (8.32)

$$\frac{3}{2}(\gamma + im\Omega_0)\hat{T} + ikc_0^2\hat{v}_r = -\chi k^2\hat{T} - ik\mu\hat{v}_\varphi - \Delta E_\sigma\hat{\sigma} - \Delta E_T\hat{T}, \tag{A3.67}$$

where μ, ΔE_σ, and ΔE_T are given by (8.33) and $\chi \equiv \chi_0/\sigma_0$. The expression for μ will be more complicated in the case of very rarefied discs.

3 Derivation of the General Dispersion Relation

We write the linearised set (8.32) in the form (assuming that $\nu \equiv \nu_0$)

$$\left(\gamma + im\Omega_0 + \left[\frac{\partial N^-}{\partial \sigma_0} - \frac{\partial N^+}{\partial \sigma_0}\right]\right)\hat{\sigma} + (ik\sigma_0)\hat{v}_r$$

$$+ (0)\hat{v}_\varphi + \left[\frac{\partial N^-}{\partial T_0} - \frac{\partial N^+}{\partial T_0}\right]\hat{T} = 0,$$

$$\left(-i\frac{2\pi G\sigma_0 - kc_0^2}{\sigma_0}\right)\hat{\sigma} + (\gamma + im\Omega_0 + \tfrac{4}{3}\nu k^2)\hat{v}_r - (2\Omega_0)\hat{v}_\varphi + ik\hat{T} = 0,$$

$$(ik\beta)\hat{\sigma} + \left(\frac{\kappa^2}{2\Omega_0}\right)\hat{v}_r + (\gamma + im\Omega_0 + \nu k^2)\hat{v}_\varphi + (ik\alpha)\hat{T} = 0,$$

$$\left(\tfrac{2}{3}\Delta E_\sigma\right)\hat{\sigma} + \left(\tfrac{2}{3}ikc_0^2\right)\hat{v}_r + \left(\tfrac{2}{3}ik\mu\right)\hat{v}_\varphi$$

$$+ (\gamma + im\Omega_0 + \chi k^2 + \Delta E_T)\hat{T} = 0. \tag{A3.68}$$

From (A3.68) we get a determinant of the following form:

$$\begin{vmatrix} \gamma + A_\sigma & A_r & 0 & A_T \\ B_\sigma & \gamma + B_r & B_\varphi & B_T \\ C_\sigma & C_r & \gamma + C_\varphi & C_T \\ D_\sigma & D_r & D_\varphi & \gamma + D_T \end{vmatrix} = 0. \tag{A3.69}$$

The evaluation of this determinant gives us the required general dispersion relation which is of the fourth degree in γ. We gave in (8.34) the dispersion relation neglecting the terms involving the external mass fluxes N^+ and N^- — when $A_T = 0$ and A_σ could be simplified.

Introducing the notation

$$\Delta N_\sigma = \frac{\partial N^-}{\partial \sigma_0} - \frac{\partial N^+}{\partial \sigma_0}, \qquad \Delta N_T = \frac{\partial N^-}{\partial T_0} - \frac{\partial N^+}{\partial T_0}, \qquad \text{(A3.70)}$$

we can write the general dispersion relation in the form

$$(\gamma + im\Omega_0)^4 + (\gamma + im\Omega_0)^3 \left[\tfrac{2}{3}(\chi k^2 + \Delta E_T) + \tfrac{7}{3}\nu k^2 + \Delta N_\sigma\right]$$

$$+ (\gamma + im\Omega_0)^2 \left\{\tfrac{4}{3}\nu^2 k^4 + \omega_0^2 + \tfrac{14}{9}\nu k^2(\chi k^2 + \Delta E_T) + \tfrac{2}{3}k^2\alpha\mu\right.$$

$$+ \left(\tfrac{7}{3}\nu k^2 + \tfrac{2}{3}\chi k^2\right)\Delta N_\sigma + \tfrac{2}{3}\left(\Delta N_\sigma \Delta E_T - \Delta N_T \Delta E_\sigma\right)\Big\}$$

$$+ (\gamma + im\Omega_0)\left\{\nu k^2 \left[\tfrac{5}{3}k^2 c_0^2 - 2\pi G\sigma_0 k\right] + \tfrac{2}{3}\left[\tfrac{4}{3}\nu^2 k^4 + \omega_*^2\right](\chi k^2 + \Delta E_T)\right.$$

$$+ \tfrac{8}{9}\alpha\mu\nu k^4 + \tfrac{4}{3}k^2 c_0^2 \Omega_0 \alpha - \tfrac{2}{3}k^2 \sigma_0 \Delta E_\sigma - \frac{\kappa^2}{3\Omega_0}k^2\mu$$

$$+ 2k^2\beta\Omega_0\sigma_0 + \left(\tfrac{4}{3}\nu^2 k^4 + \tfrac{14}{9}\nu\chi k^4\right)\Delta N_\sigma$$

$$+ \tfrac{14}{9}\nu k^2 \left(\Delta N_\sigma \Delta E_T - \Delta N_T \Delta E_\sigma\right) + \tfrac{2}{3}\mu k^2 \left(\alpha\Delta N_\sigma - \beta\Delta N_T\right)$$

$$+ \Delta N_\sigma \left(\tfrac{3}{2}k^2 c_0^2 + \kappa^2\right) - \frac{2c_0^2}{3\sigma_0}\Delta N_T \left(k^2 c_0^2 - 2\pi G\sigma_0 k\right)\Big\}$$

$$+ \tfrac{2}{3}\left\{(\chi\nu k^4 + \nu k^2 \Delta E_T + k^2\alpha\mu)(k^2 c_0^2 - 2\pi G\sigma_0 k)\right.$$

$$+ \sigma_0 k^4(2\chi\beta\Omega_0 - \beta\mu - \Delta E_\sigma\nu) + 2k^2\sigma_0\Omega_0(\beta\Delta E_T - \alpha\Delta E_\sigma)$$

$$+ \tfrac{4}{3}\nu^2\chi k^6 \Delta N_\sigma + \tfrac{4}{3}\nu^2 k^4 \left(\Delta N_\sigma \Delta E_T - \Delta N_T \Delta E_\sigma\right)$$

$$+ \tfrac{4}{3}\mu\nu k^4 \left(\alpha\Delta N_\sigma - \beta\Delta N_T\right) - \nu k^2 \frac{c_0^2}{\sigma_0}\Delta N_T \left(k^2 c_0^2 - 2\pi G\sigma_0 k\right)$$

$$+ \nu k^2 c_0^2 \Delta N_\sigma + 2k^2 c_0^2 \Omega_0 \left(\alpha\Delta N_\sigma - \beta\Delta N_T\right) + \kappa^2 \Delta N_\sigma \left(\chi k^2 - \frac{\mu k^2}{2\Omega_0}\right)$$

$$+ \frac{\kappa^2\mu}{2\Omega_0\sigma_0}\Delta N_T \left(k^2 c_0^2 - 2\pi G\sigma_0 k\right) + \kappa^2 \left(\Delta N_\sigma \Delta E_T - \Delta N_T \Delta E_\sigma\right)\Big\}$$

$$= 0. \qquad\qquad \text{(A3.71)}$$

Appendix IV.
Evaluating the Gravitational Potential Inside and Outside a Triaxial Ellipsoid

1 Potential Inside the Ellipsoid

Let us consider the potential of a solid uniform ellipsoid and its attractive force on a unit point mass inside it (Moulton 1914).

Let the equation of the surface of the ellipsoid in a ξ, η, ζ coordinate system be given by

$$\frac{\xi^2}{a^2} + \frac{\eta^2}{b^2} + \frac{\zeta^2}{c^2} = 1, \tag{A4.1}$$

and let the unit point mass be at a point P with coordinates x, y, z inside the ellipsoid. We take this point as the origin of polar coordinates ϱ, φ, θ and let the base planes of the new system of coordinates be parallel to those of the first system in such a way that the angle φ is reckoned in the ξ, η plane from $\eta = 0$, and the angle θ from the ζ-axis. The variables ϱ, φ, and θ are then related to the Cartesian coordinates through the formulæ:

$$\xi = x + \varrho\sin\theta\cos\varphi, \quad \eta = y + \varrho\sin\theta\sin\varphi, \quad \zeta = z + \rho\cos\theta. \tag{A4.2}$$

The potential of the ellipsoid in the point P is equal to

$$E = \int_M \frac{dm}{\varrho} = \sigma\int_V \frac{dV}{\varrho} = \sigma\iiint_V \varrho\sin\theta\,d\varrho\,d\varphi\,d\theta$$

$$= \sigma\int_0^\pi d\theta \int_0^{2\pi} d\varphi \int_0^{\varrho_1} \varrho\sin\theta\,d\varrho,$$

where σ is the mass density of the ellipsoid, and where the first integral is over the mass of the ellipsoid, and the second and third integrals over its volume. After integration over ϱ we have

$$E = \tfrac{1}{2}\sigma\int_0^{2\pi} d\varphi \int_0^\pi \varrho_1^2\sin\theta\,d\theta. \tag{A4.3}$$

We substitute (A4.2) into (A4.1) to write the expression for the surface of the ellipsoid ϱ_1 in terms of the polar coordinates and we find

$$A\varrho_1^2 + 2B\varrho_1 + C = 0, \tag{A4.4}$$

with

$$
\left.
\begin{aligned}
A &= \frac{\sin^2\theta\cos^2\varphi}{a^2} + \frac{\sin^2\theta\sin^2\varphi}{b^2} + \frac{\cos^2\theta}{c^2}, \\[2mm]
B &= \frac{x\sin\theta\cos\varphi}{a^2} + \frac{y\sin\theta\sin\varphi}{b^2} + \frac{z\cos\theta}{c^2}, \\[2mm]
C &= \frac{x^2}{a^2} + \frac{y^2}{b^2} + \frac{z^2}{c^2} - 1.
\end{aligned}
\right\}
\tag{A4.5}
$$

From (A4.4) we find $\varrho_1 = [-B \pm \sqrt{B^2 - AC}]/A$. It is clear that ϱ_1 is necessarily a positive quantity. As A is positive and C negative definite — because the point x, y, z lies inside the ellipsoid — we need the positive sign in front of the square root in order that ϱ_1 be positive. Substituting that value of ϱ_1 into (A4.3) we find

$$
E = \tfrac{1}{2}\sigma \int_0^{2\pi} d\varphi \int_0^\pi \frac{2B^2 - AC - 2B\sqrt{B^2 - AC}}{A^2} \sin\theta \, d\theta.
\tag{A4.6}
$$

It follows from the expression for B that if $0 \leqslant \varphi_0 \leqslant \pi$ and $0 \leqslant \theta_0 \leqslant \pi/2$, we have $B(\varphi_0 + \pi, \pi - \theta_0) = -B(\varphi_0, \theta_0)$. As the integral (A4.6) can be split into two parts corresponding, respectively, to φ_0, θ_0 and $\varphi_0 + \pi, \pi - \theta_0$ while $\sin\theta_0 = \sin(\pi - \theta_0)$, we can drop the contribution from $-2B\sqrt{B^2 - AC}/A^2$ as it will be zero. Substituting the expressions for A, B, and C into (A4.6) we can split the remaining integral into two parts:

$$
E = \tfrac{1}{2}\sigma \int_0^{2\pi} d\varphi \int_0^\pi \frac{2B^2 - AC}{A^2} \sin\theta \, d\theta = I_1 + I_2,
$$

with

$$
\begin{aligned}
I_1 &= \tfrac{1}{2}\sigma \int_0^{2\pi} d\varphi \int_0^\pi \left[\frac{\sin^2\theta\cos^2\varphi}{a^2}\left(\frac{2x^2}{a^2} - C\right) \right. \\[2mm]
&\qquad \left. + \frac{\sin^2\theta\sin^2\varphi}{b^2}\left(\frac{2y^2}{b^2} - C\right) + \frac{\cos^2\theta}{c^2}\left(\frac{2z^2}{c^2} - C\right) \right] \frac{\sin\theta \, d\theta}{A^2}, \\[3mm]
I_2 &= 2\sigma \int_0^{2\pi} d\varphi \int_0^\pi \left[\frac{xy\sin^2\theta\sin\varphi\cos\varphi}{a^2 b^2} \right. \\[2mm]
&\qquad \left. + \frac{xz\sin\theta\cos\theta\cos\varphi}{a^2 c^2} + \frac{yz\sin\theta\cos\theta\sin\varphi}{b^2 c^2} \right] \frac{\sin\theta \, d\theta}{A^2}.
\end{aligned}
$$

Let us consider the integration over φ in I_2. One shows easily that the function $A(\varphi)$ is symmetric relative to $\varphi_0 = \pi/2$ and periodic with period π. Hence, the function $1/A^2$ which occur in the integrand has the same properties. Using the properties of the function $1/A^2$ one sees that $\sin\varphi/A^2$ and $\cos\varphi/A^2$ integrated over the interval $[0, 2\pi]$, as well as $\sin\varphi\cos\varphi/A^2$ integrated over either the interval $[0, \pi]$ or the interval $[\pi, 2\pi]$ give zero. Hence I_2 is equal to zero.

To simplify I_1 we introduce the auxiliary integral

$$W = \tfrac{1}{2}\sigma \int_0^{2\pi} d\varphi \int_0^\pi \frac{\sin\theta\, d\theta}{A}, \tag{A4.7}$$

where A is given by (A4.5).

We also evaluate

$$\frac{\partial W}{\partial a} = \tfrac{1}{2}\sigma \int_0^{2\pi} d\varphi \int_0^\pi \frac{\sin\theta\, d\theta}{A^2} \frac{\partial}{\partial a}\left[-\frac{\sin^2\theta \cos^2\varphi}{a^2} \right]$$

$$= \frac{\sigma}{a^3} \int_0^{2\pi} d\varphi \int_0^\pi \frac{\sin^3\theta \cos^2\varphi\, d\theta}{A^2}.$$

Similarly, we have

$$\frac{\partial W}{\partial b} = \frac{\sigma}{b^3} \int_0^{2\pi} d\varphi \int_0^\pi \frac{\sin^3\theta \sin^2\varphi\, d\theta}{A^2},$$

$$\frac{\partial W}{\partial c} = \frac{\sigma}{c^3} \int_0^{2\pi} d\varphi \int_0^\pi \frac{\sin\theta \cos^2\theta\, d\theta}{A^2}.$$

One easily shows that

$$E = -CW + \frac{x^2}{a}\frac{\partial W}{\partial a} + \frac{y^2}{b}\frac{\partial W}{\partial b} + \frac{z^2}{c}\frac{\partial W}{\partial c}, \tag{A4.8}$$

where C is also given by (A4.5).

We bring (A4.7) to integrable form. We put

$$M - \frac{\sin^2\theta}{a^2} + \frac{\cos^2\theta}{c^2}, \qquad N = \frac{\sin^2\theta}{b^2} + \frac{\cos^2\theta}{c^2}, \tag{A4.9}$$

so that

$$M \cos^2\varphi + N \sin^2\varphi = A,$$

whence (A4.7) takes the form

$$W = \tfrac{1}{2}\sigma \int_0^{2\pi} d\varphi \int_0^\pi \frac{\sin\theta\, d\theta}{M \cos^2\varphi + N \sin^2\varphi},$$

or, if we use the symmetry properties of the integrand with regard to φ and θ,

$$W = 4\sigma \int_0^{\pi/2} d\varphi \int_0^{\pi/2} \frac{\sin\theta\, d\theta}{M \cos^2\varphi + N \sin^2\varphi}.$$

Since M and N are independent of φ we can integrate over φ, changing the integration variable to $t = \tan\varphi$:

$$W = 4\sigma \int_0^{\pi/2} \sin\theta \, d\theta \int_0^\infty \frac{dt}{M + Nt^2} = 2\pi\sigma \int_0^{\pi/2} \frac{\sin\theta \, d\theta}{\sqrt{MN}}.$$

Using (A4.9) we have

$$W = 2\pi\sigma abc^2 \int_0^{\pi/2} \frac{\sin\theta \, d\theta}{\sqrt{(a^2 \cos^2\theta + c^2 \sin^2\theta)(b^2 \cos\theta^2 + c^2 \sin^2\theta)}}. \quad (A4.10)$$

To show the symmetry in a, b, c – which is obvious in (A4.7) – we introduce a Jacobi transformation: $\cos\theta = c/\sqrt{c^2 + s}$. This means that

$$s = c^2 \tan^2\theta, \quad ds = \frac{2c^2 \sin\theta}{\cos^3\theta} d\theta, \quad \cos^2\theta = \frac{c^2}{c^2 + s}, \quad \sin^2\theta = \frac{s}{c^2 + s}.$$

Applying this substitution to (A4.10) we get

$$W = \pi\sigma abc \int_0^\infty \frac{ds}{\sqrt{(a^2 + s)(b^2 + s)(c^2 + s)}}, \quad (A4.11)$$

whence follows

$$\left.\begin{aligned}
\frac{\partial W}{\partial a} &= \pi\sigma bc \int_0^\infty \frac{s \, ds}{(a^2 + s)\sqrt{(a^2 + s)(b^2 + s)(c^2 + s)}}, \\
\frac{\partial W}{\partial b} &= \pi\sigma ac \int_0^\infty \frac{s \, ds}{(b^2 + s)\sqrt{(a^2 + s)(b^2 + s)(c^2 + s)}}, \\
\frac{\partial W}{\partial c} &= \pi\sigma ab \int_0^\infty \frac{s \, ds}{(c^2 + s)\sqrt{(a^2 + s)(b^2 + s)(c^2 + s)}},
\end{aligned}\right\} \quad (A4.12)$$

Substituting (A4.11) and (A4.12) into (A4.8) and using (A4.5) we finally find for the potential:

$$E = \pi\sigma abc \int_0^\infty \left[1 - \frac{x^2}{a^2 + s} - \frac{y^2}{b^2 + s} - \frac{z^2}{c^2 + s}\right]$$
$$\times \frac{ds}{\sqrt{(a^2 + s)(b^2 + s)(c^2 + s)}}. \quad (A4.13)$$

Hence we find for the components of the attractive force:

$$\left.\begin{aligned}
F_x &= k^2 \frac{\partial E}{\partial x} = -2\pi\sigma abcx k^2 \int_0^\infty \frac{ds}{\sqrt{(a^2 + s)^3(b^2 + s)(c^2 + s)}}, \\
F_y &= k^2 \frac{\partial E}{\partial y} = -2\pi\sigma abcy k^2 \int_0^\infty \frac{ds}{\sqrt{(a^2 + s)(b^2 + s)^3(c^2 + s)}}, \\
F_z &= k^2 \frac{\partial E}{\partial z} = -2\pi\sigma abcz k^2 \int_0^\infty \frac{ds}{\sqrt{(a^2 + s)(b^2 + s)(c^2 + s)^3}},
\end{aligned}\right\} \quad (A4.14)$$

where k^2 is the gravitational constant.

To get rid of the infinite integration limits it is useful to apply in (A4.14) the substitutions $t = a/\sqrt{a^2 + s}$, $t = b/\sqrt{b^2 + s}$, $t = c/\sqrt{c^2 + s}$, respectively. After some elementary transformations we get for the force components on a unit point mass inside the ellipsoid:

$$
\left.
\begin{aligned}
F_x &= -4\pi\sigma bck^2 x \int_0^1 \frac{t^2\, dt}{\sqrt{[a^2 - (a^2 - b^2)t^2][a^2 - (a^2 - c^2)t^2]}}, \\
F_y &= -4\pi\sigma ack^2 y \int_0^1 \frac{t^2\, dt}{\sqrt{[b^2 - (b^2 - a^2)t^2][b^2 - (b^2 - c^2)t^2]}}, \\
F_z &= -4\pi\sigma abk^2 z \int_0^1 \frac{t^2\, dt}{\sqrt{[c^2 - (c^2 - a^2)t^2][c^2 - (c^2 - b^2)t^2]}},
\end{aligned}
\right\} \quad \text{(A4.15)}
$$

where σ is the density of the ellipsoid, a, b, and c its semi-axes, and x, y, z the coordinates of the point on which the force acts, in a Cartesian coordinate system with origin at the centre of the ellipsoid.

The calculation of the force components for interior points of the ellipsoid thus reduces to evaluating integrals of the form

$$
I = \int_0^1 \frac{t^2\, dt}{\sqrt{(g^2 - et^2)(g^2 - ft^2)}},
$$

where g, e, and f are constants.

2 Potential Outside the Ellipsoid

In the case of a point which lies outside the ellipsoid the integrals are evaluated in terms of the components of the attractive force of an auxiliary ellipsoid on an interior point (Ivory method; Moulton 1914).

We are required to find the attractive force due to the ellipsoid – which we call E – given by (A4.1) on an exterior point P' with coordinates x', y', z'. We construct through P' an ellipsoid E', confocal with E, with semi-axes a', b', c' and we assume that it has the same density as E. The axes of the two ellipsoids are then connected through the relations

$$
a' = \sqrt{a^2 + \chi}, \qquad b' = \sqrt{b^2 + \chi}, \qquad c' = \sqrt{c^2 + \chi}, \qquad \text{(A4.16)}
$$

where χ can be found from the equation

$$
\frac{x'^2}{a^2 + \chi} + \frac{y'^2}{b^2 + \chi} + \frac{z'^2}{c^2 + \chi} - 1 = 0. \qquad \text{(A4.17)}
$$

One can easily show that (A4.17) has one positive and two negative roots for χ. It is clear, that we are interested in the positive root.

We establish a one-to-one connection between the points of the ellipsoids E and E′ by means of the formulae

$$\xi' \;=\; \frac{a'}{a}\xi, \qquad \eta' \;=\; \frac{b'}{b}\eta, \qquad \zeta' \;=\; \frac{c'}{c}\zeta. \tag{A4.18}$$

Let P be the point corresponding to P′. We shall show that the attractive force of E on P′ (an exterior point) can simply be expressed in terms of the attractive force of E′ on P (an interior point) which was calculated in Sect. 1 of the present appendix. Let F_x, F_y, F_z be the components of the force of E′ on P (x, y, z). According to what we found in Sect. 1 we have in terms of Cartesian coordinates

$$F_x \;=\; k^2 \sigma \iiint \frac{\partial(1/\varrho')}{\partial \xi'} \, d\xi' \, d\eta' \, d\zeta'. \tag{A4.19}$$

Let F_x', F_y', F_z' denote the components of the force of E on the point P′ which we are required to find. By analogy with (A4.19) we have

$$F_x' \;=\; k^2 \sigma \iiint \frac{\partial(1/\varrho)}{\partial \xi} \, d\xi \, d\eta \, d\zeta. \tag{A4.20}$$

Integrating (A4.19) over ξ' and (A4.20) over ξ we find

$$\left.\begin{aligned}
F_x &\;=\; k^2 \sigma \iint \left[\frac{1}{\varrho_2'} - \frac{1}{\varrho_1'} \right] d\eta' \, d\zeta', \\[2mm]
F_x' &\;=\; k^2 \sigma \iint \left[\frac{1}{\varrho_2} - \frac{1}{\varrho_1} \right] d\eta \, d\zeta,
\end{aligned}\right\} \tag{A4.21}$$

where ϱ_2 (ϱ_2') and ϱ_1 (ϱ_1') are the distances of P′ (P) from the end of the elementary column obtained by integrating over ξ (ξ').

We prove a simple relation between F_x and F_x' by means of the following lemma:

If (P, A) and (P′, A′) are two pairs of corresponding points on the surfaces of confocal ellipsoids the distances PA′ = ϱ' and AP′ = ϱ will be equal to one another: $\varrho = \varrho'$.

Indeed, if the coordinates of P and A are, respectively, equal to ξ_1, η_1, ζ_1 and ξ_2, η_2, ζ_2 and those of P′ and A′ to $\xi_1', \eta_1', \zeta_1'$ and $\xi_2', \eta_2', \zeta_2'$ we have

$$\begin{aligned}
\varrho'^2 &\;=\; (\xi_1 - \xi_2')^2 + (\eta_1 - \eta_2')^2 + (\zeta_1 - \zeta_2')^2, \\
\varrho^2 &\;=\; (\xi_2 - \xi_1')^2 + (\eta_2 - \eta_1')^2 + (\zeta_2 - \zeta_1')^2.
\end{aligned}$$

Subtracting the second equation from the first and using (A4,16) and (A4.18) we get

$$\varrho'^2 - \varrho^2 \;=\; \chi \left[\frac{\xi_2^2}{a^2} + \frac{\eta_2^2}{b^2} + \frac{\zeta_2^2}{c^2} \right] - \chi \left[\frac{\xi_1^2}{a^2} + \frac{\eta_1^2}{b^2} + \frac{\zeta_1^2}{c^2} \right].$$

Since P and A lie on the surface of the ellipsoid with semi-axes a, b, c both brackets are equal to unity. Hence, $\varrho'^2 - \varrho^2 = 0$, or $\varrho' = \varrho$. QED.

We assume that the integrals (A4.21) are evaluated in such a way that elements of the corresponding points of the two surfaces are always taken at the same time. In that case we have for the whole of the integration $\varrho_1 = \varrho_1'$ and $\varrho_2 = \varrho_2'$. Moreover, it follows from (A4.18) that

$$d\eta = \frac{b}{b'} d\eta', \qquad d\zeta = \frac{c}{c'} d\zeta'.$$

Substituting these formulæ into the second of (A4.21) we find that

$$F_x' = k^2 \sigma \frac{bc}{b'c'} \iint \left[\frac{1}{\varrho_2'} - \frac{1}{\varrho_1'} \right] d\eta'\, d\zeta',$$

and after comparing this with the first of (A4.21) we see that

$$F_x' = \frac{bc}{b'c'} F_x. \tag{A4.22}$$

Similarly we find that

$$F_y' = \frac{ac}{a'c'} F_y, \qquad F_z' = \frac{ab}{a'b'} F_z. \tag{A4.23}$$

According to (A4.14) we have for F_x, F_y, F_z

$$\left. \begin{aligned} F_x &= -2\pi\sigma a'b'c'xk^2 \int_0^\infty \frac{du}{\sqrt{(a'^2+u)^3(b'^2+u)(c'^2+u)}}, \\[2mm] F_y &= -2\pi\sigma a'b'c'yk^2 \int_0^\infty \frac{du}{\sqrt{(a'^2+u)(b'^2+u)^3(c'^2+u)}}, \\[2mm] F_z &= -2\pi\sigma a'b'c'zk^2 \int_0^\infty \frac{du}{\sqrt{(a'^2+u)(b'^2+u)(c'^2+u)^3}}. \end{aligned} \right\} \tag{A4.24}$$

Since P and P' are corresponding points we have

$$x = \frac{a}{a'} x', \qquad y = \frac{b}{b'} y', \qquad z = \frac{c}{c'} z'. \tag{A4.25}$$

Substituting (A4.24) and (A4.25) into (A4.22) and (A4.23) we find

$$\left. \begin{aligned} F_x' &= -2\pi\sigma abck^2 x' \int_0^\infty \frac{du}{\sqrt{(a'^2+u)^3(b'^2+u)(c'^2+u)}}, \\[2mm] F_y' &= -2\pi\sigma abck^2 y' \int_0^\infty \frac{du}{\sqrt{(a'^2+u)(b'^2+u)^3(c'^2+u)}}, \\[2mm] F_z' &= -2\pi\sigma abck^2 z' \int_0^\infty \frac{du}{\sqrt{(a'^2+u)(b'^2+u)(c'^2+u)^3}}. \end{aligned} \right\} \tag{A4.26}$$

From (A4.16) it follows that

$$a'^2 = a^2 + \chi, \qquad b'^2 = b^2 + \chi, \qquad c'^2 = c^2 + \chi.$$

Hence, if we also put $s = u + \chi$ we find

$$\left.\begin{array}{l}
F'_x = -2\pi\sigma abck^2 x' \displaystyle\int_\chi^\infty \frac{ds}{\sqrt{(a^2+s)^3(b^2+s)(c^2+s)}}, \\[1em]
F'_y = -2\pi\sigma abck^2 y' \displaystyle\int_\chi^\infty \frac{ds}{\sqrt{(a^2+s)(b^2+s)^3(c^2+s)}}, \\[1em]
F'_z = -2\pi\sigma abck^2 z' \displaystyle\int_\chi^\infty \frac{ds}{\sqrt{(a^2+s)(b^2+s)(c^2+s)^3}},
\end{array}\right\} \qquad \text{(A4.27)}$$

It is clear that the only difference between (A4.27) and (A4.14) lies in the lower limit on the integral. Changing variables as in the case of an interior point, namely,

$$t = a/\sqrt{a^2+s} \qquad \text{for the first expression,}$$
$$t = b/\sqrt{b^2+s} \qquad \text{for the second expression,}$$
$$\text{and } t = c/\sqrt{c^2+s} \qquad \text{for the third expression,}$$

we finally find the following expressions for the components of the attractive force on a unit mass point outside the ellipsoid:

$$\left.\begin{array}{l}
F'_x = -4\pi\sigma bck^2 x' \displaystyle\int_0^{a/\sqrt{a^2+\chi}} \frac{t^2\,dt}{\sqrt{[a^2-(a^2-b^2)t^2][a^2-(a^2-c^2)t^2]}}, \\[1.5em]
F'_y = -4\pi\sigma ack^2 y' \displaystyle\int_0^{b/\sqrt{b^2+\chi}} \frac{t^2\,dt}{\sqrt{[b^2-(b^2-a^2)t^2][b^2-(b^2-c^2)t^2]}}, \\[1.5em]
F'_z = -4\pi\sigma abk^2 z' \displaystyle\int_0^{c/\sqrt{c^2+\chi}} \frac{t^2\,dt}{\sqrt{[c^2-(c^2-a^2)t^2][c^2-(c^2-b^2)t^2]}},
\end{array}\right\} \qquad \text{(A4.28)}$$

where σ is the density of the ellipsoid, a, b, and c its semi-axes, x', y', z' the coordinates of the point on which the force acts, in a Cartesian coordinate system with origin at the centre of the ellipsoid, and χ the positive root of the third-degree equation (A4.17).

The calculation of the components of the attractive force for points exterior to the ellipsoid reduces thus to the evaluation of integrals of the same form as in the case of interior points—only the upper limit on the integral has been changed.

Appendix V. A Drift Mechanism
for the Formation of the Cassini Division[1]

1 Introduction

This appendix is an attempt to find the correct solution to the three-century-old problem of the formation of the Cassini Division. The mechanism proposed is based on a new type of quasi-stationary accretion flow in astrophysical discs when there is a density wave present. The flow appears due to the action of the Reynolds stresses caused by a non-linear wave. In this respect it is similar to "acoustic streaming", an effect which, starting from Faraday's discovery (Franklin 1831), has been observed in hundreds of different laboratory experiments and which has been described in numerous theoretical papers, starting with the pioneer work by Rayleigh (1884). However, due to the dominant role of the Coriolis forces in astrophysical discs the flow has a drift nature. This means that the direction of the flow is at right angles to the direction of the applied force. Moreover, we shall show that in order correctly to describe a non-linear accretion drift of this kind we must take into account the complex, fully three-dimensional structure of the density wave.

Under suitable conditions the density wave may produce a radial flow, carrying disc particles away from some regions, thus producing a gap. According to our model the maximum width a gap can achieve is determined by the maximum size of the zone in which the wave propagates. The latter extends from the position where the wave is generated – at the position of the orbital resonance with a neighbouring satellite – to the reflection point where the group velocity of the wave falls to zero. If we use the parameters of the Cassini Division to make numerical estimates, the result agrees with the observational data.

We have checked the applicability of the simple criterion for the formation of a gap. A gap is formed whenever our mechanism for a redistribution of the density in the disc is more efficient that the diffusion flux. In the opposite case, there is no gap and one can observe a resonant wave. We show that this criterion is sufficient to explain gaps in some parts of the rings and waves in other parts, in agreement with observations. It should be noted that the new effect studied in the appendix – radial drift in a rotating gravitating medium –

---

[1] Written with Khoruzhii as coauthor (Fridman, Khoruzhii and Gor'kavyĭ 1996).

is a universal effect and should be observed in all astrophysical discs in which density waves are present.

The important common feature of the structures which we are discussing – gaps, wave trains, and ringlets – is the fact that they are located near inner Lindblad resonances of neighbouring satellites. For instance, on the one hand, the inner edge of the Cassini Division coincides with the position of the 2:1 resonance with Mimas (Franklin et al. 1971). On the other hand, we note a strong spatial correlation between the position of narrow ringlets and resonances both in the Saturnian system (Fig. 2.3; Cuzzi et al. 1984, Rosen, Tyler, Marouf, and Lissauer 1991) and in the Uranian system (Tables 10.1 to 3 and Figs. 10.1 to 5; Gor'kavyĭ and Fridman 1985b). Indeed, this was just the reason why we could predict new Uranian satellites (Gor'kavyĭ and Fridman 1985b, c) which were observed subsequently by Voyager-2 (Smith et al. 1986, Gor'kavyĭ and Fridman 1987b, Gor'kavyĭ et al. 1988).

Is it possible to give a reason why this localisation should occur? Yes, we can, if we bear in mind that the Lindblad resonances, as we described earlier in this book in Sect 2.1, are the regions where various kinds of waves are generated. Many examples of bending waves travelling towards the planet as well as spiral density waves propagating out from the planet have been observed (see, for instance, Marouf, Tyler, and Rosen 1986 and Fig. A5.1). Moreover, the theory predicts also the existence of standing waves in a medium with a sufficiently high viscosity (Meyer-Vernet and Sicardy 1987) as well as travelling acoustic waves (Goldreich and Tremaine 1978b; Meyer-Vernet and Sicardy 1987). These waves have not been observed by Voyager 2 but it is not excluded that the wavelength of the acoustic waves is so small that their observation would require a larger resolving power of the photo-polarimeters of the Voyager spacecraft.

Goldreich and Tremaine (1978b; quoted in what follows as GT) predicted the generation, in the vicinity of the inner Lindblad resonance of a satellite, of gravitational acoustic waves and their propagation outwards from the planet. Deservingly, this important prediction occupies an important place in the scientific biography of the Saturnian rings which is so rich in discoveries. However, the main aim of their paper was, in our opinion, not achieved. It is the proof that the Cassini Division was formed through the propagation of a gravitation acoustic wave from the 2:1 resonance with Mimas.

According to GT the reason for the formation of a gap is an exchange of angular momentum between the particles in the ring and the wave. Qualitatively, their reasoning is the following. It is well known (McIntyre 1981) that the quasi-energy densities E and E_0 of a sound wave, moving with a velocity $\boldsymbol{u}$ in a medium, in the comoving and laboratory frames of references, respectively, are related to one another through the relation

$$E_0 = E \frac{\omega}{\omega - (\boldsymbol{k} \cdot \boldsymbol{u})}, \tag{A5.1}$$

where ω is the eigenfrequency and k the wavevector. For a disc, rotating with angular velocity Ω_0, we have

$$E_0 = E\frac{\omega}{\omega - m\Omega_0}, \tag{A5.2}$$

where m is the azimuthal number.

On the other hand, using the well known relation between energies in inertial and non-inertial reference frames (Landau and Lifshitz 1976), we can derive the following relation between the quasi-energies E and E_0 and the angular quasi-momentum M:

$$E_0 = E + \Omega_0 M. \tag{A5.3}$$

From (A5.2) and (A5.3) we find the following expression for the angular quasi-momentum of the wave (see also Toomre 1969, Shu 1970):

$$M = -\frac{E}{\Omega_0 - \omega/m}. \tag{A5.4}$$

For a sound wave we have $E = c^2 \varrho_1^2/\varrho > 0$, where ϱ is the total volume density, and ϱ_1 its perturbation, while c is the sound velocity (Landau and Lifshitz 1987). Hence, in the region between the inner Lindblad resonance and the corotation resonance, where $\Omega_0 > \omega/m$, the angular quasi-momentum of the wave is negative whereas the disc particles have a positive angular momentum in the same region. We should emphasise once again that the above example is given for illustrative reasons only. However, the result obtained is more general. The density waves in planetary rings, which are excited in the inner Lindblad resonance, where

$$k \to 0 \qquad \text{and} \qquad \hat{\omega}^2 = (\omega - m\Omega_0)^2 \to \kappa^2 \equiv 4\Omega_0^2\left(1 + \frac{r}{2}\frac{\Omega_0'}{\Omega_0}\right),$$

belong to the rotational rather than the acoustic branch, but (A5.2) and (A5.4) are valid for them also.

The main feature of GT's scenario is that, since the angular momentum of the wave is negative, the exchange of angular momentum between it and the particles in the ring results in the particles losing positive angular momentum and falling to the place where the wave is generated, that is, the outer edge of the B ring. According to GT's estimates this mechanism, unless restricted, could carry away the whole of Saturn's A ring to the resonance position during its lifetime (of about 5×10^9 years).

However, a simple logical argument shows that GT's mechanism cannot produce the Cassini Division. Indeed, according to this mechanism the density wave must first carry away particles from a narrow zone immediately adjacent to the exact resonance position, that is, the inner edge of the Cassini Division. The place where the wave is generated then shifts to the next non-depleted region, and so on. However, there is no wave in the empty space which has been formed as a result of the process. There is therefore no force capable of transporting matter through the empty zone up to the edge of the Cassini Division.

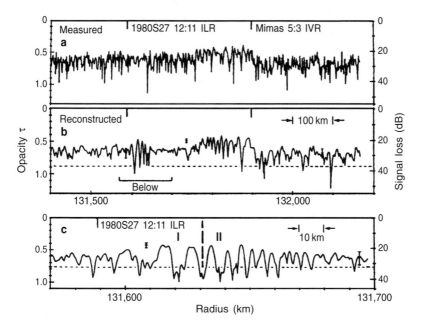

Fig. A5.1. Measured (**a**) and reconstructed (**b**) opacity profiles of a 800 km wide region in the A ring, encompassing the 1980S27 12:11 density wave and the Mimas 5:3 bending wave. The theoretical location of the corresponding resonances are as indicated. Both waves are effectively masked in the initial, diffraction-limited measurement. (**c**) Many more oscillations of the density wave are revealed at an effective resolution of 900 m. Note, however, that regions of opacity maxima are almost completely masked by noise at this resolution (from Meyer-Vernet and Sicardy 1987).

The average optical depth of the Cassini Division, which ranges from the outer edge of the B ring at a radius of 117 540 km to the inner edge of the A ring at 122 050 km, is 0.1. This is an order of magnitude less than the average optical depth of the surrounding regions of the A and B rings (compare Fig.A5.1). However, matter is distributed very non-uniformly within the Cassini Division: there are many narrow ringlets, and in the zone from 117 550 to 120 350 km there are several almost completely empty areas (see Fig.2.8). Any single one of the inner gaps must, during its own formation, interrupt the propagation of the spiral wave outwards. Moreover, these gaps should be filled rapidly – within 1000 to 10 000 years – at the normal diffusion velocity.

One can therefore reach two conclusions from the present structure of the Cassini Division: (a) particle diffusion in the Cassini Division was practically completely suppressed through the decrease of the optical depth to 0.1; (b) if the Cassini Division was formed by a wave, this wave would have emptied first the outside regions near the inner edge of the A ring, and only afterwards would the wave have emptied the inner regions, when moving back to the resonance point.

Observational data of the resonant wave trains – the density waves generated at the points of resonance – give us one more surprise. Estimates (Hord et al. 1982) similar to the ones by GT suggest that the density wave generated in the 2:1 resonance from Janus should empty a gap of a width of about 40 km within less that 10^9 years. However, this part of the ring does not show any gap (Hort et al. 1982). Why is this? To answer this question, note first of all that in recent literature all the wave characteristics connected with different aspects of the time-space symmetry have been given the prefix "quasi". This prefix is insignificant when we follow the evolution of a wavepacket and use the proper conservation laws for quasi-momentum or quasi-energy. However, there are numerous classical examples where authors find a paradox by applying these concepts too literally for the interaction between waves and a medium (McIntyre 1981). These paradoxes can all be explained by the physical differences between quasi-momentum and momentum and between quasi-energy and energy. McIntyre (1981) has given some necessary conditions when these differences can be ignored, but so far the set of sufficient conditions has not been determined. The fact that there is a contradiction between the conclusions of GT's theory and observations indicates that it could be incorrect to use the quasi-momentum rather than the momentum when studying the interactions between waves and particles in the planetary rings. It is possible that in this case some sufficient conditions for the correctness of this approach are not satisfied. For instance, McIntyre has shown that for the interactions between waves and particles it may be crucial to consider the particular geometry of the problem, that is, whether one is dealing with a three- or a two-dimensional problem. This is the reason why when in what follows we consider the interaction between a density wave and the particles in the ring we use the approach developed in the theory of acoustic streaming which is free from the above-mentioned drawback. The investigation is carried out strictly within the framework of a three-dimensional geometry and we show in Sect. 6 that when one studies the dynamics of planetary rings, it is incorrect to neglect vertical functional dependence and motions.

Let us now try to clarify the nature of the main mechanism for the formation of gaps and wavetrains.

First of all, we note that as in GT's theory, density waves are the main agents which make our mechanism operate. However, the results of the present appendix differ very markedly from those predicted by GT. The mechanism considered below is a dissipative drift which represents the modification of acoustic streaming in a rotating self-gravitating medium. Let us explain this. A wave, propagating from a point of resonance, produces a large-scale component of the Reynolds stresses – terms like $\varrho v_{i1} v_{k1}$ (see Sect. 2). The gradient of these stresses determines the force acting upon the medium. As a result, so-called "acoustic streaming" appears – a non-linear effect, known since Faraday's days (Faraday 1831). The first theoretical analysis of this acoustic streaming effect was given by Rayleigh (1884, 1945). Since then the literature on the subject has grown enormously (see, for instance, Nyborg 1969 or Zarembo 1969). Acoustic streaming moves along the gradient of the Reynolds stresses in a wave. In the case of plane-parallel inviscid flow the wave amplitude does not change and the Reynolds stresses are constant. Therefore, the gradient of the Reynolds stresses vanish, and in a dissipationless liquid there is no plane-parallel acoustic streaming. Hence, plane-parallel acoustic streaming can occur only in a dissipative medium where the wave amplitude decreases with distance and as a result the gradient of the Reynolds stresses is different from zero. This happens in an inertial frame of reference in the case of plane-parallel flow. However, we are now dealing with a rotating disc, that is, with a non-inertial frame of reference; moreover, the radial flow in the disc is divergent. The question now arises whether these two important differences change the above conclusion about the existence of acoustic streaming.

Strange as it may seem, the presence of dissipation is also required to have radial – divergent – acoustic streaming in a rotating disc. In fact, the amplitude of a resonant wavetrain propagating along the radius of the disc varies, even if there is no viscosity. Therefore in a non-rotating disc the radial gradient of the Reynolds stresses in a radially propagating wave could result in radial acoustic streaming. However, in a rotating disc, the Coriolis force plays the dominant role and this is directed at right angles to the flow velocity. Therefore, the radial force produces in a rotating disc only azimuthal drift and this cannot redistribute the density of the disc along the radius.

A redistribution of the density in the disc can be caused only by a radial flow, and this, in turn, can be produced by an azimuthal force. However, if there is no dissipation, the azimuthal gradient of the Reynolds stresses vanishes; this is a consequence of the azimuthal symmetry of an unperturbed disc. A large-scale azimuthal force, leading to radial-drift acoustic streaming, will appear only, if dissipation – viscosity – is taken into account (see Sect. 6).

In Sect. 7 we show that the radial mass flux increases with distance from the resonance; this agrees with the above-formulated conclusion which followed from the current structure of the Cassini Division. The reason for this dependence of the radial mass flux on the distance from the resonance is because it depends strongly on the wavevector. Since the wavevector of the density wave increases as it propagates from the resonance, the intensity of the radial drift also increases.

To conclude this appendix (Sect. 8) we check the applicability of the simple criterion for the formation of a gap – or wavetrain. A gap appears when the proposed mechanism for the redistributioon of the disc density is more effective than the diffusion flux. In the opposite case, there is no gap and we can observe a resonant wavetrain.

2 Statement of the Problem

Two kinds of perturbations are generated by a satellite in the resonance region: high-frequency and low-frequency ones. The first have frequencies which are multiples of the driving force and are a direct response of the medium to that force. Low-frequency perturbations of the parameters of the medium are a consequence of the interactions of high-frequency harmonics with one another and are – for moderate wave intensities – relatively weak. However, due to the slow change in these perturbations the effects caused by them may be "stored up" (the so-called secular terms) and as a result they may over a long time interval lead to a significant change in the initial state of the medium. In particular, a slow radial drift of this kind could lead to the formation of a gap in the region from which the matter is "swept away" and the formation of narrow ringlets in the region where the matter is "swept to".

When describing the change in the density of the disc over long – cosmological – times it is natural to be interested in the average of the density of the medium over a time interval Δ long as compared to the period T of the high-frequency oscillations but short as compared to the time t_{ev} for the evolution of the system,

$$T \ll \Delta \ll t_{ev}, \tag{A5.5}$$

rather than in its instantaneous average value.

The evolution of such an averaged density will satisfy an averaged equation of continuity:

$$\frac{\partial \overline{\varrho}}{\partial t} + \operatorname{div} \langle \varrho \boldsymbol{v} \rangle = 0. \tag{A5.6}$$

Here and henceforth values of a quantity averaged over a time interval Δ, satisfying condition (A5.5), are indicated by a bar or by pointed brackets:

$$\langle f \rangle = \overline{f} = \frac{1}{\Delta} \int_0^{\Delta} f \, dt. \tag{A5.7}$$

The instantaneous value of the density can be written in the form

$$\varrho = \overline{\varrho} + \varrho_1, \tag{A5.8}$$

where ϱ_1 is the high-frequency perturbation of the density caused by the satellite with an amplitude $\overline{\varrho_1}$ changing slowly with time and along the radius, a frequency ω, and an azimuthal wavenumber m:

$$\varrho_1(r, \varphi, z, t) \;=\; \overline{\varrho}_1(r, z, t) \mathrm{e}^{\mathrm{i}(m\varphi - \omega t)} + \overline{\varrho}_1^*(r, z, t) \mathrm{e}^{-\mathrm{i}(m\varphi - \omega t)}. \tag{A5.9}$$

The last term in this expression is due to the fact that the perturbed density, like any other measured physical quantity occurring in the hydrodynamical equations, must be real.

Averaging expression (A5.8) over a time interval Δ leads to an identity, provided we neglect small terms of order $(\omega \Delta)^{-1} \ll (2\pi)^{-1} \ll 1$.

Continuing along the lines developed in the preceding section we may note that in the present approach two time-scales occur at the same time, corresponding, respectively, to T and Δ. We observe the temporal and azimuthal changes of the high-frequency perturbations on the first time-scale. In this scale we are free to fix the instantaneous positions of the satellite relative to the ring, that is, the direction (azimuth) of its tidal force. The azimuthal dependence of the tidal action of the satellite is here obvious. It is impossible to notice the secular effects, caused by the action of the fast perturbations on the medium in the first, short time-scale – the medium appears here to be frozen, unchanged.

On the other hand, on the time-scale where as the minimum time interval we choose a period $\Delta \gg T$ we are unable to see any fixed position of the satellite relative to the disc – we can study only the averaged tidal interaction of the satellite with the disc, averaged over many revolutions of the satellite. It is clear that in that case we cannot speak about any azimuthal dependence – otherwise the following problem would have been solved: if there existed such a dependence one azimuthal position would somehow differ from another one and how could this occur when we average over many revolutions of the satellite relative to the disc?

All time-averages $\overline{f}$, taken over a period Δ, must therefore be independent of the angle φ with the same accuracy with which earlier we neglected, when averaging f, the dependence on the angle φ of the quantity ϱ, that is, with an accuracy of terms of order $(\omega \Delta)^{-1} \ll (2\pi)^{-1} \ll 1$.

Let us consider the factors which make a contribution to the averaged mass flow. In accordance with what we have just said we can write it as follows:

$$\langle \varrho \boldsymbol{v} \rangle \;=\; \overline{\varrho \boldsymbol{v}} + \langle \varrho_1 \boldsymbol{v}_1 \rangle. \tag{A5.10}$$

The last term in this expression is the average of the product of two high-frequency perturbations. Let us consider that quantity in some detail.

Using for the two components of high-frequency perturbations f_1 and g_1 – such as ϱ_1 and v_{i1} – expressions of the kind (A5.9) we get

$$\langle f_1 g_1 \rangle = \frac{1}{\Delta} \int_0^\Delta \left[\overline{f}_1 e^{i(m\varphi - \omega t)} + \overline{f}_1^* e^{-i(m\varphi - \omega t)} \right]$$

$$\times \left[\overline{g}_1 e^{i(m\varphi - \omega t)} + \overline{g}_1^* e^{-i(m\varphi - \omega t)} \right] dt$$

$$= \frac{1}{\Delta} \int_0^\Delta \left[\overline{f}_1 \overline{g}_1 e^{2i(m\varphi - \omega t)} + \overline{f}_1^* \overline{g}_1^* e^{-2i(m\varphi - \omega t)} \right] dt$$

$$+ \frac{1}{\Delta} \int_0^\Delta \left[\overline{f}_1 \overline{g}_1^* + \overline{f}_1^* \overline{g}_1 \right] dt$$

$$= \overline{f}_1 \overline{g}_1^* + \overline{f}_1^* \overline{g}_1 + \mathcal{O}\big((\omega \Delta)^{-1}\big). \tag{A5.11}$$

We find thus that, neglecting small terms of order $(\omega \Delta)^{-1}$, the average value of the product of two high-frequency perturbations is independent of the azimuthal angle φ:

$$\langle f_1 g_1 \rangle = \overline{f}_1 \overline{g}_1^* + \overline{f}_1^* \overline{g}_1. \tag{A5.12}$$

As a result of this the low-frequency perturbations caused by the interactions between high-frequency perturbations will be axially symmetric. We can thus in (A5.6) and later equations neglect derivatives with respect to φ of quantities which are time-averaged.

We can see from (A5.6) that the problem of the change with time of the average density of the medium reduces to evaluating the time-averaged momentum density of the medium – rather than its averaged velocity.

From a general point of view the calculation of this quantity requires a detailed kinetic description of the dynamics of the ring. However, experimental studies (Bridges et al. 1984, Hatzes et al. 1988) of collisions of icy spheres in vacuo at low temperatures have shown that the particles in the Saturnian rings collide inelastically. Moreover, the size a of the particles varies over a wide range. Observations have shown that the particle size distribution follows a power law, $n(a) \propto a^{-q}$ with $q = 2.8$ to 3.4 for $a < a_{max} \approx 5$ to 10 m and $q = 5$ to 6 for $a > a_{max}$. Both these facts make an accurate kinetic description of the ensemble of particles in the Saturnian rings too complicated to offer an adequate interpretation – from a physical point of view – of the processes involved. Indeed, all research available in the literature up to the present time uses different kinds of simplifications, sometimes of a kind which may drastically affect the results. Hence, one cannot rely on the results obtained in this way (Fridman and Gor'kavyi 1989a).

On the other hand, while the optical depth of the rings is mainly determined by the small particles, the main mass of the rings is made up of particles of meter sizes. Hence, for dynamical problems we can restrict ourselves to a kinetic equation for a distribution function of single-size particles. In doing this, the collision integral consists of two parts: one part determines the change in the distribution function due to elastic – gravitational – collisions and the other part the change caused by inelastic – contact – collisions (Fridman and Gor'kavyi 1989a).

The kinetic equation for the distribution function of the particles in Saturn's rings is similar to the kinetic equation for charged particles in an electromagnetic field. This analogy makes it possible to apply the methods of plasma physics to rotating gravitating media (Fridman and Polyachenko 1984); the epicyclic radius and frequency play here the role of the Larmor radius and frequency in plasma physics. Using this approach one can derive the transport equations by standard methods (Braginskii 1965, Fridman and Gor'kavyi 1989a) with the main parameter of the problem in the form $\kappa\tau_c$ where κ is the epicyclic frequency and τ_c the mean time between collisions.

The solution of the full linearised set of transport equations shows (Fridman and Gor'kavyi 1989a) the existence of several branches of oscillations, including ones which are unstable under certain conditions. For most ranges of the parameters these branches are independent and each can be described by a simpler set of equations. In the self-gravitating B ring near the resonance due to Mimas there exists a gravitational acoustic wave which was originally described by Lin and Shu (1964) for the case of a stellar disc. In a gas disc this wave can be described rather well using the hydrodynamic equations with a barotropic equation of state. Under those conditions the averaged momentum density can be obtained easily from the momentum transport equations. We shall derive these equations in the next section.

3 Derivation of the Non-linear Momentum Conservation Equations

Let us write down the non-linear equations of motion and of continuity (Fridman 1975, Fridman and Polyachenko 1984):

$$\frac{\partial v_r}{\partial t} + v_r \frac{\partial v_r}{\partial r} + \frac{v_\varphi}{r} \frac{\partial v_r}{\partial \varphi} + v_z \frac{\partial v_r}{\partial z} - \frac{v_\varphi^2}{r} = -\frac{\partial \chi}{\partial r}, \tag{A5.13}$$

$$\frac{\partial v_\varphi}{\partial t} + v_r \frac{\partial v_\varphi}{\partial r} + \frac{v_\varphi}{r} \frac{\partial v_\varphi}{\partial \varphi} + v_z \frac{\partial v_\varphi}{\partial z} + \frac{v_r v_\varphi}{r} = -\frac{1}{r} \frac{\partial \chi}{\partial \varphi}, \tag{A5.14}$$

$$\frac{\partial v_z}{\partial t} + v_r \frac{\partial v_z}{\partial r} + \frac{v_\varphi}{r} \frac{\partial v_z}{\partial \varphi} + v_z \frac{\partial v_z}{\partial z} = -\frac{\partial \chi}{\partial z}, \tag{A5.15}$$

$$\frac{\partial \varrho}{\partial t} + \frac{1}{r} \frac{\partial}{\partial r} (r \varrho v_r) + \frac{1}{r} \frac{\partial}{\partial \varphi} (\varrho v_\varphi) + \frac{\partial}{\partial z} (\varrho v_z) = 0, \tag{A5.16}$$

where

$$\nabla \chi \equiv \frac{1}{\varrho} \nabla P + \nabla \Phi. \tag{A5.17}$$

We add the r equation of motion (A5.13), multiplied by ϱ, to the equation of continuity (A5.16), multiplied by v_r:

$$\varrho \frac{\partial v_r}{\partial t} + \varrho v_r \frac{\partial v_r}{\partial r} + \frac{\varrho v_\varphi}{r} \frac{\partial v_r}{\partial \varphi} + \varrho v_z \frac{\partial v_r}{\partial z} - \frac{\varrho v_\varphi^2}{r}$$

$$+ v_r \frac{\partial \varrho}{\partial t} + \frac{v_r}{r} \frac{\partial}{\partial r} (r \varrho v_r) + \frac{v_r}{r} \frac{\partial}{\partial \varphi} (\varrho v_\varphi) + v_r \frac{\partial}{\partial z} (\varrho v_z) = -\varrho \frac{\partial \chi}{\partial r}.$$

As a result we find

$$\frac{\partial}{\partial t} (\varrho v_r) + \frac{1}{r} \frac{\partial}{\partial r} (r \varrho v_r^2) + \frac{1}{r} \frac{\partial}{\partial \varphi} (\varrho v_r v_\varphi) + \frac{\partial}{\partial z} (\varrho v_r v_z) - \frac{\varrho v_\varphi^2}{r}$$

$$= -\varrho \frac{\partial \chi}{\partial r}. \tag{A5.18}$$

We now add the φ equation of motion (A5.14), multiplied by ϱ, to the equation of continuity (A5.16), multiplied by v_φ:

$$\varrho \frac{\partial v_\varphi}{\partial t} + \varrho v_r \frac{\partial v_\varphi}{\partial r} + \frac{\varrho v_\varphi}{r} \frac{\partial v_\varphi}{\partial \varphi} + \varrho v_z \frac{\partial v_\varphi}{\partial z} + \frac{\varrho v_r v_\varphi}{r}$$

$$+ v_\varphi \frac{\partial \varrho}{\partial t} + \frac{v_\varphi}{r} \frac{\partial}{\partial r} (r \varrho v_r) + \frac{v_\varphi}{r} \frac{\partial}{\partial \varphi} (\varrho v_\varphi) + v_\varphi \frac{\partial}{\partial z} (\varrho v_z) = -\frac{\varrho}{r} \frac{\partial \chi}{\partial \varphi}.$$

As a result we find

$$\frac{\partial}{\partial t} (\varrho v_\varphi) + \frac{1}{r} \frac{\partial}{\partial r} (r \varrho v_r v_\varphi) + \frac{1}{r} \frac{\partial}{\partial \varphi} (\varrho v_\varphi^2) + \frac{\partial}{\partial z} (\varrho v_\varphi v_z) + \frac{\varrho v_r v_\varphi}{r}$$

$$= -\frac{\varrho}{r} \frac{\partial \chi}{\partial \varphi}. \tag{A5.19}$$

Finally, we add the z equation of motion (A5.15), multiplied by ϱ, to the equation of continuity (A5.16), multiplied by v_z:

$$\varrho \frac{\partial v_z}{\partial t} + \varrho v_r \frac{\partial v_z}{\partial r} + \frac{\varrho v_\varphi}{r} \frac{\partial v_z}{\partial \varphi} + \varrho v_z \frac{\partial v_z}{\partial z}$$

$$+ v_z \frac{\partial \varrho}{\partial t} + \frac{v_z}{r} \frac{\partial}{\partial r} (r \varrho v_r) + \frac{v_z}{r} \frac{\partial}{\partial \varphi} (\varrho v_\varphi) + v_z \frac{\partial}{\partial z} (\varrho v_z) = -\varrho \frac{\partial \chi}{\partial z}.$$

As a result we find

$$\frac{\partial}{\partial t} (\varrho v_z) + \frac{1}{r} \frac{\partial}{\partial r} (r \varrho v_r v_z) + \frac{1}{r} \frac{\partial}{\partial \varphi} (\varrho v_z v_\varphi) + \frac{\partial}{\partial z} (\varrho v_z^2) = -\varrho \frac{\partial \chi}{\partial z}. \tag{A5.20}$$

The set (A5.18) to (A5.20) comprises the well known hydrodynamical momentum transfer equations in a dissipationless medium; we shall now average these over time.

4 Time-Averaged Non-linear Momentum Conservation Equations

When we average (A5.18) to (A5.20) over the time we encounter two kinds of averages: averages of products of two functions – such as $\langle \varrho v_i \rangle$ – and averages of three functions – such as $\langle \varrho v_i v_k \rangle$. The first kind we considered already in Sect. 2. Averages of the second kind can, if we neglect small terms of order $(\omega \Delta)^{-1}$, be written as follows:

$$
\begin{aligned}
\langle \varrho v_i v_k \rangle &= \langle (\overline{\varrho} + \varrho_1)(\overline{v_i} + v_{i1})(\overline{v_k} + v_{k1}) \rangle \\
&= \overline{\varrho}\,\overline{v_i}\,\overline{v_k} + \overline{\varrho}\,\langle v_{i1} v_{k1} \rangle + \overline{v_i}\,\langle \varrho_1 v_{k1} \rangle + \overline{v_k}\,\langle \varrho_1 v_{i1} \rangle .
\end{aligned}
\tag{A5.21}
$$

In the particular case when $i = k$ we have from (A5.21)

$$
\langle \varrho v_i^2 \rangle = \overline{\varrho}\,\overline{v_i}^2 + \overline{\varrho}\,\langle v_{i1}^2 \rangle + 2\overline{v_i}\,\langle \varrho_1 v_{i1} \rangle .
\tag{A5.22}
$$

It is clear from these expressions that the averages of products of three functions can be expressed in terms of averages of products of two functions, that is, that they are axially symmetric – taking into account that the averaged values $\overline{\varrho}$ and $\overline{v_i}$ are axially symmetric – see the explanation just before (A5.10).

It is convenient for what follows to distinguish two components of the total mean mass flux – the averaged momentum density of the medium. Firstly, we have the mass flow caused by the rotation; this is necessary to maintain the equilibrium of the disc in the gravitational and pressure fields. The equilibrium rotational velocity v_{eq} can be defined by the relation

$$
\frac{v_{\mathrm{eq}}^2}{r} = \frac{\partial \overline{\chi}}{\partial r} .
\tag{A5.23}
$$

The second component of the mass flux is connected directly with the action of the time-averaged stresses arising in a density wave generated by the satellite. The averaged momentum density connected with this component has the form

$$
\Pi_r = \overline{\varrho}\,\overline{v}_r + \langle \varrho_1 v_{r1} \rangle ,
\tag{A5.24}
$$

$$
\Pi_\varphi = \overline{\varrho}\,(\overline{v}_\varphi - v_{\mathrm{eq}}) + \langle \varrho_1 v_{\varphi 1} \rangle ,
\tag{A5.25}
$$

$$
\Pi_z = \overline{\varrho}\,\overline{v}_z + \langle \varrho_1 v_{z1} \rangle .
\tag{A5.26}
$$

Because of the axial symmetry $(\partial < f > /\partial \varphi = 0)$ the first component, connected with v_{eq}, does not lead to a density redistribution in the disc and we shall therefore in what follows only be interested in the fluxes Π_i.

Using the axial symmetry of the quantities defined by (A5.24) to (A5.26), we find for the time-averaged equation of continuity

$$
\frac{\partial \overline{\varrho}}{\partial t} + \frac{1}{r}\frac{\partial}{\partial r}(r\Pi_r) + \frac{\partial \Pi_z}{\partial z} = 0 .
\tag{A5.27}
$$

We shall be interested in processes which decrease or increase the surface density in the disc. To obtain an equation describing such changes we integrate (A5.27) over the thickness of the disc, $2h$. At the boundaries we must satisfy the conditions

$$\Pi_z(-h) = \Pi_z(h) = 0. \tag{A5.28}$$

These conditions are a consequence of the fact that, in spite of the zero mass density at the disc boundaries, the energy density – for instance, $\frac{1}{2}\varrho v_z^2 (= \epsilon_z)$ – must be finite, that is, $\Pi_z^2 = \varrho^2 v_z^2 = 2\varrho\epsilon_z \to 0$. Taking the boundary conditions into account, after integrating the averaged equation of continuity we find

$$\frac{\partial \sigma}{\partial t} + \frac{1}{r}\frac{\partial}{\partial r}(r\pi_r) = 0, \tag{A5.29}$$

where

$$\pi_r = \int_{-h}^{h} \Pi_r \, dz. \tag{A5.30}$$

Thus in order to describe the slow changes in the surface density of the disc we must evaluate the average of the radial momentum component Π_r.

Time-averaging the radial momentum equation (A5.18) we find

$$\frac{\partial \Pi_r}{\partial t} - 2\Omega_{\text{eq}}\Pi_\varphi = -\frac{1}{r}\frac{\partial}{\partial r}\left(r\overline{\varrho}\langle v_{r1}^2\rangle\right) + \frac{\overline{\varrho}\langle v_{\varphi1}^2\rangle}{r}$$
$$-\frac{\partial}{\partial z}\left(\overline{\varrho}\langle v_{r1}v_{z1}\rangle\right) - \left\langle \varrho_1 \frac{\partial \chi_1}{\partial r}\right\rangle, \tag{A5.31}$$

where $\Omega_{\text{eq}} \equiv v_{\text{eq}}/R$.

We proceed similarly with the azimuthal momentum equation (A5.19):

$$\frac{\partial \Pi_\varphi}{\partial t} + \frac{\partial}{\partial t}(\overline{\varrho} v_{\text{eq}}) + \frac{1}{r}\frac{\partial}{\partial r}\left(r\Pi_r\overline{v_\varphi} + r\overline{\varrho}\langle v_{r1}v_{\varphi1}\rangle\right) + \frac{\partial}{\partial z}\left(\overline{v_\varphi}\Pi_z + \overline{\varrho}\langle v_{z1}v_{\varphi1}\rangle\right)$$
$$+ \frac{\Pi_r\overline{v_\varphi} + \overline{\varrho}\langle v_{r1}v_{\varphi1}\rangle}{r} = -\left\langle\frac{\varrho_1}{r}\frac{\partial \chi_1}{\partial \varphi}\right\rangle. \tag{A5.32}$$

We note that the quantities Π_i are of second order in the amplitudes of the high-frequency perturbations. Remaining within that order of accuracy we can replace the average azimuthal velocity $\overline{v}_\varphi$, which occurs in the equations we have just derived, by the quantity v_{eq} which differs from the former velocity by an amount of second order in the amplitude. Moreover, we note that $\Pi_z\partial v_{\text{eq}}/\partial z = 0$ (see App. II). Using the equation of continuity (A5.27) and the definition of the epicyclic frequency, $\kappa_{\text{eq}}^2/2\Omega_{\text{eq}} \equiv \partial v_{\text{eq}}/\partial r + v_{\text{eq}}/r$, we finally have

$$\frac{\partial \Pi_\varphi}{\partial t} + \frac{\kappa_{eq}^2}{2\Omega_{eq}} \Pi_r$$

$$= -\left\langle \frac{\varrho_1}{r} \frac{\partial \chi_1}{\partial \varphi} \right\rangle - \frac{1}{r^2} \frac{\partial}{\partial r} \left(r^2 \overline{\varrho} \left\langle v_{r1} v_{\varphi 1} \right\rangle \right) - \frac{\partial}{\partial z} \left(r \overline{\varrho} \left\langle v_{z1} v_{\varphi 1} \right\rangle \right). \quad \text{(A5.33)}$$

We finally write down the result of time-averaging (A5.20)

$$\frac{\partial \Pi_z}{\partial t} = -\overline{\varrho} \frac{\partial \overline{\chi}}{\partial z} - \frac{1}{r} \frac{\partial}{\partial r} \left(r \overline{\varrho} \left\langle v_{r1} v_{z1} \right\rangle \right) - \frac{\partial}{\partial z} \left(\overline{\varrho} \left\langle v_{z1}^2 \right\rangle \right) - \left\langle \varrho_1 \frac{\partial \chi_1}{\partial z} \right\rangle. \quad \text{(A5.34)}$$

5 Absence of Averaged Radial Mass Flux in a Dissipationless Disc. Large-Scale Convection

We have already noted above that the averages occurring in Eqs. (A5.31) to (A5.34) vary little over times of the order of the period of the oscillations, that is, the time-derivatives are small, and we can study the processes of the mass redistribution in the framework of the geostrophic approximation (Pedlosky 1982)

$$\frac{1}{\Omega_{eq}} \frac{\partial}{\partial t} \ll 1. \quad \text{(A5.35)}$$

This means that we can neglect the first terms in (A5.31) and (A5.33) as compared to the second ones and we can write the radial mass flow in the form

$$\Pi_r = -\frac{2\Omega_{eq}}{\kappa_{eq}^2} \left[\frac{1}{r^2} \frac{\partial}{\partial r} \left(r^2 \overline{\varrho} \left\langle v_{r1} v_{\varphi 1} \right\rangle \right) \right.$$

$$\left. + \left\langle \frac{\varrho_1}{r} \frac{\partial \chi_1}{\partial \varphi} \right\rangle + \frac{\partial}{\partial z} \left(\overline{\varrho} \left\langle v_{z1} v_{\varphi 1} \right\rangle \right) \right]. \quad \text{(A5.36)}$$

To find Π_r we need the expressions for v_{r1}, $v_{\varphi 1}$, and v_{z1}. Linearising the equations of motion (A5.13) to (A5.15) and taking the t- and φ- dependence of the perturbed functions in the form (A5.9) we find (Polyachenko and Fridman 1971)

$$v_{r1} = -\mathrm{i} \frac{\hat{\omega} \chi_1'}{D} + 2mi \frac{\Omega_{eq} \chi_1}{rD}, \quad \text{(A5.37)}$$

$$v_{\varphi 1} = -\frac{\chi_1' \kappa_{eq}^2}{2\Omega_{eq} D} + \frac{m\hat{\omega} \chi_1}{rD}, \quad \text{(A5.38)}$$

$$v_{z1} = -\frac{\mathrm{i}}{\hat{\omega}} \frac{\partial \chi_1}{\partial z}, \quad \text{(A5.39)}$$

where the prime here and henceforth indicates differentiation with respect to r, $\hat{\omega} \equiv \omega - m\Omega_{eq}$, and

$$D \equiv \hat{\omega}^2 - \kappa_{eq}^2. \tag{A5.40}$$

We can obtain an expression for ϱ_1 from the linearised equation of continuity (see also Fridman 1975, Fridman and Polyachenko 1984)

$$\varrho_1 = -\frac{i}{\hat{\omega}}\frac{\partial}{\partial r}(\bar{\varrho}v_{r1}) - \frac{i}{\hat{\omega}}\frac{\bar{\varrho}v_{r1}}{r} + \frac{m}{r\hat{\omega}}\bar{\varrho}v_{\varphi 1} - \frac{i}{\hat{\omega}}\frac{\partial}{\partial z}(\bar{\varrho}v_{z1}). \tag{A5.41}$$

Substitution of expressions (A5.37) to (A5.39) reduces it to the form

$$\varrho_1 = -\frac{1}{r}\left[\frac{r\bar{\varrho}}{D}\chi_1'\right]' + \left[\frac{2m}{\hat{\omega}r}\left(\frac{\Omega_{eq}\bar{\varrho}}{D}\right)' + \frac{m^2\bar{\varrho}}{Dr^2}\right]\chi_1 - \frac{1}{\hat{\omega}}\frac{\partial}{\partial z}\left[\frac{\bar{\varrho}}{\hat{\omega}}\frac{\partial\chi_1}{\partial z}\right]. \tag{A5.42}$$

Using (A5.37) to (A5.39) and (A5.42) we can write the time-averages occurring in (A5.36) for the radial drift in the following form:

$$\langle v_{r1}v_{\varphi 1}\rangle = \bar{v}_{r1}\bar{v}_{\varphi 1}^* + \bar{v}_{r1}^*\bar{v}_{\varphi 1} = \frac{im}{rD}\left(\bar{\chi}_1'^*\bar{\chi}_1 - \bar{\chi}_1'\bar{\chi}_1^*\right), \tag{A5.43}$$

$$\left\langle \varrho_1\frac{\partial\chi_1}{\partial\varphi}\right\rangle = -im\left(\bar{\varrho}_1\bar{\chi}_1^* - \bar{\varrho}_1^*\bar{\chi}_1\right)$$

$$= \frac{im}{r}\left[\left(\frac{r\bar{\varrho}}{D}\bar{\chi}_1'\right)'\bar{\chi}_1^* - \left(\frac{r\bar{\varrho}}{D}\bar{\chi}_1'^*\right)'\bar{\chi}_1\right]$$

$$+ \frac{im}{\hat{\omega}}\left[\frac{\partial}{\partial z}\left(\frac{\bar{\varrho}}{\hat{\omega}}\frac{\partial\bar{\chi}_1}{\partial z}\right)\bar{\chi}_1^* - \frac{\partial}{\partial z}\left(\frac{\bar{\varrho}}{\hat{\omega}}\frac{\partial\bar{\chi}_1^*}{\partial z}\right)\bar{\chi}_1\right]$$

$$= \frac{im}{r}\left[\frac{r\bar{\varrho}}{D}\left(\bar{\chi}_1'\bar{\chi}_1^* - \bar{\chi}_1'^*\bar{\chi}_1\right)\right]'$$

$$+ \frac{im}{\hat{\omega}}\frac{\partial}{\partial z}\left[\frac{\bar{\varrho}}{\hat{\omega}}\left(\frac{\partial\bar{\chi}_1}{\partial z}\bar{\chi}_1^* - \frac{\partial\bar{\chi}_1^*}{\partial z}\bar{\chi}_1\right)\right]. \tag{Λ5.44}$$

The last expression is obtained by moving $\bar{\chi}$ and $\bar{\chi}^*$ in the previous expression under the derivative sign.

Substituting (A5.43) and (A5.44) into (A5.36) we finally have

$$\Pi_r = -\frac{2\Omega_{eq}}{\kappa^2}\left\{\frac{im}{r^2}\left[\frac{r\bar{\varrho}}{D}\left(\bar{\chi}_1'^*\bar{\chi}_1 - \bar{\chi}_1'\bar{\chi}_1^*\right)\right] + \frac{im}{r}\left[\frac{r\bar{\varrho}}{D}\left(\bar{\chi}_1'\bar{\chi}_1^* - \bar{\chi}_1'^*\bar{\chi}_1\right)\right]\right.$$

$$\left. + \frac{\partial}{\partial z}\left[\frac{im\bar{\varrho}}{r\hat{\omega}^2}\left(\frac{\partial\bar{\chi}_1}{\partial z}\bar{\chi}_1^* - \frac{\partial\bar{\chi}_1^*}{\partial z}\bar{\chi}_1\right) + \left(\bar{\varrho}\langle v_{z1}v_{\varphi 1}\rangle\right)\right]\right\}. \tag{A5.45}$$

One can see that the first two terms cancel one another while the last term is a derivative with respect to z which vanishes after integration over the thickness of the disc.

We finally get from (A5.45) and (A5.30)

$$\pi_r = \int_{-h}^{h}\Pi_r\,dz = 0. \tag{A5.46}$$

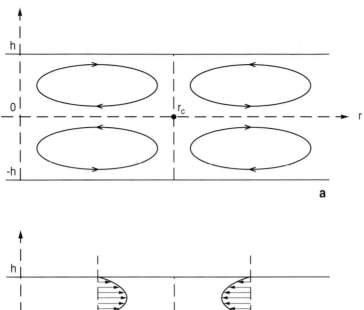

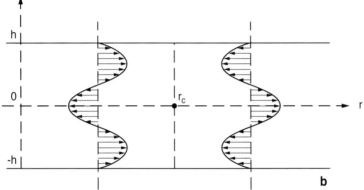

Fig. A5.2. Large-scale convection caused by a non-linear density wave. (a) The flow has the form of four toroidal vortices, separated by the vertical cylindrical surface $r = r_c$ (r_c is the corotation radius); (b) field of the flow lines in the (r, z) plane; one can see that in the vicinity of $z = 0$ the flow is directed away from the corotation circle. For details we refer to the papers by Fridman and Khoruzhii (1998, 1999).

Therefore, in a dissipationless disc the non-linear effects due to a density wave do not lead to any large-scale mass transfer when averaged over the thickness of the disc. Nevertheless, the wave generates some kind of large-scale convective flow – the last term in (A5.45). Fridman and Khoruzhii (1998) have shown recently that the non-linear convection has the form of four large-scale toroidal vortices, separated by a vertical cylindrical surface at the corotation radius and the central plane, $z = 0$, of the disc (Fig. A5.2). This convection can play a very important role in the total dynamics and the evolution of a disc.

6 Radial Mass Transfer in a Viscous Disc

The answer found in the preceding section could have been predicted beforehand by bearing in mind the analogy between the drift studied here and acoustic streaming known from non-linear acoustics (Nyborg 1969, Zarembo 1969). In both cases the large-scale components of the Reynolds stresses in the wave – terms of the form $\overline{\varrho v_{i1} v_{k1}}$ – are the cause of the excitation of flows. There is, however, a difference in that the Coriolis force plays a decisive role in the disc. The result is that whereas in the case of acoustic flows the motion of the medium is along the acting force, in the disc it will be at right angles to that direction.

The absence of flow in a dissipationless medium is a well known result in the theory of acoustic flows (Lighthill 1978). The reason for this is simple. The force induced by the non-linear self-action of the wave is a potential one which can be written as a gradient. In a dissipationless medium the amplitude of the acoustic wave is constant, and the gradient vanishes. If there is dissipation such a gradient appears. It is proportional to the damping coefficient and directed along the wavevector. Hence, the velocity of the excited flow is proportional to the damping coefficient of the wave; in particular, if the wave propagates in a viscous medium, it is proportional to the viscosity coefficient.

Let us consider what the situation in the case, considered by us, of a density wave excited by a satellite at resonance will lead to. In that case the wave amplitude changes, even when there is no dissipation as the wave parameters depend on the distance from the resonance. The change in the amplitude of the wave can be evaluated from the condition that the angular momentum flux be conserved (Toomre 1969). However, this change occurs exclusively along the radius, and as a result only the radial force is non-zero. This radial force produces solely an azimuthal drift. Because of the axial symmetry of the initial disc this drift cannot redistribute the density.

Apart from the purely radial force, described a moment ago, there must, if there is dissipation present, appear a force proportional to the damping coefficient and directed along the wavevector. This force has an azimuthal component proportional to m/k and must cause radial drift in the disc. If the velocity of this drift turns out to be non-uniform along the radius it will cause a redistribution of mass in the disc and, in particular, it may produce a region where the density is lowered – gaps – and regions where the density is enhanced – ringlets.

When evaluating the velocity of the radial drift it is thus necessary to determine the corrections to the characteristics of the density wave caused by the dissipation. Here we shall restrict our discussion to the the case of viscous dissipation. Before doing this we shall discuss a number of general conditions connected with the introduction of viscosity when studying astrophysical and, in particular, circumplanetary discs.

We have shown in App. II that in the general case the description of high-frequency perturbations in the disc is possible only on the basis of the three-dimensional equations. This is even more strongly the case when describing viscous effects in such perturbations. Indeed, the viscous force caused by the z-dependence of the velocity along the rotation axis $- F_z \propto v_z/h^2 \propto k v_r/h -$ is in general much larger – by a factor $(kh)^{-1}$ – than the viscous force caused by the radial dependence of the velocities in the plane of the disc $- F_r \propto k^2 v_r$. Hence the dissipative power connected with the vertical viscous forces $(v_z F_z)$ is of the same order of magnitude as the dissipative power connected with the viscous forces in the plane of the disc $(v_r F_r)$. We note here that even those authors (for instance, Shu, Dones et al. 1985) who have tried to give the most detailed analysis of the non-linear dynamics have mistakenly ignored z-motions in the disc. We shall see in what follows that the radial mass flux turns out to be proportional to the dissipative power (see (A5.56)). Both viscous forces must thus be taken into consideration.

Moreover, we have shown in §4 that the average radial drift is connected with the non-linear characteristics of the high-frequency perturbations, averaged over the thickness of the disc. Hence, in that case dissipative corrections may be important, even if, as a result of averaging over the thickness of the disc, they vanish in the linear approximation.

Following our discussion we start from the linearised three-dimensional equations of motion, including viscosity:

$$-i\hat{\omega} v_{r1} - 2\Omega_{\mathrm{eq}} v_{\varphi 1} = -\frac{\partial \chi_1}{\partial r} + \alpha_r, \tag{A5.47}$$

$$-i\hat{\omega} v_{\varphi 1} + \frac{\kappa_{\mathrm{eq}}^2}{2\Omega_{\mathrm{eq}}} v_{r1} = -\frac{1}{r}\frac{\partial \chi_1}{\partial \varphi} + \alpha_\varphi, \tag{A5.48}$$

$$-i\hat{\omega} v_{z1} = -\frac{\partial \chi_1}{\partial z} + \alpha_z. \tag{A5.49}$$

The α_i are here the components of the viscous forces, the actual form of which we shall give later. We shall first try to find a solution which is independent of the particular form of the viscous stress tensor.

Proceeding as in §5 we obtain expressions for the velocity and density perturbations in the wave:

$$v_{r1} = -i\frac{\hat{\omega}(\chi_1' - \alpha_r)}{D} + \frac{2\Omega_{\mathrm{eq}}}{D}\left(\frac{im}{r}\chi_1 - \alpha_\varphi\right), \tag{A5.50}$$

$$v_{\varphi 1} = -\frac{(2\Omega_{\mathrm{eq}} + r\Omega_{\mathrm{eq}}')}{D}(\chi_1' - \alpha_r) - \frac{i\hat{\omega}}{D}\left(\frac{im}{r}\chi_1 - \alpha_\varphi\right), \tag{A5.51}$$

$$v_{z1} = -\frac{i}{\hat{\omega}}\left(\frac{\partial \chi_1}{\partial z} - \alpha_z\right), \tag{A5.52}$$

$$\varrho_1 = -\frac{1}{r}\left(\frac{r\overline{\varrho}}{D}\left(\chi_1' - \alpha_r\right)\right)' + \frac{2m\Omega_{eq}\overline{\varrho}}{r\hat{\omega}D}\alpha_r$$

$$+ \frac{i}{r\hat{\omega}}\left(\frac{2r\Omega_{eq}\overline{\varrho}}{D}\alpha_\varphi\right)' + \frac{im\overline{\varrho}}{rD}\alpha_\varphi - \frac{i}{\hat{\omega}}\frac{\partial}{\partial z}\left(\overline{\varrho}v_{z1}\right) + A\chi_1, \qquad (A5.53)$$

where A is a coefficient the actual value of which is unimportant for what follows.

Using expressions (A5.50) to (A5.53) we find the time-averages occurring in expression (A5.36) for the radial flux:

$$\langle v_{r1}v_{\varphi1}\rangle = \frac{1}{D}\left[\left(\chi_1'^* - \alpha_r^*\right)\left(\frac{im}{r}\overline{\chi}_1 - \overline{\alpha}_\varphi\right) + \text{c.c.}\right], \qquad (A5.54)$$

$$\left\langle \varrho_1\frac{\partial\chi_1}{\partial\varphi}\right\rangle = \frac{1}{r}\left[\frac{r^2\overline{\varrho}}{D}\left(\overline{\chi}_1' - \overline{\alpha}_r\right)\left(\frac{im}{r}\overline{\chi}_1^* - \overline{\alpha}_\varphi^*\right)\right]' + \frac{i}{\hat{\omega}}\left(r^2\overline{\varrho}\,\overline{\alpha}_\varphi\overline{v}_{r1}^*\right)'$$

$$+ \frac{m\overline{\varrho}}{\hat{\omega}}\left(\overline{\alpha}_r\overline{v}_{r1}^* + \overline{\alpha}_\varphi\overline{v}_{\varphi1}^* + \overline{\alpha}_z\overline{v}_{z1}^*\right) + \frac{\partial G}{\partial z} + \text{c.c.} \qquad (A5.55)$$

The explicit form of the function G in the last term of (A5.55) is unimportant as this term vanishes when we integrate the radial mass flow over the thickness of the disc.

Substituting the expressions we have obtained into (A5.36) we get, neglecting terms which are derivatives with respect to z:

$$\Pi_r = -\frac{2m\Omega_{eq}\overline{\varrho}}{r\hat{\omega}\kappa_{eq}^2}\left[\overline{\alpha}_r\overline{v}_{r1}^* + \overline{\alpha}_\varphi\overline{v}_{\varphi1}^* + \overline{\alpha}_z\overline{v}_{z1}^* + \frac{i\left(r^2\overline{\varrho}\,\overline{\alpha}_\varphi\overline{v}_{r1}^*\right)'}{rm\overline{\varrho}} + \text{c.c.}\right]. \qquad (A5.56)$$

To proceed further, we need the explicit form of the components of the viscous forces.

The form of the viscous stress tensor deserves special attention. If the main mass of the disc is made up of large particles which are nearly collisionless – $\Omega\tau \gg 1$, where Ω is the rotational frequency and τ the time between collisions – it is natural to expect a suppression of the viscosity because in the plane of the disc the mean free path of the particles is restricted to the epicycle radius. It is just this result which is generally adopted in the physics of planetary rings, starting from the pioneering work on this problem by Goldreich and Tremaine (1978a).

However, one can see easily that the restrictions on the mean free path can only fully be applied to the shear viscosity, that is, the viscosity between two layers of the medium which move along one another with different velocities. In that case the transfer of momentum of one layer to the other – which is the effect which constitutes the viscosity in this case – requires the particles of one layer to penetrate into the other one and the motion of the particles along epicycles inhibits that.

On the other hand, the action of viscosity manifests itself also in processes leading to a change in volume, that is, compression or expansion of the medium. In that case the non-uniformity of the velocity occurs along the velocity itself and because of this there occur collisions between particles leading to an exchange of momentum between them. Of course, a restriction on the mean free path must affect those processes to a much lesser extent and may even not affect them at all.

The simple qualitative considerations given here are confirmed in the theory of magnetised plasmas (see, e.g., Braginskiĭ 1958, 1965). The transport theory in such a plasma has been worked out in detail and it is therefore useful to use its results as the effect of the epicyclic motion of the particles in the disc is completely analogous to the effect of the Larmor rotation of the plasma particles. The condition for rare collisions between particles, $\Omega\tau \gg 1$, corresponds for a plasma to the condition of strong magnetisation, $\omega_L\tau \gg 1$ with ω_L the Larmor frequency.

It was shown that for a magnetised plasma not only does the viscous stress tensor show a complicated non-diagonal structure, but the viscosity coefficients themselves also form a tensor (Braginskiĭ 1965). In the simplest case one introduces five viscosity coefficients, η_0, η_1, η_2, η_3, and η_4 from which only three are independent. The viscosity coefficient introduced by Goldreich and Tremaine (1978a) corresponds to η_1 and describes the shear viscosity, that is, it operates when the velocity changes across the flow. The "volume" viscosity appearing when the velocity changes along the flow is even in a strongly magnetised plasma described by the "usual" viscosity coefficient η_0 which has the same value as for an unmagnetised plasma. Moreover, when the "volume" viscosity effect is not present the viscosity η_3 may play an important role, as $\eta_3 \simeq \Omega\tau\eta_1$ (Braginskiĭ 1965); it describes the off-diagonal terms of the viscous stress tensor.

In what follows we shall mainly be interested in the rare collision case, $\Omega\tau \gg 1$, corresponding to the case of a strongly magnetised plasma, and therefore restrict ourselves to considering the contribution due to the largest viscosity coefficient, η_0 (Braginskii 1965). If the optical depth of the particles which make up the main mass of the disc becomes of the order of, or larger than, unity one must take into account also other terms but in that case all viscosity coefficients are of the same order of magnitude and one may expect that the effect of the viscosity for such a case will differ only quantitatively, but not qualitatively.

In the approximation used the expressions for the viscous stress tensor from Braginskiĭ's paper (1965) lead to the following form of the components of the viscous force:

$$\alpha_r = \frac{1}{3\varrho}\frac{\partial}{\partial r}\left[\eta_0\left(\frac{\partial v_{r1}}{\partial r} - 2\frac{\partial v_{z1}}{\partial z}\right)\right], \tag{A5.57}$$

$$\alpha_\varphi = \frac{1}{3r\varrho}\frac{\partial}{\partial\varphi}\left[\eta_0\left(\frac{\partial v_{r1}}{\partial r} - 2\frac{\partial v_{z1}}{\partial z}\right)\right], \tag{A5.58}$$

$$\alpha_z = -\frac{2}{3\varrho}\frac{\partial}{\partial z}\left[\eta_0\left(\frac{\partial v_{r1}}{\partial r} - 2\frac{\partial v_{z1}}{\partial z}\right)\right]. \tag{A5.59}$$

We shall consider the case of low-order resonances, that is, small m. In that case the azimuthal dependence of the wave characteristics is weak, and we may neglect the azimuthal component of the viscous force. Substituting (A5.57) to (A5.59) into the expression (A5.56) for the averaged radial mass flow we get, after some simple transformations,

$$\Pi_r = \frac{2m\Omega_{\rm eq}}{3r\hat{\omega}\kappa_{\rm eq}^2}\left\{\eta_0\left|\frac{\partial \bar{v}_{r1}}{\partial r} - 2\frac{\partial \bar{v}_{z1}}{\partial z}\right|^2 - \frac{\partial}{\partial r}\left[\eta_0\bar{v}_{r1}^*\left(\frac{\partial \bar{v}_{r1}}{\partial r} - 2\frac{\partial \bar{v}_{z1}}{\partial z}\right) + {\rm c.c.}\right]\right.$$
$$\left. + 2\frac{\partial}{\partial z}\left[\eta_0\bar{v}_{z1}^*\left(\frac{\partial \bar{v}_{r1}}{\partial r} - 2\frac{\partial \bar{v}_{z1}}{\partial z}\right) + {\rm c.c.}\right]\right\}. \tag{A5.60}$$

One sees easily that the first term in that expression is the dominant one. The second term is small as the external derivative with respect to the radius is taken of a function which changes little as it is an averaged function, while the third term vanishes when we integrate over the thickness of the disc. We thus are led to an expression for the total radial mass flow in the form

$$\pi_r = \int_{-h}^{h}\Pi_r\,dz = \int_{-h}^{h}\frac{2m\Omega_{\rm eq}\eta_0}{3r\hat{\omega}\kappa_{\rm eq}^2}\left|\frac{\partial \bar{v}_{r1}}{\partial r} - 2\frac{\partial \bar{v}_{z1}}{\partial z}\right|^2\,dz. \tag{A5.61}$$

This expression confirms the statement made at the beginning of the present section that in the general case one must take into account the motion in the z-direction for a correct description of dissipation. The structure of (A5.61) clearly explains why it is necessary to take into account the motion along the axis. In the case considered the viscous dissipation depends on the extent to which the velocity changes along its own direction and therefore notwithstanding the fact that the velocity along the axis is much smaller than the velocity in the plane of the disc ($v_z \simeq v_r kh \ll v_r$) its contribution to the viscous effects is comparable to the contribution from the radial velocities ($\partial v_z/\partial z \simeq \partial v_r/\partial r$).

The mass flow is proportional to the viscosity, and from this point of view we may call this effect "dissipative drift". Expression (A5.61) makes it possible at once to determine the direction of the excited drift. Since the azimuthal phase velocity of the density wave, $\Omega_{\rm p} \equiv \omega/m$, is equal to the rotational velocity at the corotation radius, which is the same as the satellite radius ($\Omega_{\rm p} = \Omega_{\rm s}$), we have for a wave excited at the inner Lindblad resonance $\hat{\omega} = m(\Omega_{\rm s} - \Omega_{\rm eq}) < 0$. This means that the radial drift in this case is directed away from the corotation circle towards the planet. As expression (A5.61) contains $\hat{\omega} \equiv \omega - m\Omega_{\rm eq}$, the total radial mass flow changes its sign on the corotation circle. Hence, outside the corotation circle the radial drift will be directed away from the planet.

7 Evolution of the Surface Density of a Disc

In order to study the action of the dissipative drift which we found in the preceding section on the profile of the average surface density in the disc we need to know how the drift velocity depends on the radius. A consideration of all possible situations goes beyond the framework of the present appendix and we shall therefore restrict ourselves to the relatively simple case of an isothermal disc when we can neglect the motion along the rotational axis (see App. I). We can then use the functional dependence of the wave parameters on the distance from the resonance which was found by GT for the two-dimensional model of the disc.

In the region where the wave amplitude is increasing (region I in Fig. A5.1c; refractive zone) we find

$$\pi_r = -\frac{16k_r^2 h\eta_0 |\bar{v}_{r1}|^2}{3\Omega_{\mathrm{eq}}^2 R} = -\frac{16k_r^2 h\eta_0 |\bar{\Psi}_1|^2}{27\Omega_{\mathrm{eq}}^4 Rx^2}. \tag{A5.62}$$

where $x = \Delta r/R$ is the dimensionless distance from the resonance, R is the radius of the resonant orbit, $\bar{\Psi}_1$ is the amplitude of the gravitational potential of the excited density wave, k_r its wavenumber, Ω_{eq} is the rotational velocity at the resonance, and we have used (A5.37) for v_{r1}. Because of refraction the wavevector changes, as follows,

$$k_r \simeq \frac{3\Omega_{\mathrm{eq}}^2}{2\pi G\sigma} x. \tag{A5.63}$$

According to GT one can express the gravitational potential of the wave in terms of the Fourier components of the forcing action of the satellite:

$$|\bar{\Psi}_1|^2 \simeq \frac{4\pi^2 G\sigma_0}{3R\Omega_{\mathrm{eq}}^2} |\bar{\Psi}_{22}^{\mathrm{s}}|^2, \tag{A5.64}$$

where

$$|\bar{\Psi}_{22}^{\mathrm{s}}| \simeq \frac{2.4GM_{\mathrm{s}}}{R_{\mathrm{s}}}. \tag{A5.65}$$

Here M_{s} and R_{s} are, respectively, the mass and the orbital radius of the satellite.

Finally, using (A5.64) and (A5.65), we find from (A5.62) the following expression for the radial mass flow:

$$\pi_r \simeq -\frac{\nu\Omega_{\mathrm{eq}}^6 R^2}{2G^3\sigma^2} \left[\frac{M_{\mathrm{s}}}{M_{\mathrm{p}}}\right]^2 x^2, \tag{A5.66}$$

where M_{p} is the mass of the planet.

Substituting this expression into the averaged equation of continuity (A5.29) we find that the rate of change of the surface density due to this drift will be of the form

$$\frac{\partial \sigma}{\partial t} = -\frac{\partial \pi_r}{\partial r} \simeq \frac{\nu \Omega_{\text{eq}}^6 R^2}{G^3 \sigma^2} \left[\frac{M_{\text{s}}}{M_{\text{p}}}\right]^2 x. \tag{A5.67}$$

Thus, the density increases in this region.

Far from resonance (region II in Fig. A5.1c) viscosity effects dominate and the wave amplitude decreases exponentially. Let us assume that in that region

$$\overline{\Psi}_1 \simeq \overline{\Psi}_{\text{max}} e^{-x/L}, \tag{A5.68}$$

where L is a characteristic scale for the wave damping and $\overline{\Psi}_{\text{max}}$ is the maximum value of the gravitational potential.

Using (A5.68) we find for the radial mass flow in this region the expression

$$\pi_r = -\frac{16 k_r^2 h \eta_0 \left|\overline{\Psi}_{\text{max}}\right|^2}{27 \Omega_{\text{eq}}^4 R x^2} e^{-2x/L}. \tag{A5.69}$$

As refractive effects are weaker in this region than the dissipative ones, we must assume that

$$L < x, \tag{A5.70}$$

and the radial dependence of the mass flow is mainly governed by the exponential. Under those conditions the rate of change of the surface density is given by the equation

$$\frac{\partial \sigma}{\partial t} = -\frac{\partial \pi_r}{\partial r} \simeq -\frac{|\pi_r|}{L}. \tag{A5.71}$$

Thus, in this region the density decreases.

8 Conditions for the Formation of Different Types of Resonant Structures: Gaps or Wavetrains?

We can easily see that the solution we have found for the whole of the resonant region has a number of unique features. Firstly, it simultaneously describes the sweeping up of matter in region I which can be considered as the formation of a ringlet and a sweeping away of matter in region II which leads to the formation of a gap. Secondly, the redistribution of the density occurs, to begin with, far from resonance. As a result, this mechanism does not suppress itself, as GT's mechanism did. Finally, the characteristics of the drift depend strongly on the properties of the disc and of the satellite – the satellite mass and the surface density of the disc. This makes it possible to explain why in some cases we observe gaps and ringlets whereas in other cases we have a continuous generation of waves without the formation of any structures.

To prove this last statement let us analyse the correlation between the predictions of the theory and the observed structures in Saturn's rings. For the process which prevents the formation of gaps we consider the diffusive – viscous – filling of regions with a reduced density.

Table A5.1 presents a set of resonant structures in the Saturnian rings: either gaps with narrow ringlets (G+R) or density waves (DW; see Table 2.2).

Table A5.1. Structures in Saturn's rings associated with Londblad resonances from different satellites

Structure number and satellite	Satellite mass (relative to mass of Mimas)	Type of the resonance and its radius [10^3 km]	Type of the structure	Ring and surface density [g/cm^2]
1. Mimas	1.0	(1:2) 117.55	Cassini Div.	(B) 70
2. Janus	0.112	(1:2) 96.25	DW	(B) 70
3. Janus	0.112	(3:4) 125.27	DW	(A) 34*
4. Prometheus	0.017	(1:2) 88.71	G+RE	(C) 4
5. Prometheus	0.017	(21:22) 135.2	DW	(A) 24*
6. Pandora	0.010	(14:15) 135.4	DW	(A) 24*

* This resonance is one of a series of similar resonances produced by this satellite in the same ring region.

We show in Fig. A5.3 in a ($\log M_s/M_p$,$\log\sigma$)-plane, where M_s and M_p are, respectively, the masses of the satellite and of the planet, the positions of the resonant structures from Table A5.1. Open circles correspond to density waves and filled circles to gaps and ringlets and the numbers correspond to those in Table A5.1.

Obviously, there are other parameters, apart from σ and M_s/M_p, which influence the formation of resonant structures, such as the optical depth τ, the order m of the resonance, and so on. Why did we restrict ourselves to considering the resonant structures only in the ($\log M_s/M_p$,$\log\sigma$)-plane, and not in an N-dimensional space, where N is the number of determining parameters? We shall show that, indeed, the two parameters M_s/M_p and σ are the most significant factors determining the efficiency of the formation of gaps. Let us assume that M_s/M_p is small and constant. For small σ even a satellite with a small mass can form a gap or a ringlet. In the region of larger σ, the diffusion is also larger, and it prevents the formation of gaps and ringlets. A satellite with a larger mass is competitive with the diffusion and can form

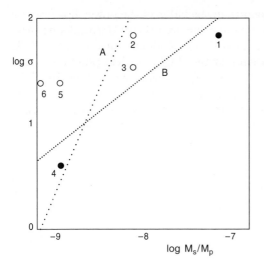

Fig. A5.3. The positions of the resonant structures from Table A5.1 in a $(\log M_s/M_p, \log\sigma)$-plane. Open circles correspond to density waves and filled circles to gaps and the numbers correspond to those in Table A5.1.

gaps in a denser medium. From this we can conclude that the relative position of the resonance structures is unlikely to change qualitatively when the other parameters $(\tau, m, \ldots)$ are varied.

Any condition which determines the kind of structure which can be produced can be represented by a curve in Fig. A5.3. Each curve will divide the plane into two parts – the region where there will be waves, and the region where there will be gaps and ringlets.

Let us first consider how we can represent the GT mechanism in Fig. A5.3. According to Goldreich and Tremaine (1980) a resonant gap is formed in optically thin discs under the condition

$$\log\tau < 2\log(\alpha M_s),\tag{A5.72}$$

where the coefficient α is given by the formula

$$\alpha = \frac{m\Omega_s R_s}{cM_p}.\tag{A5.73}$$

Here Ω_s is the rotational velocity of the satellite (or the rotational velocity on the corotation radius $m\Omega_s = (m-1)\Omega_{eq}$) and c is the velocity dispersion.

The line which divides the region of the spiral density waves (to the left) and that of the resonant gaps (to the right) according to condition (A5.72) from the GT model is marked in Fig. A5.3 by the letter A. It was assumed here that the coefficient α was a constant. Note that the line A cannot serve as a true division between the gap and the wave regions, whatever the value of the coefficient α or the connection between the optical depth and the surface density, since its slope depends on the index of the power-law function $\tau = \tau(\sigma)$. For example, if we plot a similar line in a $(\log m M_s/M_p, \log \sigma)$-plane to take into account the variation in α due to a change in the order of the resonance, this would result in a shift of the (higher-order) points 3, 5, and 6 to the right. In any case, the division of the coordinate plane by a line similar to A cannot be accepted. The position would become worse, if the disc were optically thick – more precisely, if collisions between disc particles were frequent – and this could happen for the B ring. In that case the slope of the line A would become negative and the fit with observations would become impossible.

Let us now plot the boundary between gap formation and wave generation which follows from (A5.67). It follows from that equation that the characteristic time t_{red} for the redistribution of matter is

$$t_{\mathrm{red}} \simeq C \frac{\sigma^3}{\nu M_s^2}, \tag{A5.74}$$

where C is a coefficient which depends on the resonance position and the mass of the planet. If we use the expression for the characteristic time for diffusive filling,

$$t_{\mathrm{fil}} \simeq \frac{\Delta R^2}{\nu}, \tag{A5.75}$$

we find the condition

$$3 \log \sigma = 2 \log(\beta M_s), \tag{A5.76}$$

which corresponds to line B in Fig. A5.3. We see that this line divides the coordinate plane in the appropriate manner.

We can thus conclude that the mechanism proposed by us for the formation of gaps and ringlets is more consistent with observational data than the GT mechanism.

Let us estimate the required characteristic time scales for our dissipative drift mechanism to produce some of the observed gaps. Adapting a gap width of about 1000 km, a ring density of about 70 g/cm², and a viscosity of $\nu \simeq c^2/\Omega \simeq 100$ cm²/s in the case of the Mimas 2:1 resonance we find

$$t_{\mathrm{gap}} \simeq 5 \times 10^4 \text{ years.} \tag{A5.77}$$

The time for diffusive filling for the above parameters is about 10^5 years.

In the case of Janus which generates a 300 km density wave in the B ring the characteristic time scale for redistribution is about 10^7 years, whereas the time for diffusive filling for such scales is about 10^4 years. It is thus natural that there is no gap. This example demonstrates the role of the mass of the satellite and explains why Janus with a mass about 9 times smaller than that of Mimas could not form gaps in the dense A and B rings.

9 Estimate of the Maximum Width of a Gap Produced by a Density Wave

Let us derive this estimate for the maximum width of a gap produced by a wave excited at the Lindblad resonance in the case of the Cassini Division. The maximum width of a gap to be produced by our proposed mechanism is obviously restricted to the distance the wave can propagate. In the case of the Cassini Division a gravitational acoustic wave is excited at the inner Lindblad resonance point (ILR) and propagates up to the reflection point (RP) where its group velocity vanishes:

$$\frac{d\omega}{dk} = 0. \tag{A5.78}$$

Differentiating the dispersion relation for the gravitational acoustic wave (Rolfs 1977),

$$\hat{\omega}^2 \equiv (\omega - m\Omega)^2 = \kappa^2 - 2\pi G\sigma|k| + k^2 c^2, \tag{A5.79}$$

with respect to k and using (A5.78) we find that in the reflection point we have

$$k_{\text{refl}} = \frac{\pi G\sigma_{\text{refl}}}{c_{\text{refl}}^2}. \tag{A5.80}$$

In the region between ILR and RP we have for a long-wavelength wave, $0 < k < k_{\text{refl}}$, $\hat{\omega} < 0$, and hence

$$\frac{d\hat{\omega}}{dk} = \frac{1}{\hat{\omega}} \left(k^2 c^2 - \pi G\sigma \right) > 0, \tag{A5.81}$$

which shows that a long-wavelength gravitational acoustic wave propagates outwards from the resonance. The maximum width of the Cassini Division is thus the distance between ILR and RP. Let us estimate its magnitude.

From (A5.79) and (A5.80) we have in RP, $R = R_{\text{refl}}$,

$$\hat{\omega}_{\text{refl}}^2 = \kappa_{\text{refl}}^2 - \left(\frac{\pi G\sigma_{\text{refl}}}{c_{\text{refl}}} \right)^2. \tag{A5.82}$$

Using the expression $h = \sqrt{2}c/\Omega$ for the half-thickness of an isothermal disc (Fridman and Khoruzhii 1999) and the Kepler law for rotation, $\kappa^2 = \Omega^2 = GM_{\text{p}}/R^3$, we can transform (A5.82) to the form

$$\hat{\omega}_{\text{refl}}^2 = \kappa_{\text{refl}}^2 \left(1 - \alpha^2 \right), \tag{A5.83}$$

where

$$\alpha = \frac{\sqrt{2}\pi\sigma_{\text{refl}} R_{\text{refl}}^3}{h_{\text{refl}} M_{\text{p}}}. \tag{A5.84}$$

According to observations (Zebker et al. 1985) in the Cassini Division we have

$$\sigma \approx 18.8 \pm 0.5 \text{ g/cm}^2, \qquad h \approx 10 \text{ m}, \tag{A5.85}$$

and in the inner part of the A ring

$$\sigma \approx 34 \pm 6 \text{ g/cm}^2, \qquad h \approx 5 - 15 \text{ m}, \tag{A5.86}$$

Estimating R_{refl} to be the radius of the inner edge of the A ring, $R_{\text{refl}} = 122.17 \times 10^3$ km (Cuzzi et al. 1984), and using $M_{\text{p}} \approx 5.7 \times 10^{29}$ g, we see that in any case $\alpha^2 \ll 1$. As in ILR we have the exact equality (Rolfs 1977)

$$\hat{\omega}_0 = -\kappa_0 = -\Omega_0, \tag{A5.87}$$

we can by comparing (A5.83) with (A5.87) conclude that as to order of magnitude $\delta R \equiv R_{\text{refl}} - R_0 \approx \alpha^2 R_0 \ll R_0$. To calculate δR more accurately, we expand the quantities $\hat{\omega}_{\text{refl}}^2$ and κ_{refl}^2 from (A5.83) in the vicinity of ILR. Taking into account that for a wave generated in a 2:1 resonance we have $m = 2$, we find, up to second order in $\delta R/R_0$,

$$\delta R = \tfrac{1}{3} R_0 \alpha^2. \tag{A5.88}$$

Putting numerical values into (A5.88) we have

$$\delta R = 3100 \text{ km} \left(\frac{\sigma_{\text{refl}}}{20 \text{ g/cm}^2} \right)^2 \left(\frac{10 \text{ m}}{h_{\text{refl}}} \right)^2. \tag{A5.89}$$

Let us assume that the Cassini Division was formed at the position of the A ring which until the formation of the gap started adjoined the B ring. As the acoustic drift mechanism, generated by the spiral density wave operated from the resonance with Mimas, the quantity σ_{refl} occurring in (A5.89) decreases from a value of 34 g/cm^2 – in the inner part of the A ring – to a value of 18.8 g/cm^2 – in the Cassini Division. Substituting these two values into (A5.89) we find the region – $\delta R_{\text{min}} = 2740$ km, $\delta R_{\text{max}} = 8960$ km – in which we can observe the width of the Cassini gap $\Delta \approx 4650$ km. The total width of the Cassini Division can thus be determined by the intrinsic properties of the density wave.

10 Some Additional Remarks

The distance from the resonance to the turning point of the spiral wave depends on the orbital radius as R^{-7}. The orbital distances in the C ring are reduced as compared to the Cassini gap by 35 to 40%, and this gives an eight-fold reduction in the maximum possible propagation length of the wave. This distance has already been reduced by almost two orders of magnitude due to the occurrence of the square of the surface density ($\sigma = 4$ g/cm^2). Estimates show that a typical wavelength of a spiral density wave – which does not include bending waves – in the C ring must be 20 km. Gaps of such a magnitude are in the C ring generated by the 1:3 resonance from Mimas and the 1:2 resonance of Pandora (Cuzzi et al. 1984), There are no wider gaps or density waves with a longer propagation length from satellite resonances in the C ring (for details see App. VI).

To explain why other satellites with smaller masses can produce gaps in the C ring one must take into account the role of the ring density, as well as the role of the magnitude of the rotational velocity. The first factor gives a gain in the time scale of a factor of about 5000, and the second a similar factor. This allows Prometheus to redistribute matter on spatial scales of 200 to 300 km in a time of 10^6 years. Pandora, which has the smallest mass, has difficulties in producing ringlets by itself, but the 2:1 Pandora resonance coincides with the 3:1 Mimas resonance. The joint action of these resonances can produce the observed ringlets in the C ring.

We have thus proposed a new radial dissipative drift mechanism in planetary rings. We hope that it may correctly explain the formation of some remarkable natural phenomena: wide gaps and wave trains. It is possible that this mechanism may have wider applications and that it may also operate in galactic and accretion discs.

Appendix VI. Resonance Structures in Saturn's C Ring

The aim of this appendix is to discuss the interconnection between satellite resonances and the large number of dense narrow ringlets in Saturn's C ring (see Plate 4). Although the Titan ringlet can confidently be identified with the 1 : 0 resonance from Titan (Rosen and Lissauer 1988) and two further ringlets in the outer part of the C ring correspond to resonances from Mimas, Pandora, and Prometheus (Cuzzi et al. 1984) the origin of the other ringlets has still remained a problem.

In the classical scheme of the interaction of a spiral wave caused by a satellite resonance with a dense planetary ring the redistribution of matter in the resonance region is determined by the competition between the resonant sweeping away and the diffusive filling in. If the resonant sweeping away dominates, a rarefied region is formed in the dense disc of rings (see App. V). A typical example of such a resonance interaction is the formation of the Cassini Division in a disc with a surface density of 34 to 70 g/cm^3 as the result of the action of the $1/2$ resonance from Mimas ($R_{res} = 117\,544$ km). In the case of weaker resonances one can observe spiral waves in the rings – such a wave has been produced, for instance, by the vertical $2/4$ resonance from Mimas in the B ring 1000 km from the Cassini Division ($R_{res} = 116\,715$ km).

The Uranian rings with a surface density of about 1 g/cm^3 in their proto-stage are almost two orders of magnitude less dense than the Saturnian rings. An evaluation of the maximum wavelength for wave propagation (see App. V) shows that it cannot be more than a few kilometers in the Uranian rings. We have shown in Chap. 11 that the formation of the existing dense Uranian rings on a background of the rarefied proto-rings is connected with a new kind of interaction: the capture of positively drifting ring particles in inner Lindblad resonances which arrest this drift. The appearance of such a resonance capture can be adequately described, without invoking a collective interaction, in the framework of the classical three-body celestial mechanics problem. After the formation of dense rings at the positions of resonances the collective interaction between resonant particles is amplified and the rings leave the resonance and drift away from the planet and the parent resonance.

Many Lindblad and other resonances from Saturn's satellites are situated in Saturn's C ring, and especially in its outer part. The C ring with its surface density of up to 5 g/cm^3 occupies an intermediate position between

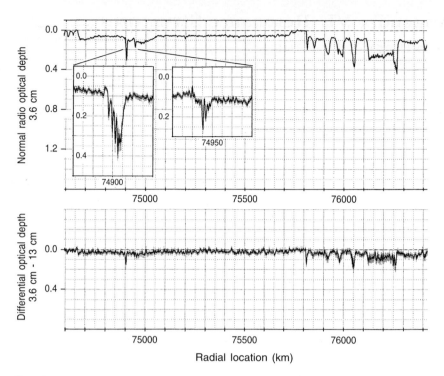

Fig. A6.1. Rosen's radio-occultation density profiles (Rosen 1989) of the Inner C Ring of Saturn near 75 000 km

the Saturnian rings, which lie further out and are more massive, and the Uranian proto-rings. Calculations show that the wavelength of a spiral density wave in the normal sections of the C ring cannot exceed 20 km.

We can, according to Chap. 11, expect in the C ring an appreciable positive ballistic particle drift caused by the erosion of the B ring by micrometeorites. It is therefore natural to assume that the mechanism for the formation of the narrow Saturnian and Uranian rings is the same and that the elliptical Titan, Maxwell, and Huygens ringlets are direct relations of the Uranian rings.

However, most of the ringlets in the C zone differ from the Uranian rings: they are broader, about 100 to 200 km wide, have zero eccentricity, and are, as a rule, not surrounded by empty space but by the normal background density of the C ring. They have an almost rectangular density profile (see Plate 4 and Fig. 2.3, and also the beautiful radio-occultation density profiles from Rosen's 1989 thesis which are called the "Inner C Ring of Saturn", the "Central C Ring of Saturn", and the "Outer C Ring of Saturn", parts of which we reproduce here as Figs. A6.1 to A6.6) and inside them one can often notice a regular structure which is characteristic for a spiral wave (see Fig. 2.7 (Cuzzi et al. 1984) and Figs. A6.1 to A6.6).

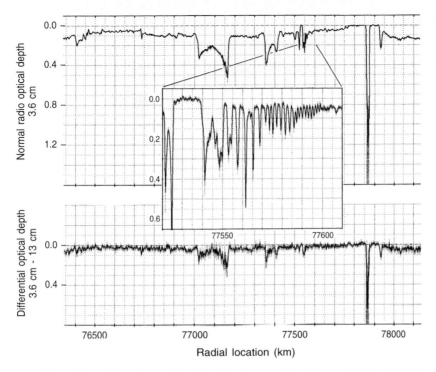

Fig. A6.2. Rosen's radio-occultation density profiles (Rosen 1989) of the Inner C Ring of Saturn near 77 000 km

It is thus necessary to consider again the problem of the origin of the ringlets in Saturn's C ring. We shall consider two problems:

1. Why are the ringlets in Saturn's C ring different from the Uranian rings, although the principal mechanism for the formation of the two systems of narrow rings must be the same?
2. Which satellite resonances led to the formation of the ringlet system in Saturn's C ring?

Probably, one of the causes for the difference between Uranus' narrow ringlets and the ringlets in Saturn's C ring is that the background density of the C ring is several times higher than the density of Uranus' proto-rings. The matter of Uranus' proto-rings was rapidly shared out between the resonance rings and the formation of rings in the resonances was stopped – each resonance produced a single narrow ring, except the innermost 1/2 resonance from Portia which successively produced three rings. In contrast to Uranus' proto-rings, in Saturn's C ring, which had an appreciable supply of matter, the drift current towards the resonance was not weakened, even after a narrow ring had been formed which left the resonance. If matter accumulated sufficiently rapidly in the resonance, till the dense ring which was formed

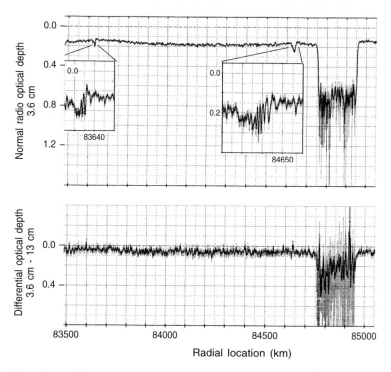

Fig. A6.3. Rosen's radio-occultation density profiles (Rosen 1989) of the Central C Ring of Saturn near 84 000 km

went further away, this would mean, in fact, rather than the formation in the resonance zone of a narrow ring, the formation of a whole dense region which was slowly drifting outwards. Indeed, the more rapidly the next ring was formed in the 1/2 resonance from Portia, or the more slowly this ring drifted outwards, the closer they – the 4, 5, 6 rings – would be situated near one another until they merged into a single dense region.

Naturally, as soon as such a dense zone was formed on the outside of the resonance, the resonance would rapidly excite in it a spiral wave with a wavelength much greater than 20 km since the wavelength of the propagating wave increases as the square of the surface density (see (A5.88) and (A5.84)). The spiral wave will efficiently retard the positive drift of the ring. In such a case there arises the possibility of a new kind of equilibrium – when the positive ballistic drift of the matter equilibrates with the negative resonance drift not only in the resonance region – for individual particles – but also in the whole of the region of the collective motion of the particles – in the whole of the spiral wave.

We can write down a condition for equilibrium:

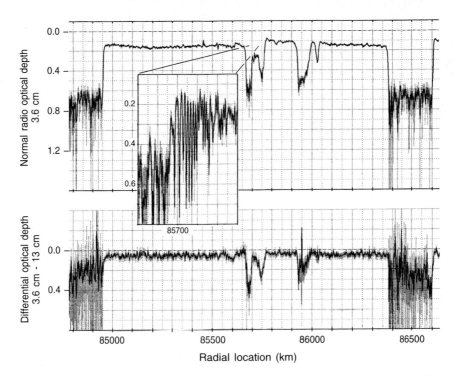

Fig. A6.4. Rosen's radio-occultation density profiles (Rosen 1989) of the Central C Ring of Saturn near 86 000 km

$$\int_{R_{\mathrm{int}}}^{R_{\mathrm{ext}}} J_{\mathrm{ball}}\, dr \;=\; \int_{R_{\mathrm{res}}}^{R_{\mathrm{max}}} J_{\mathrm{extr}}\, dr, \tag{A6.1}$$

where J_{ball} is the angular momentum (per unit radius) absorbed by the ring from ballistic injecta, R_{int} is the inner edge of the ring, R_{ext} is the outer edge of the ring, J_{extr} is the angular momentum (per unit radius) removed from the ring by the spiral resonance wave, R_{res} is the resonance radius, and R_{max} is the radius up to which the spiral wave propagates.

When the resonance removal of angular momentum is too strong, the ring may shift closer to the planet – to such a radius that the zone in which the wave propagates, which cannot be beyond the outer boundary of the dense region, is reduced so much that the balance (A6.1) can be guaranteed. In that case the upper integration limit on the right-hand side of (A6.1) will become R_{ext} rather than R_{max}.

The mechanisms for the build-up of the narrow Uranian ringlets and of the ringlets in Saturn's C ring are thus the same, but the rates of the positive drift and of the supply of matter are different so that the resonances in the Uranian system produce separate rings, but in Saturn's C ring they produce complete dense regions necessarily containing in them spiral waves. These ring

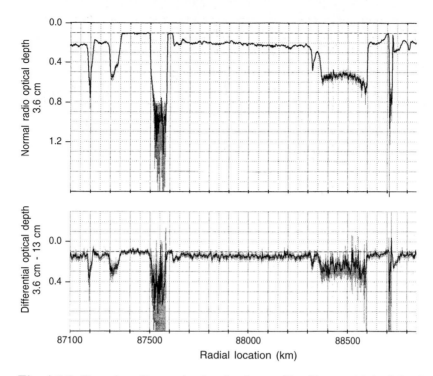

Fig. A6.5. Rosen's radio-occultation density profiles (Rosen 1989) of the Outer C Ring of Saturn near 88 000 km

regions differ noticeably from the usual resonance ringlets like the Uranian ringlets in the following features which follow from the specific mechanism for their formation and stability:

a. They must contain a strong resonance.
b. They must contain a spiral wave, which
c. plays an important role in the total balance of the angular momentum of the ring.
d. They are surrounded by matter with a lower density, but not by gaps.
e. The density of the background matter near the inner boundary of the ringlet is practically always higher than at the inner boundary – this is a clear criterion for a positive drift of the matter which remains in the ringlet.
f. They do not have a significant eccentricity although the edge of the ring can, apparently, oscillate in correspondence with the mode of the close resonance.
g. They are, as a rule, broader than the usual elliptic narrow rings and their density profile has a characteristic rectangular shape resembling a table mountain (mesa) with sharp edges.

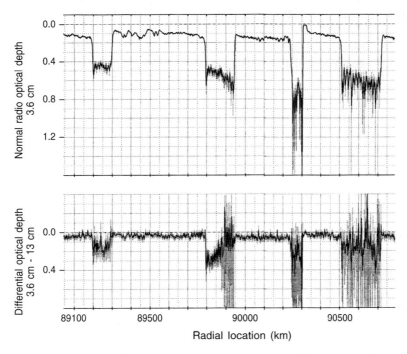

Fig. A6.6. Rosen's radio-occultation density profiles (Rosen 1989) of the Outer C Ring of Saturn near 90 000 km

In view of the last feature of these resonance rings and to emphasise the specific nature of such structures we propose to call these rings "mesa-ringlets". Under well defined conditions the interior spiral waves can considerably lower the density inside a mesa-ringlet, producing a two-hump double ringlet (as in Fig. 2.7).

The obligatory presence in mesa-ringlets of satellite resonances requires a careful analysis of the positions of resonances inside Saturn's C ring and their correspondence with observed structures. Of course, we can only propose the presence of a resonance for stationary ringlets; if, on the other hand, the ring was merely formed in a resonance and afterwards left it and is now slowly drifting or in another equilibrium state – like the inner Uranian rings – it is possible that there is no resonance in it.

In concluding this appendix we shall give a table of the correspondence of observed structures in Saturn's C ring with the positions of calculated satellite resonances. The complexity of the comparison consists in that the asphericity of Saturn's field produces a splitting of the resonances and instead of a single 2/1 resonance we find a whole set of close resonances with a shift of up to several hundred kilometers.

Table A6.1. Resonance structures in the C ring (R_{res} is the resonance position)

Structure type	Position of structure [10^3 km]	R_{res} [10^3 km]	Satellite resonance $(m+n+p)/(k-l)$	Satellite	Figure
Gap	74.65–74.67	74.680	$(0+4+0)/(2-1) = 4/1$	Mimas	
Spiral wave	74.89–74.91	74.887	$(0+1+3)/(2-1) = 4/1$	Mimas	A6.1
Spiral wave	74.94–74.95	74.950	$(0+0+4)/(2-1) = 4/1$	Mimas	
Bending wave	77.52–77.60	77.518	$(1-1-1)/(1-1) = -1/0$	Titan (V)[a]	
Eccentr. ringlet	77.871	77.853	$(1-1-1)/(1-1) = -1/0$	Titan[b]	A6.2
Spiral wave	83.63–83.64	83.631	$(0+1+1)/(2-1) = 2/1$	Pan	
Spiral wave	84.63–84.65	84.683	$(3+1+0)/(3-1) = 4/2$	Pan	
Mesa-ringlet	84.75–84.95	84.753	$(4+0+0)/(3-1) = 4/2$	Pan	A6.3
		84.836	$(0+2+0)/(2-1) = 2/1$	Pan	
Double ringlet[c]	85.67–85.77	?	?	?	A6.4
Double ringlet	85.92–86.03	86.074	$(1+1+0)/(2-1) = 2/1$	Atlas (V)	
Mesa-ringlet	86.38–86.61	86.511	$(3+1+0)/(3-1) = 4/2$	Atlas (V)	
Ringlet	87.18–87.22	87.141	$(1+1+0)/(2-1) = 2/1$	Prometheus (V)	
Ringlet	87.29–87.35	87.298	$(4+0+0)/(3-1) = 4/2$	Atlas	
Gap	87.35–87.62	87.513	$(1+1+0)/(2-1) = 2/1$	Atlas	
Maxwell ringlet	87.49	87.649	$(2+0+0)/(2-1) = 2/1$	Atlas	A6.5
Mesa-ringlet	88.35–88.60	88.349	$(4+0+0)/(3-1) = 4/2$	Prometheus	
		88.561	$(1+1+0)/(2-1) = 2/1$	Prometheus	
Gap	88.69–88.71	88.695	$(2+1+0)/(2-1) = 3/1$	Mimas (V)	
Ringlet	88.71–88.73	88.696	$(2+0+0)/(2-1) = 2/1$	Prometheus	
Mesa-ringlet	89.19–89.30	89.220	$(4+0+0)/(3-1) = 4/2$	S/1995S3	
Mesa-ringlet	89.79–89.95	89.820	$(4+0+0)/(3-1) = 4/2$	Pandora	
Mesa-ringlet	90.14–90.21	90.160	$(2+0+0)/(2-1) = 2/1$	Pandora	A6.6
Gap	90.21–90.23	90.191	$(2+0+1)/(2-1) = 3/1$	Mimas	
Mesa-ringlet	90.41–90.62	90.427	$(0+0+2)/(2-1) = 2/1$	Pandora	

[a] V indicates a vertical resonance.
[b] It is not impossible that the other structures in the outer part of the C ring may also have produced in the $-1/0$ resonance from Titan, after which drift has displaced them from the point where they were formed.
[c] McGhee et al. (1998) have recently given more accurate data about the new Saturnian satellites. Inside Atlas's orbit the satellite S/1995S1 was discovered at a radius of 137 529±87 km. Clearly, it or neigbouring as yet undiscovered (hypothetical) satellites near the A ring correspond, for instance, to the double ringlet at a radius of 85 670 to 85 770 km (for instance, an as yet undiscovered (hypothetical) satellite at 136 855 km might explain the existence of a double ringlet at this distance with spiral waves at 85 695 km caused by a strong $(2+0+0)/(2-1) = 2/1$ resonance). As a confirmation of the hypothesis of Cuzzi and Burns (1988) a whole series of satellites has been discovered near the F ring between Pandora and Prometheus: S/1995S7 (139 371±93 km or 139 382.4±0.3 km), S/1995S9 (139 445±139 km), S/1995S6 (139 594±75 km), S/1995S3 (140 742±135 km).

To be strict, our table is not the final word as far as the truth is concerned, because we have not taken into account resonances from the satellites which were recently discovered by the Hubble telescope – since the radii of their orbits are not known very exactly – and, of course, we cannot calculate the positions of resonances from possibly undiscovered satellites.

We must also bear in mind that the positions of many resonances are the same – for instance, those of the $(4+0+0)/(3-1)$ and $(2+1+1)/(3-1)$ resonances – and by indicating just one of them we may not have made the most felicitous choice.

Moreover, we have neglected such problems as: Why does one particular resonance cause the formation of a mesa-ringlet or of a spiral wave, whereas another one did not do this? The reason is that this would require a more detailed comparative analysis of the resonances and also a calculation of their competitiveness caused not only by their strength but also by the relative positions of the resonances, the strength and directions of diffusion currents in a given region – which are determined by the erosion of matter in not only the B ring but also in the mesa-ringlets themselves which are being formed – and so on. Nevertheless, we feel that this table gives conclusive indications in favour of the assumption that the observed structures in the C ring, including the mesa-ringlets, have a resonance origin.

As the most general resonance relation we use the following equations: For a horizontal resonance:

$$m\Omega_\mathrm{s} + n\mu_\mathrm{s} + p\kappa_\mathrm{s} \ = \ k\Omega_\mathrm{r} - l\kappa_\mathrm{r}, \qquad\qquad (A6.2)$$

where the index "s" corresponds to the motion of the satellite, the index "r" corresponds to the motion of the ring particles, Ω is the mean angular velocity of the orbital motion, μ is the mean angular velocity of the vertical oscillations of a particle, and κ is the mean angular velocity of the radial oscillations. These frequencies are no longer the same when we take the aspherical harmonics of Saturn's gravitational field into account (for details see Elliot and Nicholson 1984 or Rosen's 1989 thesis).

For a vertical resonance the following relation holds:

$$m\Omega_\mathrm{s} + n\mu_\mathrm{s} + p\kappa_\mathrm{s} \ = \ k\Omega_\mathrm{r} - l\mu_\mathrm{r}, \qquad\qquad (A6.3)$$

and now n must be odd (Shu 1984). To determine the resonance radius for each set of integers m, n, p, k, l we numerically solved the algebraic equations (A6.2) or (A6.3) for R_res.

Table A6.1 was constructed by comparing the results obtained with the observed structures (using the radio-occultation profiles from Rosen's 1989 thesis).

References

Abalakin, V.K., Aksenov, E.P., Grebenikov, E.A., and Ryabov, Yu.A. (1971) Handbook of Celestial Mechanics and Astrodynamics. Nauka, Moscow.

Afanas'ev, V.L., Gor'kavyĭ, N.N., Smirnov, M.A., and Fridman, A.M. (1985) Astron. Tsirkul., No 1391, p.3.

Aksness, K. (1977) Nature **269**, 783.

Alekseev, B.V. (1982) Mathematical Kinetics of Reacting Gases. Nauka, Moscow.

Andersen, D. and Benson, K. (1963) Ice and Snow (Ed. W.D. Kingeri). MIT Press, Cambridge MA.

Araki, S. (1988) Icarus **70**, 182.

Araki, S. and Tremaine, S. (1986) Icarus **65**, 83.

Avedisova, V.S. (1985) Sov. Astron. Lett. **11**, 185.

Balk, M.B. (1965) Elements of the Dynamics of Cosmic Flight. Nauka, Moscow.

Balescu, R.(1975) Equilibrium and Nonequilibrium Statistical Mechanics. Wiley-Interscience, New York.

Beust, H., Lagrange-Henri, A.M., Vidal-Madjar, A., and Ferlet, R. (1989) Astron. Ap. **223**, 204.

Bhattacharyya, J.C. and Kuppuswamy, K. (1977) Nature **267**, 331.

Binney, J.J. and Tremaine, S. (1987) Galactic Dynamics, Princeton University Press. Princeton NJ.

Bisikalo, D.V., Boyarchuk, A.A., Kuznetsov, O.A., et al. (1995) Astron. Repts. **39**, 167.

Bisikalo, D.V., Boyarchuk, A.A., et al. (1997) Ap. Space Sci. **252**, 389.

Bisikalo, D.V., Boyarchuk, A.A., et al. (1998) Mon. Not. Roy. Astron. Soc. **300**, 39.

Bisikalo, D.V. and Boyarchuk, A.A., et al. (1999) Astron. Rept. **43**.

Blitz, L., Binney, J., Lo, K.Y., and Ho, P.T.P. (1993) Nature **361**, 417.

Bobrov, M.S. (1970) The Rings of Saturn. Nauka, Moscow (English translation: NASA TTF-701).

Bogorodskiĭ, V.V. and Gavrilo, V.P. (1980) Ice. Gidrometeoizdat, Leningrad.

Bonnor, W.B. (1957) Monthly Not. Roy. Astron. Soc. **117**, 104.

Borderies, N., Goldreich, P., and Tremaine, S. (1984) Planetary Rings (Eds R. Greenberg and A. Brahic). Univ. Arizona Press, Tucson AZ, p.713.

Borderies, N. (1987) Dynamics of the Solar System (Ed. M. Sidlichovsky). Publ. Astron. Inst. Czech. Acad. Sci. **68**, 151.

Borderies, N., Goldreich, P., and Tremaine, S. (1982) Nature **299**, 209.

Borderies, N., Goldreich, P., and Tremaine, S. (1985) Icarus **63**, 40.

Borderies, N., Goldreich, P., and Tremaine, S. (1989) Icarus **80**, 344.

Braginskiĭ, S.I. (1958) Sov. Phys. JETP **6**, 358.

Braginskii, S.I. (1965) Rev. Plasma Phys. **1**, 205 (Ed. M.A. Leontovich), Consultants Bureau, New York.

Brahic, A. (1982) Formation of Planetary Systems (Ed. A. Brahic). Copadues, Toulouse, p.651.

Brahic, A. (Ed.) (1984) Planetary Rings. Copadues, Toulouse.

Brahic, A. and Sicardy, B. (1986) Bull. Am. Astron. Soc. **18**, 778.

Brahic, A. and Sicardy, B. (1987) Dynamics of the Solar System (Ed. M. Sidlichovsky). Publ. Astron. Inst. Czech. Acad. Sci. **68**, 159.

Bridges, F.G., Hatzes, A, and Lin, D.N.C. (1984) Nature **309**, 333.

Broadfoot, A.L., Herbert, F, Holberg, J.B., et al. (1986) Science **233**, 74.

Brophy, T.G. (1989) The Dynamics of the Observed Clumps in 1989N1R.

Brophy, T.G. and Esposito, L.W. (1989) Icarus **78**, 181.

Brophy, T.G., Stewart, G.R., and Esposito, L.W. (1990) Icarus **83**, 133.

Brouwer, D. and Clemence, G.M. (1961) Methods of Celestial Mechanics. Academic Press, New York.

Burns, J., Lamy, P.L., and Soter, S. (1979) Icarus **40**, 1.

Burns, J.A. (1986) Satellites (Eds J.A. Burns and M.S. Matthews). Univ. Arizona Press, Tucson AZ, p.1.

Camichel, H. (1958) Ann. Ap. **21**, 231.

Cassini Mission (1989) Saturb Orbiter (NASA; A.O.No OSSA-1-89) Houston TX.

Chandrasekhar, S. (1943) Principles of Stellar Dynamics. Dover, New York.

Chandrasekhar, S. and ter Haar, D. (1950) Ap. J. **111**, 187.

Chapman, S. and Cowling, T.G. (1953) The Mathematical Theory of Non-Uniform Gases. Cambridge Univ. Press.

Cherepanov, G.P. (1974) Mechanics of Brittle Fraction. Nauka, Moscow.

Cherepashchuk, A.M. (1981) Stars and Stellar Systems (Ed. D.Ya. Martynov). Nauka, Moscow, p.38.

Chew, G.F., Goldberger, M., and Low, F. (1956) Proc. Roy. Soc. **A236**, 112.

Churilov, S.M. and Shukhman, I.G. (1981) Astron. Tsirkul. No 1157, p.1.

Clark, R.N., Fanale, F.P., and Zent, A.P. (1985) Icarus **56**, 233.

Coates, A. (1997) Astronomy Now, October 1997, p.41.

Colbeck S.C. (Ed.) (1980) Dynamics of Snow and Ice Masses. Academic Press, New York.

Cantopoulos, G. (1975) Ap. J. **201**, 566.

Cantopoulos, G. (1979) Astron. Ap. **71**, 229.

Cantopoulos, G. (1981) Astron. Ap. **104**, 116.

Cantopoulos, G. and Mertzanides, C. (1977) Astron. Ap. **61**, 477.

Cooke, M.L., Nicholson, P.D., Matthews, K., and Elias, J. (1985) Bull. Am. Astron. Soc. **17**, 719.

Cosmogony (1951) Proc. I All-Union Meeting on Cosmogony Problems. Nauka, Moscow.

Covault, C.E., Glass, I.S., French, R.G., and Elliott, J.L. (1986) Icarus **67**, 126.

Cuzzi, J. (1983) Rev. Geophys. Space Phys. **21**, No 2, p.173.

Cuzzi, J.N. and Burns, J.A. (1988) Icarus **74**, 284.

Cuzzi, J.N. and Durisen, R.Y. (1990) Icarus **84**, 467.

Cuzzi, J.N., Durisen, R.Y., Burns, J.A., and Hamill, P. (1979) Icarus **38**, 54.

Cuzzi, J.N., Lissauer, J.J., Esposito, L.W., et al. (1984) Planetary Rings (Eds R. Greenberg and A. Brahic). Univ. Arizona Press, Tucson AZ, p.73.

Cuzzi, J.N. and Scargle, J.D. (1985) Ap. J. **292**, 707.

Davis, D.R., Weidenschilling, S.J., Chapman, C.R., and Greenberg, R. (1984) Science **224**, 744.

Dermott, S.F. and Gold, T. (1977) Nature **267**, 590.

Dermott, S.F., Gold, T., and Sinclair, A.T. (1979) Astron. J. **84**, 1225.

Dermott, S.F., Jayaraman, S., Xu, Y.L., Gustafson, B.A.S., and Liou, J.C. (1994) Nature **369**, 719.

Di Prima, D.H. and Swinney, H.G. (1981) Hydrodynamic Instabilities and the Transition to Turbulence (Eds H.G. Swinney and J.P. Gollub). Springer, Berlin, p.163.

Dobrovolskis, A.R., Borderies, N.J., and Steiman-Cameron, T.Y. (1989) Icarus **81**, 132.

Doyle, L.R., Dones, L., and Cuzzi, J.N. (1989) Icarus **80**, 104.

Durisen, R.H., Cramer, N.L., Murphy, B.W., et al. (1989) Icarus **80**, 136.

Elliot, J.L. (1984) Planetary Rings (Ed. A. Brahic). Copadues, Toulouse, p.197.

Elliot, J.L., Dunham, E.W., and Mink, D.J. (1977) Nature **267**, 328.

Elliot, J.L., Mink, D.J., Elias, J.N., et al. (1981) Nature **294**, 526.

Elliot, J.L. and Nicholson, P.D. (1984) Planetary Rings (Eds R. Greenberg and A. Brahic). Univ. Arizona Press, Tucson AZ, p.25.

Esposito, L.W. (1986) Icarus **67**, 345.

Esposito, L.W. and Colwell, J.E. (1989) Nature **339**, 605.

Esposito, L.W., Cuzzi, J.N., Holberg, J.B., et al. (1984) Saturn (Eds T. Gehrels and M.S. Matthews). Univ. Arizona Press, Tucson AZ, p.463.

Esposito, L.W., Harris, C.C., and Simmons, K.E. (1987) Ap. J. Suppl. **63**, 749.

Faraday, M. (1831) Philos. Trans. Roy. Soc. London **121**, 229.

Feigenbaum (1978) J. Stat. Mech. **19**, 25.

Ferrari, C. and Brahic, A. (1994) Icarus **111**, 193.

Ferziger, J.H. and Kaper, H.G. (1972) Mathematical Theory of Transfer Processes in Gases, North-Holland, Amsterdam.

Foryta, D.W. and Sicardy, C. (1996) Icarus **123**, 129.

Flynn, B.C. and Cuzzi J.N. (1989) Icarus **82**, 180.

Franklin, F.A., Colombo, G., and Cook II, A.F. (1971) Icarus **15**, 80.

Franklin, F.A., Cook II, A.F., Barrow, R.T.F., et al. (1987) Icarus **69**, 280.

French, R.G., Elliot, J.L., French, L.M., et al. (1988) Icarus **73**, 349.

French, R.G., Elliot, J.L., and Levine, S.E. (1986) Icarus **67**, 134.

Fridman, A.M. (1975) Equilibrium and Stability of Collisionless Gravitational Systems. VINITI, Moscow.

Fridman, A.M. (1986) Sov. Astron. **30**, 525.

Fridman, A.M. (1989) Sov. Astron. Lett. **15**, 487.

Fridman, A.M. (1990) Dynamics of Astrophysical Discs (Ed. J.A. Scllwood). Cambridge Univ. Press, p.185.

Fridman, A.M. and Gor'kavyi, N.N. (1987) Dynamics of the Solar System (Ed. M. Sidlichovsky). Publ. Astron. Inst. Czech. Acad. Sci. **68**, 175.

Fridman, A.M. and Gor'kavyi, N.N. (1988) Plasma Theory and Nonlinear and Turbulent Processes in Physics (Eds V.G. Bar'yakhtar *et al.*). World Scientific, Singapore, p.275.

Fridman, A.M. and Gor'kavyi, N.N. (1989a) Sov. Sci. Rev. **12**, 289.

Fridman, A.M. and Gor'kavyi, N.N. (1989b) Physics of Nonlinear Waves, Vol.2 (Eds A.V. Gaponov-Grekhov et al.). Springer, Berlin, p.156.

Fridman, A.M. and Gor'kavyi, N.N. (1992) Chaos, Resonance, and Collective Dynamical Phenomena in the Solar System (Ed. S. Ferraz-Mello). Kluwer, Dordrecht, p.75.

Fridman, A.M., Gor'kavyi, N.N., and Ozernoy, L.M. (1992) Proc. 180th Meeting Am. Astron. Soc., July 11–17, Columbus OH.

Fridman, A.M. and Khoruzhiĭ, O.V. (1996) Chaos in Gravitational N-Body Systems (Eds. Muzzio et al.). Kluwer, Dordrecht, p.197.

Fridman, A.M. and Khoruzhiĭ, O.V. (1998) in "Nonlinear Dynamics and Chaos in Astrophysics; Festschrift in Honor of George Contopoulos" (Eds J.R. Buchler et al.), Annals of the New York Academy of Sciences **867**, 156.

Fridman, A.M. and Khoruzhii, O.V. (1999a) Introduction to Classical Graviphysics (to be published).

Fridman, A.M. and Khoruzhiĭ, O.V. (1996) in "Astrophysical Discs" (Eds. J.A. Sellwood, and J. Goodman) ASP Conference Series **160**, 341.

Fridman, A.M. and Khoruzhiĭ, O.V., and Gor'kavyi, N.N. (1996) Chaos **6**, 334.

Fridman, A.M., Morozov, A.I., and Polyachenko, V.L. (1984) Ap. Space Sci. **103**, 137.

Fridman, A.M. and Ozernoy, L.M. (1991) Testing in AGN Paradigm, Proc. 2nd Ann. Oct. Ap. Conf., Maryland.

Fridman, A.M., Palous, J., and Pasha, I.I. (1981) Monthly Not. Roy. Astron. Soc. **194**, 705

Fridman, A.M. and Polyachenko, V.L. (1984) Physics of Gravitating Systems. Springer, New York.

Fridman, A.M., Polyachenko, V.L., and Zasov, A.V. (1991) Dynamics of Galaxies and their Molecular Cloud Distribution (Eds F. Combes and F. Casoli). Kluwer, Dordrecht, p.109.

Friedson, A.J. and Stevenson, D.J. (1983) Icarus **56**, 83.

Froidevaux, L., Matthews, K., and Neugebauer, G. (1981) Icarus **46**, 18.

Geguzin, Ya.I. (1969) Physics of Sintering. Nauka, Moscow.

Gehrels, T. (Ed.) (1978) Protostars and Planets. Univ. Arizona Press, Tucson AZ.

Ginzburg, I.F., Polyachenko, V.L. and Fridman, A.M. (1971) Astron. Zh. **48**.

Gold, T. (1975) Icarus **25**, 489.

Goldreich, P. and Lynden-Bell, D. (1965) Monthly Not. Roy. Astron. Soc. **130**, 7.

Goldreich, P. and Nicholson, P. (1977) Nature **269**, 783.

Goldreich, P. and Tremaine, S. (1978a) Icarus **34**, 227.

Goldreich, P. and Tremaine, S. (1978b) Icarus **34**, 240.

Goldreich, P. and Tremaine, S. (1979a) Nature **277**, 97.

Goldreich, P. and Tremaine, S. (1979b) Astron. J. **84**, 1638.

Goldreich, P. and Tremaine, S. (1980) Ap. J. **241**, 425.

Goldreich, P. and Tremaine, S. (1981) Ap. J. **243**, 1062.

Goldreich, P. and Tremaine, S. (1982) Ann. Rev. Astron. Ap. **20**, 249.

Goldreich, P., Tremaine, S., and Borderies, N. (1986) Astron. J. **92**, 490.

Goldsmith, W. (1960) Impact. Arnold, London.

Gor'kavyĭ, N.N. (1985a) Sov. Astron. Lett. **11**, 28.

Gor'kavyĭ, N.N. (1985b) Sov. Astron. Lett. **11**, 195.

Gor'kavyĭ, N.N. (1986a) On the Dynamics of Planetary Rings. Thesis, Moscow.

Gor'kavyĭ, N.N. (1986b) Nauchn. Inform. Astrosov. Akad. Nauk SSSR, No 61, p.132.

Gor'kavyĭ, N.N. (1989a) Sov. Astron. Lett. **15**, 370.

Gor'kavyĭ, N.N. (1989b) Astron. Tsirkul. No 1538, p.31.

Gor'kavyĭ, N.N. (1990a) Sov. Astron. Lett. **16**, 77.

Gor'kavyĭ, N.N. (1990b) Nauchn. Inform. Astrosov. Akad. Nauk SSSR, No 68, p.66.

Gor'kavyĭ, N.N. (1991) Sov. Astron. Lett. **17**, 428.

Gor'kavyĭ, N.N. (1994) Physics of the Gaseous and Stellar Discs of the Galaxy (Ed. I.R. King). ASP Conference Series **66**, 117.

Gor'kavyĭ, N.N. and Fridman, A.M., (1985a) Sov. Astron. Lett. **11**, 264.

Gor'kavyĭ, N.N. and Fridman, A.M., (1985b) Sov. Astron. Lett. **11**, 302.

Gor'kavyĭ, N.N. and Fridman, A.M., (1985c) Astron. Tsirkul. No 1391, p.1.

Gor'kavyĭ, N.N. and Fridman, A.M., (1986) Sov. Phys. Uspekhi **29**, 1152.

Gor'kavyĭ, N.N. and Fridman, A.M., (1987a) Nonlinear Waves: Structures and Bifurcation (Eds A.V. Gaponov-Grekhov and M.I. Rabinovich). Nauka, Moscow, p.86.

Gor'kavyĭ, N.N. and Fridman, A.M., (1987b) Sov. Astron. Lett. **13**, 96.

Gor'kavyĭ, N.N. and Fridman, A.M., (1988a) Astron. Tsirkul. No 1531, p.25.

Gor'kavyĭ, N.N. and Fridman, A.M., (1988b) Proc. VIII Soviet-American Workshop on Planetology, GEOKhI Akad. Nauk SSSR, Moscow, p.35.

Gor'kavyĭ, N.N. and Fridman, A.M., (1989) Nonlinear Waves: Dynamics and Evolution (Eds A.V. Gaponov-Grekhov and M.I. Rabinovich). Nauka, Moscow, p.376.

Gor'kavyĭ, N.N. and Fridman, A.M., (1990) Sov. Phys. Uspekhi **33**, 95.

Gor'kavyĭ, N.N., Minin, V.A., and Fridman, A.M., (1989a) Proc. II All-Union Conf. on Classical Graviphysics, Volgograd, p.48

Gor'kavyĭ, N.N., Minin, V.A., and Fridman, A.M., (1989b) Study of Planetary Physics in the USSR and the USa (Preprint SAO Akad. Nauk SSSR, No 38). Moscow.

Gor'kavyĭ, N.N., Minin, V.A., and Fridman, A.M., (1994) Astron. Zh. **71**.

Gor'kavyĭ, N.N., Morozov, A.I., and Fridman, A.M., (1986) Sov. Phys. Tech. Phys. **31**, 711.

Gor'kavyĭ, N.N. and Ozernoy, L. (1999) Ap. J. (in press).

Gor'kavyĭ, N.N., Ozernoy, L., Mather, J., and Taidakova, T. (1998) Earth, Planets and Space **50**, No 6-7.

Gor'kavyĭ, N.N., Polyachenko, V.L., and Fridman, A.M., (1989) Proc. II All-Union Conf. on Classical Graviphysics, Volgograd, p.41.

Gor'kavyĭ, N.N., Polyachenko, V.L., and Fridman, A.M., (1990) Sov. Astron. Lett. **16**, 79.

Gor'kavyĭ, N.N. and Taĭdakova, T.A. (1989) Sov. Astron. Lett. **15**, 234.

Gor'kavyĭ, N.N. and Taĭdakova, T.A. (1991a) Sov. Astron. Lett. **17**, 462.

Gor'kavyĭ, N.N. and Taĭdakova, T.A. (1991b) Proc. Internat. Conf. on Origin and Evolution of the Solar System, 27–31 August 1991, IFZ Akad. Nauk SSSR, Moscow, p.27.

Gor'kavyĭ, N.N. and Taĭdakova, T.A. (1992) Izv. Crimean Ap. Obsevr. **90**, 82.

Gor'kavyĭ, N.N. and Taĭdakova, T.A. (1993) Astron. Lett. **19**, 142.

Gor'kavyĭ, N.N., Taĭdakova, T.A., and Fridman, M.A. (1988) Sov. Astron. Lett. **14**, 441.

Gor'kavyĭ, N.N., Taĭdakova, T.A., and Gaftonyuk, N.M. (1991) Sov. Astron. Lett. **17**, 457.

Gradshteyn, I.S. and Ryzhik, I.M. (1965) Tables of Integrals, Series, and Products. Academic Press, New York.

Grebenikov, E.A. and Ryabov, Yu.A. (1984) The Search for Planets. Nauka, Moscow.

Greenberg, R. (1973) Astron. J. **78**, 338.

Greenberg, R. (1988) Icarus **75**, 527.

Greenberg, R. and Brahic, A. (Eds.) (1984) Planetary Rings. Univ. Ariz. Press, Tucson AZ.

Greenberg, R., Hartmann, W.K., Chapman, K., and Baker, D. (1978) Protostars and Planets (Ed. T. Gehrels). Univ. Arizona Press, Tucson AZ.

Gresh, D.L. (1989) Voyager Radio Occultation by the Uranian Rings. Thesis, Palo Alto CA.

Gresh, D.L., Rosen, P.A., Tyler, G.L., and Lissauer, J.J. (1986) Icarus **68**, 481.

ter Haar, D. (1948) Proc. Roy. Dan. Acad. Sc. **25**, Nr 3.

ter Haar, D. (1950) Ap. J. **111**, 179.

ter Haar, D. (1967) Ann. Rev. Astron. Ap. **5**, 267.

ter Haar, D. and Cameron, A.G.W. (1963) Origin of the Solar System (Eds R. Jastrow and A.G.W. Cameron). Academic Press, New York, p.1.

Hämeen-Anttila, K.A., Hanninen, J., and Verronen, M. (1988) Earth, Moon, and Planets **43**, 61.

Hamilton, D.P. (1994) Icarus **109**, 221.

Harris, A.W. (1976) Icarus **24**, 190.

Harris, A.W. (1984) Planetary Rings (Eds R. Greenberg and A. Brahic). Univ. Arizona Press, Tucson AZ, p.641.

Hartmann, W.K. (1978) Icarus **33**, 50.

Hartmann, W.K. (1984) Planetary Rings (Ed. A. Brahic). Copadues, Toulouse, p.407.

Hartmann, W.K. (1985) Icarus **63**, 69.

Hatzes, A.P., Bridges, F.G., and Lin, D.N.C. (1988) Monthly Not. Roy. Astron. Soc. **231**, 1091.

Hayashi, C., Nakazawa, K., and Nakagawa, Y. (1985) Protostars and Planets (Eds D.C. Black and M.S. Matthews). Univ. Arizona Press, Tucson AZ, p.1101.

Hayley, J. and Smarr, L. (1988) Numerical Astrophysics (Eds J. Centrella et al.). Jones and Bartlett, Boston MA.

Hirschfelder, J.O., Curtiss, C.F., and Bird, R.B. (1954) Molecular Theory of Gases and Liquids. Wiley, New York.

Holberg, J.B., Forrester, W.T., and Lissauer, J.J. (1982) Nature **297**, 115.

Horanyi, M. and Porco, C.C. (1993) Icarus **106**, 525.

Hord, C.W., West, R.A., Esposito, L.W., et al. (1982) Science **215**, 537.

Hubbard, W.B., Brahic, A., Bouchet, P., et al. (1985) Lunar Planet. Sci. **16**, 368.

Hubbard, W.B., Brahic, A., Sicardy, B., et al. (1986) Nature **319**, 636.

Hunten, D.M., Tomasko, M.G., Flasar, F.M., et al. (1984) Saturn (Eds T. Gehrels and M.S. Matthews). Univ. Arizona Press, Tucson AZ, p.671.

Hunter, C. (1972) Ann. Rev. Fluid Mech. **4**, 219.

Jackson, A.A. and Zook, H.A. (1989) Nature **337**, 629.

Jaffe, L.D. (1967) Icarus **16**, 75.

Jeans, J.H. (1929) Astronomy and Cosmology. Cambridge Univ. Press.

Jewitt, D. (1985) Jupiter's Moons (Ed. D. Morrison). Mir, Moscow, p.57.

Jeffreys, H. (1947) Monthly Not. Roy. Astron. Soc. **107**, 260.

Johnson, K. (1985) Contact Mechanics. Cambridge University Press.

Kalnajs, A.J. (1972) Ap. J. **175**, 63.

Kato, S. (1970) Publ. Astron. Soc. Japan **22**, 285.

Kerr, R.A. (1985) Science **229**, 1376.

Kessler, D.J. (1985) Icarus **48**, 39.

Kuhn, T. (1962) The Structure of Scientific Revolutions. Univ. Chicago Press, Chicago IL.

Kuiper, G.P., Cruikshank, D.P., and Fink, U. (1970) Sky Telesc. **39**, 14.

Kumar, S.S. (1960) Publ. Astron. Soc. Japan **12**, 552.

Landau, L.D. and Lifshitz, E.M. (1970) Theory of Elasticity. Pergamon, Oxford.

Landau, L.D. and Lifshitz, E.M. (1975) Classical Theory of Fields. Pergamon, Oxford.

Landau, L.D. and Lifshitz, E.M. (1976) Mechanics. Pergamon, Oxford.

Landau, L.D. and Lifshitz, E.M. (1987) Fluid Mechanics. Pergamon, Oxford.

Lane, A.L., West, R.A., Hord, C.W., et al. (1989) Science **246**, 1450.

de Laplace, P.S. (1789) Mém. Acad. Sci, Mécanique Céleste, Vol.3, Part 6.

Lifshitz, E.M. (1946) Zh. Eksp. Teor. Fiz. **16**, 587.

Lighthill, J. (1978) Waves in Fluids. Cambridge Univ. Press.

Lin, C.C. and Shu, F.H. (1964) Ap. J. **140**, 646.

Lin, D.N.C. and Bodenheimer, P. (1981) Ap. J. **248**, L83.

Lin, D.N.C., Papaloizou, J.C.B., and Ruden, S.P. (1987) Mon. Not. Roy. Astron. Soc. **227**, 75.

Lissauer, J.J. (1985) Nature **318**, 544.

Lissauer, J.J. and Cuzzi, J.N. (1985) Protostars and Planets (Eds D.C. Black and M.S. Matthews). Univ. Ariz. Press, Tuscon AZ, p. 921.

Lissauer, J.J., Goldreich, P., and Tremaine, S. (1985) Icarus **64**, 425.

Lissauer, J.J., Shu, F.S., and Cuzzi, J.N. (1981) Nature **292**, 707.

Longaretti, P.-Y. (1989) Icarus **81**, 51.

Longaretti, P.-Y. and Rappaport, N. (1995) Icarus **116**, 376.

Lynden-Bell, D. and Pringle, J.E. (1974) Monthly Not. Roy. Astron. Soc. **168**, 603.

Maddox, J. (1985) Nature **318**, 505.

Manfroid, J., Haefner, R., and Bouchet, P. (1986) Astron. Ap. **157**, L3.

Marouf, E.A. and Tyler, G.L. (1986) Nature **323**, 31.

Marouf, E.A., Tyler, G.L., and Eshleman, V.R. (1982) Icarus **49**, 161.

Marouf, E.A., Tyler, G.L., and Rosen, P.A. (1986) Icarus **68**, 120.

Marow, M.Ya. (1987) Die Planeten des Sonnensystems. Kleine Naturwissenschaftliche Bibliothek, Leipzig.

Maxwell, J.C. (1859) Scientific Papers, Vol.1, p.287.

McDonald, J.S.B., Hatzes, A.P., Bridges, F.G., and Lin, D.N.C. (1989) Icarus **82**, 167.

McGhee, C.A., Nicholson, P.P., French, R.G., and Hall, K.J. (1998) BAAS **30**, 1140.

McIntyre, M.E. (1981) J. Fluid Mech. **106**, 454.

Mellor, M. (1980) Physics and Mechanics of Ice (Ed. P. Tryde). Springer, Heidelberg.

Meyer-Vernet, N. and Sicardy, B. (1987) Icarus **69**, 157.

Millis, R.L., Wasserman, L.H., and Birch, P.V. (1977) Nature **267**, 330.

Mishurov, Yu.N., Peftiev, V.N., and Suchkov, A.A. (1976) Sov. Astron. **20**, 152.

Mitton, S. (1976) Exploring the Galaxies. Faber and Faber, London.

Molnar, L.A. and Dunn, D.E. (1995) Icarus **116**, 397.

Morozov, A.G. and Fridman, A.M. (1986) Sov. Phys. Tech. Phys. **21**, ??.

Morozov, A.G., Torgashin, Yu.M., and Fridman, A.M. (1985) Sov. Astron. Lett. **11**, 94.

Morozov, A.G., Torgashin, Yu.M., and Fridman, A.M. (1986) Nauchn. Inform. Astrosov. Akad. Nauk SSSR, No 61, p.110.

Morrison, D.D., Owen, T., and Soderblom, L.A. (1986) Satellites (Eds J.A. Burns and M.S. Matthews). Univ. Arizona Press, Tucson AZ, p.764.

Moulton, F.R. (1914) An Introduction to Celestial Mechanics. Macmillan, New York.

Müller, G. (1893) Publ. Potsdam Observ. **8**, 193.

Murray, J.D. (1977) Nonlinear Differential Equation Models in Biology. Clarendon, Oxford.

Nazarenko, A.I. and Skrebushenvskiĭ, B.S. (1981) Evolution and Stability of Satellite Systems. Mashinostroenie, Moscow.

Nicholson, P.D. and Jones, T.J. (1980) IAU Circ. N 3515.

Nicolis, G. (1986) Dynamics of Hierarchic Systems. Springer, Heidelberg.

Nicolis, G. and Prigogine, I. (1977) Self-Organization in Nonequilibrium Systems. Wiley, New York.

Nicolis, G. and Prigogine, I. (1989) Exploring Complexity. Freeman, New York.

Nyborg, W.L. (1969) Physical Acoustics (Ed. U.M. Mason), Vol. 2B.

Owen, W.M. and Synnot, S.P. (1987) Astron. J. **93**, 1268.

Pandey, A.K. and Mahra, H.S. (1987) Earth, Moon and Planets **37**, 147.

Panovko, Ya. G. (1977) Introduction to the Theory of Mechanical Shocks. Nauka, Moscow.

Pasha, I.I. (1983) Zemlya i Vselennaya, No 6, p.42.

Pedlosky, J. (1982) Geophysical Fluid Dynamics. Springer, New York.

Pollack, J.B. (1978) The Saturn System (NASA SP 2068; Eds. D.M.Hunten and D.Morrison). Houston TX, p.9.

Polyachenko, V.L. and Fridman, A.M. (1971) Sov. Astron. **15**, 396.

Polyachenko, V.L. and Fridman, A.M. (1972) Sov. Astron. **16**, 123.

Polyachenko, V.L. and Fridman, A.M. (1988) Sov. Phys. JETP **67**, 1.

Porco, C.C. (1991) Science **253**, 995.

Porco, C.C., Danielson, G.E., Goldreich, P., et al. (1984) Icarus **60**, 17.

Porco, C.C. and Goldreich, P., (1987a) Astron. J. **93**, 724.

Porco, C.C. and Goldreich, P., (1987b) Astron. J. **93**, 730.

Porco, C.C. and Nicholson, P.D. (1987) Icarus **72**, 437.

Porco, C.C., Nicholson, P.D., Borderies, N. et al. (1984) Icarus **60**, 1.

Potter, D. (1973) Computational Physics. John Wiley, New York.

Prentice, A.J.R. (1978) The Origin of the Solar System (Ed. S.F. Dermott). Wiley, London, p.111.

Prentice, A.J.R. and ter Haar, D. (1979) Nature **280**, 300.

Pringle, J.E. (1981) Ann. Rev. Astron. Ap. **19**, 137.

Rayleigh, Lord (1884) Philos. Trans. Roy. Soc. London **171**, 1.

Rayleigh, Lord (1941) Theory of Sound. Dover, New York.

Reach, W.T., Franz, B.A., et al. (1995) Nature **374**, 521.

Roddy, D.J., Pepin, R.O., and Merrill, R.B. (Eds.) (1977) Impact and Explosion Cratering. Pergamon, New York.

Rolfs, K. (1977) Lectures on Density Wave Theory. Springer, Berlin.

Rosen, P.A. (1989) Waves in Saturn's rings probed by radio occultation, Sci. Rept. Stanford Univ., N d845-1989-1.

Rosen, P.A. and Lissauer, J.J. (1988) Science **241**, 690.

Rosen, P.A., Tyler, G.L., and Marouf, E.A. (1991) Icarus **93**, 3.

Rosen, P.A., Tyler, G.L., Marouf, E.A., and Lissauer, J.J. (1991) Icarus **93**, 25.

Roy, A.E. (1978) Orbital Motion. Adam Hilger, Bristol.

Ruskol, E.L. (1986) Natural Satellites of the Planets. VINITI, Moscow.

Safronov, V.S. (1960) Sov. Phys. Dokl. **5**, 13.

Safronov, V.S. (1969) Evolution of the Preplanetary Cloud and the Formation of the Earth and Planets. Nauka, Moscow (English translation: Safronov (1972)).

Safronov, V.S. (1972) Evolution of the Preplanetary Cloud and the Formation of the Earth and Planets. NASA TTF-677.

Safronov, V.S. and Vityazev, A.V. (1983) The Origin of the Solar System. VINITI, Moscow.

Salo, H., Lukkari, J., and Hanninen, J. (1988) Earth, Moon, and Planets **43**, 33.

Schwarzschild, K. (1907) Nachr. Kgl. Ges. Wiss. Göttingen, 614.

Science News (1990) **127**, N.1728, p.22.

Seeliger, H. (1887) Abh. Bayer. Akad. Wiss., Math. Naturw. Kl. II **18**, 1.

Seeliger, H. (1895) Abh. Bayer. Akad. Wiss., Math. Naturw. Kl. II **16**, 405.

Shakura, N.I. and Sunyaev, R.A. (1973) Astron. Ap. **24**, 337.

Shchigolev, B.M. (1969) Mathematical Processing of Observations. Nauka, Moscow.

Shkarovskiĭ, I., Johnson, T., and Bachinskiĭ, M. (1966) The Particle Kinetics of Plasmas. Addison-Wesley, Ontario.

Showalter, M.R., Cuzzi, J.N., Marouf, E.A., and Esposito, L.W. (1986) Icarus **66**, 297.

Showalter, M.R., Burns, J.A., Cuzzi, J.N., and Pollack, J.B. (1985) Nature **316**, 526.

Shu, F.H. (1970) Ap. J. **160**, 99.

Shu, F.H., Dones, L., Lissauer, J.J., et al. (1985) Ap. J. **299**, 542.

Shu, F.H. and Stewart, R.G. (1985) Icarus **62**, 360.

Shu, F.H., Yuan, Ch., and Lissauer, J.J. (1985) Ap. J. **291**, 356.

Shukhman, I.G. (1984) Sov. Astron. **28**, 574.

Sicardy, B. (1987) Dynamics of the Solar System (Ed. M. Sidlichovsky). Publ. Astron. Inst. Czech. Acad. Sci. **68**, 169.

Sicardy, B. (1991) Icarus **83**, 197.

Sicardy, B., Brahic, A., Bouchet, P., et al. (1985) IAU Circ. N 4100.

Silkin, B.I. (1982) In the World of the Many Moons: the Satellites of the Planets. Nauka, Moscow.

Smith, B.A., Soderblom, L.A., Banfield, D., et al. (1989) Science **246**, 1422.

Smith, B.A., Soderblom, L.A., Beebe, R., et al. (1986) Science **233**, 43.

Smith, B.A. and Terrile, R.J. (1984) Science **226**, 1412.

Spahn, F. and Sponholz, H. (1989) Nature **339**, 607.

Spaute, D. and Greenberg, R. (1987) Icarus **70**, 289.

Spehalski and D.Matson (1998) The Cassini Mission, 3-19-93 (information from Internet).

Spitzer, L. (1962) Physics of Fully Ionized Gases. Interscience, New York.

Steigmann, G.A. (1978) Nature **274**, 454.

Steigmann, G.A. (1984) Monthly Not. Roy. Astron. Soc. **209**, 359.

Stewart, G.R. and Kaula, W.M. (1980) Icarus **44**, 154.

Stewart, G.R., Lin, D.N.C., and Bodenheimer, P. (1984) Planetary Rings (Eds R. Greenberg and A. Brahic). Univ. Arizona Press, Tucson AZ, p.447.

Stewart, G.R. and Wetherill, G.W. (1988) Icarus **74**, 542.

Stone, E.C. and Miner, E.D. (1989) Science **246**, 1417.

Subbotin, M.F. (1968) Introduction to Theoretical Astronomy. Nauka, Moscow.

Syer, D. and Clarke, C.I. (1992) Mon. Not. Roy. Astron. Soc. **255**, 92.

Tabor, D. (1948) Proc. Roy. Soc. **A192**, 247.

Taĭdakova, T.A. (1990) Nauchn. Inform. Astrosov. Akad. Nauk SSSR, No 68, p.72.

Taidakova, T.A. and Gor'kavyi, N. (1998) The Dynamics of Small Bodies in the Solar System: A Major Key to Solar Systems Studies (Eds. A.E. Roy and B.A. Steves). Proc. NATO Advanced Study Institute, Acquafredda di Maratea, 1999.

Thomas, P., Weitz, C., and Veverka, J, 1989 Icarus **81**, 92.

Thompson, D.E. 1984 Sci. News **126**, 133.

Thompson, D.E. 1987 Sci. News **121**, 403.

Thompson, W.T., et al. (1981) Icarus **46**, 187.

Toomre, A. (1964) Ap. J. **139**, 1217.

Toomre, A. (1969) Ap. J. **158**, 899.

Trulsen, J. (1971) Ap. Space Sci. **12**, 329.

Trulsen, J. (1972) Ap. Space Sci. **17**, 330.

Turing, A.M. (1952) Phil. Trans. Roy. Soc. London **B237**, 37.

Tyler, G.L. (1987) Proc. IEEE **75**, 1404.

Van Flandern, T.C. (1979) Science **204**, 1076.

Veverka, J., Thomas, P., Johnson, T.V., *et al.* (1986) Satellites Eds J.A.Burns and M.S.Matthews). Univ. Arizona Press, Tucson AZ, p.342.

Vityazev, A.V., Pechernikova, G.V., and Safronov, V.S. (1990) Planets of the Terrestrial Group; Origin and Early Evolution. Nauka, Moscow.

Voĭtkovskiĭ, K.F. (1977) Mechanical Properties of Snow. Nauka, Moscow.

Ward, W.R. (1981) Geophys. Res. Lett. **8**, 641.

Ward, W.R. and Harris, A.W. (1984) Planetary Rings (Ed. A. Brahic). Copadues, Toulouse, p.439.

Weizsäcker, C.F.von (1943) Zs. Ap. **22**, 319.

Weidenschilling, S.J. and Jackson, A.A. (1993) Icarus **104**, 244.

Wilford, J.N. (1989) N.Y.Times, Dec. 15, p.A22,

Zarembo, L.K. (1969) Physics and Technology of High-Power Ultrasound (Ed. L.D. Rozenberg). Nauka, Moscow, p.87.

Zebker, H.A. (1984) Analysis and Interpretation of the Voyager-1 Radio Occultation Measurements of Saturn's Rings with Emphasis on Particle Size Distribution (Sci. Repts. Stanford Univ. D841-1948-4). Palo Alto CA.

Zebker, H.A., Marouf, E.A., and Tyler, G.L. (1985) Icarus **64**, 531.

Index

Springer
and the
environment

At Springer we firmly believe that an international science publisher has a special obligation to the environment, and our corporate policies consistently reflect this conviction.

We also expect our business partners – paper mills, printers, packaging manufacturers, etc. – to commit themselves to using materials and production processes that do not harm the environment. The paper in this book is made from low- or no-chlorine pulp and is acid free, in conformance with international standards for paper permanency.

 Springer

Myne Awne good cardinall I recommende me unto yow
w all my hart and thanke yow for the greate payne
and labour that yow do take in my bysynes and maters
desyryng yow (that wen yow have well establyshd them)
to take summe pastyme and cofort to the intente yow may
the lenger enduze to serue vs for allways payne can nott
be induryd / suerly yow have so substancyally ordeyd oure
maters bothe off thys syde the see and by onde that in myne
oppynyon lytyll or no thyng can be addyd / neverthelesse accordyng
to your desyre I do send yow myne oppynyon by thys berar
the reformation wherof I do remytte to yow and the
remnante off our trusty coselleurs whych I am suze wyll
substantially loke on hyt / as tochyng the mater that sir
wylliam pirs broght answar off I am well cotentyd
w what ordez so euer yow do take in att the quene
my wyff hathe desyry'd me to make haz most hasty
recomendatiouns to yow as to hym that she louyth
very well and bothe she and I wolde knowe fayne
wen yow wyll repayre to vs no more to yow
att thys tyme but that w godis helpe I truste we shall
desapoynte oure enymys off theyre intendyd pporpose
wryttyn w the hand off your louyng master

HENRY R

To my lord cardinall.

THE DIVORCE

by Marvin H. Albert

SIMON AND SCHUSTER · NEW YORK

1/6/68

for my beloved

MELBA

I would like to express my appreciation to:
The British Museum
for the courtesy extended to me during the year spent in England doing the research for this book, and to
The New York Public Library
for allowing me to use the Frederick Lewis Allen Memorial Room while writing the final version.

—M.H.A.

Contents

Illustrations

Introduction

THE TRIANGLE

HENRY VIII, as seen by Thomas More:
"If a lion knew his strength, it were hard
for any man to rule him."

CATHERINE OF ARAGON, as seen by Henry VIII:
"She is a woman of most gentleness, humility
and buxomness; yea, and of all good qualities
pertaining to nobility she is without comparison."

ANNE BOLEYN, as seen by Wyatt, the poet who loved her:
"There is written her fair neck round about
Noli me tangere; for Caesar's I am
And wild for to hold, though I seem tame."

HISTORY knows no triangle so fateful as this one. At its center
was a common emotional entanglement: two women in a contest of wills for one man. A husband tied to his wife but infatuated with a new love; a wife clinging to her marriage with
all her strength; another woman exploiting the husband's passion for her in a gamble for the wife's position. But in this

instance the husband was Henry VIII; his wife, Catherine of Aragon; the other woman, Anne Boleyn. Never has the consummation of an illicit love been so far-reaching in its consequences.

King Henry VIII's long battle for a divorce gave birth to side controversies which grew to fantastic proportions. It brought to England both Reformation and counter-Reformation, in which Protestants and Catholics waged savage war upon each other. It broke the bond between Rome and England, changed the religion of the British people, and brought to the surface a power struggle between the rusting machinery of the Church and the emerging strength of the State. It involved all Europe in a clash that altered British history as drastically as the Battle of Hastings, and influenced the course of world history more pointedly than the Battle of Waterloo.

Without it, there might have been no Drake plundering the Spanish Main, no Invincible Armada, no Dutch or English colonization of America; England might never have risen from a third-rate power to the strongest nation on earth, and Holland never have won its independence; Spain might not have declined and the United States of America never have been born.

These events were determined by the divorce controversy, which visited the sins of Henry VIII upon his daughter by Catherine, his daughter by Anne Boleyn, and the children of his two sisters. It forced his daughter Mary to destroy Lady Jane Grey, the granddaughter of one of his sisters; it drove his daughter Elizabeth to dispatch Mary of Scotland, granddaughter of his other sister. For from the hotly debated question of the validity of the divorce arose even more momentous matters: the succession to Henry's throne, the religion of England, and the position of Britain in world affairs.

This is the story of the struggle fought out by Catherine, Henry and Anne Boleyn; of four remarkable men named Thomas—Wolsey, More, Cromwell and Cranmer—who became fatally enmeshed in it; and of two helpless, terrorized little girls who somehow survived its dangers to become known to history rather one-sidedly as Bloody Mary and Good Queen Bess.

Above all, this is the story of a clash of opposing faiths, in which is seen once more that sincere and decent men, when acting to protect their faiths, are capable of martyrdom—and also capable of committing acts immeasurably more vile than any twisted criminal would commit for gain or lust.

—M.H.A.

1964

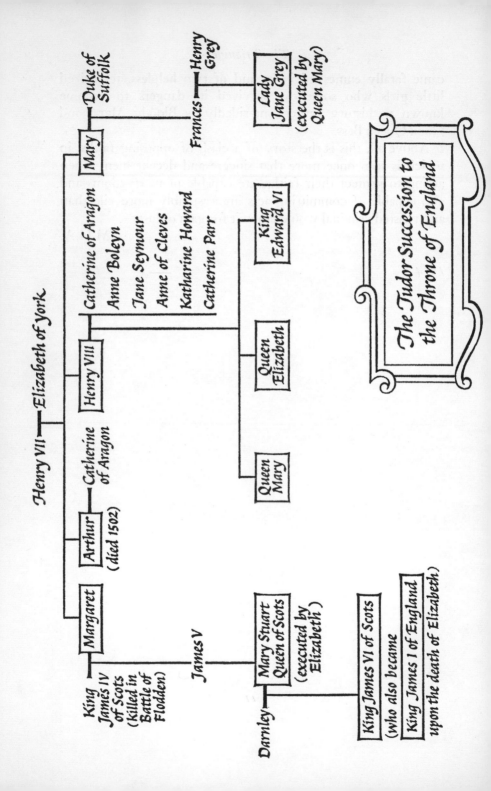

The Tudor Succession to the Throne of England

Henry VII — Elizabeth of York

Margaret | Arthur — Catherine of Aragon | Henry VIII | Mary — Duke of Suffolk

Arthur (died 1502)

King James IV of Scots (Killed in Battle of Flodden)

James V

Mary Stuart Queen of Scots (executed by Elizabeth)

Darnley

King James VI of Scots (who also became King James I of England upon the death of Elizabeth)

Henry VIII married:
Catherine of Aragon
Anne Boleyn
Jane Seymour
Anne of Cleves
Katharine Howard
Catherine Parr

Queen Mary

Queen Elizabeth

King Edward VI

Frances — Henry Grey

Lady Jane Grey (executed by Queen Mary)

Part One

THE WIFE

One's heart must either harden or break.
—CHAMFORT

Chapter One

CATHERINE OF SPAIN

CATHERINE MET the future King Henry VIII on the day she married his elder brother, Arthur. She first saw Henry as a plump ten-year-old boy richly costumed in gold and white, waiting to lead her into St. Paul's Cathedral. It was this boy who took her hand and escorted her past staring crowds of nobles to the platform upon which the wedding between her and Arthur was solemnized by the Archbishop of Canterbury.

Less than eight years later this same boy became her second husband.

It was to be neither the first nor the last of the shocks which England held in store for the Princess from Spain, youngest child of Ferdinand of Aragon and Isabella of Castile. And through all the difficulties that awaited her, Catherine was to suffer from her inability to adjust to the basic change: from the Spain of her first fifteen and a half years to the England where she would spend her remaining thirty-five; from a country of dry-hot sunshine and burning religious fanaticism to an island of damp mists and thin sunlight and cool political intrigues.

Catherine had the warmth, strength of character and devotion to duty to have made her the perfect royal wife and mother. Reasons physical, personal and political denied her this role. Instead, there were flung at her challenges which she was ill equipped to meet. It was Catherine's misfortune to inherit the wrong qualities from her parents for coping with her fate. She had her mother's militant piety and rigid courage in the face of enemy forces, unleavened by her father's tactical instincts in the battle.

Catherine's youth was dominated by her mother's impressive character and direct mind. Isabella, called "the Queen with a man's heart," had the extreme virtues and faults of the unquestioning champion of any faith. She was brave, dedicated, dignified, bigoted, and without mercy. As she saw it, outside the authoritative mantle of the Church there was no salvation—and no one who need be treated as human. It was with this religious fire that Isabella rallied the Catholic troops of Spain to her "last crusade." Clad in shining armor and flashing a sword, Isabella led them against the Moors who had overrun Spain seven hundred years before, and who still held the southern kingdom of Granada when Catherine was born.

The motives of Catherine's father in this "crusade" had little to do with religion. King Ferdinand was a more complex character than his wife. He believed that religion had its uses, he was normally pious, and he tended to be circumspect in dealings with Church authority. Bold in war and devious in diplomacy, he used whatever methods the occasion called for to reach his goal. It was his ambition to control Italy and become the most important monarch in Europe. But to attain this he needed a unified Spain behind him. His marriage to Isabella had joined his kingdom of Aragon with her kingdom of Castile. He regarded his wife's crusade simply as a means of bringing the kingdom of Granada into this new unity.

Catherine was born during this war, on December 15, 1485. She spent her childhood in the midst of events that were forging Spain into one of Europe's most powerful nations. While

Queen Isabella was giving birth to her, King Ferdinand was winning the Moorish city of Ronda by storm. And once the child was born Isabella wasted little time in getting back into action. Catherine grew from babyhood in the battle camp of the Spanish troops besieging the last great strongholds of the Moors. She witnessed the Spanish attacks on Moorish ramparts and the answering raids of the Moors against the Spanish positions. One night she almost perished in flames when the massed tents of her parents' siege camp were consumed by fire. She watched the rebuilding of the camp as a stone city named Sante Fé.

Catherine was six years old during the most eventful year of her life in Spain: 1492.

That was the year Granada finally surrendered, in exchange for the sworn promise of Ferdinand and Isabella that the Moslems would be allowed freedom of worship—which promise they broke after Granada was safely in their hands. With her crusade won, Isabella could begin her campaign for a totally Catholic state. Moslem men, women and children were sold into slavery. A quarter of a million Spanish Jews fled the country, leaving their homes and belongings behind— rather than be forced to choose between becoming Catholics and being slaughtered. The Inquisition seized heretics, suspected heretics and rumored heretics. Terror, torture and thousands of slow deaths were used to stamp out all independence of thought—with such success that eventually Spain was to have too few independent thinkers left to be able to compete with the rest of Europe.

It was in 1492 also that Catherine's mother sent Columbus on his voyage across the Atlantic which opened a New World to Spain—for which Spain was to repay him with disgrace and poverty, drawing from him the bitter comment: "If I had stolen the Indies and given them to the Moors, Spain could not have shown me greater enmity."

Through these historic events, Catherine was being raised

and educated as a royal princess, acquiring a pride in her heritage and an understanding of her role in furthering that heritage. She learned other things required of her—from embroidery to Latin—from the best tutors in Europe. And always, she knew that her future awaited her in England.

She was only three years old when her parents concluded a marriage treaty between Catherine and Arthur, Prince of Wales, son and heir to the throne of England's Henry VII. The two babies represented pawns in a game from which both parental parties expected important political advantages.

France, expanding its power into weak and divided Italy, was the main threat to Spain's position in the Mediterranean. So Ferdinand used all five of his children to strengthen Spain's position against France, surrounding France with enemies allied to him. He protected his own vulnerable west flank by marrying two of his daughters to Portugal. Another daughter and his only son became weapons against France's northern and eastern borders by their marriage contracts with children of Emperor Maximilian, head of the Holy Roman Empire—Germany. West of France lay England; so it was to England that his youngest child, Catherine, was given.

Given—in contract and treaty; but not in person. And England's Henry VII was soon urging that she be sent to him. For not until Catherine was physically within his realm could he be absolutely certain the marriage would take place; and he had reasons for desiring this alliance perhaps more strongly than Ferdinand.

Henry Tudor's right to the throne of England was most tenuous. In the general exhaustion at the end of the protracted War of the Roses he had won the throne in battle from Richard III, who was killed in the fighting. The other royal families of Europe were not impressed. The throne Henry VII had won in war he continued to hold in peace by strength of purpose, shrewd political maneuvering and cold-blooded financial chicanery. But there were other men who had a clearer title to the throne by birth. The Earl of Warwick, for one; for another, there was a Perkin Warbeck, who claimed to be the son of Edward IV, and was recognized as such by various important personages on the Continent.

Catherine's marriage to Henry Tudor's elder son would change all this. By bringing to the marriage the undoubted royal blood of Aragon and Castile, Catherine would make the Tudor line respectable and respected. The various royal families allied to Ferdinand and Isabella would support the Tudor right to the crown of England against all contenders.

But if Henry VII was anxious to have Catherine in England to buttress the Tudors' shaky claim to the crown, Ferdinand and Isabella were not anxious to let her go to England until the Tudor crown was less insecure.

There were other reasons for the delay. There was squabbling over the size of the marriage portion which Catherine should bring with her to England. And there was the fact that from start to finish Henry and Ferdinand were each scheming to cheat the other. Ferdinand wanted to keep his daughter as a lure to persuade Henry VII to go to war against France— while secretly keeping open the possibility of marrying Catherine into some other family more solid than the Tudors. Henry, for his part, used the promise of making war on France as a lure to get Catherine to England—while actually continuing his policy of keeping both Spain and France forever courting his friendship.

The years of Catherine's young life slipped by and she grew out of childhood—still in Spain. Henry VII continued to urge that she be sent to him; Ferdinand and Isabella continued to delay.

A Spanish ambassador wrote from England to warn Their Spanish Majesties that it might be unwise to wait till Catherine was too old to adjust to the drastic change of countries: "Though it is not my business to give advice, I take the liberty to say that it would be a good thing if she were to come soon, in order to accustom herself to the way of life in this country and learn the language.

"On the other hand," he hedged, to qualify an opinion that might prove unpopular with his rulers, "when one sees and knows the manners and the way of life of this people on this island, one cannot deny the grave inconveniences of her coming to England before she is of age. . . .

"But," he added, with the traditional caution of international diplomats, "the Princess can only be expected to lead a happy life through not remembering those things which would make her less enjoy what she would find here. It would, therefore, still be best to send her directly, and before she has learned fully to appreciate our habit of life and our government."

Since Ferdinand and Isabella had other matters on their minds than Catherine's future happiness, she remained in Spain, learning "fully to appreciate" the Spanish way of life. Her birthdays rolled by: twelve, thirteen, fourteen. She was addressed as the "Princess of Wales." But still her royal parents waited and evaded, simultaneously using her as a tool to induce King Henry to submit to their will and as bait in quiet alliance negotiations with other powers.

Her husband-to-be, young Arthur, joined his father in urging her to wait no longer: "Most illustrious and most excellent lady, my most entirely beloved spouse," he wrote to Catherine when he was twelve and she was thirteen, "I have read the most sweet letters of Your Highness lately given to me, from which I have easily perceived your most entire love for me. Truly those letters, traced by your own hand, have so delighted me, and have rendered me so cheerful and jocund, that I fancied I beheld Your Highness and conversed with and embraced my dearest wife. I cannot tell you what an earnest desire I feel to see Your Highness, and how vexatious to me is the procrastination about your coming. . . . I cherish your sweet remembrance night and day . . . let your coming to me be hastened, that instead of being absent we may be present with each other, and the love conceived between us and the wished-for joys may reap their proper fruit. . . . Your Highness' most loving spouse, Arthur, Prince of Wales."

This was supported by assurances and flattery from Henry VII's wife, Elizabeth of York, to Queen Isabella: ". . . Although we before entertained singular love and regard to Your Highness above all other queens in the world . . . for the eminent dignity and virtue by which Your Majesty so shines and excels that your most celebrated name is noised abroad and diffused everywhere; yet much more has this our love increased and accumulated by the accession of the most noble affinity which has recently been celebrated between the most illustrious Lord Arthur, prince of Wales, our eldest son, and the most illustrious princess the Lady Catherine, the infanta, your daughter. Hence it is that . . . we wish and desire from our heart that we may often and speedily hear of the health and safety of Your Serenity, and of Lady Catherine, whom we think of and esteem as our own daughter . . ."

In the end it was not boyish supplications and maternal assurances that changed the situation, but the shifting political circumstances: Over the years Henry VII grew in value as an ally, by gradually strengthening his position in Europe and by steadily improving the stability of his reign in England. Finally, he insured the future of the Tudors by hanging Perkin Warbeck and chopping off the head of the Earl of Warwick—thus eliminating the two men most likely to contest the right of his son to the throne.

The Spanish ambassador sent Ferdinand and Isabella the good news: "Now it has pleased God that all should be thoroughly purged and cleansed, so that not a doubtful drop of royal blood remains in this kingdom, except the true blood of the king and queen, and above all, that of the lord prince Arthur."

If this was not quite true, it was close enough. The rulers of

Spain wrote back that they were sending the ratification made by their daughter of her marriage with King Henry's elder son, and hoped that "great advantages to both countries will flow from this matrimonial union."

In later years Catherine was to say from the depths of misery that perhaps her marriage to the house of Tudor was doomed as "a judgement of God" because she had "achieved it through blood and murder."

2

CATHERINE was in her fifteenth year when she sailed from Spain, never to return. She was at that age a plain but not unattractive girl with long russet hair and a lovely, fair complexion. Her precocious figure was strong and pleasingly plump, she moved with dignity upon tiny, delicate feet, and her graceful gestures showed to advantage her small, shapely hands. People found her learned but not clever, good-natured but humorless; reserved, pious and proud. Her favorite pastime was needlework. Though she had a warm love of life her nature was devoid of frivolity.

Catherine was always deeply conscious of the fact that she was a royal princess, personifying a great complex of rights, privileges and duties, and destined to be a queen. But she was also a human being, and at the time of her departure from Spain a very young one with her feelings still close to the surface. She felt not only the wrench of leaving the familiarity of her own land, but also the fright of separation from a mother to whom she was deeply attached.

Catherine had seen little of her busy father during her lifetime. But Isabella, with her crusade won, the Inquisition successfully launched and illness curbing her activities, had finally found time to devote herself to her family. By then the only one of her children still at home was her youngest. For several years Catherine had absorbed the full strength and heat of Isabella's dominating, bigoted, passionate personality.

Now, abruptly, the land of Catherine's childhood and the mother who had cherished her were left behind. Her fears as she journeyed toward the unknown were heightened by a terrifying, storm-tossed sea voyage during which the ship carrying her almost sank beneath the waves. She arrived at Plymouth ill and shaken, and went immediately to church to thank God for preserving her life.

Still not over the horrors of the ocean passage, she next experienced the strange sensation of being for the first time in an alien land whose citizens spoke a language she did not understand. But she was cheered by the obvious delight with which the English people greeted her as she and her attendants made their way toward London. Her mother learned of this in one of the first reports she received from England: "The Princess could not have been received with greater joy were she the Savior of the world."

Added to the openhearted welcome of the people was a message of greeting Catherine received from Henry VII: "Your arrival is to us so very agreeable, that we cannot say or express the great pleasure, joy, and consolation which we have from it, and especially from the expectation of seeing your noble presence, which we have often desired. . . .

"Madam, it has likewise been most pleasing to us that you have escaped and passed the great dangers of the sea, and have arrived here at a port of safety, you and your fair company . . . and we give thanks to God for all. I beg that it may please you to regard us henceforward as your good and loving father, as familiarly as you would do the king and queen your parents; for on our part we are determined to treat, receive and favor you like our own daughter . . ."

This welcome from king and people was the only warmth Catherine found as she proceeded toward her new life. England was damp and cold. The blazing fireplaces in the buildings in which she spent her nights as she journeyed north roared most of their heat straight up the chimneys. October rains turned the roads to deep mud that slowed the progress of her procession to a crawl.

The King finally grew impatient to see what she looked like and rode south to meet her, taking along the fourteen-year-old bridegroom. They were met along the way by the Spanish ambassador, who had irritating news. The leading personages in Catherine's retinue—her duenna and the Archbishop of Santiago—had decided to adhere to the rigid letter of Castilian custom: Neither the King nor the Prince of Wales might see the Princess until the wedding ceremony.

Henry VII was not a man used to being thwarted in his own land. He felt the first straining of his fatherly affection toward Catherine—an affection that he was soon to forget had ever existed. He held a hasty conference with his advisers. They saw his anger and were quick to assure him that Catherine of Aragon, being now in his country, was his subject and must put his wishes before those of anyone else.

Thus reinforced in his own belief, the King advanced on the village where Catherine had taken refuge for the night. He flatly informed the heads of Catherine's retinue that he intended to meet her. They tried delaying tactics: "The Princess has retired to her chamber and can see no one." To which Henry VII responded with growing heat: "Tell the lords of Spain that the King of England will see the Princess even if she is in her bed!"

The Spaniards wilted. Catherine was brought out to meet her future father-in-law and the Prince of Wales. Henry VII studied her and decided that she appeared to be healthy enough to bear him the grandchildren he needed for the continuance of the Tudor line upon the throne of England. He also declared that he was impressed by her "agreeable and dignified manners."

For her part, Catherine looked for the first time upon that tough warrior and slippery politician, the lean, shrewd-eyed, tight-lipped, gaunt-cheeked King of England—and upon the boy who was to be her husband.

At fourteen, Arthur was blond, tall for his age and delicately handsome. Unfortunately, he was also exceedingly fragile, and had seldom known good health since birth. His brother Henry, at ten, was already more solidly built and stronger than Arthur, and a great deal more athletic. But in terms of practical politics young Henry was a nonentity; a younger brother, of not much value except as a reserve for emergency use. Arthur was the elder brother, and as such heir to the throne of Britain, be he ever so puny and sickly.

Catherine found it difficult to understand them and to make herself understood. They finally resorted to an interpreter. With translated speeches, gestures and manner the King and the Prince assured her they were delighted to have her in England. After they parted for the night Arthur dutifully wrote to Catherine's parents that he had never felt so much joy in his life as when he "beheld the sweet face" of his bride. He swore that no woman in the world could be more agreeable to him, and he promised to be a good husband to her.

They were married on November 14, 1501. The festivities over, Catherine and Arthur set off to begin their life together in the grim Welsh Ludlow Castle. There were forebodings, even that early.

Henry VII wrote to Ferdinand and Isabella: "We recently despatched into Wales the most illustrious Arthur and Catherine, our common children. For although the opinions of many were adverse to this course by reason of the tender age of our son, yet were we unwilling to allow the Prince and Princess to be separated at any distance from each other. Thus much we wished to show unto you by this our letter, that you may understand our excessive love towards the most illustrious

Lady Catherine, our common daughter, even to the danger of our own son."

There had indeed been much debate about whether marital intimacy should not wait till Prince Arthur was older, and perhaps stronger. The final decision was influenced by the fact that the binding legality of the marriage would remain unsure until it was physically consummated.

The question of this consummation was destined to be publicly debated across Europe. The future of England's history was to hang on the answer.

There were those who would swear that on the morning after his wedding night, young Arthur came out of his bride's chamber demanding wine and boasting: "Last night I was in Spain. . . . Marriage is thirsty work."

Catherine herself swore that she emerged from her marriage to Arthur "as intact as when I emerged from my mother's womb."

Consummated or not, the marriage ended less than five months after it began. In April of 1502 Arthur was felled by an unknown illness. He died at Ludlow Castle—leaving Catherine a sixteen-year-old widow with a most uncertain future.

3

CATHERINE had hardly seen her young husband into his grave before she found herself once more the pawn in a renewed game of political maneuvering between her parents and her father-in-law. With one immense difference: this time she was in Henry VII's hand. And this king knew how to use whatever came to hand to his advantage.

That Henry VII was unscrupulous in no way differentiates him from the other rulers of his time. That he was successful at it is a great deal more revealing, considering the odds against him. A sickly child, he had survived a rigorous, dangerous youth of dodging assassins. An exile, he had invaded his own country at the head of two thousand ill-assorted fighting men, killed Richard III at the Battle of Bosworth, and made himself king. In a country that had been torn time and again by civil conflicts and violent changes of monarchs, he smashed conspiracies against him and quelled armed uprisings; he established his authority, kept it, and made it recognized. All this, while making England felt as a power upon the European stage, despite the fact that she only possessed some three million people—against Spain's eight million, France's twelve million, the Holy Roman Empire's twenty million in its German population alone.

Henry VII played dynastic power politics with relish and patience. He concealed his thoughts while reading those of the other players with intuitive accuracy. He bluffed constantly and cheated wherever possible, as did his opponents. It was a game for deceivers, and Henry VII wove as tangled a web as any.

Bacon summed him up as "a wonder for wise men . . . sad, serious and full of thoughts. . . . His wit increased upon the occasion; and so much the more, if the occasion were sharpened by danger."

This was the king who decided after Arthur's death that as long as he held Catherine—and kept her future in suspense—she would remain a useful lever which he could employ at will in negotiating with Spain.

Ferdinand and Isabella were as aware of this as King Henry; while he sought to use their daughter as a weapon, they concentrated on trying to turn that weapon in his hand.

They moved quickly on learning of Arthur's death. A special ambassador, Estrada, was sent to England with two opposing sets of orders—one open, the other secret. In their name Estrada demanded of King Henry three things which he obviously would not want to do: First, that he return the hundred thousand scudos which he had been paid as the first half of Catherine's marriage portion. Secondly, that he immediately turn over to Catherine the lands, manors and revenues promised to her by Henry as a dowry. Thirdly, that he send Catherine back to Spain "in the shortest time possible."

The object of these demands was to pressure King Henry into suggesting, himself, the real purpose behind the sending of the special ambassador. The secret commission Estrada carried with him gave him the power to conclude a marriage between Catherine and King Henry's remaining son.

For the eleven-year-old boy, Henry, was now Prince of Wales—sole remaining heir to the throne of England.

King Henry turned a deaf ear to Estrada's open demands. He was too tight-fisted to disgorge even part of the dowry he had promised Catherine. He had no intention of returning the first installment of the marriage portion sent by Spain; in fact, he pointed out that the second installment was now due him. As for Catherine, he never considered sending her back to Spain.

He knew that Ferdinand and Isabella did not really want her returned; he understood Estrada's secret mission. He could always delve into the mind of Spain's regular ambassador to England, Puebla, who had become more devoted to Henry than to Ferdinand and Isabella, his master and mistress.

The notion of his elder son's widow becoming the wife of his younger son had already occurred to King Henry. He was not against it. But he was in no hurry. He could marry her to Prince Henry whenever he wished. Meanwhile there was time for a long, careful look at the map of Europe—to see if there might not be an even stronger match.

Catherine, still confined at Ludlow Castle, became ill. Terrified of following Arthur to the grave, she pleaded for King Henry to remove her from that unhealthy pile of stone with its clinging damp and memory of death. He graciously allowed her to move to more comfortable quarters near London. And there he let her wait, while he studied the possibilities at length.

Isabella soon began to exert pressure on King Henry to settle her daughter's future. She instructed Estrada in detail:

Press much for the departure of the Princess of Wales, my daughter, so that she may immediately come here. You must say that the greater her loss and affliction, the more reason is there for her to be near her parents. . . . Besides, the Princess of Wales can show the sense of loss she entertains better here in Spain, and give freer vent to her grief, because the customs of this country better permit it than do those of England.

You shall say to the King of England that we cannot endure that a daughter whom we love should be so far from us when she is in affliction, and that she should not have us at hand to console her; also that it would be more suitable for a young girl of her age to be with us than in any other place. . . .

You shall, moreover, tell him that you have commandment from us to freight vessels for her voyage. To this end make such a show of giving directions and setting about preparations for the journey, that all persons may believe it is true . . . show all other signs of approaching departure.

Moreover, you shall speak without delay about the restitution of the 100,000 scudos of the marriage portion. . . .

If, while urging these things, they should speak to you about the betrothal of the Prince of Wales with the Princess, you shall hear what they have to say, not showing any desire for it . . .

. . . the one object of this business is to bring the betrothal to a conclusion as soon as you are able. For then all our anxiety will cease, and we shall be able to seek the aid of England against France.

But King Henry continued to delay. When his wife, Elizabeth of York, died some ten months after Arthur, he was presented with a unique opportunity. Though he had cared for Elizabeth, he did not waste time in mourning her. In that age, and at that lofty level, a man's interest in his family's future usually took precedence over his concern for any immediate member of it.

As soon as his wife was buried, Henry—who was forty-six now, though the ravages of illness made him seem older— offered to marry his son's widow himself. In this way he could achieve the wished-for alliance between Spain and the Tudor name—and still leave his remaining son free to be married to some other powerful nation, perhaps France or the Austrian house of Hapsburg. He persuaded Puebla to communicate his offer to Spain.

Isabella flatly rejected his offer: "This would be a very evil thing—one never before seen, and the mere mention of which offends the ears . . . we would not for anything in the world that it should take place . . . a thing not to be endured . . . on no account would we allow it, or even hear it mentioned."

She ordered Estrada to leave no doubt that she meant what she said, "in order that the King of England may lose all hope of bringing it to pass, if he have any. For, the conclusion of the betrothal of the Princess, our daughter, with the Prince of Wales, his son, would be rendered impossible if he were to nourish any such idea."

Isabella saw in King Henry an unhealthy, aging man, with little hope of enabling Catherine to bear future kings of England. She emphasized that there were only two choices: either Catherine became betrothed to the young prince immediately, or she returned to Spain.

King Henry gave up his notion of becoming Catherine's husband—but parried both of Isabella's alternatives. He assured the Queen of Castile that he desired what she desired; however, this marrying of one son to another son's widow was no simple matter. God forbade it, as the Holy Bible warned: "And the Lord spake unto Moses, saying . . . thou shalt not uncover the nakedness of thy brother's wife . . . if a man shall take his brother's wife, it is an unclean thing . . ."

Isabella had anticipated this. The question was: *Had* Catherine ever been Arthur's wife? If the marriage had been physically consummated the answer was yes; if Catherine was still a virgin the answer was no. To Estrada was entrusted a delicate mission: "Be careful to get at the truth as regards whether the Prince and Princess of Wales consummated the marriage . . . use all the flattering persuasions you can to prevent them from concealing it from you."

Estrada inquired, as discreetly as he could; but Spaniards tend to be blunt in such vital matters.

Catherine was also a Spaniard, and her answer was equally blunt. Whether it was also truthful is one of the unknowns of history. She knew the answer expected of her. She gave it, firmly: she was still as intact as when she had come from her mother at birth.

But the English were skeptical. Arthur's post-wedding-night boast had become common knowledge among the nobility.

Catherine stood her ground: She was a virgin. Young Arthur had died without once in their short span together having intercourse with her.

Still, Arthur's boast had been heard, and Isabella was too clearheaded to ignore its effect. While maintaining that her daughter was a virgin, she at the same time petitioned for a papal dispensation which would allow Catherine to marry Prince Henry even if she had had intercourse with Arthur.

Ferdinand explained to his ambassador at Rome: "Though they were wedded, Prince Arthur and the Princess Catherine never consummated the marriage. It is well known in England that the Princess is still a virgin. But as the English are much disposed to cavil, it has seemed to be more prudent to provide for the case as though the marriage had been consummated."

King Henry was asked to prove his good faith by also ask-

ing the Pope for such special dispensation. Seeming at last to
bend before the steady pressure, he petitioned Rome.

"Ferdinand and Isabella, as well as Henry VII," read the
treaty all three signed, "promise to employ all their influence
with the Court of Rome, in order to obtain the dispensation
of the Pope necessary for the marriage of the Princess Catherine with Henry, Prince of Wales. The Papal dispensation is
required . . . because her marriage with Prince Arthur was
solemnized according to the rites of the Catholic Church, *and
afterwards consummated.*"

The last three words seemed to Isabella, when she signed
the treaty, to insure her daughter's future—since they meant
that once the Pope gave his dispensation Catherine's marriage
to Prince Henry would be valid no matter what had occurred
between her and Prince Arthur.

It was on those terms that the Pope issued his dispensation,
and that Catherine was finally betrothed to Henry, Prince of
Wales. Catherine was now eighteen; the prince was just turning twelve. It was agreed that their actual marriage would not
occur until he became fourteen. Catherine settled down to
wait out the two years. Two years that were to stretch into
six.

Chapter Two

BRIDE-IN-WAITING

CONSIDERING THE GHASTLY MARRIAGES into which princesses were likely to be plunged for dynastic reasons, Catherine had reason to be delighted with her husband-to-be. True, Prince Henry was some years younger than she; but that would become less of a drawback with time. And from what she saw of the young Prince, it was obvious that he was going to grow up to be a magnificent man: big, strong, athletic; intelligent and learned, handsome and amiable. Many observers from foreign courts were reporting these qualities in the Prince; he was already being described as a young god.

But Catherine saw little of him. The King made sure of that. He did not want his son becoming overfond of his bride. The marriage seemed settled; but it could still be undone. A better arrangement might yet arise before the two years were up. Still, the King was not unkind to Catherine; not yet. He took her with him as he moved about the country—from Richmond to Windsor to Westminster to Greenwich—attending to his business as king. During one twelve-day stay at Windsor, she accompanied him on his daily hunts through the nearby forests.

This life ended for Catherine when she fell prey to a series of illnesses brought on by her inability to adjust to the English climate. The King went on his way without her. Despite purgings and bleedings by the physicians assigned to her, Catherine remained ill for months. Each time she seemed to be getting better, she was struck down again. She suffered from alternating colds, fevers and stomach upsets, experienced coughing fits and spells of shivering. She lost all appetite, her formerly clear complexion broke out in blemishes, and weakness brought on bouts of depression.

In this condition she missed the protective presence of her mother terribly, and began to realize how alone and friendless she actually was in England. This persisted when she became well, and with reason. She saw the Spanish Ambassador, Puebla, more attached to King Henry than to her interests or those of her parents. Her young husband-to-be was kept away from her most of the time deliberately. And the King now became too busy with other affairs to pay any attention to her.

One of the King's prime concerns at that moment was to find a bride for himself. Though getting old, as lives went in those days, he believed himself still able to produce more sons. With fatal illness likely to strike even the healthiest specimens of boyhood, additional male heirs were sensible insurance that his line would continue on the throne. So Henry VII sought a prospective breeder of boys. There were other qualifications: a desirable alliance and a large marriage portion. Also—old or not—Henry VII had some highly personal requirements for any woman to whom he would offer his bed.

One possible bride had been suggested by Isabella, when she had refused to let him have her daughter. The substitute was to be Isabella's niece, the young Queen of Naples, who Isabella felt would "suit and comfort the King of England." King Henry sent three observers to meet with the young Queen of Naples and her mother, and report back on the young Queen's qualifications as a bride. The observers took with them a document containing a list of detailed questions from King Henry, with spaces left between the questions for them to fill in the answers in equal detail. The final report, containing the King's instructions and the replies of his observers, indicates that aging King Henry was far from disinterested in the physical qualifications of a prospective wife:

Instructions given by the King's Highness to his trusty and well beloved servants, Francis Marsin, James Braybroke, and John Stile, showing how they shall order themselves when they shall come to the presence of the old Queen of Naples and the young Queen her daughter . . .

Item, specially to mark the favor of the young queen's visage, whether she be painted or not, and whether it be fat or lean, sharp or round, and whether her countenance be cheerful and amiable, frowning or melancholy, stedfast or light, or blushing in communication.

As to this article as far as we can perceive or know, the said queen is not painted, and the favor of her visage is after her stature, of a very good compass, and amiable, and somewhat round and fat, and the countenance cheerful, not frowning . . . and not light nor bold-hardy in speech, but with a demure womanly shamefaced countenance, and of few words . . .

Item, to note the clearness of her skin.

As to this article, the said queen is very fair and clear of skin.

Item, to note well her eyes, brows, teeth, and lips.

. . . the eyes of the said queen be of color brown, somewhat grayish; and her brows of a brown hair and very small, like a wire of hair; and her teeth fair and clean . . . and her lips somewhat round and thick . . .

Item, to mark well the fashion of her nose . . .

. . . her nose is a little rising in the midward, and a little coming or bowing towards the end, and she is much nosed like the queen her mother.

Item, specially to note her complexion.

. . . the said queen is of a very fair sanguine complexion and clean.

Item, to mark her arms, whether they be great or small, long or short.

. . . the arms of the said queen be somewhat round and not very small . . . and as to the length of her arm . . . it is of good proportion according unto her personage and height.

Item, to see her hands bare, and to note the fashion of them, whether the palm of her hand be thick or thin, and whether her hands be fat or lean, long or short.

As to this article, we saw the hands of the said queen bare at three sundry times that we kissed her said hands, whereby we perceived . . . they be somewhat fully and soft and fair and clean skinned.

Item, to note her fingers, whether they be long or short, small or great, broad or narrow.

. . . the fingers of the said queen be right fair and small, and of a meetly length and breadth, according unto her personage very fair handed.

Item, to mark whether her neck be long or short, small or great.

As to this article, the neck of the said queen is fully and comely, and not misshapen, nor very short nor long . . . but her neck seemeth for to be the shorter because that her breasts be fully and somewhat big.

Item, to mark her breasts and paps, whether they be big or small.

. . . the said queen's breasts be somewhat great and fully, and inasmuch as that they were trussed somewhat high, after the fashion of the country, the which causeth her grace for to seem much the fullyer and her neck to be the shorter.

Item, to mark whether there appear any hair about her lips or not.

As to this article, as far as that we can perceive and see, the said queen hath no hair appearing about her lips or mouth.

Item, that they endeavor to speak with the said young queen fasting . . . and to approach as near unto her mouth as they honestly may, to the intent that they may feel the condition of her breath, whether it be sweet or not, and to mark at every time when they speak with her if they feel any savor of spices, rose-water, or musk by the breath of her mouth or not.

To this article: we . . . have approached as nigh unto her visage as that conveniently we might do, and we could feel no savor of any spices or waters, and we think verily . . . that the said queen is like for to be of a sweet savor.

Item, to inquire of the manner of her diet . . .

As to this article . . . the said queen is a good feeder.

Unfortunately for Henry VII, the royal ladies about whom he inquired in such detail also inquired about him. And those ladies he considered eligible did not consider him eligible. *Their* observers reported that Henry VII was disintegrating physically; that death had already marked him as an old man soon due for cutting down. The only future for a royal lady in England now lay in marriage with the King's son. And Catherine of Aragon appeared to be certain of that honor.

But the certainty of this prospect collapsed abruptly with the death of her mother in 1504. Emotionally, Catherine felt deprived of the only person on earth she could count on. Practically, the death of her mother made Catherine a much less desirable match for King Henry's son.

While Isabella lived, Catherine represented an alliance with all Spain. With her death her kingdom of Castile—the largest part of Spain—passed on to Catherine's sister Juana, and thus into the control of Juana's husband Philip, of the Austrian house of Hapsburg. Catherine now represented an alliance only with Ferdinand's kingdom of Aragon. Henry VII was certain he could do better than that for his son.

Just before Prince Henry turned fourteen—on which day his marriage was to have become a fact—his father had him make a formal protest that he had been betrothed to Catherine without his consent.

This did not mean that the marriage agreement was definitely broken. What the protest did was to postpone the marriage, until such time as the Prince—or rather, his father—made up his mind. Catherine's father, King Ferdinand, deprived of the strength of Castile, was too enmeshed in his own troubles with Italy and France to counter this move. He received Catherine's appeals for help with considerable irritation.

The result was that Catherine found herself kept dangling, to be used as a last resort should King Henry fail to find a better match for his son. She was to go on waiting like this for four more years; years of miserable loneliness, neglect and degradation.

2

"I ENTREAT YOUR HIGHNESS," Catherine pleaded with her father by letter, *"consider that I am your daughter."*

She was desperately in need of money. None was left for her living expenses; her servants went unpaid. She had asked King Henry to help her. Henry had sidestepped adroitly: He would certainly pay her servants—except that they were not English; as all in her household were Spanish, and had come with her from Spain, they were the responsibility of King Ferdinand. It would be presumptuous of the King of England to interfere.

At last Catherine ordered her father's ambassador, Puebla, to inform Ferdinand about the "misery in which she lived"; to "tell him in plain language" that it would reflect dishonor on his character "if he should entirely abandon his daughter"; to make plain to him that she had been forced to borrow because she had "nothing to eat."

When no help came in response to this, Catherine became convinced that Puebla was entirely at fault; that he was not letting Ferdinand know how desperate her situation had become, and that he was acting in the interests of King Henry against her and her father.

Finally she wrote directly to her father of this, urging him to get rid of Puebla and send a better ambassador to England, and begging him personally to help her out of her money difficulties:

Most high and most puissant lord,

Hitherto I have not wanted to let Your Highness know the affairs here, that I might not give you annoyance, and also thinking that they would improve; but it appears that the contrary is the case, and that each day my troubles increase . . . since coming into England I have not had a single maravedi except a certain sum which was given me for food, and this such a sum that it did not suffice without my having contracted many debts in London . . .

That which troubles me more is to see my servants and maidens so at a loss, and that they have not wherewith to get clothes. . . .

I entreat Your Highness that you will consider that I am your daughter. . . . I have had so much pain and annoyance that I have lost my health in a great measure; so that for two months I have had severe tertian fevers, and this will be the cause that I shall soon die.

King Ferdinand was battling for his holdings in Italy at that moment. He himself was too hard-pressed financially to help Catherine; too hard-pressed politically to do without Puebla close to the King of England; and too hard-pressed militarily to make any threatening demands on King Henry.

Instead, he continued to act as though he believed the marriage of his daughter and Prince Henry was still a certainty, as had been previously agreed.

"Oftentimes when I am troubled in my mind for the death of my queen," he told representatives sent by King Henry to Aragon, "it right greatly rejoiceth me that I am assured that my daughter shall be married unto so noble a prince, and that she shall have so noble a father-in-law as my brother of England."

In the same vein, Ferdinand ordered Puebla to tell Catherine that all her troubles would be over if only she would do a better job of making King Henry love her: "You may say from me to the princess my daughter, that it seems to me that in all these things she should be very conformable and pay much respect and obedience to the King of England my brother, her father . . . because by this means he will more love her, and do more for her. Since it so happens that she has always to be in that land, and spend her life with the King of England . . . the expense of her and her house, and the salaries of her people, are and must ever be at the charge of the said King of England . . ."

Ferdinand wrote this, of course, knowing Puebla would show it to King Henry.

Catherine tried her best. She broke down and wept before King Henry when she reminded him again of her circumstances. But when she begged him to pay her bills, the King of England told her savagely that her father should pay her expenses and complained that Ferdinand still owed him 100,000 scudos—the unpaid second half of the marriage portion promised by Spain on the marriage of Catherine and Arthur.

Catherine came to believe that she would be released from her degrading position and married to the Prince, if only her father would send King Henry the rest of the marriage portion. She wrote letter after letter to King Ferdinand, begging him to send the 100,000 scudos, to send her money to pay her mounting debts, at the very least to answer her letters. She told her father in one such letter:

Louis XII's France was the strongest and richest nation in Europe. Maximilian's Holy Roman Empire looked greater on the map—but he had very little control over much of these lands; and those he did control strained his finances severely. Alone, Spain, for all its growing strength, was no match for France.

Italy was the principal bone of contention between Ferdinand, Maximilian and Louis XII. It was divided into so many mutually hostile parts that almost anyone with the energy and will could seize one of them. Ferdinand already had his portion; so did Maximilian. Louis XII at this moment did not, and he wanted one; in fact, if he could contrive to get it, he wanted all of Italy.

France's recurrent lunges across the southern Alps threatened Maximilian's holdings in northern Italy, the Papal States in central Italy, and Ferdinand's kingdom of Naples in southern Italy. Ferdinand pinpointed the anxiety about French ambitions: "Should the King of France gain Naples he would be sole monarch of all Italy; and if Italy were joined to France it is easy to see how much it would threaten all other princes."

But there was a more immediate object of dispute between Ferdinand and Louis XII at the moment: the little kingdom of Navarre, squeezed between the borders of Spain and France. Ferdinand wanted it, and decided to use England to help him get it—over French opposition.

Guienne, in southern France, had once belonged to England. Ferdinand persuaded the unsuspecting Henry to send an English army to Spain to recapture Guienne for England.

But once the English troops were in Spain Ferdinand used them instead for his real purpose. He positioned them as a shield against any French attempt to interfere, while he overran unprotected Navarre.

With Navarre in his pocket, Ferdinand had no further use for the English army; he certainly saw no point in helping them to regain Guienne for England. The English troops waited idly through a sweltering summer, their food going bad, their morale collapsing, their organization disintegrating. Finally they mutinied and went back to England.

Europe, which saw only that they had quit the field without fighting, jeered that because the English had kept away from war so long they had forgotten how to fight. Henry was too stung by this humiliation to be properly angry with Catherine's father. He saw that England could achieve the respect of Europe only by proving itself, in battle, an ally to cherish and a foe to fear.

3

HENRY PREPARED FOR WAR with a will, recruiting and equipping his armies, enlarging his navy—all out of his father's seemingly inexhaustible treasure. In 1513 he was ready. He crossed the Channel to Calais, and led his troops inland to win glory, land, and a proper regard for English worth.

Before sailing Henry revealed how he felt about his wife after four years of marriage: He left his country in Catherine's hands, naming her regent of England, governor of the realm until his return. She was also left in complete charge of the armed forces remaining in England, and soon proved how well she could use them.

On the Continent, Henry's troops, driving forward in search of the foe, got bogged down by stormy weather and muddy roads. Henry spent most of the night wandering among his troops in the rain, steadying them with his vitality, firing them with his own enthusiasm.

"Well, comrades," he told them cheerfully, "now that we have suffered in the beginning, fortune promises us better things, God willing."

So it proved. Henry advanced; the French, not eager to play for high stakes, avoided battle and tried to stop him with little skirmishes which the English brushed aside. Henry moved deeper into French territory and lay siege to the fortified town of Thérouanne. Maximilian arrived with his troops, offering to serve under Henry's command—for a considerable payment. Henry dug deeper into his father's treasure, and used the combined armies to tighten the pressure around Thérouanne.

The French sent a force of cavalry to deliver food and supplies to the besieged town. Henry's army lunged at this relief force. The enemy cavalry turned and fled with such alacrity that the English laughingly named it "the Battle of Spurs."

Ten days later, the besieged town surrendered to Henry, who thus became the first English king since Henry V to take a French city. He marched on and captured another: Tournay. By mid-September he was writing happily: "Since our entry into France we have in every encounter been victorious . . ."

Meanwhile, Catherine found herself involved in a more serious war than her husband. Scotland had long been an ally of France and a threat to England's northern border. To achieve peace with Scotland, Henry's older sister, Margaret, had been married to the Scottish King, James IV. But, though she bore James a son who was to become the next king of Scotland, Margaret proved unable to influence him in any way. When Henry invaded France, James decided that he was committed by the terms of his alliance with France to invade England.

Margaret tried to deter him and failed. Henry wrote him an indignant letter pointing out that James had a peace treaty with England. This had as little effect as the persuasions of Margaret. With the bulk of the English Army in France, James IV started south across the Tweed, with between sixty thousand and a hundred thousand men-at-arms.

It was with this possibility in mind that Henry had left Catherine in command of all the forces remaining in England;

and had posted an old soldier, the seventy-year-old Earl of Surrey, in the north with about twenty thousand troops. As the Scots came across the border, Catherine responded with all the vigor of her mother. She rallied the nation to resist the invader, swiftly collecting thousands of men from all over southern England and sending them north to reinforce Lord Surrey. She prepared to go north herself.

Until the Scottish invasion distracted her, Catherine had been nervously waiting for her husband to return, worrying about his health, writing anxious letters to France, begging those close to Henry to restrain him from being too foolhardy in battle. She had also to bear alone the painful misfortune of another miscarriage, news of which was not publicly announced.

Now she had a war of her own to take up her thoughts, and into which she could fling her pent-up energies. Her letters to the Continent took on a joyful, eager—even rather boastful—tone:

"Ye be not so busy with the war as we be here encumbered with it. I mean touching mine own self, for going where I shall not so often hear from the King." She expressed her own relief at having more to do than worry about Henry, when she wrote that all his people in England "be very glad (I thank God) to be busy with the Scots, for they take it for a pastime. My heart is very good to it, and I am horribly busy with making standards, banners and badges."

Henry had sent Catherine some French banners and a noble French prisoner as tokens of his triumphs. Catherine swore to send him, in return, a token of a victory of her own; perhaps so great a token as a captive king of Scotland. She started north, but was stopped by news from Surrey.

Reinforced by the troops gathered so quickly by Catherine, Surrey met the Scots at Flodden, destroyed their army, cut down the fathers and sons of their proudest families, and left King James himself an arrow-pierced, sword-hacked corpse on the field. In all, the English lost fewer than two thousand men while killing over ten thousand of the enemy. In a single battle, Surrey had stopped an invasion and won a war.

It was Catherine's proudest moment. She wrote to her husband of "the great victory that Our Lord hath sent your subjects in your absence. . . . To my thinking this battle has been to Your Grace, and all your realm, the greatest honor that could be, and more than should you win all the crown of France."

From the corpse of James IV was taken a memento which Catherine sent on to Henry to "keep my promise, sending you for your banners a king's coat. I thought to send himself to you, but our Englishmen's hearts would not suffer it. It should have been better for him to have been in peace, than to have this reward."

Henry received the news of his brother-in-law's death a bit sadly: "He has paid a heavier penalty for his treachery than we would have wished."

His own war, in France, continued without any battle to compare with Flodden. But this in itself was a triumph for the prestige of Henry and England. When the Scots had invaded England, the English had rushed forward to meet and stop them. When Henry invaded the Continent, the French armies retreated without a decisive stand.

It was the coming of winter that finally halted Henry's advance. Postponing his march on Paris till summer, he left the Continent. Less than three months of war had accomplished exactly what he had wanted. The martial determination the English had shown in attacking the French on their own ground, coupled with the stunning victory over the Scots, had forced the rest of Europe to take England seriously. Ferdinand and Maximilian signed an agreement to join Henry in attacking France with the coming of warm weather.

Henry returned to England looking forward eagerly to spring. He was confident that with the help of Spain and Germany he would slash his way across France and capture Paris. He did not know that less than two weeks after Ferdinand

and Maximilian signed the anti-French treaty with him, Ferdinand was betraying England in secret negotiations with France.

Maximilian was warned not to follow Ferdinand's example. His daughter had met Henry before he left for England, and she gave her father a thoughtful appraisal: "Unless you give him cause to the contrary, this young King, be well assured, will help you both with his person and his purse, without deceit (like France) or hypocrisy (like Ferdinand). Therefore, he should be treated in the same way, and promises made to him should not be broken."

Ignoring her advice, Maximilian joined Ferdinand in listening to French proposals. His daughter told him darkly that this was "very much opposed to my judgment. I know not how the King of England will accept it, considering the great preparations he has made for war."

Louis XII of France had also been impressed with Henry's fighting spirit. He did not relish trying to halt an English attack while simultaneously parrying Ferdinand and Maximilian. So he made considerable concessions: Ferdinand was to keep Navarre; the French claims in Italy might be put aside for the grandsons of Maximilian and Ferdinand.

In return, Ferdinand and Maximilian agreed to ignore their treaty with England. If Henry attacked next spring, he would learn to his astonishment that he was attacking alone, without allies.

Henry did not have to wait that long to find out: Louis XII deliberately revealed what Ferdinand and Maximilian had done.

4

HENRY'S FIRST REACTION was self-righteous shock: "I see no faith in the world, save in me."

Shock was followed by fury. Catherine was caught in the difficult position of trying to be loyal to her father and her husband at the same time. She felt that Henry's anger was out of proportion, and typical of the overemotional reactions of his countrymen. "There is no people in the world more influenced by the good or bad fortunes of their enemies than the English," she wrote to her father. "A little adversity of their antagonists renders them overbearing, and a small success of their enemies prostrates them."

But Henry's anger was not the final aspect of his reaction to being duped. He was following an emotional progression that was to mark his later personality—first shock, then fury; and then the vengeful, brutal blow of retaliation.

Ferdinand and Maximilian had treated with France behind his back. Henry's retaliation was to link England to France with a blood tie.

Henry's younger sister, Mary, was unmarried and exceptionally lovely. France's King Louis XII was an aged widower with a taste for tender young beauty. The reluctant Mary, in love with the dashing Duke of Suffolk, was abruptly pushed into marriage with the doddering Louis. The French King received her with feverish delight.

It was the turn of Ferdinand and Maximilian to be shocked. This alliance-by-marriage left Louis free to hit again at their possessions with all the forces at his disposal. And if Mary were to bear Louis a son, it might bind England and France together permanently—against Spain and Germany's Holy Roman Empire.

"The penance," complained Maximilian's daughter, "is too great for our offense toward England."

Henry himself was quite pleased with his revenge. He had tried being the opposite of his father, and had been treated as an easy tool because of it. Now he would emulate his father's trickiness.

He did try, but he never really made a success of it. He

could be savage and treacherous; but he had no talent for the devious. In later years, for example, he attempted to trick the Venetian ambassador into believing that a friend of Venice was actually its enemy. He told the ambassador of reports that he claimed his spies had brought to him. The Venetian listened carefully to all Henry had to say, thanked him warmly for the information, and passed it on in full detail to his ruler in Venice. At the end of his dispatch the ambassador added: "It has seemed fit to me to mention the whole to Your Sublimity, because I deem the words of Kings worthy of consideration—not that I believe anything he told me."

The marriage of Mary and King Louis turned England's back on Spain and made Catherine's position most uncomfortable. Through her father she represented Spain; through her husband she was queen consort of England. The two countries were now at odds, and she was in the middle.

The Spanish ambassador warned Ferdinand that his daughter was showing herself loyal to England in this difficulty. But whenever Henry looked at her he saw Spain—and her father who had used her to play him for a fool. Though Maximilian had betrayed him too, Henry directed most of his anger against Ferdinand. For Maximilian was not his father-in-law; Maximilian's daughter was not his wife.

Henry struck at his wife through the one person who had been faithful to her during her worst troubles—her confessor, Fray Diego. Without warning, the King suddenly lashed out at the friar before witnesses, bringing up his evil reputation and charging him with being a "fornicator." Brushing aside Catherine's attempts to defend her confessor, Henry had him stripped of position and arrested.

Fray Diego well understood that he was merely the scapegoat for the King's anger against Catherine: "If I am badly used, the Queen is still more badly used." Banished from England, he returned to Spain and from there wrote Henry: "I left my father and mother, the country of my birth, and the sacred order to which I belong, to serve the Queen, my mistress. For nine years I have served her faithfully, enduring every evil for her sake, even lack of meat and drink, of clothes and fire. . . . Your Grace has called me a fornicator. By the holy Gospel I swear this charge is false; never, within your kingdom, have I had to do with woman. . . . I was condemned unheard. The witnesses brought against me were not only personal enemies, but disreputable rogues. Ascrutia is a perjurer and traitor to the Queen. Pedro, the keeper of your chapel, has a bastard son. Vadillo is unclean of life. Such men as these should not have been permitted to give evidence against the confessor of so great a Queen. . . . If you desire me to return to the Queen, I will; but only on condition that I be heard by honest judges. . . . Wherever I go, I shall pray that you may have sons."

Wherever he did go, Fray Diego was not invited to return to England. Catherine never saw him again. Without him, she was once more alone, adrift within the English court; and her father found that she was of no use at all in attempting to placate King Henry.

"The Queen may mean the best," Ferdinand was informed by his ambassador, whom Catherine distrusted, "but she has no one near her person who will tell her how she may be useful to her father. Fray Diego is to blame for what has gone so wrong. He told her that she ought to forget her country and her family, in order to secure the love of Henry and his people. She has grown so accustomed to this idea that she will do nothing else, unless some person is about her who can tell her sharply what to do."

Apparently, Catherine was concentrating all her powers on trying to win back her husband's affection. In time she was helped by outside occurrences. The coldness between Catherine and Henry began to thaw in 1515 with the death of Louis XII—brought on, it was said, because the beauty of his young bride wrung from him more passion than was sensible for a man of his advanced age.

His widow, Mary, acted on Henry's earlier promise that whenever King Louis died she might wed whom she pleased. She quickly broke her legal ties with France by marrying the Duke of Suffolk. Young Francis I became the new king of France, and there were no treaties between Francis and Henry. Thus the link to France was no more. Henry, seeing that it might become necessary to realign himself with Maximilian and Spain, cooled his righteous indignation.

The following year the healing of the personal rift between Henry and his wife was hastened by the birth of Catherine's first healthy child.

The child was a girl, and was named Mary. Catherine and Henry were united in their happiness. True, being a girl, Mary could not be considered a proper heir to the throne. English experience and custom indicated that a woman could not rule England. But the fact that Mary was an obviously sound baby was a good omen for the future.

"We are both young," Henry said happily. "If it was a daughter this time, the sons will follow."

With the death of Ferdinand in the same year, Henry no longer had any reason at all to resent Catherine as representing her father. And if she still represented Spain in some measure, that was now all to the good. Spain and the kingdom of Naples fell to Ferdinand's grandson Charles V, son of Catherine's sister Juana and Maximilian's son Philip. This Charles was already ruler of the Netherlands, and when Maximilian died in 1519 Charles also took over all his German holdings and his title, Emperor of the Holy Roman Empire.

All this made Charles a most attractive ally; and Charles was Catherine's nephew. So, what with one thing and another, Catherine regained her husband's affection and regard. But she was never to regain her old influence over him. That had been squandered irretrievably by her late father.

In the next decade of constantly shifting European alliances the position of influence next to the King was to be occupied, to the exclusion of everyone else, by an astonishing man of low birth who became one of the most powerful statesmen of Europe—Thomas Wolsey.

5

WOLSEY was a plump, robust man with a full-fleshed face, an impressive manner, and a voice by turns hearty and heavy. Eighteen years older than Henry, he used for all they were worth a rare combination of talents: an agile, subtle mind, an ability to deal with others, and an enormous capacity for work.

His driving force was ambition—ambition to control the destiny of England and influence the fortunes of Europe; ambition to enjoy pomp and luxury; ambition to be treated with the worshipful respect of a king or pope.

Wolsey's father was a village merchant—a butcher, it was said by some, especially by his enemies. He was to have many enemies, most notably among the highborn gentlemen who resented his place of prominence above them. Shakespeare has one nobleman sum up what all the nobility came to feel about Thomas Wolsey in his years of success:

> This butcher's cur is venom-mouth'd, and I
> Have not the power to muzzle him.

For Wolsey was to become England's supreme religious power under the Pope and supreme civil power under the King—thus grasping in his thick hands more combined authority than any man before in the history of his country. By 1518 he was able to write to Rome: "Never has the kingdom been in greater harmony and repose than now; such is the effect of *my* administration of justice and equity." He might even have eventually achieved his ultimate ambition to become pope, if he had not fallen into the pit dug for him by the determination of Anne Boleyn and the stubbornness of Queen Catherine.

Nothing tells more about Wolsey than the steadiness—the inevitability, almost—of his climb from the bottom of society to the top. He got his B.A. at Oxford at the age of fifteen, went on to become a fellow of Magdalen College, its bursar, and master of Magdalen College school—being ordained as a priest as soon as he was old enough, two years before the turn of the century. He was impressive enough even then, in his youth, to attract the patronage of important men. From one of them he gained the rectory of Limington in Somerset. He quickly won a papal dispensation enabling him to hold two other incompatible benefices.

By 1501, just three years after becoming a priest, he was appointed chaplain to Henry Deane, Archbishop of Canterbury. By the time the Archbishop died, two years later, Wolsey was able to secure his appointment as chaplain to England's deputy-lieutenant of Calais.

The deputy, Sir Richard Nanfan, was getting old and tired. He was ready to delegate much of his authority to a man whose ability and faithfulness he could trust. Wolsey's critics grant that he knew how to be faithful. Even in his personal life, though he ignored, as did many priests of his time, the vow of celibacy, he remained loyal to a single mistress and to his children by her. As for ability, Nanfan soon found that he could rest easy with his business affairs in his chaplain's hands.

So well did Wolsey serve the deputy-lieutenant that on Nanfan's retirement he recommended Wolsey to Henry VII. The old Henry valued ability above inherited rank; at the age of thirty-four the "butcher's cur" found himself chaplain to the King—within reach of the heights to which he aspired. Sent on the King's business first to Scotland and then to the Continent, Wolsey gained a reputation for the swift efficiency with which he carried out his assignments. He also acquired an insight into how great nations conducted their affairs with each other.

"There is here," he discovered on the Continent, "so much inconstancy, mutability, and little regard of promises and causes, that . . . there is little trust or surety; for things surely determined to be done one day are changed and altered the next."

When Henry VIII mounted the throne he was eager to play the king, but impatient of details of state business that kept him from the fun of jousting, hunting, and presiding with his bride over lavish parties. It was this that gave Wolsey his opportunity.

Wolsey was appointed Henry VIII's royal almoner—dispenser of the king's charity. It was not a very high position, but it sufficed to keep Wolsey near to the King, and the young King was soon more than aware of Wolsey's presence. Other members of the King's Council annoyed Henry with demands that he study problems at hand, consider alternatives, weigh decisions, arbitrate debates, immerse himself in the pressing matters of conducting the government. Wolsey, on the other hand, took it upon himself to understand the King's mind and will, and carry out himself all the detailed work that was necessary.

After Wolsey's death his gentleman-usher, George Cavendish, wrote a biography of his master in which he explained how Wolsey, "perceiving a plain path to walk in towards promotion . . . daily attended upon the king in the court . . . he was most earnest and readiest among all the council to advance the king's only will and pleasure, without any respect to the case; the king, therefore, perceived him to be a meet instrument for the accomplishment of his devised will and pleasure, called him more near unto him, and esteemed him so highly that his estimation and favor put all other ancient counselors out of their accustomed favor. . . .

"The king was young and lusty, disposed all to mirth and pleasure. . . . The which the almoner perceiving very well, took upon him therefore to disburden the king of so weighty a charge and troublesome business." Wolsey's success in taking more and more of the weight of government off the King's shoulders delighted Henry "and caused him to have the greater affection and love to the almoner. Thus the almoner ruled all them that before ruled him . . ."

By the time Henry made his decision to invade France in 1513, he automatically turned to Wolsey as the man to make all the preparations for the expedition. Wolsey proved again that no work load was too much for him. He supervised everything—from the readying of a fleet to the recruiting and arming of troops, from the butchering and salting of meat to the stocking of a keg of nails that might be needed for building a siege platform. It was a job for twelve men, not for one. No other single man could have kept track of it all. "I pray God . . ." Bishop Fox wrote Wolsey, "soon deliver you of your outrageous charge and labor."

So well did he carry out his multitudinous tasks that by the time Wolsey sailed with Henry for France, the King considered him the one indispensable man in the kingdom—and everyone knew he had become the most influential. Even those members of Henry's family who were closest to the King now turned to Wolsey when they themselves could not influence Henry.

Queen Catherine, worried about Henry's health and safety during the campaign against the French, wrote most of her letters to Wolsey, begging him to take care of her husband. Henry's older sister, Margaret, widow of the Scottish King slain at the Battle of Flodden, wrote to Wolsey, ". . . next to the King's Grace, my next trust is in you, and you may do me most good of any." Henry's younger sister, Mary, fearing Henry's wrath when she hastily married the Duke of Suffolk after the death of Louis XII, placed her future in Wolsey's hands: ". . . my trust is in you . . . for now I have no mother to put my trust in but only the King my brother, and you."

After the successful French campaign, Henry proceeded to shower lands, incomes and offices of government upon Wolsey. It was Wolsey who arranged the marriage of Princess Mary and Old Louis XII. It was Wolsey who was named godfather of Catherine and Henry's daughter, Mary. He built himself the most splendid home in the country; his servants, in gorgeous livery, waited upon him on their knees; he became Henry's lord chancellor.

Church positions began to descend upon him, one after another: He became bishop of Tournai, bishop of Lincoln, and archbishop of York; he later added to these the bishoprics of Bath, of Wells, of Durham, of Winchester. The Pope made Wolsey a cardinal under pressure from Henry, who wrote to Rome: "Such are his merits that I esteem him above my dearest friends, and can do nothing of importance without him."

The head of the church in England was normally the archbishop of Canterbury. Since the incumbent showed some reluctance to make way for Wolsey by dying, King Henry solved the problem by asking the Pope to name Wolsey his papal legate in England, with powers that took precedence over those of Canterbury.

Within England Thomas Wolsey was the supreme authority in the Church, and in the civil government under the king. To many he appeared to *be* the government of England. The Venetian ambassador marveled at the variety of Wolsey's powers and activities: "He alone transacts the same business as that which in Venice occupies all the magistracies, offices, and councils, both civil and criminal; and all State affairs are managed by him, let their nature be what it may."

The heads of other governments came to realize that in order to influence England they must influence Wolsey first. And Wolsey was quick to use this, in turn, to influence the rest of Europe to England's advantage. The international scene had changed. Henry was no longer contending with older and more experienced rulers. The throne of France was occupied by Francis I, five years younger than Henry, a strong-nerved, wiry, impetuous, direct fellow, much fussed over and sometimes dominated by mother and sister, a glory hunter on the battlefields of Europe, a writer of poetry in the privacy of his chamber. Spain, Naples, Germany, Austria and the Netherlands were ruled by Charles V, ten years younger than Henry—a stolid, uncommunicative, solemn, none-too-healthy young man who struggled stoically for a long time to fulfill the demands of his many crowns.

Both Francis and Charles acknowledged Wolsey's influence by sending him regular payments to prevent his ill will. For the rest of Europe sometimes believed that it was Wolsey, not King Henry, who really ruled England.

They were wrong, of course, as are those historians who have pictured Henry as Wolsey's tool. There may have been some truth in the charge in the first few years of the reign. But not after Henry grew into manhood. Wolsey was the tool, and the tool was held firmly in the hand of Henry.

The truth is that Henry, as he matured, became a man quick to seize on the right tools for his purposes. When he delegated authority to Wolsey, Henry observed him in action and learned from what he observed. He had a questing, absorbent mind. He also had the ability to draw his own conclusions. It was not Wolsey who realized the importance of sea power to England's future. It was Henry. It was Henry who encouraged the building of warships and royal dockyards, who studied the new types of cannon aboard foreign ships which entered English ports, and who founded the standing Royal Navy with which a daughter yet unborn would one day break the might of Spain.

Those who try to cut down Henry's stature confuse his character with his talent. As Acton says, "Judge talent at its best and character at its worst." By this yardstick, Henry's character as a mature man was to become utterly corrupt. But his talent for achieving and consolidating power was prodigious. He had almost a genius for seizing on exactly the right men, events and popular feelings to gather the power he wanted into his own hands. So it was as a tool that Wolsey achieved the authority toward which his ambition had driven him. And a tool can be discarded when it ceases to function properly.

It was to Wolsey that Henry was to hand the dangerous problem of the divorce.

For by 1525 the relationship between Catherine and Henry was undergoing another change—because of circumstances over which neither had any control. People said of Catherine that she was "as beloved as if she had been of the blood royal of England." But, despite the optimism of 1516 when their daughter had been born healthy, Catherine had borne Henry no further children in the years that had followed. There had been many more pregnancies. All had ended in miscarriages. Mary remained an only child. There was no male heir to King Henry's throne.

A male heir was essential to prevent the kind of uncertainties about succession that had led to civil war before. That Henry was deeply troubled by this practical problem was hinted to Catherine's nephew, Charles V, in the summer of 1525. Henry's representatives reported back to him that they had told Charles, as instructed, that "my lady Princess was your only child at this time, in whom Your Highness put the hope of propagation of any posterity of your Lady, seeing the Queen's Grace hath been long without children; and albeit God may send her more children, yet she was past that age in which women most commonly are wont to be fruitful and have children."

Henry *did* have a son, Henry Fitzroy—an illegitimate child to whom one of the king's mistresses, Elizabeth Blount, had given birth in 1519. Finally, in 1525, Henry took the hard decision to groom this illegitimate son to inherit the crown of England. It was a move of desperation. For a bastard succeeding to the throne could cause as much trouble as the accession of a girl. But it was the only protective measure Henry could think of—for the moment.

The boy was created Duke of Richmond—the title held by Henry's father before becoming King of England. He was also named Lord High Admiral of England, given the revenues of large tracts of land, and placed over all other nobility, including Henry's legitimate daughter, Mary.

The Venetian ambassador recorded: "It seems that the Queen resents the earldom and dukedom conferred on the King's natural son, and remains dissatisfied . . . but the Queen was obliged to submit and have patience."

Patience was the only thing left to Catherine now in dealing with her husband. He ceased to visit her bed, and not only because he believed she had become too old to bear another child. The years since Mary's birth had dealt harshly with Catherine. She was aging fast, becoming heavy of face and body, taking on the too-solid, unattractive appearance of her later portraits.

Henry was changing, too. By his late twenties he was no longer quite the dazzling youth; he had grown a red-gold beard to compensate for the thinning hair of his head, his face was assuming a heaviness, and his body was turning to fat at a rapid rate. But he was so tall and strongly built that his bulk did not yet show to disadvantage. Instead, it rather added to the general appearance of a mature and powerful king.

When Henry and Catherine visited France together, King Francis I looked them over and gave his opinion of what he saw: "The King of England is young—but his wife is old and deformed."

Yet not all was changed. Henry might enjoy sporting with his mistresses; but no single one of them ever managed to keep his favor for long. He shed them as carelessly as old suits of clothing. Even Elizabeth Blount, mother of his illegitimate son, had been discarded by the time the boy was born. And she was sent away to live at some distance from court to prevent her having any influence upon her son who might possibly grow up to be the next king.

Though certainly not happy with developments around her, Catherine could still cling to this: She *was* the queen. Nothing and no one, it seemed, could change that.

Part Two

THE
OTHER WOMAN

No man can guess in cold blood what he may do in a passion.
—H. G. Bohn

Chapter Four

ENTER ANNE BOLEYN

ANNE BOLEYN's appearance—like that of Catherine—is blurred by religious controversy. Most descriptions of both are based on prejudice.

Because Catherine represented the tie between England and Rome, Protestant writers point to a painting of her in her late thirties or early forties—after she had suffered the ravages of illnesses and miscarriages—and claim that this was what she looked like when she arrived in England as a girl of fifteen. Because Anne Boleyn was the tangible reason for the break between England and Rome, Catholic writers descend to ludicrous descriptions of her as a physical, mental and moral monster.

Many of the more recent descriptions of Anne Boleyn are taken from the work of the Jesuit writer, Nicholas Sanders, who loathed her. He wrote that she was deformed of body and blemished of face, that her hand possessed an extra finger, and so on. Yet he grudgingly concluded: "She was handsome to look at, with a pretty mouth, amusing in her way."

Through the fog of religious and political spite one glimpses a slim, spirited Anne Boleyn with flashing dark eyes, a saucy flirtatiousness acquired during a stay at the French court, and raven hair which she wore unbound and flowing down her back. Judged by the tastes of her day she was not a beauty, not even very pretty. Her figure was unfashionably slender; she was dark when a fair complexion like Queen Catherine's was considered more attractive.

But her eyes—they dominate descriptions by those who met her. Thomas Wyatt, the poet who cherished a hopeless love for her, wrote:

> The lively sparks that issue from those eyes,
> Sunbeams to daze men's sight . . .

And an Italian who was not well disposed toward her reported from the English court: "Madame Anne is not one of the handsomest women in the world. She is of middling stature, swarthy complexion, long neck, wide mouth, not a very full bosom, and in fact has nothing but the passion of the King, *and her eyes*, which are black and beautiful, and have a great effect."

This was the young woman who captured the imagination and desire of an easily irritated absolute monarch used to having his whims obeyed—and who for a number of dangerous years managed to stave off his lust without losing his love.

It was during this period that another of Wyatt's poems about Anne Boleyn was penned:

> There is written her fair neck round about
> *Noli me tangere*, for Caesar's I am
> And wild for to hold, though I seem tame.

The secret lay in her eyes; and in something else, inside her: a will as strong and supple as a blade of spring steel. With this blade a royal marriage would be slashed apart, England would be severed from the Roman Catholic Church, and her own head would be cut away from her body before she reached thirty.

2

ANNE BOLEYN was about fifteen years younger than Henry, and over twenty years younger than Catherine. She was born into a family close to the court, and its position rose during her youth. Her mother Elizabeth was a Howard, daughter of the Earl of Surrey who had won the Battle of Flodden. The Howards had family connections to royalty which, though remote, helped to inflate their political pretensions to sometimes dangerous proportions. Anne Boleyn's uncle, the third duke of Norfolk, was one of the most powerful noblemen in the land—a brave, jealous, snobbish brute of a man; a blunt soldier scornful of scholars and priests; Thomas Wolsey's natural enemy, though for years discretion forced him to curb his hatred.

Her father's family was less distinguished. But her great-grandfather, a wealthy merchant, had managed to become lord mayor of London, and her grandfather had been knighted by Richard III. Her father, Thomas Boleyn, was left a considerable inheritance, which he used to increase his position in society by marrying Elizabeth Howard. Sir Thomas then dedicated his life to increasing both inheritance and position in a cold-blooded manner which provoked the Bishop of Worcester to say of him: "He would sooner act from interest than from any other motive." He was a climber, willing to use his daughters as rungs in his ladder, ready to abandon them when they were in trouble.

Sir Thomas Boleyn rose, slowly at first, by making himself useful to Henry VIII. He was made governor of Norwich Castle in 1511 and ambassador to the Netherlands the next year. He served as an officer under Henry during the invasion of the Continent in 1513, was sent on various foreign missions afterward, and became ambassador to France in 1519. Besides his daughter Anne, he had a son named George and another daughter, Mary. The fact that his daughter Mary was for a time one of King Henry's mistresses helped to speed Boleyn's rise. And when, sometime after discarding Mary, the King fell in love with Anne, Thomas Boleyn swiftly outpaced his rivals at court—being made in turn comptroller, treasurer, vice-chamberlain, lord chamberlain, Viscount Rochford, earl of Wiltshire, knight of the Order of the Garter, and lord privy seal. Eventually he became one of the most influential members of the King's Council.

Perhaps the sharpest comment on the character of Anne Boleyn's father is the fact that he continued to hold royal honors after Anne's death; and not one word of protest against the beheading of his daughter comes down to us—not even a hint of open sorrow.

Anne Boleyn was, as near as can be determined, somewhere between twelve and fourteen when her father was sent as ambassador to France. She became part of the French court, where her youthful gaiety would naturally have made her popular, and she had learned much by the time she returned to England in 1522, at the age of fifteen or sixteen.

An enthusiastic account of the young Anne Boleyn was written by a member of the French court, Viscount Chateaubriant. His remembrance of her during her time in France:

"She possessed a great talent for poetry, and when she sang, like a second Orpheus, she would have made bears and wolves attentive. She likewise danced the English dances, leaping and jumping with infinite grace and agility. Moreover, she invented many new figures and steps, which are yet known by her name, or by those of the gallant partners with whom she danced them. She was well skilled in all the games fashionable at courts. Besides singing like a siren, accompanying herself on the lute, she harped better than King David, and handled cleverly both flute and rebec. She dressed with marvelous taste, and devised new modes, which were followed by the fairest ladies of the French court. But none wore them with her gracefulness, in which she rivaled Venus."

With her saucy wit, penchant for music, fluent French and manner learned in France, Anne Boleyn was an immediate success at the English court. She was made part of Queen Catherine's household. There, according to Wolsey's servant Cavendish, "for her excellent gesture and behavior, she did excel all others."

But the time was not yet ripe for Anne and Henry to fall in love with each other. Instead, she fell in love with young Lord Henry Percy, the Earl of Northumberland's son, who was at that moment completing his worldly education by serving as a member of Cardinal Thomas Wolsey's household.

Young Percy often accompanied Wolsey to court. There, according to Cavendish, while Wolsey attended to business, "the Lord Percy would then resort for his pastime unto the queen's chamber, and there would fall in dalliance among the queen's maidens." Soon Percy was ignoring the other young ladies and concentrating all his attentions on Anne Boleyn. Anne was flattered and amused—then much more than amused. Their interest in each other increased with his every visit. Soon they were deeply in love.

For a time they kept their love a secret, but their youthful feelings became too intense to be satisfied by their occasional, semi-clandestine meetings. Finally they announced to the court that they wanted to get married—and their love affair was abruptly destroyed.

Before Anne's return from France the King and Wolsey had decided, without her knowledge, to use her to settle a quarrel between her father and the powerful Irish family of Butler over which of them was entitled to the earldom of Ormond. Both were faithful and useful to the crown; whichever one the King settled the earldom on, the other would be offended. So the King was considering marrying Anne Boleyn to Sir Piers Butler, thus joining the two families and ending the dispute. Nothing ever came of this project. But it was still under consideration when Anne and Percy announced their wedding plans.

Henry did not like having his own projects thus thwarted. We cannot be sure if there was another consideration behind his opposition to the marriage of Anne and Percy—whether or not Henry's fancy had already secretly been caught by the young Anne Boleyn. There were those at court who later said this was the real reason for his anger. Whatever the reason, he *was* angry.

As usual, it fell to Cardinal Wolsey to handle the problem. Wolsey quickly summoned Percy to him. "I marvel not a little at thy peevish folly," he told Percy harshly, "that thou would tangle thyself with a foolish girl yonder in the court. I mean Anne Boleyn."

He warned Percy that he was endangering the position he expected to inherit: "For after the death of thy noble father, thou art most like to inherit and possess one of the most worthiest earldoms in this realm. Therefore it had been most meet . . . for thee to have sued for the consent of thy father . . . and to have also made the King's Highness privy thereto."

Wolsey pointed out that Percy, instead of conducting a secret love affair with Anne Boleyn, should have told his father and the King that he would like to get married. Then the King would have chosen for him a bride more suitable to his future station in life; and everyone would have been pleased. "But now behold what ye have done through your willfulness. Ye have not only offended your natural father, but also your most gracious sovereign lord, and matched yourself with one such as neither the King nor your father will be agreeable with . . ."

When young Lord Percy protested that he loved Anne Boleyn, Wolsey threatened him: "I will send for your father, and at his coming, he shall either break this unadvised contract, or else disinherit thee forever." And he revealed that the King had already picked out another husband for Anne.

Through tears, Percy pleaded with Wolsey: "Sir, I knew nothing of the King's pleasure therein, for whose displeasure I am very sorry." But then, instead of bowing instantly to the King's displeasure, Percy put up a fight for his right to the girl he loved. He begged Wolsey to help him persuade the King to let him have her: "I considered that I was of good years, and thought myself sufficient to provide me of a convenient wife where my fancy served me best. . . . And though she be a simple maid, and having but a knight to her father, yet is she descended of right noble parentage. As by her mother she is nigh of the Norfolk blood; and of her father's side lineally descended of the Earl of Ormond. . . . Why should I then, sir, be anything scrupulous to match with her, whose estate of descent is equivalent with mine when I shall be in most dignity? Therefore I most humbly require Your Grace of your especial favor herein; and also to entreat the King's Most Royal Majesty most lowly on my behalf."

His love for Anne, Percy declared, was something "which I cannot deny or forsake."

Wolsey turned with exasperation to the other members of his household assembled around them. "Lo, sirs, ye may see what conformity or wisdom is in this willful boy's head." Turning back to Percy he said, "I thought that when thou heardest me declare the King's intended pleasure . . . thou wouldest have relented and wholly submitted thyself."

"Sir, so I would," Percy explained, "but in this matter I have gone so far, before many so worthy witnesses . . ."

Such obstinacy was unheard-of. Wolsey sent for Percy's father to help him turn Anne's beloved from his "hasty folly." When the Earl of Northumberland arrived, Wolsey made plain to him the danger of the situation. The Earl in turn descended upon Percy:

"Son," he told him, "thou hast always been a proud, presumptuous, disdainful, and a very unthrift waster, and even so hast thou now declared thyself. Therefore, what joy, what comfort, what pleasure or solace should I conceive in thee, that . . . hath so unadvisedly ensured thyself to her for whom thou hast purchased the King's displeasure. . . . But that His Grace of his mere wisdom doth consider the lightness of thy head, and willful qualities of thy person, his displeasure and indignation were sufficient to cast me and all my posterity into utter subversion and desolation."

He pointed out to his son that the King and Wolsey were, out of kindness, willing to forgive and forget "thy lewd fact" if only Percy would promise never to see Anne Boleyn again. And he swore to disinherit his son if he persisted in being stubborn about it: "For I do not purpose, I assure thee, to make thee mine heir; for, praises be to God, I have more choice of boys who, I trust, will prove themselves much better . . . among whom I will choose and take the best and most likeliest to succeed me."

Young Percy finally broke down before the combined wrath of his father, Cardinal Wolsey and King Henry. He swore to forget Anne Boleyn, and proved his submission by contracting, instead, to marry the daughter of the Earl of Shrewsbury.

As for Anne, she was sent away from the court to spend a season with her father—making certain that she would not see Percy again before he went north, and giving her time to get over her infatuation with him. She went, but according to Cavendish she was so angry "she smoked." Wolsey's servant said that she blamed the whole thing on Wolsey, "saying that if it lay ever in her power she would work the Cardinal as much displeasure."

But she was still young, knowing well how exciting many other eligible men found her, aware that her future held further delights for her, resilient enough for time to heal her wounds and cool her anger. In time she returned to the royal court. She got over her love for Percy and the loss of him. Soon she found herself seeing a great deal of the King of England.

Henry flirted with her. Anne Boleyn knew how to flirt as pleasantly as any young woman at court. Finding her delightful, he sought her company. A king might expect any woman, married or single, highborn or low, to yield rather quickly to his advances. Anne Boleyn did not yield, but there was nothing in her manner of refusal to offend him. She knew how to say no gracefully.

Intrigued by such delicious resistance, Henry spent more time with her, found his interest in her growing, pursued her with increasing fervor. Anne could not have failed to be flattered; he was the King. And she could not fail to find him attractive and let him know it. Henry was now in his thirties, in his prime, practiced in courtliness, entertaining, skillful with women—and very much a man.

"In this eighth Henry," declared a foreign ambassador, describing the King in his thirties, "God has combined such corporeal and intellectual beauty as not merely to surprise but astound all men. . . . His face is angelic, rather than handsome; his head imperial and bold; and he wears a beard, contrary to English custom. Who would not be amazed when contemplating such singular beauty of person, coupled with such bold address, adapting itself with the greatest ease to every manly exercise. He sits his horse well, manages him yet better. He jousts, wields the spear, throws the quoit, and draws the bow admirably. He plays at tennis most dexterously. . . ."

A member of the Venetian embassy said of Henry, even later, at forty: "Never have I seen a man so handsome, elegant, and well proportioned, as this English King. Tall, agile, strong, with flesh all pink and white, graceful in his mien and his walk. It seems to me that Nature, in creating such a prince, has done her utmost to present a model of manly beauty to these modern days."

And this King whom Anne found courting her knew how to augment his natural looks with elegant apparel. A description of his dress on one occasion: "He wore a cap of crimson velvet, in the French fashion, and the brim was looped up all round with lacets of gold enamelled tags. His doublet was in Swiss fashion, striped, alternately with white and crimson satin, and his hose was scarlet, and all slashed from the knee upwards. Very close round his neck he had a gold collar, from which there hung a rough cut diamond, the size of the largest walnut I ever saw, and to this was suspended a most beautiful and very large round pearl. His mantle was of purple velvet lined with white satin. . . . Beneath the mantle he wore a pouch of cloth of gold, which covered a dagger; and his fingers were one mass of jewelled rings."

But attractive though she might find Henry, still she did not yield to his passionate persuasions. She had a special kind of pride different from that of most ladies of the court. She would be no man's plaything; not even the King's.

Frustrated, Henry's desire for her became feverish. In time he realized that what he felt was more than an itch for another bedchamber conquest. He was wildly in love with her. In 1526 he had one of his ships named the *Anne Boleyn*. By then she already knew, as Cavendish puts it, "the great love that he bare her in the bottom of his stomach."

But she continued to thwart his desire—without, however, denying that she returned his love.

3

"I HAVE BEEN in great agony about the contents of your letters," a confused and anxious Henry wrote to Anne in 1527, "not knowing whether to construe them to my disadvantage or to my advantage. I beg to know expressly your intention touching the love between us.

"Necessity compels me to obtain this answer," he continued, "having been more than a year wounded by the dart of love, and not yet sure whether I shall fail or find a place in your affection."

It was a role with which Henry was unfamiliar. He was used to commanding and demanding. He was not used to pleading—to no avail. It was difficult to keep his dignity under such conditions, for Anne Boleyn gave him no firm ground on which to make a stand. She did not spurn him. In fact she appeared to find him as exciting as he found her. But she did not allow their mutual attraction to achieve its natural physical consequences.

"This has prevented me naming you my mistress . . ." he wrote, seeking to assure her that what he sought from her now was not merely a night of passing, careless pleasure. "But if it please you to do the office of a true, loyal mistress, and give yourself, body and heart, to me, who have been and mean to be your loyal servant, I promise you not only the name, but that I shall make you my sole mistress, remove all others from my affection, and serve you only."

Anne Boleyn declined to be his mistress, and suggested instead that he consider her his servant—in all but lovemaking. Henry could not understand her attitude. His answer was bitter:

"Though it is not for a gentleman to take his lady in the place of a servant, nevertheless, according to your desire, I shall willingly grant it if thereby I may find you less ungrateful in the place chosen by yourself than you have been in the place given you by me; thanking you most heartily that you are pleased still to have some remembrance of me."

But his bitterness only added to his lust for her. He did all in his power to prove to her that she reigned alone in his heart. Her father was heaped with honors. Her brother

George was brought to court and assigned to Henry's privy chamber, being daily with the King. And when Anne was away from the court, Henry let her know how deeply he missed her. In writing to her he began to refer to her as his mistress in the sense that she was mistress of the situation between them, and he her servant in love:

"Although, my mistress, you have not been pleased to remember your promise when I was last with you, to let me hear news of you and have an answer to my last, I think it the part of a true servant to inquire after his mistress' health and send you this, desiring to hear of your prosperity. I also send by the bearer a buck killed by me late last night, hoping when you eat of it you will think of the hunter. Written by the hand of your servant, who often wishes you in the place of your brother."

Men close to the royal court said later that Queen Catherine knew of her husband's affection for Anne Boleyn, and was not alarmed by it. She had reason for her lack of alarm. Catherine had seen other pretty ladies kindle the King's fancy, satisfy it, and then lose it. She could not have known that Anne had skillfully refrained from satisfying Henry's passion, and by her elusiveness had awakened in him an adoration such as he had never before experienced. Not even Wolsey guessed at this. The King was suffering the pangs of genuine love, for perhaps the only time in his life. And when the woman who had opened up in him this new capacity for love delayed her return to court, he felt a sharpening anxiety.

"I and my heart put ourselves in your hands," he wrote to Anne. "Let not absence lessen your affection; for it causes us more pain than I should ever have thought, reminding us of a point of astronomy that the longer the days are, the further off is the sun, and yet the heat is all the greater. So it is with our love, which keeps its fervor in absence, at least on our side. Prolonged absence would be intolerable, but for my firm hope in your indissoluble affection. As I cannot be with you in person, I send you my picture set in bracelets."

Anne Boleyn knew well how to keep him in her control, at any distance. Though she did not immediately return to him, she sent him a present, assurances of her love for him, and begged him to forgive her any unhappiness she was causing him.

This response seemed to encourage Henry to hope that he would soon obtain that which he had so long desired. He was raised to the heights of happy anticipation, forgetting the number of times she had raised him to such heights only to plunge him again into frustrated despair.

"For a present so beautiful," he wrote to her expansively, "that nothing could be more so, I thank you most heartily, not only for the splendid diamond and the ship in which the solitary damsel is tossed about, but also for the pretty interpretation and too humble submission made by your benignity. I should have found it difficult to merit this but for your humanity and favor, which I have sought and will seek to preserve by every kindness possible for me. . . ."

It is a pity that none of the early letters from Anne to Henry were preserved. But that she knew how to play on the King's emotions even from afar is plain in his reactions to what she wrote: "Your letter, and the demonstrations of your affection, are so cordial that they bind me to honor, love and serve you. I desire also, if at any time I have offended you, that you will give me the same absolution that you ask, assuring you that henceforth my heart shall be devoted to you only. I wish my body also could be. God can do it if he pleases, to whom I pray once a day that it may be, and hope at length to be heard."

It did not take long for Henry to fall from this pinnacle of expectant ecstasy.

"The time seems so long," says his next letter to Anne, "since I heard of your good health and of you, that I send the bearer to be better ascertained of your health and your purpose; for since my last parting from you I have been told you have quite given up the intention of coming to court, either with your mother or otherwise. If so, I cannot wonder sufficiently; for I have committed no offense against you, and it is very little return for the great love I bear you to deny me the presence of the woman I esteem most of all the world."

Finally, Henry descended to that old, familiar lover's lament: "If you love me as I hope you do, our separation should be painful to you. I trust your absence is not wilful on your part; for if so, I can but lament my ill fortune, and by degrees abate my great folly."

Anne Boleyn knew better than to keep Henry too long in such a state of suspense. Returning to him, she saw to it that he was left in no doubt about two things: first, that she was passionately in love with him; second, that she would never plunge herself into the uncertainties of an illicit affair, not even as mistress of a king she loved.

Henry could not deny that she had good reason for her obstinacy. She could point to a series of mistresses whom he had loved for a time, each of whom he had dropped and utterly forgotten. She had one example close to her: her sister Mary. Once Mary had had both the security of a marriage to Sir William Carey and the pleasure of basking in the King's attentions. Now she was no longer the King's mistress, and when her husband died in 1528 she was left with nothing, not even money. Her father, seeing no further use in Mary, turned a deaf ear to her pleas for aid.

In desperation, Mary was forced to beg Anne for help. Anne asked Henry to intercede for his ex-mistress. It was only because of his love for Anne that he gave a few moments' attention to his ex-love's problems.

"As touching your sister's matter," he assured Anne, "I have caused Walter Walshe to write to [Thomas Boleyn] my mind therein . . . for surely what soer is said, it cannot so stand with his honor, but that he must needs take her his natural daughter, now in her extreme necessity."

Anne's father bowed grudgingly to the King's wishes, doling out a small income to his now-useless daughter. It was a lesson that was not lost on Anne Boleyn. She made it quite clear to Henry that much as she wanted him, she would continue to exercise all her willpower to deny herself to him—under the present circumstances.

She was gambling for the ultimate prize: She wanted to be the King's wife, queen of England.

But it was not Anne Boleyn who first put into Henry's mind the idea of divorcing Catherine. Anne merely sensed the thoughts he already had, played on them, and served to bring them to a head. The ground had been prepared by Catherine's failure to bear Henry a male heir.

As Henry grew older, his anxiety over this problem grew. It was a very practical problem, and likely to breed very great disasters. After Catherine became too old for further childbearing it became a problem that could no longer be evaded. But before Henry could do anything, he first had to reassure himself that his action was morally justified. He began to dwell on the Bible's warning against any man having carnal knowledge of his brother's wife.

God's punishment for such an enormous sin was clearly stated: childlessness. It was true that Henry and Catherine had a child, the Princess Mary. But for a king that was the same as having no child at all. A king needed a son to carry on his line. Though great honors had been heaped on Henry's illegitimate son, Henry knew that a bastard was a makeshift, shaky base on which to attempt to set a crown. Only a legitimate royal son could prevent the possibility of civil war and assure the continuance of the Tudor line upon the throne of England.

Henry and Catherine had no son, and would have none. Henry became convinced that this was God's punishment for his having married his brother's wife. This in turn convinced him that the late Pope who had granted the dispensation for his marriage to Catherine had had no right to do so. If this was true, then Henry had never really been married to Catherine, in the eyes of God; he was not now married.

Other kings, in similar difficulties, had in the past been able to obtain papal permission to shed one wife and take another. Henry saw this as the only solution to his problem. So righteous had he come to feel about it, and so pressing was the nation's need for a male heir to the throne, that Henry even entertained hopes that Queen Catherine herself would help him obtain what he wanted.

Henry had great respect and affection for Catherine; but he did not know her so well as he thought he did. What he wanted, as his need for a legitimate son and his desire for Anne Boleyn became linked together, was called by everyone of the time a divorce. And as "the divorce" it has been known ever since. But it was not a divorce. What Henry actually sought was an annulment. In other words, what he hoped for from Catherine was that she would acknowledge that they had never been legally married—that for some twenty years she had lived with him not in holy matrimony as his wife, but in sin as his concubine.

So sensible, politically, did this plan seem, that when Henry first broached the subject to Cardinal Wolsey, Wolsey saw it as nothing but a practical political move. Wolsey immediately began maneuvering to marry Henry to a French princess, not knowing that Henry had already chosen the woman who would be his new wife.

At the start only two people knew it: Henry and Anne Boleyn. But it was not kept secret from Wolsey for long. Soon all of Europe knew it. The battle lines began to form quickly. Anne Boleyn, her nerves under rigid control, launched her campaign to win the King of England.

It was a very high and thin wire she thus ventured across,

and far below she could see the deadly spikes waiting for her. One misstep and she would fall; even if her skill was equal to the feat, the wire might snap under her weight at any moment. As soon as the matter was out in the open, enormous forces began to combine against her. There were Queen Catherine and the political groups allied to her within the country. There were the people of England, who loved Catherine and included a generation that had grown up knowing no other queen. There was Catherine's nephew, Charles V, with the united power of Spain, Naples, Germany and the Netherlands behind him. And before long Anne was also facing the wrath of the Roman Catholic Church all over Europe.

But even Anne's worst enemies granted her one virtue: courage. Even the man who came to hate her perhaps more than anyone else—Charles V's special ambassador to England, Eustache Chapuys—said of Anne Boleyn at a moment when he had more reason than ever to hate her:

"The lady . . . is braver than a lion."

Chapter Five

THE POPE AND
THE KING'S CONSCIENCE

IT WAS in May of 1527 that Catherine first learned there was a plot to destroy her marriage. The frightening news was divulged to her through secret channels: On the previous day Cardinal Wolsey had summoned King Henry to appear before him and Warham, the aging Archbishop of Canterbury—to hear a charge that he was living in sin with his dead brother's wife.

From the start the horrified Catherine jumped to the conclusion that Wolsey was the instigator of this plot against her. Her husband, she was certain, was as much an innocent victim as herself. Wolsey had instilled doubt into him, confused him, duped him into walking a path which the King himself did not really want to follow. Or so she chose to believe; and continued to believe long after others acknowledged the obvious truth.

There was to be much guessing about the role Wolsey actually played in launching the struggle for the divorce. The guessing game has continued to this day. Some have claimed that it was Wolsey who first whispered the idea of divorcing

HENRY VIII
By gracious permission of H. M. The Queen

CATHERINE OF ARAGON
Kunsthistorisches Museum, Vienna

ANNE BOLEYN

The Metropolitan Museum of Art, Whittelsey Fund, 1951

MARY, IN 1544, AT THE AGE OF TWENTY-EIGHT
National Portrait Gallery

ELIZABETH, AT THE AGE OF THIRTEEN
By gracious permission of H. M. The Queen

EDWARD, AT THE AGE OF FOURTEEN
By gracious permission of H. M. The Queen

SIR THOMAS MORE
By gracious permission of
H. M. The Queen

CARDINAL WOLSEY
National Portrait Gallery

Queen Catherine into Henry's ear; hoping to wed Henry to a French princess so that they might bear a son to rein over both France and England. Others have believed that Wolsey favored the divorce until he discovered that Henry planned to marry Anne Boleyn; by which time matters had progressed too far for him to back out. Still others have said that Wolsey opposed the divorce from the start, pleaded with Henry to change his mind, and finally went along with it only reluctantly, and thus ineffectually.

What Wolsey felt in his heart will never be known. Guessing at it is pointless. Whatever he himself felt could not and did not influence his subsequent actions. From the moment Henry decided to get rid of Catherine, Wolsey was committed to the divorce. There was no question of choice on his part.

One advantage of greatness, Montaigne was to write later in the century, is that a man can usually relinquish it whenever he chooses: "For one does not fall from every height; there are more from which one can descend without falling."

Whether this was as generally true as Montaigne believed, it was not true in the case of Cardinal Wolsey. He could not step down from his greatness. Only his exalted position protected him from the wrath of his enemies.

The nobility hated him because his influence excluded them from positions of power which they felt should belong to them. The heads of other nations resented the highhanded manner in which he dealt with them. Members of Parliament hated Wolsey for his contemptuous way of issuing them orders. The people had come to hate him because it was Wolsey who raised money for the King by extortion and heavy taxes. What he did, he did for his king. But it was Wolsey who was blamed; Wolsey who bore the burden of unpopularity.

He had no friends left; only those who feared him and those who hated him. His enemies had long memories, and nurtured them with bitter patience while they awaited their time of vengeance. They were like a gathering pack of hungry wolves and jackals warily circling a bear. So long as the bear remained strong and healthy they feared to attack him; they even paid him homage. But all the while they watched for any sign that he was losing his strength.

The summons Wolsey issued to Henry had been prearranged between them. The King obeyed the summons, presenting himself before Wolsey and Warham. Wolsey informed him that the legality of his marriage to Catherine was being questioned. There were charges that the dispensation granted by the late Pope for their marriage was not valid. If so, Henry and Catherine were not man and wife. Their living together as such was an offense against God and religion.

Wolsey humbly asked his king's permission to consider the question. Henry gave permission. He took the position of not agreeing with the charge. He considered himself Catherine's true husband; otherwise, he would not have lived with her as such for the past eighteen years. But he was a pious man with a sensitive conscience. If living with Catherine as her husband *was* an offense to God, it certainly would explain Catherine's failure to bear him a male heir. The problem must of course be looked into. Whatever the truth, he must know it so his conscience would not be troubled.

As Henry could spare no more time for the matter at the moment, a Dr. Bell was appointed as Henry's advocate, to argue against the charge in his place. Another theologian, Dr. Wolman, was made prosecutor, to state the charge against the marriage of Henry and Catherine. And prove it.

In the days that followed, Cardinal Wolsey held "court" while Bell and Wolman argued the case back and forth. They finally finished presenting their arguments. Wolsey ruled that there was some substance in the charge against the validity of the marriage. Enough to justify asking the opinion of other doctors of theology and experts in canon law. He set himself to the task of insuring that a majority of the experts questioned came up with the right answer.

All this was supposed to have been conducted in secrecy. But word of it had leaked out, and was spreading fast. Catherine learned of the proceedings the day after they started. So did Spanish Ambassador Mendoza, who immediately sent word to Catherine's nephew.

The letter told Charles V that Wolsey, "as the final stroke to all his iniquities, has been scheming to bring about the Queen's divorce." Catherine, Mendoza wrote, was "full of apprehension." The danger to her marriage was also a danger for Charles. Wolsey had for some time been working to ally England and France against the growing power of Catherine's nephew. Knowing nothing about Henry's secret passion for Anne Boleyn, Mendoza believed Wolsey was now trying to free Henry from Catherine so he could join England and France through a new marriage. Charles must, in his own interest, do everything he could to counter Wolsey's efforts.

Mendoza did not, like Catherine, absolve King Henry from complicity in this business: "The King is so bent on this divorce that he has secretly assembled certain bishops and lawyers to sign a declaration that his marriage with the Queen is null and void, on account of her having been his brother's wife."

Mendoza then went to the heart of the struggle: "It is therefore to be feared that either the Pope will be induced by some false statement to side against the Queen, or that the Cardinal [Wolsey], in virtue of his legatine powers, may take some step fatal to the said marriage. . . . It would be very advisable if, with all possible secrecy, the Pope were to be put on his guard . . . and by having the cause referred entirely to himself, should prevent him [Wolsey] from taking part in it. . . . Should the King see that he cannot succeed, he will not run the risk."

Catherine was frightened, but kept both her fright and her

awareness of her danger from her husband. For over a month she pretended she did not know what was going on behind her back; as though hoping that if she kept silent the threat would go away.

At last, on June 22, it became impossible to pretend any longer.

On that day, Henry told her that their marriage had been called into question. It had been said that the Pope who had granted the dispensation for them to marry had not had the right to do so. Because of this, his conscience had been deeply troubled. He had questioned canonists and theologians about the matter. In the opinion of these experts, they were living together in mortal sin. So now his conscience was more troubled than ever.

To ease his conscience, Henry told her, they must separate. He asked her to choose a place to which she would like to go to live.

All these weeks Catherine had borne her knowledge with quiet strength. But now that she saw her husband, as she thought, so deluded by Wolsey's lies that he actually believed them, she could bear it no longer. She broke down and wept. Tears were the only answer she gave him.

Her grief unnerved Henry. He could not bring himself to press her further in the face of it. Instead, he begged her not to be so distressed, assuring her that "all should be done for the best." No more was said, for a time, about Catherine separating from her husband. Henry begged her, for her part, to say nothing to anyone else about the matter.

This was an attempt to delay any appeal for help that Catherine might make to her nephew. Henry wanted the matter settled before Charles could stick his weighty thumb into it. For something had occurred which put Charles V in a more strategic position than anyone had anticipated.

Henry and Wolsey had originally foreseen no unconquerable obstacle to getting the marriage declared invalid. The arguments in Wolsey's "court," followed by the solicited opinions of English theologians, would give them the formal, legal groundwork upon which to submit the question to the present Pope, Clement VII. There was every reason to expect that Pope Clement would decide the way King Henry wanted him to decide—that the dispensation granted by the previous Pope was invalid; that Henry was actually still a bachelor, free to marry again.

The rulers of nations supporting the Church were usually accommodated in such matters. King Henry IV of Castile, for example, had in 1437 obtained a papal dispensation to discard his first wife because she had failed to produce an heir. The dispensation granted him the right to try out a second wife for a time, discard her in turn if no heir was produced by this new match, and then marry his first wife again if he wished. There were other such precedents.

Pope Clement VII could surely be counted on to act as the English King wished when the problem was presented to him. Henry and Wolsey began their proceedings with confidence.

Then, after they had already begun and the secret was out, England was stunned by fantastic news from Italy: Charles V's Imperial troops had taken and sacked Rome. The Pope was virtually a prisoner of Catherine's nephew.

2

THE SACK of Rome climaxed for Wolsey a series of failures in his attempt to wrest advantage for England from the continuing power struggle between Charles V and Francis I. Before Henry fell in love with Anne Boleyn, Charles had one advantage over Francis: His aunt was Henry's wife. In loyalty to this blood tie, Catherine naturally took Charles's part at every turning of the international game. Charles had strengthened this special position by promising to marry Henry's and

Catherine's daughter, Mary. This had brought England to his side for a time. The alliance had kept France at bay while Charles strengthened his grip in Italy.

The Italian rulers, frightened by the danger to their territories of Charles's increasing power, tried to counter by allying themselves with France. As alarmed as any of them was Pope Clement VII, a man of excellent taste, keen mind, large faith, small courage and shaky principles. He had territorial and political ambitions of his own—both for the international Church of which he was the head, and the powerful Medici family of which he was a member. In 1524 he joined France and Venice in a league to block Charles.

Armed with Italian support and papal favor, Francis led his army south across the Alps to take on the forces of Charles. For a time in 1525, it seemed that Francis would prevail. Then his string of victories ran out. The Imperial army smashed the French at the Battle of Pavia, and the twenty-five-year-old Charles took Francis prisoner.

It had appeared to Henry and Wolsey to be the opportunity of the century. With Francis a defeated captive, and the victorious Emperor Charles England's ally, what was to prevent Henry from from invading France, conquering it, and declaring himself ruler of France as well as England? "Now," Henry exulted, "is the time for the Emperor and myself to devise the means of getting full satisfaction from France. Not an hour is to be lost."

But the hour *was* lost. Wolsey's strenuous efforts to raise English money for the invasion petered out against a wall of public resistance and pleas of poverty. And Charles showed no inclination to assist, or even to encourage Henry.

After the Battle of Pavia, Charles no longer needed England. He had wrung enormous concessions from Francis in exchange for his release from captivity. With his great enemy tamed, there was no one left strong enough to oppose Charles. He could now do whatever he wanted, ignoring England. As though to drive this point home, Charles, ignoring his agreement to wed Princess Mary, married the wealthy Isabella of Portugal. He then announced that he intended to invade Central Italy and revenge himself on those who had sided with Francis against him—"most especially on that villain Pope."

Pope Clement retaliated by absolving the French King from all promises he had made to Charles while a prisoner. Francis began once more to gather his forces to oppose Charles. He started by allying France with Pope Clement, Venice, Milan and Florence—and negotiating for a similar alliance with England.

Henry and Wolsey were, by then, most receptive to such an alliance. They had been stung by Charles's failure to honor his promise to Princess Mary, annoyed by his reluctance to aid England's territorial ambitions, and worried by his seemingly limitless growth of power. So it was that in 1527—the same year that Henry began proceedings to discard Catherine—he signed a treaty with France to "prevent Charles climbing any higher." At the same time, a considerable sum of English money was sent to help the Pope.

But more than money was needed to withstand the force descending through Italy. It required solid military intervention. Henry was unwilling to commit himself that far, and Francis was not yet fully prepared to do so. Charles's troops captured Milan and swept south toward Rome.

The Imperial army reached Rome on the first Saturday in May, led by the Duke of Bourbon. It consisted of some twenty thousand Germans and Spaniards in a vicious mood, being long unpaid, hungry, ragged, and blaming their privation on the papal enemy. It was because of their condition and temper that Bourbon had decided to abandon his siege of Florence and hurry south to attack Rome instead. The reason, an Imperial officer wrote to Charles, was that Florence was too well defended by the forces of the Holy League. The attacking army's meager provisions would run out before they could possibly take the city.

"On the other hand," he explained to Charles, "Rome was defenseless, and by plundering it and putting the Pope to great extremity we should gain all the rest."

On May 6th, under cover of a heavy morning mist, Bourbon led the ravenous, hate-filled, near-mutinous German and Spanish troops toward a section known as the Borgo, where the walls would be easiest to scale. At the first assault, Bourbon was shot in the stomach, dying almost immediately. After that, there was no leader left strong enough to exercise any measure of control over the attacking forces. Charles's army broke into the city, slaughtered the few defenders, and went wild—giving itself up to an orgy of looting, rape and blood lust.

The report Charles received later from one of his officers told him of what followed: "Having entered, our men sacked the whole Borgo, and killed almost every one they found, making only a very few prisoners. . . . The Trans-Tiberine was stormed and taken, and shortly afterwards the bridges of Sisto and S. Maria, by which the army made an entrance into Rome. This was early in the evening of the 6th . . . persons of every nation, age, sex and degree were taken prisoners, and not one escaped. All the monasteries were rifled, and the ladies who had taken refuge in them carried off. Every person was compelled by torture to pay a ransom, not according to his condition, but according to the will of the soldiers, after being stripped of all his goods. The greater part were unable to pay, and remained in prison, subjected to ill treatment. . . . The ornaments of all churches were pillaged, and the relics and other sacred things thrown into sinks and cesspools. Even the holy places were sacked. The church of St. Peter and the Papal palace, from basement to the top, were turned into stables for horses. I am convinced Your Majesty, as a most Christian prince, will be grieved . . . but every one considers it has taken place by the just judgment of God, because the court of Rome was so ill ruled. . . ."

Cardinal Como wrote: "Rome was taken on the 6th. They began to sack the city the same day. The sacking and taking of prisoners continued for 12 days, and it would have lasted longer still if there had been anything to sack, or any more prisoners to take. . . .

"All the churches and monasteries, both of friars and nuns, were sacked. Many friars were beheaded, even priests at the altar; many old nuns were beaten with sticks; many young ones violated, robbed and made prisoners. . . . In short, there is not a house in Rome, either of cardinals' or others, not a church or monastery . . . which has not been sacked; even the houses of the water-carriers and porters.

"Cardinals, bishops, friars, priests, old nuns, infants, dames, pages, and servants,—the very poorest,—were tormented with unheard-of cruelties,—the son in the presence of his father, the babe in the sight of its mother. Fathers were separated from sons, husbands from wives, so that they knew nothing about each other; menservants and maidservants tortured to reveal hidden treasures . . . ."

While the armed horde sacked his city, Pope Clement huddled within the heavily fortified Castle San Angelo, to which he had fled for his life on the first day of the sack. There he remained, safe for the moment but with no way to escape. The massive walls were a shelter—and a prison.

The attitude of many of Charles's associates toward the Pope kept his future in jeopardy. The sacking of Rome was not yet finished when a letter to Charles informed him of the view of one cardinal allied to the Imperial cause: "Cardinal Colonna is of opinion that on no account ought the Emperor to trust the Pope, whatever may be his promises or the securities he offers. He knows his disposition to be so fickle and changeable that no reliance whatever can be placed on him."

One of Emperor Charles's representatives in Italy wrote to

him before the month was out: "The sack of Rome must be looked upon as a visitation from God. He permitted that the Emperor—who is his most devoted servant and true Catholic prince—should become the instrument of his vengeance, to teach his Vicar on Earth and the rest of the Christian Princes that their wicked purposes shall be defeated, the unjust wars they have raised against the Emperor cease, and peace be restored to Christendom, so that the exaltation of Faith and the extirpation of heresy may be accomplished. . . .

"Should the Emperor think that the Church of God is not what it should be, and that the Pope's temporal power emboldens him to promote rebellion . . . it will not be a sin, but on the contrary a meritorious action to reform the Church in such a manner that the pope's authority may be confined exclusively to his own spiritual duties, and temporal affairs left to be decided upon by the Caesar."

Charles himself expressed regret over what his troops had done to Rome. But Pope Clement remained trapped in the Castle San Angelo.

This was the news that reached Henry and Wolsey—and Queen Catherine. At the very moment when Henry was preparing to ask the Pope to free him from Catherine, Catherine's nephew held the Pope's fate in his hand.

3

IT WAS a staggering blow. But Henry was a resilient, tenacious fighter. In tournaments his charge might be stopped, his lance shattered harmlessly against his opponent's armor—but he was seldom unhorsed. He sat his saddle too firmly for that, and his powers of recovery were quick and strong. It did not take Henry long to reassess his opponent, shift to different weapons, and wheel his mount back into the lists.

Henry began by sending Wolsey to France. The first part

of his mission was to enlist help in freeing the Pope. If Pope Clement could not be freed, Wolsey was to try to get the other cardinals of Europe to declare that, since the Pope was no longer in a position to give impartial consideration to Henry's problem, it should be turned over to Wolsey for final judgment. Henry wrote to a number of cardinals on the Continent to explain the first part of Wolsey's mission, and to gain their good will in case it came to the second part.

"No one can receive the news of the disaster at Rome without grief and indignation," Henry told them. As a champion of the Faith, he said, he was "determined to resist this storm, and leave nothing undone to restore the Pope to liberty, and vindicate the dignity of the Church . . . no doubt the Cardinal, and those of his brethren who have been spared this degrading bondage, burn with the same zeal."

Catherine felt the net closing around her. In her panic, she decided to send a personal messenger to her nephew; to warn Charles of the latest developments and beg him to make the Pope remove Wolsey's power to harm her. She chose Francisco Felipez, one of her oldest servants, as her messenger. And knowing that her husband did not want her to communicate with Charles, she concocted a ruse.

Felipez went to King Henry with a worried face and a delicate problem. His aged mother, he told Henry, was sick in Spain. Felipez wanted to see her before she died. But Queen Catherine refused to let him go. Felipez did not want to disobey his mistress; neither did he wish his mother to die without one last look at her beloved son. Was there anything a merciful king could do to help him out of this dilemma?

Henry was not fooled. He realized that it was Charles whom Felipez intended to see in Spain. But if he prevented him from going, Catherine might be driven to other methods of seeking legal and political aid. She might even break her silence and cry out her cause to the country.

Henry pretended to be taken in by the ruse. He went to

Catherine and told her it would be humane to give Felipez permission to go to his dying mother. Catherine argued against depriving herself of a servant she was so accustomed to. Henry set about persuading her. She allowed herself to be persuaded.

Henry gave Felipez more than the Queen's permission. The best way to Spain was through France. But with Francis and Charles in conflict this could be a dangerous route. Henry promised to use his influence to help Felipez get through without harm or hindrance. Felipez was sent on his way. So was a letter from Henry to Wolsey: Felipez was to be seized in France and kept prisoner, without letting it be known that anyone English had anything to do with it.

But if Henry was not fooled, neither was Felipez. He went to France, and vanished. Whether he used a disguise, an unexpected route, or a combination of both, those sent to intercept him failed. When he reappeared, it was at the court of Charles in Spain.

"Madame and my Aunt," read the letter Catherine received from Charles, "Your letter by Francisco Felipez has come duly to hand. I have perfectly well understood the verbal message he brought from you respecting the affair, and the reason why you sent him to me. . . . You may well imagine the pain this intelligence caused me, and how much I felt for you. I cannot express it otherwise than by assuring you that were my own mother concerned, I should not experience greater sorrow. . . . I have immediately set about taking the necessary steps for the remedy, and you may be certain that I shall omit nothing to help you in your present trouble."

But he had other troubles to contend with: his entanglements in Italy, the enmity of Francis, the spread of Lutheranism in his German territories. He wanted to stop the divorce without further alienating King Henry, if at all possible. "To this end," he wrote Ambassador Mendoza in England, "it seems to us that this matter ought to be treated with all possible moderation, having recourse to kind remonstrances alone for the present."

In keeping with his desire to try friendly persuasion first, Charles communicated his feelings about the divorce to Henry: "We cannot be persuaded to believe in so strange a determination. . . . In fact, we do not believe it possible, considering the good qualities of His Serenity and of the Queen, his wife . . . the Queen herself being so good and virtuous, loving His Serenity so much, having always conducted herself towards him in so irreproachable a manner, and being of such Royal blood. . . . To which we may add that having so sweet a Princess for their daughter, it is not to be believed that His Serenity would consent to have her and her mother dishonored . . ."

Charles earnestly entreated Henry, "for the honor and service of God, put an end to this scandalous affair."

At the same time, Charles wrote to the Pope, urging him to revoke Cardinal Wolsey's legatine power—or to command that neither Wolsey nor anyone else in England do anything about the divorce. The case must be settled in Rome by the Pope and the sacred college of cardinals.

The divorce was coming irresistibly out into the open. Henry's and Wolsey's attempts at subterfuge had failed. And the cardinals on the Continent had failed to respond to Wolsey's plan to circumvent the Pope. They would not fulfill Clement's function—though the Pope was still trapped inside the Castle San Angelo, unharmed but in no position to disregard Emperor Charles's requests or threats.

Henry decided the time had come to send Clement a request and threat of his own. As usual, Henry sent the request, in the most pleasant of terms, and it was left to Wolsey to deliver the threat. Since the sack of Rome, the devotion of England was more vital than ever as a support for what was left of the Pope's safety and prestige. This was part of what Pope Clement must be made to keep in mind when he considered the problem of Henry's divorce—which was beginning to be referred to as "the King's great matter."

Three men—Knight, Lee and Sir Gregory Casale—were delegated to represent Henry's position in Rome. Casale was to disguise himself, pretend to be in someone else's employ, slip into the Castle San Angelo and obtain a secret interview with the Pope. Henry gave Casale a brief note of introduction to Clement:

"Sends Gregory Casale to offer consolation, and to request the Pope's indulgence in a matter of great moment. . . . Greenwich, 6 Dec. 1527.

"P.S.—The matter which Gregory has to speak about is of the deepest interest to the King, and therefore he implores the Pope's prompt kindness."

Wolsey briefed Casale on the details: "I have told you already how the King . . . has found his conscience somewhat burdened with his present marriage. . . . He has made diligent inquiry whether the dispensation granted for himself and the Queen as his brother's wife is valid and sufficient, and he is told that it is not.

"To this the King attributes the death of all his male children, and dreads the heavy wrath of God if he persists. . . . He is resolved to apply for his remedy to the Holy See, trusting that, out of consideration of his services to the Church, the Pope will not refuse to remove this scruple out of the King's mind, and discover a method whereby he may take another wife, and, God willing, have male children."

After which, Wolsey announced his objective: "You shall then request the Pope, all fear and doubt aside, to . . . freely grant a special commission . . . granting me a faculty to summon whom I please to inquire into the sufficiency of the dispensation."

Finally, Wolsey ordered Casale to deliver the threat: "The King's friendship is of the utmost moment to the Pope, as his enmity is fraught with the most terrible consequences . . . the

King is absolutely resolved to satisfy his conscience. . . . If he [the Pope] refuses, I can do nothing hereafter in his behalf."

Meanwhile, Charles had been considering the disadvantages of keeping Pope Clement almost a prisoner in the Castle San Angelo. Any decision the Pope made that was favorable to Charles would appear to have been forced out of him—and might be treated by the rest of Europe as invalid. And it was unseemly for an emperor who was the first prince in Christendom to be at war with God's vicar on earth.

For these reasons, Charles finally allowed Pope Clement to escape from the castle, slip out of Rome in disguise, and make his way to Orvieto. There Pope Clement would once more appear to be a free agent. But Charles could have his every move watched and exercise a measure of less obvious control over what he said and did.

So it was in a bishop's threadbare palace in Orvieto, rather than the Castle San Angelo, that the English representatives found the Pope. They immediately began trying to persuade him to grant what King Henry wanted. But representatives of Charles, and cardinals allied to him, were also at Orvieto. These warned the Pope against each English demand. Pope Clement found himself in a fierce tug of war, with himself as the rope.

It was a situation which dredged up all of Pope Clement's considerable timidity. To be fair to him, his fears were not only for himself. With the flames of Lutheranism licking at the walls of Catholicism, he could not afford to turn either King Henry or Emperor Charles against the Holy Roman Church. He struggled desperately, and a bit peevishly, against being forced to make any decision at all. For one decision would alienate Henry; the other would alienate Charles.

When the English kept prodding him, the Pope finally appeared to give in. Knight happily wrote Wolsey that Pope Clement told him "that when he was in the Castle of San Angelo, the general of the Observants in Spain required of him in the Emperor's name to grant no act whereby the King's divorce should be judged in his own dominions. . . . He urges, as he is still in captivity in effect, he cannot grant this commission without evident ruin; but he is willing to run the hazard, rather than that the King and you should suspect him of ingratitude."

But the Pope was only playing for time, hoping that the divorce issue would vanish or be solved by someone else before he could be maneuvered into taking a definite stand. What he granted the English turned out to be merely a dispensation allowing Henry to marry another women *if* his marriage to Catherine turned out to be invalid. He continued to delay giving what Henry really wanted: a commission enabling Wolsey—or someone else in Henry's camp—to decree whether the marriage was valid or not.

Clement's timidity over the divorce has been stressed by many historians to the point of making his actions appear ridiculous. Protestants have contemptuously dismissed his qualms as mere camouflage for cowardice and self-interest. Catholics have harshly blamed his indecision for the loss of England to the Church—which ultimately defeated efforts to stamp out Protestantism.

It is undoubtedly true that Pope Clement showed little integrity and less courage in his handling of the divorce dispute. But it is wrong to exaggerate his failure to the point of believing his dilatory actions were unreasonable. It was in Emperor Charles's power to crush, cripple or remold the papacy; just as it was in his power, if he chose, to ignore the growth of Lutheranism in Germany. Clement's own future, and the future of his family, the Medicis, hung on Charles's decisions. If the Imperial army should seize Clement and kill him, Charles could again express regret, as he had after the sack of Rome; but Clement would be beyond hearing him. More likely was the possibility that Charles might use the fact that Clement had been born illegitimate, and elected pope through the manipulations of the Medici interests, to unseat him from the papal throne.

On the other hand, England, if not so powerful as the Empire, was, in alliance with France, almost so. For Clement to lose England's friendship was to lose an ally which had consistently supported the Church against its enemies—and not only with cash and political support. Six years before, Wolsey had presided over the burning of Luther's books in the churchyard of St. Paul's. In the same year, Henry had written a learned, spirited defense of the Church in answer to Luther's heresies. For this, Pope Leo X had called Henry "Defender of the Faith" and Luther had called him an assortment of abusive names. Now the papacy was being asked to prove its appreciation.

At this point there was only one aspect of "the King's great matter" which was still a secret: that it was Anne Boleyn whom King Henry wanted to marry.

By the beginning of the next year, 1528, even that item was no longer hidden. In February, France sent its ally, Venice, a concise summation of the facts: "The English King intends to repudiate the Queen his consort, saying that the dispensation given by the Pope . . . is defective and invalid, and also because the Queen is of such an age that he can no longer hope for offspring from her. For the maintenance and welfare of his realm, he purposes marrying Sir Thomas Boleyn's daughter, who is very beautiful."

Wolsey found it necessary to write to Clement, admitting what Clement had already heard. But, Wolsey added, the rumors about Anne Boleyn which had reached Orvieto were grossly distorted: "The Pope has been laboring under some misapprehension, as if the King had set on foot this cause, not from fear of his succession, but out of a vain affection or undue love to a gentlewoman of not so excellent qualities as she is here esteemed."

Henry had nothing at all against Catherine, Wolsey assured

the Pope: "The King's desire is grounded upon justice, and not from any grudge or displeasure to the Queen, whom the King honors and loves, and minds to love and to treat as his sister, with all manner of kindness. . . . But as this matrimony is contrary to God's law, the King's conscience is grievously offended."

However, there *was* another woman, and her name *was* Anne Boleyn; and Wolsey had to tell the Pope about her. This he did, compiling a list of every womanly virtue he could think of.

"On the other side," Wolsey wrote, "the approved, excellent virtuous qualities of the said gentlewoman, the purity of her life, her constant virginity, her maidenly and womanly pudicity, her soberness, chasteness, meekness, humility, wisdom, descent of right noble and high thorough regal blood, education in all good and laudable qualities and manner, apparent aptness to procreation of children, with her other infinite good qualities . . . be the grounds on which the King's desire is founded . . ."

In this same month, King Henry sent off two more men—his almoner and Wolsey's secretary, Fox and Gardiner—to deal with the Pope. On their way they stopped to visit Anne Boleyn and give her a letter from King Henry.

"The bearer and his fellow," Anne read, "are dispatched with as many things to compass our matter and bring it to pass as wit could imagine; which being accomplished by their diligence, I trust you and I will shortly have our desired end. This would be more to my heart's ease and quietness of my mind than anything in the world.

"Keep him not too long with you," Henry told her, "but desire him, for your sake, to make the more speed; for the sooner we shall have word from him, the sooner shall our matter come to pass. And thus, upon trust of your short repair to London, I make an end of my letter, mine own sweetheart.

"Written with the hand of him which desireth as much to be yours as you do to have him."

4

"THE AIR OF THIS CITY," Henry's representatives wrote home on reaching Orvieto in March, "is very contagious, and the weather so moist that . . . it will be of little consequence who are lords of this country, unless for penance you would wish it to the Spaniards as being unworthy to die in battle."

They found Orvieto impoverished, foul-smelling, dreary and chronically discouraged. Pestilence from the river a mile away wafted through its dirty streets on the breath of the south wind.

As for the man they had come to see: "Cannot tell how the Pope should be described as at liberty here, where hunger, scarcity, bad lodgings, and ill air keep him as much confined as he was in Castle San Angelo. The Pope occupies a decayed palace of the bishop of Orvieto. Before reaching his privy chamber we passed three chambers, all naked and unhanged, the roofs falled down, and, as we can guess, 30 persons, riff raff and other, standing in the chamber for a garnishment. The furniture of the Pope's bed chamber was not worth 20 nobles, bed and all. It is a fall from the top of the hill to the lowest part of the mountain . . ."

Pope Clement received Gardiner, Fox and Sir Gregory Casale with apparent pleasure. He told them—at length—how much he appreciated Henry's services and devotion to the Church. He intended, he promised, to do everything he could to satisfy Henry in the matter of his marriage—adding "that if the Emperor complained he would show him and the world that in administering justice he was bound to show favor to one so meritorious as the King of England."

These were brave words. But Clement delayed acting on

them. "His Holiness," Casale wrote home, "is not adverse to pleasing the King and Wolsey, but fears the Spaniards more than he ever did, as they hold all the lands of the Church."

It was not only fear that caused Pope Clement to stretch out the discussions in Orvieto. With Charles pressing him one way, and Henry in alliance with Francis pressing him the other, he found power being fed back into his hands. Both sides needed him to achieve their purposes. So long as the controversy continued, both sides would continue to court his favor. To delay was sensible and safer, for as soon as the Pope made a decision he would offend one of the opposing sides.

Clement's tactics in subsequent meetings with Henry's men revealed his talent as an actor. When they pleaded for an end to delay he sighed and wiped his eyes and complained that he did not understand the fine points of the case; he must wait on the opinions of men more learned than he. When they began to doubt his sincerity he shifted to cheerful confidence, assuring them that he expected to shortly satisfy the wishes of their king. When they became openly angry at his failure to do so, he sat with his head bowed in downcast silence; then leaped up and strode around the chamber in great agitation, flinging his arms to heaven and crying out that they must do what they wanted without him.

Outside events cut this knot. The French army reasserted itself in Italy and began winning victories over the forces of Emperor Charles. Pope Clement appeared to give in. He granted a commission, as the English had wished. Cardinal Lorenzo Campeggio would go to England and judge "the King's great matter" together with Cardinal Wolsey.

"Before the Pope would grant this brief," Casale wrote home, "he said, weeping, that it would be his utter ruin."

Casale soothed Clement with promises that King Henry would protect him from the wrath of Emperor Charles, and use his influence with Francis to assist the papal cause. This was what the Pope was hoping for. He asked that the French should march immediately to liberate the States of the Church from Charles.

After further assurances of aid Pope Clement signed the long-sought commission. He was trusting his future entirely to King Henry and Wolsey, he said; placing himself "in the King's arms." Henry was overjoyed. "The approach of the time which has been delayed so long," he wrote Anne Boleyn, "delights me so much that it seems almost already come."

The source of his joy was a mirage. Pope Clement's decision was, in reality, a means for further delay. He informed Henry that Cardinal Campeggio would have the power to act together with Wolsey to settle the divorce matter in England. At the same time he assured Charles that Campeggio had no such power; the final decision would be rendered not in England, but in Rome. Campeggio was instructed to take his time preparing for his journey; and to make the trip a slow one.

Wolsey waited in France for the arrival of Campeggio. He now seemed to be at the very pinnacle of his power and royal favor. After all, it was on him, and his influence on Cardinal Campeggio, that Henry and Anne Boleyn were depending to achieve their freedom to marry.

Anne Boleyn wrote to Wolsey on June 11th of 1528: "My Lord, in my most humble wise I desire you to pardon me that I am so bold to trouble you with my simple and rude writing. . . . The great pains you take for me, both day and night, are never likely to be recompensed, but only in loving you, next unto the King's Grace, above all creatures living, as my deeds shall manifest. I long to hear from you news of the Legate [Campeggio], and hope they will be very good . . ."

Henry added some words of his own to this letter: "The writer of this would not cease till she had called me likewise to set to my hand. Both of us desire to see you, and are glad to hear you have escaped the plague so well, trusting the fury of it is abated, especially with those that keep good diet, as I trust you do. The not hearing of the Legate's arrival in France causeth us somewhat to muse; but we trust by your diligence shortly to be eased of that trouble."

Wolsey waited. The legate, Campeggio, did not appear. The days passed. Wolsey sent assurances to England, along with expensive tokens of his esteem. In the beginning of July he received another loving letter from Anne Boleyn: "In my most humble wise that my poor heart can think, I thank Your Grace for your kind letter and rich present, which I shall never be able to deserve without your help."

She swore to Wolsey that "all the days of my life I am most bound, of all creatures, next the King's Grace, to love and serve Your Grace. I beseech you never to doubt that I shall ever vary from this thought while breath is in my body."

To which Anne added what Wolsey already knew: "I much desire the coming of the Legate, and, if it be God's pleasure, I pray Him to bring this matter shortly to a good end, when I trust partly to recompense your pains."

Wolsey continued to wait. Still there was no word of Campeggio's coming.

For Pope Clement continued to play for time—still hoping that the divorce problem would vanish, absolving him of the necessity to choose sides. Failing a settlement between Henry and Charles themselves, there were other possibilites. Queen Catherine or Anne Boleyn might die—either at the hands of their enemies, or from the plague that Henry had mentioned in his note to Wolsey.

Called "the sweat," this sickness was virulent and extremely contagious. It struck England in June of 1528 and spread swiftly through the more populous areas, felling thousands in the first days of its assault. King Henry fled from the infected area to Waltham, taking Catherine with him for appearances' sake. His desire to break with his wife and marry Anne Boleyn was now common knowledge. But he clung to the pretense that he had no intention of separating from Catherine unless the Church decided that he *must* do so. For this same reason he did not take Anne with him, but sent her to the Boleyn manor at Hever. Deprived of her presence, Henry occupied himself with hunting and hard riding, saw his wife as little as possible, and thought much of the woman he loved.

Then word reached him that Anne Boleyn had caught the disease. "There came to me in the night," the worried King wrote Anne, "the most afflicting news possible. I have to grieve for three causes: first, to hear of my mistress's sickness, whose health I desire as my own, and would willingly bear the half of yours to cure you; secondly, because I fear to suffer yet longer that absence which has already given me so much pain,—God deliver me from such an importunate rebel!; thirdly, because the physician I trust most is at present absent when he could do me the greatest pleasure. However, in his absence, I send you the second, praying God he may soon make you well, and I shall love him the better. I beseech you to be governed by his advice, and then I hope to see you soon again."

His anxiety was cured days later, with the news that Anne was entirely over whatever had ailed her. "The doubt I had of your health troubled me extremely," Henry wrote her in relief, "and I should scarcely have had any quiet without knowing the certainty; but since you have felt nothing, I hope it is with you as with us. . . . I think if you would retire from Surrey, as we did, you would avoid all danger . . ."

A few days later Henry wrote her again to inquire after her health and "praying God (and it be His pleasure) to send us shortly together, for I promise you I long for it."

The plague subsided. Anne Boleyn was still alive. So were Catherine and Henry. Despite Pope Clement's delays and evasions, the problem would not go away. Having returned with Catherine to Bridewell Palace, Henry arranged to get a house for Anne on the Strand. There he could reach her easily and privately by merely taking a short boat ride on the Thames at night.

Many had expected the passing of time to "abate the King's great appetite" for Anne Boleyn. Instead, his desire for her seemed to be growing more tumid. When he shot a hart during one day's hunting, he sent it to Anne for her dinner—accompanied by this note: "Seeing my darling is absent, I . . . send her some flesh representing my name, which is hart's flesh for Henry . . . hereafter, God willing, you must enjoy some of mine."

In another letter, after a separation of a few days, he told her how long it seemed since she had gone, and that he could not have imagined so short an absence would so upset him. He closed the letter: "Wishing myself an evening in my sweetheart's arms, whose pretty dubbys I trust shortly to caress."

Under these circumstances even so skillful and desperate a delayer as Pope Clement could not procrastinate forever. Cardinal Campeggio finally left Italy and began making his way toward England.

Lorenzo Campeggio was a restrained and steady man devoted to the twin ideals of cleansing the Church and crushing the Lutherans. He had the mind and manner of a good, solid lawyer. He had been an eminent Italian jurist and teacher of law before becoming a priest. The father of five children, three of them legitimate, he had taken his vows after the death of his wife in 1509, the year Henry became king of England and married Catherine. Since then he had become a distinguished papal diplomat. For two years he had been rallying the authorities in Austria and Germany against Lutheranism.

Coming as he did from an influential family markedly successful in real estate and politics, Campeggio had a healthy respect for possessions and position. Rome was regularly reminded that he appreciated and expected tangible rewards for his industry on its behalf. At the same time he was unswervingly loyal to the Church, undeviatingly alert against any infringement of its power or prestige. He never agreed with Erasmus that Lutherans and other heretics should be tolerated, their fate left to the workings of time and the will of God. But then, in the gathering war between Luther and Rome, very few people *did* agree with Erasmus.

Campeggio was an aging man. He suffered from recurrent bouts of the gout which caused him so much agony that he often could not walk, having to be carried about on a litter.

He began his journey in July, traveling north and west at a pace so slow as to be almost imperceptible. This was due, Campeggio said, first to the danger and difficulty of moving through Italy unsettled by war, and then to an especially excruciating attack of gout. Since he might understandably be reluctant to face the problem awaiting him in England, some referred to his ailment as "diplomatic gout."

Campeggio did not reach Paris until the latter part of September. By then Pope Clement had reason to be grateful for the two months Campeggio had consumed. The English situation had not changed during this time; but much had altered in Italy. The initial victories of the French and their allies against Charles had again given way to utter defeat. Pope Clement had to resign himself to the fact that Emperor Charles would not be evicted from Italy. The Pope's secretary sped fresh instructions after Campeggio:

". . . as the Emperor is victorious, and has made overtures for peace, the Pope must not give him any pretext for a fresh rupture, lest the Church should be utterly annihilated. As soon as you can do so . . . proceed on your journey to England, and there do your utmost to restore mutual affection between the King and Queen. You are not to pronounce any opinion without a new and express commission hence."

Five days later the nervous Pope had his secretary send another warning to Campeggio: "I am ashamed of repeating the same thing so many times . . . but every day stronger reasons are discovered which compel the Pope to remind you that you are to act cautiously, and to use your utmost skill and address in diverting the King from his present desire, and restoring him to his former love toward the Queen."

At almost that same moment Henry was writing to Anne Boleyn: "The Legate which we most desire arrived at Paris . . . so that I trust by next Monday to hear of his arrival at Calais, and then I trust within a while after to enjoy that which I have so longed for. . . .

"I would you were in mine arms or I in yours, for I think it long since I kissed you. . . ."

5

IT WAS October 7th when Campeggio at last reached London. He came for the avowed purpose of joining with Wolsey in examining the validity of Henry's marriage, conducting a divorce trial, and pronouncing judgment. But this last act, the Pope's secretary warned him again and again by mail, was a goal which Campeggio must never reach:

"If in satisfying His Majesty the Pope would incur merely personal danger, his love and obligations to the King are so great that he would content him unhesitatingly; but as this involves the certain ruin of the Apostolic See and the Church, owing to recent events, the Pope must beware. If so great an injury be done to the Emperor, all hope is lost of the universal peace, and the Church cannot escape utter ruin, as it is entirely in the power of the Emperor's servants. You will not, therefore, be surprised at my repeating that you are not to proceed to sentence, under any pretext, without express commission [from the Pope]. Protract the matter as long as possible. . . ."

Riding from Canterbury toward London, Campeggio discovered that the people of England knew what their king had in mind to do, and were violently opposed to it. Especially the women. A crowd of more than a thousand gathered along the road and chanted at him as he went by: "No Nan Boleyn for us! No Nan Boleyn for us!"

By the time Campeggio entered the suburbs of London he was too exhausted and tortured by gout to ride any farther. The nobles escorting him were forced to call a halt, though Henry and Wolsey were waiting eagerly to receive the legate in state. Campeggio was carried in a litter into the Duke of Suffolk's home and put to bed. There he lay through the night and most of the following day.

Henry and Wolsey had waited too long to be delayed still further by Campeggio's illness. If the road was too painful for him, there was the more comfortable water route. Campeggio was carried from Suffolk's house to the Thames. A river barge floated him the rest of the way into London, to lodgings prepared for him at Bath House. There he was allowed a night to recover. It was not enough. The next morning he found that he was unable to rise from his bed.

His impatient hosts were forced to give him more time to recuperate. He was allowed to remain in bed—but not in peace. That same morning Wolsey paid a visit to Bath House. Settling himself by Campeggio's bedside, Wolsey welcomed him to the city and got down to business—the business of annulling Henry's marriage to Catherine. Campeggio accepted the inevitable. Propped up on his pillows, he discussed the problems involved. He found that they were even thornier than he had anticipated.

Pope Clement had instructed Campeggio to enlist Wolsey's help in persuading King Henry to remain married to Catherine. This Wolsey said he could not—and *would* not—do.

"He alleged," Campeggio wrote later, "that if the King's desire were not complied with . . . the speedy and total ruin would follow of the kingdom, of His Lordship [Wolsey], and of the Church's influence in this kingdom."

Campeggio countered by warning Wolsey that a divorce would result in mortal war with Catherine's nephew. Wolsey denied this emphatically. He was certain Emperor Charles would not take it so much to heart. With Campeggio's co-operation, the divorce could be conducted with dignity, Catherine treated respectfully, and both she and her husband come out of it honorably. Surely Charles, burdened with so many problems all over Europe, would not choose to burden himself further with a great war merely to avenge an aunt.

Campeggio brought out every other argument he could think of, but he could make no dent in Wolsey's decision. "I have no more moved him," Campeggio informed the Pope, "than if I had spoken to a rock."

In the days that followed, Wolsey lost no opportunity to impress upon Campeggio the absolute necessity of giving in to King Henry. He led him toward it with the lure of an England deeply grateful and forever faithful to Pope and Church. He drove him toward it with the threat of Lutheranism. He reminded Cardinal Campeggio that "the greater part of Germany, owing to the harshness and severity of a certain Cardinal, has become estranged from the Apostolic See and from the Faith." And he warned: "Beware lest, in like manner . . . it may be said that another Cardinal has given the same occasion to England with the same result."

Because of his gout, Campeggio was able to confine himself to Bath House for a time, putting off his first audience with Henry. But the King soon reached the end of his patience. Campeggio was compelled to take the short but painful ride to the King's palace.

Meeting Henry for the first time, Campeggio could not fail to be impressed, as other Italians had been impressed before him. At thirty-seven he was still a handsome, towering man with enormous vitality.

Anne Boleyn was unseen but very much felt at this first interview between king and prelate, her waiting presence goading Henry and reinforcing his resolve. Campeggio's delays had made her difficult to handle. Despite Henry's repeated reassurances, her impatience had broken through her self-restraint. She had become irritable and frightened, angrily doubting that Henry would ever be able to marry her.

Henry had assured her that Campeggio's illness was not a pretense, that the Cardinal had come to help them. Anne allowed herself to be calmed. Her relieved lover told her "what joy it is to me to understand of your conformableness to reason, and of the suppressing of your inutile vain thoughts and fantasies with the bridle of reason. . . . Wherefore, good sweetheart, continue in the same, not only in this but in all your doings hereafter; for thereby shall come, both to you and me, the greatest quietness that may be in this world."

But neither Anne nor Henry could bear to wait much longer. Henry made this plain to Campeggio.

"The King," Campeggio informed the Pope after their first meeting, "exhibited a most ardent desire for this divorce; and he seems to me to be so persuaded of the nullity of the marriage, and so firmly to believe it, that I have come to the conclusion that it will be impossible to persuade him otherwise."

This was a device of diplomatic self-protection for Campeggio. It was written to make sure the Pope understood the difficulty of Campeggio's assignment, and the possibility that he might fail through no fault of his own. But in spite of Henry's stubbornness, Campeggio did not concede defeat on that first day. Too much hung in the balance. And Campeggio was not lacking in courage. After a night's rest, he prepared to try again.

King Henry, eager to sweep aside all legal complications that stood between him and Anne Boleyn, came to Campeggio in private the following evening after dinner. He shut himself up alone with the Cardinal for four straight hours. Campeggio begged him not to proceed with the divorce. Henry said that he was forced to, in order to clear his conscience. Campeggio offered to sooth his conscience by obtaining a new dispensation from Pope Clement. Henry was not interested. Campeggio attempted to confuse Henry by debating the fine points of canon law. They thrashed out divine prohibition against marrying the wife of one's brother; also the pope's right to grant a dispensation which removed the stain from doing so. It was a debate which Campeggio lost.

"His Majesty," he discovered, "has so diligently studied this matter, that I believe in this case he knows more than a great theologian and jurist."

But Henry did agree with Campeggio on one aspect of the matter: The scandal and controversy involved in an open battle over divorce would do no one good. He preferred, if possible, to separate himself from Catherine in a friendly manner. If she would only allow him to do so, he would reward her with whatever she wished. Especially, he would swear to see to it that their daughter Mary sat upon the throne of England after his death—*if* he failed to have a son by another wife.

In the end a limp Campeggio found himself discussing one solution that pleased Henry: Suppose Catherine renounced her claims as Henry's wife by entering a nunnery and taking vows of chastity? By the time Henry left, he had Campeggio's promise to help Wolsey persuade Catherine to do so.

It was, if it could be done, as good a way out for the Church as any. The Pope had already considered such a solution, and given Campeggio leave to suggest it. If Henry would not give in for the sake of his soul, Catherine must give way for the safety of her religion.

The next day Wolsey took Campeggio to meet Catherine. He found himself facing a stubborn, aging woman grown thick and slow with years, her face ravaged by illnesses and betrayals and the carrying of dead children, armed only with the weapon of pride. But it was the pride of a Spanish princess and English queen. Campeggio was to find it as unshakable as Henry's determination.

Chapter Six

THE TRIAL AT BLACKFRIARS

"I BEGAN," Campeggio wrote to Clement of his first meeting with Catherine, "by telling her that as the Pope could not refuse justice to any one who demands it, he had sent the Cardinal of York and myself hither to understand the state of the question between Her Highness and the King's Majesty: but as the matter was very important and full of difficulty, His Holiness . . . counselled her, confiding in her prudence, that rather than press it to trial she should take some other course . . ."

Catherine did not give him time to come slowly to the point. She already knew what "other course" he had in mind. She was no longer without friends at court. Rumors reached her quickly. She informed Campeggio and Wolsey that she had heard they had come to persuade her to renounce her worldly status and enter a convent. Was this true?

Campeggio did not deny it, but added that it was entirely up to her. No one else could make such a decision. However, he could not refrain from pointing out that if she freed Henry by taking vows of chastity, she would lose nothing that she had not already lost. He was now convinced, he told her, that her husband would never again regard her as his wife. It would be much wiser, he told her, to yield to the King's wishes and avoid a trial. For if the sentence went against her she would lose much—even her dowry.

"I begged her to consider the scandals and enmities which would result," Campeggio informed the Pope. He told Catherine that if she would only give in to Henry she would keep her dowry, the guardianship of her daughter, her rank as a princess, plus anything else she cared to ask of the King—"and she would offend neither God nor her own conscience. Then I alleged the example of the queen of France who did a similar thing, and who still lives in the greatest honor and reputation with God and all that kingdom."

When at length he ran out of arguments, Wolsey spoke, repeating essentially the same warnings and inducements. Catherine listened to him with growing anger. Losing her composure, she charged him bitterly with having originated her troubles, by first instilling doubts about the marriage in her husband's mind. Then she controlled herself once more and turned back to the Italian Cardinal.

She was sorry Henry's conscience was disturbed, she told Campeggio. She held her husband's conscience and honor above all else in the world. But there was no reason for his conscience to be troubled. She had absolutely no scruple about their marriage, knowing herself to be Henry's true and legitimate wife in the sight of God. As for the proposal that she enter a convent, made in the name of the Pope, she could not even consider it.

Campeggio replied lamely that the proposal was only intended to solve the problem with a minimum of difficulty for herself—and for all Europe. But of course, she could depend upon the Pope to deal justly with so dutiful and obedient a daughter of the Church. Catherine told him pointedly that all she asked was that she not be condemned before she was heard. And on that note she dismissed them.

With the Pope's special representative against her, Catherine now seemed utterly alone. But she knew she was not alone. And her ally was in a position to see to it that the Pope did nothing to harm her.

She had already spoken to the Imperial ambassador, who had passed on her words to her nephew, Emperor Charles: "The Queen relies entirely on Your Majesty, and has no other consolation in her troubles than the hope of being powerfully assisted. She begs me to remind Your Majesty of one thing, which must greatly benefit her case, namely, the promise of another pressing letter to His Holiness, making him understand how very dangerous it would be for the peace of Christendom if he were to allow this King's will to be done."

After Campeggio and Wolsey were gone, Catherine went to see Henry and spoke about what they had said to her. Her husband insisted again that he was motivated only by disturbed conscience, and must abide by whatever the religious authorities decided. These authorities, Henry told her, were already of the opinion that she was not his true wife. Even the Pope, he claimed, was of this opinion. This was the reason he had sent Campeggio to England: to accomplish the divorce.

"How can the Pope condemn me without a hearing?" Catherine demanded.

Henry replied that Clement already knew all the evidence, and had decided on it. The sensible thing now would be for her to follow Campeggio's advice and retire to a convent. Otherwise she might be compelled to do so.

"May God forbid," Catherine cried, "that I should be the cause of that being done which is so much against my soul, my conscience, and my honor! I know very well that if the judges are impartial, and I am granted a hearing, my cause is gained. For no judge will be found unjust enough to condemn me."

She begged Henry for councilors, proctors and advocates to advise and defend her. He granted her a number of them, including her confessor, the Archbishop of Canterbury, three bishops, and a proctor and advocate from Flanders.

She also asked if she might make her confession the next day to Campeggio. Henry gave his permission—and on leaving her he ordered Wolsey to forewarn Campeggio. Wolsey did so at break of day the next morning, while the gout-ridden Campeggio was still in his bed. An hour later Catherine arrived.

She spoke to Campeggio privately under the seal of confession, but insisted that he pass on what she said to Pope Clement. To begin with, she told him the story of her life in England, from the day she had arrived as a bride to this moment. One thing she stressed: Henry had no reason to be troubled in his conscience about marrying his brother's wife. The papal dispensation allowing Henry to marry her was valid. Besides, she had never been Arthur's wife. In all, she had not slept with Arthur more than seven nights, and on none of those nights had they had sexual intercourse. Arthur had died leaving her as he had found her—a virgin. And it was as a virgin that she had married Henry.

Campeggio begged her to stop dwelling on this point, to accept the inevitable, enter a convent, and content herself with all the gifts a grateful Henry would bestow upon her for doing so.

Catherine said flatly that she would never do so. She intended to live and die in the estate of matrimony into which God had called her. She would defend to the last her honor and the honor of her husband. Campeggio must understand that all attempts to make her take the veil in a convent would fail. This was her decision; it always would be her decision.

Campeggio tried to dissuade her with the arguments he had used the day before. Everything he said to weaken her determination seemed instead to strengthen it. Catherine kept repeating, each time more fiercely, that nothing could ever make her change her mind.

"She insists," Pope Clement was informed, "that everything shall be decided by sentence. If that should go against her, she would then remain as free as His Highness. She says that neither the whole kingdom on the one hand, nor any great punishment on the other, although she might be torn limb from limb, could compel her to alter this opinion; and that if after death she should return to life, rather than change her opinion, she would prefer to die over again."

Her words were strong, and she intended them to be so to make her refusal emphatic. But having said them, Catherine softened her tone. She begged Campeggio to help her remove from her husband's mind "this fantasy." They had *not* lived in sin all these years. She even offered Campeggio bait: In return she would use her influence to persuade her nephew, Emperor Charles, to lower his demands on the international scene, so that there could be universal peace.

After she left him, a disheartened and rather irritated Campeggio wrote: "I do not despair of success in persuading the Queen to religion, though I see it is difficult and more than doubtful. . . . As she is nearly 50 and would lose nothing whatever, and as so much good would ensue, I cannot see why it should be impossible to induce her to take this course."

Campeggio tried to induce her again the following day. And when he failed, Wolsey fell to his knees before Catherine, praying and pleading that she accept Campeggio's advice. She told the two cardinals firmly that she would not.

It is understandable that her opponents came to consider her unreasonable and almost irresponsible. In her inflexible refusal to capitulate, Catherine was putting her own name and honor above the good of the Church, the peace of Europe, and the future of England. And in doing so Catherine emerges as one of the few incorruptible, decent and steadfastly human figures in this history. What her Church and her husband were doing to her, they felt compelled to do for the greater good. But it led them into what Ruskin termed "the fatal error of supposing that . . . the necessity of offenses renders them inoffensive."

The people of England understood this in their hearts. Catherine's plight called forth all their sympathy. This was the reason for their growing hatred of Anne Boleyn. They liked Henry too much to turn against him, and they needed someone to blame for what was being done to Catherine.

Henry in turn understood the temper of his people, and knew he must walk softly. These were a people he ruled, but also a people upon whom he depended. He had no huge standing army to enforce his will. In war it was a yeomanry that owned arms, knew how to use them, and were willing to leave home for the battlefield that formed the bulk of England's troops. Such men could oppose a monarch's will—even overthrow him—if sufficiently aroused.

That they were becoming aroused was obvious. Londoners shouted encouragement to Queen Catherine whenever they saw her. They growled in taverns and street-corner gatherings about the undeserved indignities inflicted on her. To prevent them from becoming further enflamed by Catherine's appearances in public, Henry sent her back to Greenwich. And though he was with Anne Boleyn more and more, he was careful not to be seen with her in public.

These measures were not enough. Henry sensed the need of doing something to keep his popularity from being drowned in this tide of public passion.

To keep his people from turning against him, Henry decided to assure them he shared their own feelings about the divorce. He called the foremost citizens of London to his great hall at Bridewell Palace and made an impassioned speech. He told them it would make him miserable if the Church forced him to part with Catherine—even though the succession to the throne and the peace of the kingdom would become more secure if he took another wife.

On the other hand, he swore, "if it be adjudged by law of God that she is my lawful wife, there was never anything more pleasant and more acceptable to me. . . . She is a woman of most gentleness, of most humility and buxomness, yea and of all good qualities appertaining to nobility she is without comparison."

If it were found that his marriage with Catherine was not a sin he would be delighted. For "if I were to marry again, if the marriage might be good, I would surely choose her above all other women." But he *must* be governed, as a religious man, by the decision of the Church.

Henry felt quite safe in speaking thus because he had utter confidence in what that decision would be. With Catherine refusing to remove herself to a convent, he was eager for Campeggio to bring the question to trial. He could not envision the Pope's alienating him by allowing the sentence to go against him. He had been assured that Campeggio had with him a secret decretal from Clement, allowing the sentence to be decided in England.

But if Campeggio did have such a decretal, he contrived to avoid producing it. Just as he contrived to keep putting off the trial. Weeks slipped by. He alternated between assuring Wolsey that Catherine might yet be persuaded to give in, and urging him to persuade King Henry that a divorce trial was too dangerous for the peace of Europe.

At one point Campeggio toyed with another solution to the problem. If what was really troubling Henry was fear that civil war would ensue if his daughter inherited the throne, her succession might be strengthened by marrying her to Henry's illegitimate son, under special dispensation from the Pope. The notion of Princess Mary's marrying her half brother was not repugnant to the Church under these circumstances. When Campeggio finally rejected it, it was not because of morality or human decency. The reason, he explained to Pope Clement, was only "that I do not believe this design would suffice to satisfy the King's desire."

Once again he concentrated upon avoiding trial by trying to persuade either King Henry or Queen Catherine to give way.

Just as he tried to use Wolsey to influence Henry, so he tried to use Emperor Charles's ambassador to influence Catherine. He failed with Wolsey. And he failed with the ambassador.

"I spoke some time ago to the legate," the ambassador reported to Charles. "He ended by expressing his opinion that the best way to avoid the embarrassments that might arise would be for the Queen of her own free will to take the veil before her case was submitted for trial."

In reply the ambassador coldly told Campeggio that Emperor Charles would never allow his aunt to be unjustly treated. Campeggio backed down hastily, saying he realized how close this affair was to Charles's heart. Pressing harder, the ambassador pointed out that Catherine wished for nothing more than her right to be heard in her own defense. But since the case was of such delicacy and importance she naturally wanted it decided at Rome, *not* in England.

Campeggio promised that he did not intend to act without fresh instructions from Rome. "This, however," he warned Charles's ambassador, "must be a secret between Your Reverence and me. I have just dispatched a trusty messenger to the Pope to inform him how the affair stands, and what the Queen has said to me, and also to inquire what are His Holiness' real intentions and wishes respecting this case. Until an answer comes from Rome I shall not move to step in the affair."

Nor did he. Cardinal Wolsey tried to move him and failed, though his efforts were intensified by a growing sense of terror. Henry was fuming over the delay. And behind Henry an irritated, anxious Anne Boleyn was becoming more influential. She was drawing to herself a group of men who wanted power and who hoped to use her to cut down Wolsey.

"The lady who is the cause of this King's misconduct," the Imperial ambassador wrote to Charles, "perceiving that her marriage, which she considered certain, is being put off, begins to suspect that the Cardinal of England is preventing it as much as he can, from fear of losing his power the moment she becomes Queen. This suspicion has been the cause of her forming an alliance with her father, and with the two dukes of Norfolk and Suffolk, to try to see whether they can conjointly ruin the Cardinal. . . . The Cardinal is no longer received at Court as graciously as before, and . . . now and then King Henry has uttered angry words respecting him."

2

IN DESPERATION, guided by Henry's doubts about his ability and persistence, Wolsey wrote to Sir Gregory Casale, England's papal ambassador. He ordered Casale to tell the Pope that Campeggio would not produce his papal commission to try the divorce in England:

"You shall say Campeggio has taken a course entirely different from his instructions and attempts to dissuade the King and Queen from the divorce. . . . What is worse, though I am his colleague, he will not entrust me with his commission. . . .

"The King feels his honor touched by this, especially considering what a benefactor he has been to the Church. . . . I see ruin, infamy, and subversion of the whole dignity and estimation of the See Apostolic if this course be persisted in. You see in what dangerous times we are. . . . Without the Pope's compliance I cannot bear up against the storm."

There was no trickery behind this. The terror Wolsey expressed, he genuinely felt, and it was to prove prophetic. He warned that unless Clement cooperated, Lutheran influence in England and English opposition to the Church, "which have been extinguished with such care and vigilance, will blaze forth.

"It is useless," he went on, "for Campeggio to think of reviving the marriage. . . . If the Pope wishes to preserve his honor, to show his gratitude and his sincerity, to preserve the dignity of the Church and the safety of this kingdom, now is the time. Let him, then, command Campeggio to proceed to sentence."

By this time Pope Clement, having acknowledged the fact of Emperor Charles's power, had returned to Rome under Charles's protection. Though he still felt obligated to King Henry and afraid of Henry's anger, he was more afraid of Charles. This was the situation when, Gregory Casale being ill, his brother John went in his place to see the Pope and carry out Wolsey's instructions. Clement's replies to the pleas of Wolsey were confusing, and intended to confuse; or at least to spin out the delay still further. He implied that Campeggio had the power to render sentence in England—and then again that he did not; that he had such a commission— but that it was merely to soothe Henry, not for actual use.

John Casale asked if anything could induce the Pope to change his mind so the divorce could proceed. Angrily, Clement said that he now realized the commission he had given Campeggio would be the ruin of him, and that he was determined not to make any more concessions.

"But," Casale pleaded, "let Your Holiness consider what ruin and what heresy will be occasioned in England upon the alienation of the King's mind by this resolution on the part of Your Holiness."

"I do consider the ruin which now hangs over me," cried the agitated Pope, flinging his arms about. "I repent what I have done. If heresies arise, is it my fault? My conscience acquits me. None of you have reason to complain. . . . The King and the Cardinal have never asked anything in my power which I have not yielded with the utmost promptness. But I will do no violence to my conscience."

Casale tried to pin him down. "Well, is Your Holiness *un-*

willing that proceedings should be taken by virtue of the commission?"

Clement saw he had gone too far. No, he said, he was not unwilling.

"But then, Campeggio opposes your wish, and dissuades the divorce."

"Well, I commissioned him to dissuade the King from the divorce, and to persuade the Queen. But he is to execute his commission."

This reluctant, vacillating capitulation on the part of Clement was meaningless. The next letter Campeggio received from Rome read: "The Pope has been highly satisfied with your negotiations hitherto, and thinks you have acted very prudently . . . hold out for the love of God."

Campeggio was told to disabuse Wolsey of the notion that Charles would not be much upset if the Church aided the divorce. Nothing the Pope could do would anger Charles more. On the other hand, Wolsey was to be reassured that the Pope had not lost his great affection for Henry.

But Clement was beginning to blame Wolsey for having involved the Vatican in the divorce—just as Catherine blamed Wolsey for instituting the divorce action, and Anne Boleyn blamed him for impeding it. "Would to God," Clement's secretary wrote to Campeggio, "the Cardinal had allowed the matter to take its course. Because if the King had come to a decision without the Pope's authority, whether wrongly or rightly, it would have been without blame or prejudice to His Holiness."

Clement would be far from unhappy if Henry became impatient enough to divorce Catherine without the papal permission. This was hinted to Gregory Casale's brother. "Let them if they like," Clement suggested hopefully, "send the legate back again, on the pretext that he will not proceed in the case, and then do as they please. Provided they do not make me responsible for injustice."

That it was not injustice Clement feared was made plain in letters approving Campeggio's alternate solution: "It would greatly please the Pope if the Queen could be induced to enter some religion because . . . it would involve injury to only one person."

But Catherine would not so injure herself. And Henry was adamant that any son he had by Anne Boleyn must be his legitimate heir, recognized as such by the Church.

Steel was added to Catherine's stubbornness and Henry's determination in February of 1529, by a letter from Charles in Spain: "We are sadly grieved at an affair so scandalous . . . the matter being so injurious to the honor of the said Queen, our aunt, and of the illustrious Princess, her legitimate daughter, and our most beloved cousin.

"We firmly believe this conduct to emanate not from the King's own free will, but solely from the sinister persuasions and intrigues of some persons who, for the sake of their own private interest and wicked purposes and from their lust for power, have deceived him. . . .

"The above reasons . . . oblige and compel us to hinder the said divorce, and strenuously to uphold and maintain the rights of the Queen. To this end our Ambassadors have in our name requested and besought our most Holy Father, the Pope, to have the cause summoned, tried and sentenced at Rome . . . not elsewhere."

Such a letter could not fail to infuriate a man like Henry. It challenged his authority in his own country, and his influence with the Church. The battle lines were clearly drawn now: Henry, lured by Anne Boleyn, demanded the divorce; Catherine, backed by Emperor Charles, was fighting for her marriage; Campeggio, instructed by Pope Clement, was delaying the decision. Wolsey, increasingly alone, was caught in the crossfire.

Campeggio contrived to drag his heels for a few more months. But it became increasingly difficult. Emperor Charles had flaunted his strength too blatantly, for all Europe to witness. Henry could not relent now, even if he should wish to. Charles had made it obvious that he controlled the Pope, and that if the decision were rendered at Rome he would dictate it. He had also intimated that his influence reached even into England; that he could oppose Henry's will in Henry's own kingdom, and prevent the decision from being made there.

Much of Henry's energy was now channeled into proving Charles wrong. Forced forward by this relentless will, Campeggio finally agreed to bring the divorce to trial at Blackfriars, in June of 1529.

3

THERE ARE hinges upon which historical revolutions turn. The great drama of the divorce between Henry and Catherine—and between England and the Roman Catholic Church —turned on the hinge of the Blackfriars trial.

To Catherine, the announcement that the trial would be held in England meant that Clement was preparing to give in to Henry. The advisers and advocates Henry had assigned her were unlikely to endanger themselves by opposing their king. It appeared that her cause was lost. Catherine began to plan, by herself and in secret, exactly how she would conduct herself in the face of this outrage.

The other principals approached the trial with different feelings. Henry and Anne Boleyn saw their desired goal within grasp. Campeggio regarded the trial as a last chance for a compromise: Henry's determination might weaken before the enormity of his act; or Catherine, seeing that her ruin was inevitable, might yet retreat to the safety of a convent life.

Cardinal Wolsey approached the trial with the most uncertainty. His power, his career, perhaps his life were at stake.

And with his position at its most precarious, it was also at its most ineffective. He could not control the decision. Though he was to be one of the two legates at the trial, Campeggio was the *presiding* legate.

The trial began on June 21. The great personages of England crowded into the Chapter House of Blackfriars as spectators. An assembly of English bishops entered, led by Warham, the aged Archbishop of Canterbury, to form the court. Above them upon a dais Wolsey and Campeggio, wearing their scarlet robes, were ceremoniously ushered in to act as judges. King Henry's counselors were grouped on one side of the officers of the court. On the other side were Queen Catherine's—among them John Fisher, the old, devout Bishop of Rochester.

Catherine made her appearance, taking her seat under a cloth-of-gold canopy to the left of the legates. Only she knew what she intended to do that day. Only she knew what it cost to control her nerves, to maintain her composure and wait for the strategic moment to do it.

Henry was last to arrive. He took his seat far from his wife, to the right of Wolsey and Campeggio.

The court was called to order and the great hall to silence. The crier commenced the proceedings: "King Henry of England, come into the court . . ."

"Here, my lords," answered Henry.

"Catherine, Queen of England, come into the court!"

Her moment was upon her. She did not reply. Instead she rose from her seat and made her way solemnly to where her husband sat. She stood before him, looking into his face. Then she lowered herself to her knees before him.

"Sir," she cried out in halting English, that all in the hall might hear and understand her, "I beseech you for all the loves that have been between us, and for the love of God, let me have justice and right! Take of me some pity and compassion. For I am a poor woman and a stranger born out of your dominion. I have here no assured friend, and much less in-different [impartial] counsel. I flee to you as to the head of justice within this realm. Alas, Sir, wherein have I offended you?"

Henry stared at his wife in embarrassed silence, unable to think of any courteous way to silence her.

Her words continued to strike at him and echo around the vast chamber: "I take God and all the world to witness that I have been to you a true humble and obedient wife, ever conformable to your will and pleasure, that never said or did anything of the contrary thereof, being always well pleased and contented with all things wherein you had any delight or dalliance. Whether it were in little or much, I never grudged in word or countenance, or showed a visage or spark of discontent. I loved all those whom you loved only for your sake, whether I had cause or no; and whether they were my friends or my enemies. This twenty years I have been your true wife or more, and by me ye have had divers children—although it has pleased God to call them out of this world, which hath been no default in me."

Every person in the Blackfriars hall was listening. Everyone heard. Catherine came to the heart of the matter: Henry's belief that he had sinned because he had married his brother's true wife.

"When ye had me at the first, I take God to be my judge, I was a true maid without touch of man. And whether it be true or no, I put it to your conscience."

Having thus skillfully implied that Henry, above all men, had reason to know that she had come to him a virgin, Catherine shifted to the ground that their fathers, who had arranged their marriage, had been wise men. "Also, as me seemeth, there was in those days as wise, as well-learned men, and men of as good judgment as be at this present in both realms, who thought then the marriage between you and me good and lawful. Therefore it is a wonder to hear that new inventions are now invented against me, that never intended but honesty. And cause me to stand to the order and judgment of this new court, wherein you may do me much wrong, if ye intend any cruelty."

This struck at Henry's image, in his own and the public's eyes, as a just, good-natured man. Now she struck at the court's claim to impartial judgment, and its consequent right to try her:

"For ye may condemn me for lack of sufficient answer, having no indifferent counsel, but such as be assigned me, with whose wisdom and learning I am not acquainted. Ye must consider," Catherine reminded her husband, "that they cannot be indifferent counsellors for my part which be your subjects . . . and dare not, for your displeasure, disobey your will and intent. . . . Therefore I most humbly require you, in the way of charity, and for the love of God, who is the just judge, to spare me the extremity of this new court, until I may be advertised what way and order my friends in Spain will advise me to take.

"And if ye will not extend to me so much indifferent favor, your pleasure then be fulfilled. And to God I commit my cause!"

That was it. She had said what she had come to say. Now it was time for the final gesture. She rose to her feet, made a low curtsey to her husband, and left him.

At first Henry assumed she intended to return to her seat. Instead Catherine went past it and, leaning on the arm of one of her attendants, headed for the door of the hall. Henry ordered the crier to summon her back.

"Catherine, Queen of England, come into the court!"

She heard, but did not heed it. Walking away with tired dignity, she left the court, never to return. As she emerged from Blackfriars a throng of London women who had gathered outside cheered her. They called to her to resist her enemies and not worry. Catherine entered her coach and had herself driven back to Bridewell Palace, there to await her fate.

Inside the hall at Blackfriars, King Henry spoke up hastily to counter the impression Catherine had made as a wife unjustly spurned. Once more he assured everyone that he had no desire to shed Catherine: "For as much as the Queen is gone, I will, in her absence, declare unto you all . . . she hath been to me as true, as obedient, and as comformable a wife as I could in my fantasy wish or desire."

They had come here before this court, not because he wished to discard his wife, but because doubt had been cast by others upon the validity of their marriage. This had caused Henry to wonder if the death of all their male children might be a sign of God's indignation.

"Thus being troubled in waves of a scrupulous conscience," he told the assembly at Blackfriars, "and partly in despair of any male issue by her, it drove me at last to consider the estate of this realm, and the danger it stood in for lack of male issue to succeed me in this imperial dignity. I thought it good therefore . . . to attempt the law therein, and whether I might take another wife in case that my first copulation with this gentlewoman were not lawful. Which I intend not for any carnal concupiscence, nor for any displeasure or mislike of the queen's person or age."

It was up to the bishops assembled here, Henry said, to try the validity of his marriage. They had all, on being consulted earlier, signed a document declaring that the marriage was in doubt, and that the doubt must be resolved by trial.

The Archbishop of Canterbury quickly agreed: "That is truth if it please Your Highness. I doubt not but all my brethren here present will affirm the same."

It was now that John Fisher, Bishop of Rochester, spoke up: "No, Sir, not *I*. You have not *my* consent thereto."

Henry was stunned. "No?" He held up the document in

question. "Look here upon this. Is not this your hand and seal?"

"No, Sire, it is *not* my hand nor seal!"

In all that assembly only the aged Fisher could have dared to challenge Henry. He had seen the King through his childhood; had been confessor to Henry's grandmother, and his father's most trusted adviser.

Henry turned on the Archbishop of Canterbury. "Sir, how say *ye*, is it not his hand and seal?"

"Yes, Sir," Warham assured him unhappily.

"That is not so," Fisher objected stubbornly. Warham had indeed asked him to sign the document as the other bishops had done, "but then I said to you that I would never consent to no such act, for it were much against my conscience."

He knew the danger of the thing he was doing against his King. But the speech Catherine had made on her knees had pricked him deep. And he was, after all, an old man, not far from death.

Warham admitted that Fisher had spoken such words to him. But he maintained that afterward Fisher had agreed that Warham could sign Fisher's name to the document for him.

"All which words and matter," Fisher snapped, "there is no thing more untrue."

Henry did not want to be forced into harshness toward the old bishop who had been for so long attached to his family. Checking his rage, Henry brushed aside the quarrel as insignificant.

"Well, well, it shall make no matter," he told Fisher. "We will not stand with you in argument herein, for you are but one man."

One man against many. And it was to the many, who had already agreed to give Henry what he wanted, that Henry turned for judgment of his marriage.

Though ordered by the court to present herself during the days of hearings that followed, Catherine did not appear. The divorce trial continued without her. But it did not go as smoothly as Henry expected. For Fisher, unlike Catherine, did not absent himself from the hearings. He listened to Henry's counsel bringing forth alleged proofs that the King was not really married, that Catherine had been known carnally by Henry's brother, that the papal dispensation for the marriage was not valid. He listened to Cardinal Wolsey acting more and more openly as Henry's chief advocate, rather than an impartial judge. The fact that his was the only voice of determined opposition to the divorce seemed to fire Fisher with ever more courage.

For, he explained to the court at Blackfriars, "I know the truth." And he went on to explain: "I know that God is truth itself . . . this marriage was made and joined by God . . . which cannot be broken or loosed by the power of man."

"So much do all faithful men know," countered Wolsey. "Yet . . . the King's counsel doth allege divers presumptions, to prove the marriage not good at the beginning. *Ergo,* say they, it was *not* joined by God at the beginning, and therefore it is not lawful."

Fisher begged to differ. He considered Henry and Catherine married in the eyes of the law and the Lord, based on his long and careful study of the matter. "I devoted more attention to examining the truth of it, lest I should deceive myself and others, than to anything else in my life."

All the other bishops were with Wolsey, eager to show Henry their desire to be helpful. But it was up to Campeggio to declare an end to the arguments and pronounce sentence. And this he did not do. Instead he listened to the arguments day after day, apparently insatiable in his appetite for hearing the same claims and counterclaims interminably repeated. The first week of the trial ended. A second week slipped by.

Henry, boiling with impatience, summoned Wolsey to Bridewell late one morning. From eleven till noon they were closeted in the King's privy chamber, while Wolsey tried to explain his failure to force the trial to a conclusion. When he left the palace, Wolsey met the Bishop of Carlisle.

"Sir," the Bishop said, wiping his perspiring face, "it is a very hot day."

"Yea," Wolsey agreed with bitter humor, "if ye had been as well chafed as I have been within this hour, ye would say it were very hot."

The emotionally exhausted Wolsey was conveyed in his barge along the Thames to his house at Westminster. There he cast himself down on his bed to regain some strength in sleep. Two hours later he was awakened by Anne Boleyn's father. King Henry had been thinking it over, Boleyn said, and decided that Wolsey and Campeggio should make another attempt—immediately—to persuade Catherine to stop being obstinate. Wearily, Wolsey got up and took his barge to Bath Place to pick up Campeggio. The two cardinals were rowed back to Bridewell, went straight to the Queen's quarters, and had themselves announced.

Catherine had been supervising embroidery work in her privy chamber. Emerging with a skein of white thread about her neck, she greeted the cardinals courteously: "Alack, my lords, I am very sorry to cause you to attend upon me. What is your pleasure with me?"

Wolsey glanced at the members of her household around them. "If it please you to go into your privy chamber, we will show you the cause of our coming."

"My lord," Catherine told him, "if you have anything to say, speak it openly before all these folks, for I fear nothing that ye can say or allege against me . . . I pray you speak your minds openly."

Wolsey began to speak to her in Latin.

"Nay, good my lord," Catherine interrupted, "speak to me in English I beseech you; although I understand Latin." She understood also how uncomfortable it was for Wolsey to have all her servants hear his words. But she had been made more than uncomfortable, and had borne it. For the moment the shoe was on the other foot.

"Madam, if it please Your Grace," Wolsey began again in English, "we come to know your mind . . . and also to declare secretly our opinions and our counsel unto you."

"I had need of good counsel in this case," Catherine acknowledged pointedly. But "any counsel or friendship that I can find in England are nothing to my purpose or profit. Think you, will any Englishman counsel or be friendly unto me against the King's pleasure, they being his subjects? Nay . . . my counsel in whom I do intend to put my trust be not here. They be in Spain, in my native land. . . . I am a simple woman, destitute and barren of friendship and counsel here in a foreign region."

Then she relented a bit, and allowed the cardinals to go with her into her privy chamber. What was said between them there is not known. But Wolsey emerged with no reason for new hope. And Campeggio was finally certain that there was absolutely nothing left that could be done to bend Catherine to her husband's wishes.

July 23 was set as the day for the trial to end. Henry came to Blackfriars that day and took a gallery seat to hear Campeggio hand down his decision. He was to receive a shock. Campeggio, with all eyes on him, again evaded rendering sentence—this time permanently. In Rome, he pointed out, the courts recessed between July and October. Therefore, in adherence to that custom, he was adjourning *this* court for two months.

No one present misunderstood Campeggio's true meaning: the trial was over; there would be no sentence. The case would be revoked to Rome, where Pope Clement would eventually decide it.

Henry stormed out of the hall.

Catherine was quickly informed, and rose in an instant from stoic defiance to overconfidence. Anne Boleyn's peak of anticipation crumbled, leaving her, it was said, filled with apprehension that she was "wasting her time and youth to no purpose."

As for Wolsey, who had failed to achieve what Henry wanted, his fall was signaled by the Duke of Suffolk's ominous growl at the trial's termination: "It was never merry in England whilst we had cardinals among us."

Henry did not turn on Cardinal Wolsey. He merely turned *from* him and walked away, leaving him, as Shakespeare has Wolsey lament, "naked to mine enemies."

4

WOLSEY suddenly found himself outside the King's circle. It became difficult to see the King, impossible to speak with him in private. For two months he was kept in a state of unbearable suspense.

Henry still had one last hope of using him to obtain his divorce. Campeggio was supposed to have brought with him a papal authorization for the decision to be made in England. If it could be found, Wolsey might even now act on that authority.

In the beginning of October Campeggio prepared to sail away from England. Pope Clement's decretal, if it still existed—if it had ever existed—would leave with him. King Henry's agents at Dover resorted to an extreme last-gap measure: They seized and searched all of Campeggio's baggage.

What they sought, they did not find. The last hope was lost—to Henry, and to Wolsey. Campeggio left England. Wolsey's enemies fell upon him.

They charged him with everything from common theft to manipulating political, religious and financial policy behind the King's back—forty-four charges in all. Wolsey went down under their assault so swiftly and completely that all Europe was amazed.

"Cardinal Wolsey, after so much long continued prosperity, has at last found fortune irate and hostile beyond measure, bringing him to ruin which may be said to exceed his late fame and elevation." So read a secret Italian report. "He has been forbidden to act as legate, and has lost the chancellorship, the bishopric of Winchester, the abbacy of St. Albans, and all his other revenues . . . in a moment and unexpectedly, he has lost everything. In truth a memorable example for those who believe worldly prosperity to be stable and true happiness."

Four days later the French ambassador was writing: "On Tuesday the great seal was taken from him [Wolsey] and an inventory made of his goods. Everyone who had been in his service these 20 years was commanded to render an account of all they have touched. . . . While writing, I have heard that Wolsey has just been put out of his house, and all his goods taken into the King's hands. Besides the robberies of which they charge him, and the troubles occasioned by him between Christian princes, they accuse him of so many other things that he is quite undone."

King Henry appropriated Hampton Court, the magnificent palace Wolsey had built for himself. The Duke of Norfolk succeeded him as head of the King's Privy Council, with the Duke of Suffolk as second in charge. Sir Thomas More, having become lord chancellor of England in Wolsey's place, delivered a savage attack on Wolsey to a meeting of Parliament.

More compared the people of England to a multitude of sheep, and called King Henry their shepherd. As for Wolsey: "As you see that amongst a great flock of sheep some be rotten and faulty, which the good shepard sendeth away from the good sheep, though the great wether which is of late fallen, as you all know, so craftily, so scabbedly, yea, and so untruly juggled with the King, that all men must guess that he thought in himself that the King would not see or know his fraudulent juggling and attempts. But he was deceived, for His Grace's sight was so quick and penetrable that he saw him—yea, and saw through him . . . and according to his desert he hath had a gentle correction."

What was "gentle" about Wolsey's correction was that he was still alive and not in prison—probably because Henry wanted him available in case he should ever be needed again. Alive, but utterly crushed. The French ambassador reported that Wolsey broke down while lamenting his fall to him in private: "For heart and words entirely failed him. He wept much . . . his enemies, even if English, could not but feel sorry for him."

Wolsey did not fight the charges brought against him— knowing it would be useless; fearing it would goad his enemies into more vicious retaliation. Instead of attempting to defend himself, he surrendered to self-abasement and self-effacement—desperately trying to make himself into so small an object as to be overlooked, or at least to be looked upon as an object of pity.

He confessed in writing to the offenses charged against him, signed over his revenues and possessions to the King, retired to a house in Esher, and threw himself completely on Henry's mercy.

"I daily cry to you for mercy," he wrote the King. "I covet nothing so much in this world as your favor and forgiveness. The rememberance of my folly, with the sharp sword of your displeasure, have so penetrated my heart that I cannot but cry: it is enough; now stay, most merciful King, your hand."

Henry stayed his hand to the point of allowing Wolsey to move from unhealthy Esher to Richmond Lodge. From there he watched, in subdued silence, as Norfolk, Thomas More, and Suffolk conducted his former offices.

These three seemed about to inherit the spoils of Wolsey's power. But this rich inheritance was to be too heavy a burden to be borne by men such as these. Norfolk and Suffolk would prove unequal to it; More would be unwilling to assume it at the cost of his identity. Wolsey's religious and civil powers would descend instead upon the only man in England strong enough to carry them: Henry, the King who was at last coming to know his own strength, and who was at last willing to throw aside his now threadbare cloak of appealing boyishness and reveal that strength.

The business of administering his religious power would devolve upon a still obscure priest named Thomas Cranmer. The administration of Henry's civil power, fallen from Wolsey's hands, fumbled by Norfolk and Suffolk, thrown away by More, would then be skillfully caught and adroitly managed by a man who was at the moment one of Wolsey's legal advisers—Thomas Cromwell.

It was this same Thomas Cromwell, at the moment working hard for the alleviation of his master's misery, who received a warning from Wolsey's enemies: Wolsey was still too close to court; it was time for him to move farther away.

"Show him," Norfolk told Cromwell threateningly, "that if he will go not away shortly, I will, rather than he should tarry still, tear him with my teeth."

Cromwell passed on the warning and Wolsey abjectly obeyed it. He rode two hundred miles north to his remaining diocese of York. There, the Milanese ambassador shortly reported, "we hear that he makes good cheer and is gentle as a little lamb with everybody."

This picture of the overbearing Wolsey as a gentle little lamb was astonishing—but accurate. As a priest ministering to his new flock he began to show the same unflagging devotion to the welfare of people who came to him as he had once given to the conduct of his high office. There is one instance of his having a journey interrupted by hordes of parents bringing their children to him to be confirmed. Wolsey delayed for two entire days while confirming every single child—several hundred in all.

"Who was less beloved in the north than my Lord Cardinal before he was amongst them?" read a pamphlet written not long after this. "Who better loved after he had been there a while? He gave bishops a right good example how they might win men's hearts."

But behind his new humility Wolsey began to be emboldened as time went by and no further moves were made against him. This was what his enemies had been waiting for. The most implacable of them was Norfolk, who feared the day when Henry might feel a renewed need for the Cardinal. He dug deep and found what he was looking for: Wolsey was secretly in contact with the French envoy—and, much worse, with the envoy of Emperor Charles.

Norfolk quickly conveyed this intelligence to Henry. Without doubt, Wolsey was once again attempting to pull international strings. Norfolk did not care to openly consider the possibility that Wolsey might be trying to regain favor by working to achieve what the King wanted. He dwelt, instead, on the rumor that Wolsey was acting for Catherine, vying for Charles's favor by urging international opposition to his King's desire. Henry's furious response was all that Norfolk could have wished.

Wolsey was finishing dinner in the midst of his household when the Earl of Northumberland entered with a number of men. The Earl was the same Sir Percy whose romance with Anne Boleyn Wolsey had once so brutally broken. This was the Earl's moment of revenge. But he was not quite up to it.

His voice became weak with awe as he placed his hand on Wolsey's arm and spoke in the King's name: "My lord, I arrest you of high treason."

Wolsey's stunned silence gave way to the sure knowledge that his end had come. Slowly gathering all the dignity he could muster, he surrendered himself, assembled all his household and bade them farewell. The shocking suddenness of it drained him of strength and hope. He was a sick old man when he left to begin his journey to London.

It was a very slow journey, with a number of long stop-overs. For Norfolk wanted Wolsey's fate sealed before he arrived. Wolsey's physician, Agostino—who had been his contact with the French after his fall from power—was being rushed to London bound underneath a horse. He arrived before Norfolk in a state of terror, knowing that to save himself from torture and execution as a traitor, he must denounce his master.

"Since they have had the Cardinal's physician in their hands they have found what they want," Emperor Charles was informed by his ambassador. "He lives in the Duke of Norfolk's house like a prince, and is singing the tune they wished."

The tune Agostino sang sealed Wolsey's doom, and Norfolk conveyed it to the King. The Constable of the Tower of London, Sir William Kingston, was sent out with twenty-four soldiers to bring Wolsey the rest of the way under guard.

Kingston met Wolsey at Sheffield Park. As soon as he saw the Constable of the Tower, Wolsey understood that his worst fears were justified. He brushed aside an attempt to persuade him that the soldiers were merely a guard of honor: "I perceive more than you can imagine or can know. Experience," he said sardonically, "has taught me."

He knew that he was being taken to the Tower; that Norfolk, Suffolk, and the whole group led by Anne Boleyn's family had no intention of allowing him to remain there long as a *live* prisoner. The Tower would be too close to the King for their comfort. Wolsey's nerve broke completely with the arrival of the Constable. By the time they reached Leicester he was so weak he could ride no farther, and had to be helped to bed. He never rose from it.

His dying words to those around him were of Henry, and they dripped with bitterness:

"If I had served God so diligently as I have done the King, he would not have given me over in my gray hairs. . . .

"He is sure a prince of royal courage, and hath a princely heart. And rather than miss any part of his will or appetite he will put the loss of half his realm in danger. . . .

"I warn you to be well-advised what matter ye put in his head. For ye shall never put it out again."

He died, some said, from sheer terror of being put to death. In France it was rumored that he had killed himself by swallowing a deadly powder. Another rumor hinted that he had indeed been killed by such a powder, but not knowingly.

Whether his death was caused by suicide, an assassin, or the sickness of fear and despair, one fact remained: Wolsey was gone forever, and could no longer influence what was about to happen.

"The serpent being dead," Milan's ambassador to England wrote, "his poison also died. He has at length disappeared, nor is there any longer remembrance of him."

Chapter Seven

"DOWN WITH THE CHURCH"

ANNE BOLEYN had felt her chance of becoming Queen of England almost within her grasp—only to see it shrink away from her, leaving her frightened and doubtful. Now she was allowing herself to hope once more, but cautiously this time. She tried to protect herself in advance from any further disappointment by keeping as tight a rein as she could on her expectations.

Gardiner was off for another attempt to persuade Pope Clement to allow the divorce. Anne thanked him for his faithfulness. She assured him he would not have reason to repent of it. But she could only hope, she added, that the end of this journey would be more pleasant to her than his first ". . . for that was but a rejoicing hope, which causing of it does put me to the more pain . . . I do trust that this hard beginning shall make the better ending."

Catherine was also caught on the twin horns of cautious hope and painful disappointment. She was careful never to let Pope Clement forget it: "I humbly beg Your Holiness to have pity on me, and accept as though I had been in purgatory the penance I have already suffered for so many years. The remedy lies in the sentence and determination of my case without any delay."

Everything could be repaired, Catherine insisted, by Clement's "commanding the King, my Lord, to dismiss and cast away from him this woman with whom he lives. . . . If I could only have him two months with me, as he used to be, I alone should be powerful enough to make him forget the past."

On the surface of it, Catherine did still have the King with her. But so did Anne Boleyn. During this period Henry found himself living with both while sleeping with neither. Anne accompanied him on his sporting jaunts and was much with him in private. Catherine continued to be part of his court and appear with him in public.

It was a situation which amazed foreign observers, such as the ambassador from Venice: "His Majesty is still at Hampton Court, enjoying his usual sports and royal exercises, and the Queen remains constantly with him. Nor does she at all omit to follow her lord and husband. So much reciprocal courtesy is being displayed in public that anyone acquainted with the controversy cannot but consider their conduct more than human."

If the situation put a strain on the credulity of others, it soon put a much severer strain on Henry himself. What was going on in private, behind this public masquerade, was secretly passed on to Emperor Charles by Eustace Chapuys, the new ambassador sent to England to help Charles's aunt:

"On St. Andrew's Day the Queen, having dined with the King, said to him that she had long been suffering the pains of purgatory on earth; and that she was very badly treated by his refusing to dine with and visit her in her apartments."

Henry, according to Chapuys, told Catherine "that she had no cause to complain of bad treatment, for she was mistress in her own household, where she could do what she pleased. As to his not dining with her for some days past, the reason was that he was so much engaged with business of all kinds, owing to the Cardinal having left the affairs of government in a state of great confusion."

But with Catherine pointedly taking him to task, Henry's politeness suddenly began to fray badly. He became unpleasantly blunt with her: "As to his visiting her in her apartment and partaking of her bed, she ought to know that he was not her legitimate husband." He went on to remind her angrily that many theologians and canonists were of this opinion, including his own almoner, Dr. Lee—and that if the Pope failed to declare their marriage null and void, Henry would denounce the Pope as a heretic and marry whom he pleased.

The Queen's own anger flared: "You know perfectly well yourself, without the help of doctors, that you are my husband and the principal cause claimed for the divorce does not really exist. I came to you as much a virgin as I came from my mother's womb. You have often said so yourself.

"As to your Almoner's opinion in this matter, I care not a straw," she went on heatedly. "He is not my judge in the present case. It is for the Pope, not for him, to decide. . . . Indeed, if you give me permission . . . for each doctor or lawyer who might decide in your favor and against me, I shall find a thousand to declare that our marriage is good and indissoluble."

After a protracted session of this bickering between Henry and Catherine, Chapuys related, "the King left the room suddenly . . . disconcerted and downcast."

Henry's black mood was still on him when he joined Anne Boleyn for supper that night. When she learned the reason, it became her turn to heap bitterness on him: "Did I not tell you that whenever you argue with the Queen she is sure to have the upper hand? I see that some fine morning you will succumb to her reasoning, and cast me off."

With this, Anne's bitterness gave way to despair:
"I have been kept waiting a long time, and might in the

meanwhile have contracted some advantageous marriage—out of which I might have had issue, which is the greatest consolation in this world. But alas!—farewell to my time and youth, spent to no purpose at all."

"Such, I am told," Chapuys wrote Charles, "was the Lady Anne's language to the King on the evening of the day that he dined with his Queen."

It was a situation which could not continue. And Henry did not intend it to.

After Wolsey's fall, what had been a confusing, ever-shifting tangle of complicated maneuverings, evasions and probings finally became an open contest—with Henry and Anne on one side, Emperor Charles and Catherine on the other side, and the Pope still caught in the middle. Anne prodded Henry to action; Catherine prodded Charles to counteraction. But both Henry and Charles had reasons for their actions which were more important than their personal attachments to these two women.

Charles acted to secure England as a willing tool for his international power struggle; if not through Catherine, then through her daughter Mary. With Spain, Austria, Germany, the Netherlands and much of Italy his, all he needed was England to make the war chain around France complete. That this was a practical possibility was to be shown when Mary finally married some years after Henry's death.

Henry, on the other hand, was acting to secure a legal male heir to his throne; and to solidify the power of that throne in England. To accomplish this he now proceeded to advance on several different levels at the same time:

His representatives in Rome fought to counterbalance the influence of Charles with the Pope. Henry sent other representatives to persuade important religious leaders in England and the Continent to declare that his marriage to Catherine was by nature invalid. His third move, the most important, was to summon a meeting of Parliament. It was in his handling of this Parliament over the next few years that Henry finally revealed himself as the superbly matured politician which opposition, controversy and setbacks were making of him.

On the morning of November 3, 1529, shortly after the fall of Wolsey, King Henry left both Anne and Catherine behind at Greenwich and set out to open what was to evolve into "the Reformation Parliament." The people of London crowded into boats and massed along the banks of the Thames to watch as the King's barge carried him up the river to his palace of Bridewell.

There Henry and his nobles, as reported in *Hall's Chronicle*, "put on their robes of Parliament, and so came to the Blackfriars Church, where a mass of the Holy Ghost was solemnly sung by the King's chaplain; and after the mass, the King, with all his Lords and Commons which were summoned to appear on that day, came into the Parliament. The King sat on his throne or seat royal, and Sir Thomas More, his chancellor, standing on the right hand of the King, made an eloquent oration, setting forth the causes why at that time the King so had summoned them."

Thomas More, the new lord chancellor of England, eulogized the King who sat beside him as the mighty and benevolent shepherd of his flock. This shepherd-king, said More, felt that various ancient laws had become outmoded with changing times; and that the changing times had brought forth new ills within his realm for which there were no existing laws. It was to bring the outmoded laws up to date, and to institute other laws to reform the new ills, that Henry had now summoned the Parliament.

If this opening speech was purposely vague in not naming specific problems to be dealt with, there was nothing vague about Henry's intent—or his Parliament's response to it. It was through this Parliament that Henry began to skillfully apply a gradually increasing pressure on the Roman Catholic

Church to grant him his divorce. The pressure would grow until the Church bent to Henry's will—or something broke under the strain.

Though few could have predicted it on the fateful day when Henry convened the Reformation Parliament, what was to break, ultimately, was the link between England and the Church of Rome. Nothing could have been more opposed to the desires and religious convictions of the man whose speech opened the Parliament—Sir Thomas More.

2

"WHAT did nature ever create milder, sweeter or happier than the genius of Thomas More?" So wrote Erasmus.

Wit, scholar, lawyer, philospher, writer, devoted father—all these Thomas More was; and much else besides. There were piquant contradictions in his nature. He was a passionate, lusty man with a patiently inquiring mind and a deep, abiding religious faith. A gregarious, humorous man with a quiet, secret core to him. When he jested his face was solemn; when he spoke from the heart he grinned as though mocking himself.

He was of average height and build, with a strong, good-humored, thoughtful face. He had the steady gray-blue eyes of a man much given to studying himself, yet never failing to exercise an ability to read the minds of others. One who knew him, Robert Whittinton, summed him up thus during his lifetime: "More is a man of an angel's wit and singular learning. I know not his fellow. For where is the man of that gentleness, lowliness, and affability? And, as time requireth, a man of marvellous mirth and pastimes, and sometime of as sad gravity. A man for all seasons."

Born to the gentry, son of a successful lawyer who rose to become a judge, Thomas More was educated to follow in his father's footsteps. But a life of lonely study and piety in a monk's cell always had great appeal for him. While at Oxford he came close to entering the Church. He discovered, however, that he was not fashioned to be celibate. So he returned to his law studies and went on into the legal profession. And he married, becoming a man who enjoyed to the fullest his family, friends and the world around him. But for the rest of his life, as though to atone for his failure to enter the priesthood, he wore a hair shirt hidden under his garments, ceaselessly tormenting his offending flesh.

These contradictions in his personality, this duality in his life's purpose, continued. While raising his family and becoming a successful London lawyer, his piety and religious studies also led him into becoming one of the most prominent "men of new learning" that the Renaissance was bringing to the fore all over Europe.

For religion still permeated everything else that a man might think of—from philosophy to politics to war. But recent years had brought a bewildering variety of new things to think about, along with new ways of thinking about them. The long-stale incense of the medieval centuries was dissipating in a whirlwind of change and challenge.

The quasi-unity of Europe under the all-embracing Church through the Middle Ages had been broken up by the inexorable growth of large, mutually antagonistic nations. States like France, Spain, Germany and England were seizing unto themselves the powers and privileges which had belonged to the boundary-spanning Church—a Church whose working parts had become clogged with the rust of dogma and bloated with the fat of worldly wealth.

At the same time a new land had been discovered on the other side of the Atlantic Ocean; the world and the universe were turning out to be quite different from what everyone had believed; thinking men were delving into the works of the ancient Greeks and Romans for answers that could not be found in contemporary writings. Learning and teaching, long regarded as the exclusive territory of the clergy, were being grasped by men outside the Church. Gold from the mines of America and trade with the Far East was pouring into Europe, painfully wrenching the static economic structure of its society. A new class of merchants and workingmen was on the rise, while the peasantry was sunk into homeless poverty by swiftly rising prices and changes in land usage.

"Old things were passing away," in the words of Froude the historian, "and the faith and the life of ten centuries were dissolving like a dream."

The effect of Henry's desire for Anne Boleyn was swelled to fantastic proportions by this time of monumental change. It made of the divorce an instrument which helped to bury the Middle Ages and give birth to modern history.

Men like Thomas More were thinking and writing about the need to deal with what was happening by making sweeping changes in a society that was still Church-oriented. These men, often referred to as "men of the new learning" in their time, have since been called Christian Humanists. They sought to improve the condition of humanity by applying pure Christian idealism to the present working of society.

In a visionary book called *Utopia*, which became popular all over civilized Europe, More created an imaginary land where everyone lived a unified life of sweet reason, peaceful learning and simple piety. In this Utopia he even allowed heretics to live, so long as they kept their doubt about the established religion to themselves and did not try to preach it.

Other men of the new learning were thinking and writing along similar lines. Greatest of them all was Erasmus of Rotterdam, who has been variously remembered as "the soul of the Christian Humanism" and "the King of the Renaissance." It was Erasmus who, without meaning to, supplied much of the ammunition fired in the war between the Catholic and Protestant faiths, and in the power struggle between Church and State. And the controversy over Henry's divorce was to become an integral part of both the religious war and the political struggle.

"Wherever you encounter truth, look upon it as Christianity," Erasmus wrote, attacking the Church's unbending preoccupation with dogma. He warned the clergy that all their devotion to ritual and regulations, fastings and prayer, would not help them on Judgment Day. For when that day came, Jesus would remind them: "I left you only one precept, to love one another; and none of you pleads that he has done that."

One of the countries that this famous writer visited a number of times was England. There he spent much of his time with Thomas More. They found each other to be kindred souls, men who combined a soul searching opposition to the practices of Church authorities with an absolute loyalty to Catholicism. It was while living in More's house in Chelsea that Erasmus wrote the most famous of his books, *In Praise of Folly*. It was inspired by a previous work of More's, and in turn it inspired More to write his *Utopia*.

"He has the quickest sense of the ridiculous of any man I ever met," Erasmus wrote in describing More while he lived in his home. "He dresses plainly . . . holds forms and ceremonies unworthy of a man of sense. His talk is charming, full of fun. . . . He is fond of animals of all kinds, and likes to watch their habits. All the birds in Chelsea come to him to be fed."

Of More when younger, Erasmus learned quite a bit during his prolonged stay in Chelsea: "He had his love affairs when young, but none that compromised him; he was entertained by the girls running after him. . . . His original wish was to be a priest. . . . He gave it up because he fell in love, feeling that a chaste husband was better than a profligate cleric."

By the time Erasmus became a guest in his home, Thomas More had achieved considerable success in public life—and survived a personal tragedy. His wife had died, after seven years of marriage, leaving him with three daughters and a son. It was chiefly to provide his children with a mother that he startled his friends by marrying again within weeks of his first wife's death.

What startled them as much as the swiftness of his remarriage was the woman he chose: plump, plain Alice Middleton. She was, More himself acknowledged with a smile, "neither a pearl nor a girl." A robust widow with a child of her own, Alice was both tart-tongued and affectionate, unlearned and full of common sense, vain and loyal. More had chosen well. Theirs became a warmhearted duel of wits, a bantering, loving relationship that lasted to the steps of the scaffold.

More, Erasmus noted, "lives as pleasantly with her as if she were the loveliest of maidens. He rules her with jokes and caresses better than most husbands do with sternness and authority. . . . He controls his family with the same easy hand: no tragedies, no quarrels. . . . His whole house breathes happiness."

Outside this cheery household of books, birds, family and friends, in the harsher doings of public affairs, Thomas More was also doing very well for himself. His success as a lawyer, writer and speaker had won him a seat in Parliament by the age of twenty-six. Since Parliament seldom sat, he had plenty of time to further his legal career. He rose to the position of undersheriff—garnering so much praise and popularity for his handling of civil cases that he caught the attention of Wolsey and King Henry. He was forty when he became a member of the King's Privy Council, and was knighted three years later.

More's scholarship, reputation, public ability and the charm of his conversation combined to make him most attractive to Henry, who began seeking his advice quite often. It was More who helped Henry write his reply to Luther's charges against the Church, the reply which earned Henry his title from the Pope of "Defender of the Faith."

Henry sent More on a foreign mission, and he did well. Erasmus wrote: "The King would not afterwards part with him, and dragged him into the circle of the Court. 'Dragged' is the word, for no one ever struggled harder to gain admission there than More struggled to escape. But the King was bent on surrounding himself with the most capable men in his dominions . . . and he now values More so highly, both as a companion and as a Privy Councillor, that he will scarcely let him out of his sight."

That Henry acquired a real affection for More there can be no doubt. One evening the King's barge brought him unexpected and unannounced up the Thames to More's home, where he invited himself to dinner. Afterward, strolling in the garden with his arm around More's shoulders, Henry discussed animals, science, scholarship and religion with him, as one friend to another.

More knew Henry's friendship for him was real. He also knew exactly how far the friendship could be stretched: "If my head could win him a castle in France, it should not fail to go."

It was such knowledge that made Thomas More extremely reluctant when Henry, after dropping Wolsey into the depths, insisted he become the new Lord Chancellor. He accepted because he had to; and also because in this position he might yet help steer the nation through the stormy waters of "the King's great matter" without its smashing upon the reefs of religious controversy.

Before More, the lord chancellor had always been drawn from the Church. In appointing More, Henry was slapping the clergy's wrist. Also, he was attempting by this appointment to enlist the support of More's influential friends in England and the Continent: the men of the new learning, the Christian Humanists of whom More was a distinguished member. At the same time he assumed that in Thomas More he had an ally.

Henry was wrong. But his mistake was based on a reasonable assumption. More had spoken and written much about the need for reforming the Church—as had so many of his friends.

More's spiritual adviser had been the late John Colet, dean of St. Paul's Cathedral, who had preached against the "devilish pride" and worldliness creeping into the Church: "We are nowadays grieved of heretics, men mad with marvellous foolishness. But the heresies of them are not so pestilent and pernicious unto us and the people, as the evil and wicked life of priests."

It was in More's house that Erasmus had written, concerning the worldliness of even the popes: "How many treasures would the Holy Fathers have to forfeit if wisdom were suddenly to subdue their minds!"

But though Thomas More might concur with this, he could also concur with a judgment which Erasmus pronounced upon the *kings* of the day: "Of all birds, the eagle alone has seemed to wise men the type of royalty—not beautiful, not musical, not fit for food; but carnivorous, greedy, hateful to all, the curse of all, and with its great powers of doing harm, surpassing them all in its desire of doing it."

In choosing the new lord chancellor, King Henry had for once picked the wrong tool for his purpose.

3

THOMAS MORE was no longer the same man he had been when he wrote *Utopia*. He who had dabbled in the ideal of religious tolerance had developed into a harsh prosecutor (some would say persecutor) of heretics. He who had been a sharp critic of the Church had become its defender. The reason can be summed up with a single word: Luther.

A dozen years earlier, Martin Luther had been merely a

lone German monk joining his voice to a chorus of others calling for a reform of Church abuses—most particularly the abuse of granting indulgences in exchange for money. The Church claimed that a man, no matter how evil his life, might by these indulgences buy advance protection against his soul's being sent to Hell.

"Will you grudge a quarter of a florin to bring your soul that is immortal safely to the fatherland that is Paradise?" demanded John Tetzel, who preached the Mainz Indulgence, intended to pay for rebuilding St. Peter's Cathedral. Even more, it was claimed that a man could, by contributing enough money, get the souls of his dead loved ones who were aleady in Hell transferred to Heaven: "Can you not hear the voices of your dead father and mother pleading with you? 'A tiny alms,' they are saying, 'and we shall be free from this torment.'"

Most of the Christian Humanists agreed heartily with Luther in finding repugnant the notion that the road to salvation was paved with hard cash, rather than a good life. But Luther went on to deny that even living a good life helped a person toward salvation: A Christian's salvation was to be found, not in "good works," but only in faith.

He went on, also, to deny the authority and purpose of the Roman Catholic Church: God did not speak to man through the pope, but through the Bible and each person's individual conscience. The papacy, past and present, was a source of only one thing: disgust. "If Rome is not a brothel above all other brothels imaginable, I know not what a brothel is."

Since then Martin Luther had risen from a hounded, excommunicated priest to become the agitator of a raging German nationalism—and the leader of a religious revolution that would evolve into a new, Protestant, religion. He had taken the criticisms of men like Erasmus, who wanted to reform the Catholic Church from within, and forged them into a sword with which to destroy Catholicism.

By the time Henry began struggling to divorce Catherine, it was being said all over Europe: "Erasmus laid the egg, and Luther hatched it."

It was possibly true in part; and the possibility alarmed Thomas More. It was possibly true; and Erasmus shrank back in horror from what Luther had hatched from his egg. Eramus had sought a peaceful unity of the Christian world under a revitalized Church. Luther's preachings were bringing war, a division of Europe into even more violently antagonistic camps, and an undermining of the foundations of the Church.

No two men could have been more different in aims and personalities than the meek, kindly, delicate Erasmus and the courageous, savage, brawny Luther. There is no final common ground for the humanist who searches for truth and the prophet who knows the truth.

Luther denounced Erasmus: "Human considerations outweigh the divine with him." Erasmus countercharged: "Wherever Lutherism is dominant the study of literature is extinguished."

"True religion," Erasmus said, "is peace, and we cannot have peace unless we leave the conscience unshackled on obscure points on which certainty is impossible."

"Without certitude, Christianity cannot exist," Luther snapped back, substituting new dogmas for the dogmas of Catholicism. "A Christian must be sure of his doctrine and his cause, or he is no Christian."

As for Erasmus' plea for peace in unity, Luther's answer to him was blunt: "Let be with your complaining and clamor; against such a fever no medicines can prevail. This war is our Lord God's war. He has unchained it, and never will it cease raging until all the enemies of His word have been wiped from the face of the earth."

Luther was not one to shrink from bloodshed. The peasants of Germany took Luther's call for reform to include reforming the government, and revolted to obtain relief from oppression and starvation. Luther turned his wrath on them in shocked indignation. Civil obedience was necessary for order, Luther realized, and he could see only one way to obtain it: "The donkey needs a thrashing, and the brute populace must be governed by brute force." He demanded that the military authorities "strike down and strangle these miscreants." The rebellious peasants were crushed without pity; and when it was over Luther announced: "I, Martin Luther, have slain all the peasants who died during this rebellion; for I goaded authority to the slaughter. Their blood be on my head."

For a long time Erasmus refused to choose sides between Luther and the Church, seeing right and wrong on both sides, hoping for a reconciliation. "I neither approve Luther nor condemn him," he told a leader of the Church. "If he is innocent, he ought not to be oppressed by the factions of the wicked; if he is in error, he should be answered, not destroyed."

Luther had no such hesitations or scruples. When Erasmus at last came out with a mild challenge to some of Luther's beliefs, and refused to support his cause, Luther reacted with a spasm of cold ferocity: "He who crushes Erasmus cracks a bug which stinks even worse when dead than when alive."

For Luther there was only *one* truth, he was its voice, and anyone who doubted anything he said was an enemy of the Almighty. "The anger of my mouth is not my anger," he explained, "but God's anger." And: "I have delivered the pure and unadulterated word of God." And again: "I teach and preach the veritable word of God. . . . If thou believest this thou art blessed; if thou dost not, thou art damned."

The worst inclinations in the personality of much of the German people were turned loose by the savage certainty of Luther. And German princes began accepting Lutheranism as a means of establishing their independent power over their realms. Reformers like Thomas More were aghast at what they might have helped bring about.

Like More, the King regarded Lutheranism with revulsion, and would continue to do so as long as he lived. But he was prepared to use some of the side effects its energy was generating to help shape coming events to his design. For what Henry meant by reform was subjugation of the Church to his authority within his kingdom.

It was not a new idea for an increasingly nationalistic Europe. In the rigidly Catholic countries of Spain and France the Pope had some time ago lost his authority to the rulers of the states. Now Catherine's obstinacy had cornered Henry into trying the same thing for England. He could not know, when he started, that England was about to leap far past both Spain and France in this direction.

4

"ENGLAND was the largest and most important political unit to secede at one blow from Catholic Christiandom," E. H. Harbison points out in *The Age of Reformation*. And this secession was the decisive one for the Protestant cause: "If England had remained Catholic, it seems fairly certain that Protestantism later on could not have maintained itself in the Netherlands, and it is at least arguable that in this case all of Europe except for a few innocuous Lutheran states in Germany might have been won back to Roman Christianity by the end of the century."

But this secession was still a long way off when Henry opened the first meeting of the Reformation Parliament. It was to be approached slowly and hesitantly over a great many years, a few short steps at a time with long pauses between them. The earlier pauses were called by Henry each time to allow Rome another chance to avoid the breach by giving him what he wanted: freedom from Catherine and permission to marry Anne Boleyn.

Henry proceeded cautiously, with masterful understanding of his people and shrewd manipulation of Parliament. The bulk of Englishmen were against his divorce. But they would not tolerate foreigners interfering with them in their own country. Emperor Charles was ruler of Spain, Germany, the Netherlands, much of Italy and most of the New World—but not of England. The English people were incensed that he should presume to give their king orders.

The fact that he gave these orders through the Pope did not lessen their indignation. It merely reminded them that a pope in Rome was as much a foreigner as an emperor in Germany. By forging this feeling to the long-smoldering resentment against the power, wealth and abuses of the English clergy, Henry had what he needed to increase his own authority in England.

For the first few years of his manipulation of the Reformation Parliament Henry used no overt act to push his unpopular divorce in England. Instead he dropped the divorce into the current of popular feeling against foreign interference and clerical authority—and then proceeded to stimulate that current.

The Parliament which met late in 1529 was ready and eager for the measures suggested to them by King Henry. In England, as elsewhere, the Church had grown much too fat—fat on land, cash, special privileges and political influence. The nonclerical men who sat in this Parliament could expect to acquire at least part of any of this fat they managed to slice from the body of the Church.

Behind such practical reasons there was, unquestionably, a genuine popular anger, long frustrated, against the decadence of churchmen. It was not that the clergy were any worse than the rest of society around them; but in their failure to be *better* than the rest, they were no longer justifying their special privileges.

The Church had become the largest landowner in the country. On these Church lands, as on others, the peasantry suffered harshly from evictions and steadily rising rents as the landlords grasped for higher profits through sheep raising. With poverty, starvation and beggary stalking the land, the Church continued to drain enormous amounts of wealth from the general population through Church taxes and fees.

Members of the clergy could not be haled into civil courts for any crimes they committed. Since there were at this time from one to two members of the clergy for every hundred Englishmen, the resentment this caused can be imagined. Especially when it is considered that in certain large areas of public conduct the Church had the power to bring people to trial in its own courts; and had almost as much to do with running the country's internal affairs and imposing taxation as the civil government.

The vow of chastity taken by every priest had been broken or ignored repeatedly by churchmen, from the lowliest monks to the highest bishops—and even by some of the popes. This situation had become so common as to be almost accepted, or at least winked at, with prominent churchmen openly bringing forth their children for advancement in society. It had also become usual for a politically agile priest to acquire the incomes from a number of different parishes, none of which he ever troubled to visit; for a prominent churchman to become bishop over a plurality of sees without once setting foot in any of them. Money and power, ran the complaint, were all that concerned the Church; and all that the Church had to offer was for sale.

The outcry against all this had been growing louder and more eloquent for a long time in England. Two centuries before Luther, an influential English preacher named Wyclif had challenged the Church practice of selling absolution: "Thus sin might be bought for money as one buys an ox or a cow; and so rich men had occasion to dread not for to sin, when they might for a little money be thus assoiled of all their sins; and poor men might despair, for they had naught to buy thus sin . . ."

Intensifying resentment of clerical power and immunities had erupted into the open in the 1514 case of Richard Hunne, a London merchant. The Church had placed Hunne in the Bishop of London's prison to be questioned on a charge of heresy growing out of his refusal to pay a priest the mortuary fee demanded for the burial of his dead baby. Two days later Hunne died in his cell, of a broken neck. The Bishop declared the death a suicide; a clerical court found Hunne to have been a heretic, had his body burned, and made paupers of his family. An aroused London public cried that Hunne had been murdered, and demanded that the clergymen they accused be tried for the murder in civil court.

The alarmed Bishop sought to prevent such a trial on the grounds that the public was so hostile to the Church that a civil jury would find any member of the clergy guilty "be he innocent as Abel." A civil jury went right ahead and charged the Bishop's chancellor, jailer and bell ringer with murdering Hunne. Though the charge was later dropped, popular wrath ran so high that the next session of Parliament came close to stripping many members of the Church of their immunity from civil prosecution.

But the Church weathered this crisis, retained all its power, and continued to deal harshly with English "heretics." The most notable of these was William Tyndale, who was forced to become a fugitive exile on the Continent. In defiance of Church opposition, he translated portions of the Bible into English, and his translations were smuggled into England. Tyndale's version of the Bible, printed by Coverdale, was soon destined to be legalized in Britain and become the first version that most Englishmen could read for themselves. But Tyndale did not live to see this. He was captured by agents of Emperor Charles and executed for his faith—by slow strangulation.

In the very year that the Reformation Parliament began,

another exile, an English lawyer named Simon Fish, wrote a
pamphlet, "Supplication of Beggars," to King Henry. In it
Fish summed up the complaints against the clergy in terms
that were sometimes ridiculously exaggerated. But he reflected
an equally intense feeling in many of the English people about
the vast quantity of clergymen among them, and the inferior
quality of so many of them:

> In the times of your noble predecessors past, [there] craftily
> crept into your realm . . . holy and idle beggars and vagabonds
> . . . bishops, abbots, priors, deacons, archdeacons, suffragans,
> priests, monks, canons, friars, pardoners, and summoners. And
> who is able to number this idle ravinous sort, which . . . have
> gotten into their hands more than a third part of your Realm?
> The goodliest lordships, manors, lands and territories are theirs.
> Besides this, they have the tenth part of all the corn, meadow,
> pasture, grass, wool, colts, calves . . . of every servant's
> wages. . . .
>
> Every man and child that is buried must pay somewhat for
> masses and dirges to be sung for him, or else they will accuse the
> dead's friends and executors of heresy.

Thomas More quickly countered with a pamphlet he titled
"Supplication of Souls," in which he denied emphatically that
the numbers of beggars in the country were increasing, or
that a "ravinous clergy" was to blame for any poverty that
did exist. He smelled heresy in Simon Fish's claim that the
Church had no power to shorten the pains of purgatory, and
no right to exact payment for it.

While King Henry took no open position in this controversy, he could not fail to draw aid and comfort from Fish's
book. Especially the part where Fish asked him what the
"holy thieves" did with all the wealth they had acquired; and
answered: "Truly nothing but exempt themselves from
obedience to Your Grace. Nothing but translate all rule,

power, lordship, authority, obedience and dignity from Your Grace unto them!"

Henry quietly let Fish know that it was now safe for him to come home. As for Thomas More, despite his loyalty to the Church he had already let himself be used as a weapon against the Church. He had done so by becoming the first nonchurchman to receive the position of lord chancellor. It was Henry's way of informing the lay members of Parliament that they could expect to benefit from whatever the Church lost.

That Parliament was "a bow only Henry could bend" was true. It was also true that he could not have bent it against its will. Parliaments had denied Henry his wishes in the past, especially in the matter of taxes for war. But this time Henry was pushing the members in a direction toward which the majority of them wanted to go. Bishop Fisher observed angrily: "Now with the Commons is nothing but 'Down with the Church!' "

What the Commons produced, in this first session of the Reformation Parliament, were several relatively mild restrictions on the clergy. These were measures that a man like Thomas More might accept as no more than long-overdue reforms. The clergy thought otherwise—and blocked the measures in the House of Lords, where half the members were bishops.

But not for long. With the incisive skill of a born politician, Henry circumvented the rejection. He managed to get the final decision turned over to a committee composed of sixteen members of Parliament, equally divided between the two Houses. Since the members from the House of Commons had been in favor of the measures from the start, and since half of the members from the House of Lords were lay peers, Henry thus obtained a clear majority.

These measures regulated the amount that churchmen could charge for funerals and wills; they restricted the number of parishes or sees that a single member of the clergy could acquire in the future; and they made the absentee parish priest punishable by law.

That was all that Henry had wanted, for now. Having gotten it he ended the first session of this Parliament. Then he sat back to wait and see if Pope Clement would recognize the warning—and release him from his wife so that he could marry Anne Boleyn.

Part Three

THE HUSBAND

The glory of princes is in their self-esteem and in undertaking great peril; all principal forces meet at a small point, which is called pride.

—Chastellain

Chapter Eight

CONCUBINE OR QUEEN?

IN THE YEAR that followed the Reformation Parliament's first session, Anne Boleyn concentrated every fiber of her being on her incredibly difficult relationship with the King: She continued to stimulate Henry's desire for her, without allowing him to make of her what many thought she was already—his concubine.

Henry peevishly pointed out that she owed him *something* in return for the fact that he was making enemies everywhere for her sake. Anne retorted that his difficulties could not compare with the danger she was in for his sake—even if she finally became his queen: "For it is foretold in ancient prophecies that at this time a queen shall be burned." She claimed to believe that she was to be the queen to whom this prophecy of execution applied.

"But even if I were to suffer a thousand deaths," Anne told Henry passionately, "my love for you will not abate one jot."

With infinite skill, Anne thus continued to keep Henry at bay without letting him slip from her leash. But it was nerve-racking. The terrifying prospect of failure and utter ruin was always around the next corner, and bound to affect her personality. She became irritable, imperious toward other members of the court. She acted as though she were already queen—an easily ignited, quick-tempered queen.

She became merciless toward all real or imagined enemies, from whom she knew she could expect no mercy if she lost her gamble. When Cardinal Wolsey had failed to obtain the divorce, Anne had decided he was conspiring against it and turned her wrath on him. Before his death Wolsey saw her as the chief instrument of his downfall, and privately referred to her as "the night-crow . . . the enemy that never slept, but studied and continually imagined, both sleeping and waking, his utter destruction." Later, on hearing that Guilford, the King's comptroller, was not on her side, Anne warned him that when she became queen she would ruin him. Guilford promptly went to Henry and offered his resignation. Henry told him soothingly that he should pay no attention to "woman's talk"—but accepted his resignation.

Anne did retain a warmer, more delightful side to her personality. Perhaps she showed it only to Henry. But that was enough to hold his love. More than a year after the first session of this Parliament ended, the ambassador from Milan, learning that another representative was being dispatched to England, wrote home: "The one sent should not fail to . . . appease the most illustrious and beloved Anne with some presents."

Every crumb of gossip concerning fluctuations in Henry's feeling for Anne was seized upon and studied on the Continent as the months of maneuvering went on. No such rumor, however unimportant, could be ignored. "The King of England has lately quarrelled with his Lady [Anne]," Charles was informed by his ambassador at one point, "owing to her illtreating a gentleman of his household in his presence. But it appears that they soon made it up again. As usual in such cases, their mutual love will be greater than before."

As for Queen Catherine, reports of her deepening distress kept reaching Charles from his new representative in England, Eustace Chapuys: "Indeed, she has daily more and more ground for complaint, the King's indifference to her and neglect of her increasing rapidly, in proportion to his passionate attachment for the Lady."

Soon after his arrival in England, Chapuys had come to realize that Henry could not be moved by persuasion to give up Anne Boleyn. But the plight of Catherine caught at his emotions. His letters to Charles became a constant stream of honest sympathy for the Queen, indignation at Henry, and abuse of Anne—all of whose male relatives were rapidly being exalted in rank to befit her anticipated role as the new queen.

In the week that the Reformation Parliament ended, Chapuys informed Charles that King Henry had given a grand fete: "Several ladies of the Court were invited . . . the Lady Anne taking precedence of them all, and being made to sit by the King's side, occupying the very place allotted to a crowned queen. . . . After dinner there was dancing and carousing, so that it seemed as if nothing were wanting but the priest to give away the nuptial ring and pronounce the blessing.

"All the time," he added, "and whilst the carousal was going on, poor Queen Catherine was several miles away from this place holding her own fete of sorrow and weeping."

A year later Chapuys was reporting that Henry was still possessed by "his blind, detestable and wretched passion for the Lady." And: "Lady Anne said the other day to one of the Queen's ladies-in-waiting that she wished all Spaniards were at the bottom of the sea. The lady attendant observed that she should not for the sake of the Queen's honor express such sentiments. She [Anne] replied that she cared not for the Queen or any of her family, and that she would rather see her hanged than have to confess that she was her Queen."

Catherine, meanwhile, knew there was only one remedy for her situation: "Nothing will suffice except a final decision about my marriage." She kept writing to Rome pleading for this decision, and to Charles begging that he force the Pope to render it, in her favor.

"You may be sure," Charles wrote her, "that I have greatly at heart this affair of yours, and that as much care shall be bestowed upon it as if it were my own."

But in his heart Charles already believed Catherine's cause was hopeless. "I cannot tell yet how this affair will end," he wrote to his brother, "but of this I am quite certain, namely that the King will commit this folly, and with or without the Pope's consent marry the Lady . . ."

There was nothing he could do to prevent it, short of declaring war on Henry—which was what his old enemy, the King of France, was hoping it would come to. Charles had no intention of making war on England for his aunt's sake. He had other problems on his hands.

In any case, it was too late to help Catherine. Charles was now working for the future. At present the only legitimate heir to the throne of England was his cousin, Princess Mary. No matter how many children Anne Boleyn had by Henry, if the Church did not recognize their marriage then Princess Mary would *remain* the only legitimate heir on Henry's death. It was for this that Charles maintained pressure on Pope Clement: to prevent him from giving permission for the divorce and remarriage.

This pressure, Henry wrote to the Pope, made it impossible for the matter to be decided fairly in Rome. He demanded that Clement allow the case to be determined in England, and warned that he would not brook denial. "If the Pope desires his own rights to be respected, let him not interfere with those of Henry."

Pope Clement placated Charles by prohibiting Henry from remarrying until the case was decided in Rome—under threat of excommunication. At the same time he tried to placate Henry: "We will speak with you as a friend, and beg of you to put away false suspicion you have conceived of us . . .

especially in that reiterated taunt that we are governed by the Emperor."

Clement said that since Catherine had appealed to him, he had no choice but to have the case decided in Rome. And having written this, Clement continued to put off making the decision.

Henry's representative in Rome, Gardiner, had earlier warned Pope Clement that if this procrastination continued Henry and his nobles would conclude that God had taken from the papacy the key of knowledge; and would listen to those who said that "those papal laws which neither the Pope himself nor his council can interpret, deserve only to be committed to the flames."

Clement had replied coolly that he knew that all of God's laws were supposed to be locked within his breast, but that God had as yet not given him the key to open that lock. He prayed for time to solve the matter for him.

2

HENRY decided to increase the pressure. He called another session of Parliament. He had the entire clergy of England charged with violation of the statutes of praemunire—a centuries-old law which forbade anyone to exercise any papal authority within England which infringed upon the jurisdiction of the King. The application of this ancient law was extremely vague; which gave Henry the scope to use it now.

Wolsey had confessed to violating this law when he became papal legate in England. The fact that King Henry had wanted him to be papal legate did not change the fact that in accepting the appointment he had violated the law. If Wolsey was guilty, then the Church in England was also guilty because it had recognized his legatine powers. Again, the fact that King Henry had wanted the Church to do so failed to remove its guilt according to the letter of the law—whose interpretation, as Chapuys saw it, "lies solely in the King's head."

"His use of the statutes of praemunire," remarks Tudor expert A. F. Pollard, "was very characteristic. It was conservative, it was legal, and it was unjust."

The guilty clergy were left with only one way to avoid losing all their worldly goods and going to prison for their crime. They were forced to purchase the King's pardon—for a hundred and twenty thousand pounds. But there was more to the price of pardon: The clergy also had to recognize King Henry as the "Supreme Head of the Church and Clergy in England"—though they did manage to add the weak hedge: ". . . so far as the law of Christ allows."

Henry had the weapon he needed now, to be used or not used as he saw fit; but there, in any case, as a hint to Clement of what could happen.

"Immediately after the above act was passed," the horrified Chapuys wrote, "the woman of the King [Anne] made such demonstrations of joy as if she had actually gained Paradise."

Catherine was as frightened as Anne was joyful. All that Henry was doing, she wrote Charles, was "on account of the shameless life he leads with the woman whom he keeps with him." She begged Charles again to extract a decision from the Pope, "and to order testimony to be sent from Spain as to my virginity when I married the King."

Charles had already sent representatives to find women who had been members of Catherine's intimate household when she was married to Arthur. They were wanted to answer such questions as "whether it be true that the said Arthur was very young and thin, delicate, and of a weak complection, and unfit for a woman; and whether he looked as if he were impotent." Charles had instituted a search for the woman who had made Catherine's bed during her first marriage—and could testify that judging from the condition of the sheets, Catherine had not lost her virginity to Arthur.

But this was all a waste of time and effort. For Henry had solicited and obtained opinions from universities in England and the Continent—opinions that he was not legally married to Catherine, since the Pope had had no right to give them a dispensation to marry. Of course other universities had given a contrary opinion, but these Henry ignored. Bolstered by the opinions favorable to him, he at last summoned the courage to separate himself from Catherine's physical presence—while still waiting for the Pope to separate him from her legally. At the same time Catherine was separated from her daughter Mary, so she could exercise no further influence on her.

Catherine was sent to live henceforth at Moore Park, once Wolsey's country house. She went, protesting that she would rather be sent to the Tower, swearing she was Henry's lawful wife—and not realizing that she would never again, as long as she lived, look upon the face of her husband.

What she did realize was that her situation had sunk lower than she could ever have believed it would. Yet she continued to conduct herself with the combination of pride and humility which befitted a true queen. From her place of banishment from the court, she wrote to Henry saying that she was sorry that he was angry at her without cause; all she had done to block the divorce she had done for the honor of both of them.

Henry answered that she need not bother writing to him any more; she might better employ her time in seeking witnesses to prove her "pretended virginity" when she married him. He sent a delegation of nobles and churchmen to make another try at persuading her to withdraw her appeal to the Pope—and agree to allow the case to be decided elsewhere.

Catherine refused. She denied their claim that Henry had been motivated only by conscience in seeking the divorce: "The King's plea of conscience is not honest. He is acting out of passion, pure and simple."

Some of the delegation fell to their knees and begged her to obey the King. Whereupon she fell to *her* knees and begged them to persuade the King to return to her. The delegation retreated in defeat.

"I am the King's lawful wife, and while I live I will say no other." On this Catherine stood firm, and on her belief that she had done nothing to bring this suffering upon herself. "God knows what I suffer from these people, enough to kill ten men, much more a shattered woman who has done no harm."

If only the Pope would finally declare that her marriage was legal, perhaps Henry would come to agree with her. "I hope," she wrote Charles, "that the King will acknowledge that God has enlightened him when he sees himself loosed from the bondage in which he now is. . . . Those he employs continually irritate him like a bull in a circus, by giving him empty hopes and false reasons."

She closed this letter with the words: "At the Moore, separated from my husband without having offended him in any way, Catherine, the unhappy Queen."

"The Pope has caused all this," Catherine charged, "by his refusal to render justice."

Still Pope Clement delayed.

King Henry summoned Parliament a third time. Through two stormy sessions that lasted four months, against growing opposition, he tightened the screws on the Church. From Parliament he obtained a Petition of the Commons against a long list of clerical abuses and special privileges—and a request that in the future the legal actions of the Church in England be regulated by the King and Parliament.

The clergy protested that this would be a violation of their obligations to Church and God. "We think," Henry told the Commons on hearing this, "that their answer will smally please you." He retaliated with a veiled warning that the churchmen were violating their obligations to *him* and the laws of the land: "We thought that the clergy of our realm were our subjects wholly; but now we have well perceived that they be but *half* our subjects. For all the prelates . . . make an oath to the Pope clean contrary to the oath they make to us, so that they seem his subjects and not ours."

The resistance of the churchmen, which had been propped up by Thomas More, collapsed. Henry obtained a "Submission of the Clergy." By it they gave up much of their right to prosecute laymen; and their immunity, in turn, from civil prosecution. They also agreed to reform their ecclesiastical laws, and to submit those laws to the judgment of men appointed by Henry.

At the same time Parliament passed the Act of Annates, which drastically reduced the percentage of Church fees going to Rome. Having obtained all this, Henry let it be known that the decision to enforce these acts still depended on him. Once more he awaited Pope Clement's reaction.

Meanwhile, isolated at Moore Park, Catherine knew quite well that all these weighty acts of government were aimed at a single person, herself: "The storms of this land do not cast thunderbolts except to strike at me."

But they struck at others, too. Notably Sir Thomas More. He had done his utmost to obtain the rejection of these concessions by the Church to the King, and had failed. If he continued as lord chancellor he would be forced to openly defy the King, or secretly deny his conscience. To escape the choice he resigned his office, determined to keep his mouth shut, his thoughts to himself, and his person out of public sight. But he would soon learn that the choice must still be made, nevertheless—and that it would be a choice, also, between life and death.

Still the Pope would not render a decision.

3

THE MANEUVERING which had gone on for so many years might have gone on for many more, except for one person. It was Anne Boleyn herself who singlehandedly brought the maneuvering to an abrupt end. For over six years she had based her gamble on simultaneously keeping Henry's passion flaming and her virtue intact. Now, sometime during 1532, she played her last card; she let the King have her.

Henry was overjoyed. He bestowed the title of Marquess of Pembroke on her, with an appropriate income. When Henry went to the Continent for a conference with the King of France, Anne went with him to meet King Francis, as though she were already representing England as its queen. When Anne insisted that she must have Catherine's royal jewels, in addition to her own, she got what she wanted. "The companion the King now takes everywhere with him," Catherine wrote, "and the authority and place he allows her have caused great scandal and the most widespread fear of impending calamity." Before long, Chapuys had an ominous little incident to report:

Anne Boleyn had startled members of the court with whom she was speaking by suddenly exclaiming: "I have such a longing to eat apples! Do you know what the King says?" Her voice grew louder as she went on: "He says it means I am with child! But I tell him no. No! It couldn't—no!"

Before anyone else could say anything Anne burst into a fit of wild laughter and ran off.

Later, Henry explained to Chapuys that he had to have children by a new marriage to ensure the succession to his throne. The Ambassador pointed out that he already had a marriageable daughter who could soon give him grandchildren—and that the King could not be sure he would have any more children even if he got a new young wife.

"Am I not a man like others?" Henry demanded angrily.

And then, simply, "I need not give proofs to the contrary, or let you into my secrets."

By which Chapuys understood Henry to mean "that his beloved Lady is already in the family way."

Chapuys was right. Anne Boleyn was pregnant. It was an end to protracted maneuvering. Henry had to act swiftly and decisively now, if the son he was expecting Anne to bear him was to be born legitimate.

4

FORTUNATELY for the pregnant Anne Boleyn, Henry had already found the man to do what had to be done: Thomas Cranmer. At about the time that Anne had finally given in to Henry and become his mistress, the aged Archbishop of Canterbury, Warham, had died. Henry had made a mistake in choosing Thomas More to replace Wolsey as chancellor. He made no such mistake when he picked Thomas Cranmer to succeed Warham.

Cranmer, destined to become father of the Reformation Church of England, was a perfect tool for what Henry and Anne now needed. He was extremely pliable. He was submissive to authority. His detractors have blamed this quality on a detestable lack of a will of his own. His defenders have seen it as a practical ability to adjust to changing conditions. Certainly, though he was a man who accepted little on faith, he consented to much out of fear; and though he devoted much thought to making up his mind about any difficult question, pressure could make him change his mind.

A short, strong, nearsighted man with a dignified manner, Cranmer was an excellent sportsman—and though never a brilliant scholar, he was a methodical and dedicated one. He had quit his studies for the priesthood to marry at the age of twenty-five. But his wife died in childbirth within the year, together with his child. Cranmer returned to his religious studies, and was ordained a priest at the age of thirty. He went on to a career as a lecturer in divinity at Cambridge—a career that was utterly undistinguished until he came to the notice of Henry.

This happened only a few months before the first session of the Reformation Parliament. Cranmer knew Fox and Gardiner, the King's almoner and secretary. Chancing to spend an evening with them, he suggested that a decision from the Pope was not necessary to determine whether Henry's marriage to Catherine was valid. It was a theological matter that could be decided by the theologians of Europe, most notably by those at the various universities.

Fox and Gardiner asked what Cranmer, as one of these theologians, would himself decide. He replied that he had not yet given the question enough thought to have an opinion on the validity of the marriage. It was suggested that he *give* the matter sufficient thought—and he did.

Those theologians who believed the marriage was invalid and those who believed it valid both based their arguments on the Bible.

Those who felt that the Pope had had no right to give a dispensation for the marriage argued from Leviticus: "Thou shalt not uncover the nakedness of thy brother's wife . . . And if a man shall take his brother's wife, it is an unclean thing . . . they shall be childless."

Those who believed the Pope had acted within his powers in granting the marriage dispensation could point to Deuteronomy: "If brethren dwell together, and one of them die, and have no child, the wife of the dead shall not marry without unto a stranger: her husband's brother shall go in unto her, and take her to him to wife . . ."

Whether Cranmer's study of the question was determined by worldly ambition, honest cogitation, or a combination of the two—he finally decided in favor of Leviticus and against Deuteronomy; for Henry and against Catherine. And when a group of Cambridge and Oxford theologians decided that Henry and Catherine *were* lawfully married, Cranmer argued with them so effectively that a number of them reversed their decision. From that moment, the hitherto obscure Cranmer was catapulted onto the stage of world history.

Henry summoned him to an interview and suggested he write a book explaining why the marriage was not valid. Cranmer wrote the book in the home of Anne Boleyn's father. Next Henry sent Cranmer to Italy to solicit the opinions of the universities there. Henry himself could bring the English universities to heel, and his ally King Francis could be counted on to extract favorable opinions from the French universities. Italy was more difficult. But even there, aided by bribes and politics, Cranmer managed to obtain a respectable number of opinions that the marriage was invalid.

Returning to England, he became Anne's chaplain, was absorbed into the Boleyn family circle, and was made one of Henry's chaplains. Henry also set him the task of persuading Sir Thomas More to favor the divorce. Cranmer tried with More, and failed.

He also failed—when he returned to the Continent as ambassador to Emperor Charles—to persuade the Lutherans whom he met in secret to pronounce in favor of Henry's divorce. Luther and the other Protestant leaders considered Henry a Catholic; and they had no desire to further irritate Charles.

But Cranmer's meetings with Lutherans did bear fruit, of a different sort. When he returned again to England he took with him two things—which he was to keep secret for a number of years: a leaning toward certain aspects of Lutheranism, and a wife.

This was the man whom King Henry chose as a successor to the dead Warham. Pope Clement, eager to do anything short of granting the divorce that would placate Henry, quickly agreed that Cranmer was the one most qualified to become the new archbishop of Canterbury. So Cranmer was already settled into the key position when Anne informed Henry, early in 1533, that she was carrying his child—the child who would certainly be a boy.

5

THERE WAS ONLY one more thing that was needed before Cranmer could perform what was now urgently necessary. Henry made the leaders of the English Church declare, in Convocation, that the Pope had no power to grant dispensation for a man to marry his brother's wife; that Arthur and Catherine *had* consummated their marriage and consequently *had* been man and wife. From Parliament Henry obtained an Act in Restraint of Appeals to Rome—which decreed that the final decision on all English ecclesiastical questions must be made within England. This made it treason to appeal to the Pope against any such decision—including the decision which Cranmer would soon render that Catherine was not Henry's wife.

Now there was nothing left to prevent Archbishop Cranmer from giving Henry what he wanted. He formally asked for, and got, Henry's permission to conduct a trial to determine if the King's marriage was valid. With the decision a foregone conclusion, Chapuys wrote hastily and heatedly to Emperor Charles, insisting he must now invade England at the head of an army: "Considering the very great injury done to Madame, your aunt, you can hardly avoid making war upon this king and kingdom. For it is to be feared that the moment this accursed Anne sets her foot firmly in the stirrup she will try to do the Queen all the harm she possibly can; and the Princess also. Indeed, I hear she has lately boasted that she will make of the Princess a maid of honor in her Royal household, that she may perhaps give her too much dinner on some occasion, or marry her to some varlet . . ."

Chapuys assured Charles that an invasion of England would be "the easiest thing in the world just now, for this King has neither cavalry nor well trained infantry; besides which the affection of his subjects is entirely on Your Majesty's side." Charles knew better, and would not consider such an invasion.

Cranmer went on with the trial.

It was held quietly some distance from London, where there would be less chance of popular disturbances. Catherine, summoned to appear before Cranmer's court, refused to come —which made it all the easier for Cranmer. On May 23, Cranmer rendered his decision: Henry's marriage to Catherine was not valid; therefore they had never been lawfully married.

Cranmer immediately followed this decision with an announcement: King Henry had secretly married Anne Boleyn some time earlier. Since Catherine was not Henry's wife there was no reason why he should not have done so. Anne was King Henry's wife.

Now all that was needed to make it official was her public coronation. The date was set for the first day of June—one week away. All the notables in the land would be present to watch Anne Boleyn crowned queen of England.

Almost all. Thomas More refused to go. Though he had been careful to make no public statements of his beliefs, it was known that he considered Henry's divorce from Catherine and his marriage to Anne illegal—on the grounds that the Pope had the final decision in such matters. Anne Boleyn was believed to hate him, for this reason, as much as she had once hated Wolsey. More's friends warned him that his very life would be in jeopardy if he did not go to the Abbey.

But More felt that to appear at her coronation would be to proclaim his agreement with the King's action. He explained why he could not attend by reminding his friends of the story in Tacitus about an emperor who wanted to execute a certain woman—but could not because she was a virgin and by law no virgin could suffer the death penalty. The Emperor's problem was solved by an adviser's suggestion: "Let her first be deflowered, and after that she may be devoured."

"They may devour me," Thomas More said wryly, "but they shall never deflower me."

For Anne Boleyn, the days before her coronation were one long, lavish ceremony. It began with a water procession on the Thames that carried her from Greenwich to the Tower of London. Anne left the palace at Greenwich in a barge painted with her own colors, festooned with flags and banners and bangles to catch the rays of sunlight and reflect them. Her court of women attendants were with her in the barge, which was accompanied up the river by several hundred vividly decorated boats carrying England's great bishops and noblemen. From these boats came the crackle of fireworks, the music of flutes, trumpets and drums, the singing voices of young ladies. From other boats lining the shores, cannons boomed in tribute as Anne went past.

As she arrived at the Tower, the fortress guns roared a salute to her. She was met at the landing by the Lord Chamberlain and the officers of arms, who conducted her to the water gate where King Henry stood awaiting her. Henry kissed her, waited while she turned to thank the Lord Mayor and the assembled citizens of London, and then took her inside with him.

Anne emerged from the Tower the day before the coronation to show herself to the people in a great procession through London. She was preceded through the tapestry-hung streets by a dozen French gentlemen in purple velvet, each with one sleeve bearing Anne's colors as the new queen, riding horses covered with violet taffeta decorated with white crosses. These were followed by over two hundred bishops and noblemen. Bringing up the rear of the procession were twelve ladies on hackneys, dressed in cloth of gold; a chariot bearing Anne's mother and the Duchess of Norfolk; twelve gentlewomen on horseback arrayed in crimson velvet; three gilded coaches bearing more young ladies; and twenty or thirty others on horseback, dressed in black velvet.

Anne was dressed in white, a delicate circlet of gold and diamonds around her head and her long hair flowing down over her slender shoulders. The open litter in which she rode was covered inside and out with white satin; a cloth-of-gold canopy was borne over it by four knights. All the bells in London were ringing as she moved past the people crowding every inch of the way for a glimpse of her. Pageants were performed as the procession moved through the city, music played, children cast flowers and read poetry, fountains poured forth wine—all in her honor.

The next morning Anne emerged from her lodging wearing purple velvet furred with ermine. England's great bishops and noblemen conducted her along a road carpeted with cloth to Westminster Abbey. Inside, she heard Mass and mounted upon a platform covered with red cloth, before the great altar. There she went through the long, solemn ritual of coronation. The Archbishop of Canterbury anointed her, placed the golden scepter in her hand—and the crown upon her head.

Anne Boleyn was Queen of England—according to King Henry and the laws of his realm.

But Catherine was still Queen of England—according to Pope Clement and the laws of the Roman Catholic Church. The Pope had finally been stung hard enough to retaliate: Henry was excommunicated—conditionally. His divorce from Catherine and his marriage to Anne were pronounced invalid. Catherine was Henry's wife and Anne Boleyn must be put away. Henry was given three months to think it over. If he failed to comply with the papal order by then, the Pope would openly publish the sentence of excommunication.

Chapter Nine

THE FATE OF THOMAS MORE

For as long as he could, Henry kept from Anne Boleyn the news that Rome had declared their marriage illegal and pronounced him excommunicated unless he took Catherine back. Such news might injure the unborn son she carried within her. Henry concentrated on forestalling the papal action.

First, he sent the Pope an appeal against the excommunication. This had the effect of delaying actual sentence while the appeal was considered. It allowed time, still, for the Pope to declare in Henry's favor as regards the divorce—which would in turn nullify the sentence of excommunication.

Meanwhile, Henry did all he could to hurry English acceptance of Anne as the new queen. And to make Catherine acknowledge that she was no longer the queen.

His affection and respect for Catherine were gone. Whether or not he had really persuaded himself that his marriage to Catherine was invalid in the sight of God may be questioned. But it is certain that he *did* believe the welfare of his country depended upon his having a legitimate male heir. In his eyes, Catherine, by her refusal to let him have that heir by another wife, had been deliberately endangering the future of England.

For years he had tried to obtain what he felt was necessary by persuasion and diplomacy. He had been repaid by increasing frustration; and now the conditional excommunication. In this he read a lesson: A ruler who cannot obtain his will from a subject must inflict it. The reluctance of the Church had drawn from Henry all his latent cunning. Catherine's resistance pushed him to a brutality that was to become a larger part of his character the more he exercised it.

He could not have two queens of England; that much was obvious. Anne Boleyn was Queen. If Catherine had never been his wife, she was still the widow of Prince Arthur. So from now on she would bear the title of Princess Dowager.

Henry had a proclamation to this effect nailed to the door of every parish church in England. Since most of the citizens who read it could remember no queen but Catherine, it was received with little enthusiasm. There were isolated flare-ups of vocal disgust. In Yorkshire, smoldering with pro-Church resentment, a "lewd and naughty priest" named James Harrison was arrested for stating out loud: "I will take none for Queen but Queen Catherine. Who the devil made that whore, Nan Boleyn, Queen?"

In another attempt to make Catherine accept what Henry considered an established fact, he sent a deputation to her. His instructions to them read, in part: "They shall say to her that the King . . . was lawfully divorced, and married to the Lady Anne, who has been crowned Queen. As the King cannot have two wives he cannot permit the Dowager to persist in calling herself by the name of Queen. . . . Finally, that as the marriage is irrevocable, and has passed the consent of Parliament, nothing that she can do will annul it, and she will only incur the displeasure of Almighty God and of the King."

Catherine was propped up on a sofa when Henry's deputation arrived. She had a bad cough and could not stand comfortably because she had pricked her foot with a pin. But her spirit was as iron-shod as ever. The King's representatives began to read aloud Henry's instructions to the "Princess Dowager." Catherine immediately objected that it was not addressed to her, since she was the Queen.

They read on, nevertheless, informing her that Anne Boleyn had been judged the lawful Queen.

"We know by what authority it has been done—more by power than justice," Catherine retorted. "I think I might have been judged more impartially in hell . . . I think the devils themselves do tremble to see the truth in this case so sore oppressed."

They offered her bribes, in the form of revenues, estates and more respectful treatment—if she would acknowledge her new rank. Catherine told them that they misunderstood if they thought it was rank or riches that mattered to her: "I would rather be a poor beggar's wife, and be sure of heaven, than Queen of all the world."

She was being obstinate, she said, "not for vain glory, but because I know myself the King's true wife. . . . I came not into this Realm as merchandise, nor to be married to some merchant. . . . If I should agree to your persuasions, I should slander myself, and confess to having been the King's harlot for twenty-four years."

The next day the deputation let her read the report of this interview that would be presented to Henry. Catherine seized a pen and went through it crossing out the title "Princess Dowager" wherever she found it.

Two months later, in the palace at Greenwich, Anne Boleyn reached the end of her pregnancy. "The King," wrote Chapuys four days earlier, "believing in the report of his physicians and astrologers that his Lady will certainly give him a male heir, has made up his mind to solemnize the event with a pageant and tournament."

On September 7, 1533, between three and four in the afternoon, Anne Boleyn gave birth—to a daughter.

The fair-haired baby girl was named Elizabeth.

Gleefully, Chapuys reported to Charles that Anne had been "delivered of a girl, to the great disappointment and sorrow of the King, of the Lady herself, and of others in her party; and to the great shame and confusion of physicians, astrologers, wizards, and witches—all of whom had affirmed that it would be a boy. . . . It must, therefore, be concluded that God has entirely abandoned this king, and left him a prey of his own misfortune and obstinate blindness, that he may be punished and completely ruined."

In Rome, the Milanese ambassador agreed: "The King of England has had a daughter by his new wife, which shows that God disapproves of his unholy designs and appetites."

But King Henry choked down his disappointment. He chose to read a different meaning in it as he first gazed upon Elizabeth. The girl was healthy. Anne had come through the delivery well and was also healthy—and young. The birth of this baby meant only that he could have children through Anne. This was the first. The next, surely, *would* be a boy.

It was reported to Charles that Henry had been overheard to say that he loved Anne more than ever, and would rather be reduced to beggary, and ask alms from door to door, than abandon her. Anne's daughter was christened with great pomp at the Friars' church in Greenwich, Archbishop Cranmer being named godfather. The title of Princess of Wales was stripped from Henry's other daughter, Mary, and bestowed upon his new daughter, Elizabeth. Elizabeth was to be recognized as the true heir to the throne, until Anne Boleyn produced the eagerly anticipated son. Mary, at the age of seventeen, suddenly found herself officially declared illegitimate.

2

FOR AS LONG as Mary could remember, she had been surrounded by love, respect and adulation. She had grown up on the attendance and flattery of the English court, the praise of great visitors from the Continent. Her mother, Catherine, had lavished all the warmth of her Spanish nature upon Mary, supervising her early training and education. And from the moment she had been born, the affection and pride with which her father regarded her had been obvious. "By God," Henry had boasted to the Venetian ambassador, "this baby never cries." As she grew older Henry remained a kindly, teasing, indulgent father who was always showing her off to foreign visitors, praising her for her beauty, learning and manners.

She matured into a spirited princess with reddish hair, a quick wit and a regal bearing. When she was eleven a marriage with the royal house of France was contemplated for her. The French ambassador danced with Mary and found her "very pretty, and admirable by reason of her great and uncommon mental endowments; but so thin, spare and small as to render it impossible for her to be married for the next three years." Four years later an Italian nobleman visited Mary's household at Richmond and recorded: "This Princess is not very tall, has a pretty face, and is well built, with a beautiful complexion. She speaks Spanish, French, and Latin, besides English and Greek; and understands Italian but does not venture to speak it. She sings excellently, and plays on several instruments."

But by fifteen her world was disintegrating. For Henry's campaign to marry Anne Boleyn was as much a menace to Mary as it was to Catherine. The projected divorce threatened her status and future. If Catherine was not Henry's lawful wife, then Mary was a bastard.

This common peril, and the natural feeling between mother and daughter, drew Mary and Catherine closer together than ever as Henry pushed on toward the divorce. Their closeness of spirit, sympathy and intentions only grew stronger after Henry separated them permanently. Mary dug in her heels, resisted her father's will with a stubborn courage that matched Catherine's, and revealed herself a true granddaughter of Isabella of Spain. She also began to suffer from attacks of what her doctors called "hysterical illness."

Despite his daughter's resistance, Henry undoubtedly continued to care for her. Chapuys believed that Anne Boleyn hated Mary more than Catherine, because she knew Henry could still be influenced by his affection for Mary. And the Italian nobleman who visited Mary during this period wrote: "The Princess is much beloved by her father."

It was this same father who two years later, when Mary was seventeen, declared her to be illegitimate—and no longer to be called "Princess." That title, the King's herald proclaimed in front of the church in which Elizabeth was christened, henceforth belonged to Henry's new daughter.

In the months that followed, Mary was sorely in need of the encouragement and consolation which her mother wrote to her. In one letter Catherine advised Mary on how to conduct herself in this humiliating and dangerous situation:

Daughter:
I heard such tidings today that I do perceive, if it be true, the time is come that Almighty God will prove you . . . be sure that, without fail, He will not suffer you to perish if you beware to offend Him. I pray you, good daughter, to offer yourself to Him . . . for then you are well armed.

[She warned Mary to be careful not to provoke Henry]:

Answer with few words, obeying the King, your father, in everything save only that you will not offend God and lose your own soul; and go no further with learning and disputation in the matter. And wheresoever, in whatsoever company you shall come, observe the King's commandments. Speak you few words and meddle nothing. . . .

But one thing especially I desire you, for the love that you do owe unto God and unto me: Keep your heart with a chaste mind, and your body from all ill and wanton company . . . neither determine yourself to any manner of living until this troublesome time be past. For I am certain that you shall see a very good end . . . I perceive very well that God loveth you . . .

By your loving mother,
Catherine, the Queen.

Henry now found himself dealing with *two* stubborn women. Less than a month after the birth of Elizabeth, Mary was writing to her father: "This morning my chamberlain came and informed me that he had received a letter from Sir Will. Paulet, controller of your House, to the effect that I should remove at once to Hertford Castle. I desired to see the letter; in which was written 'the lady Mary, the King's daughter,' leaving out the name of Princess. Marvelled at this, thinking Your Grace was not privy to it, not doubting but you take me for your lawful daughter, born in true matrimony. If I agreed to the contrary I should offend God; in all other things you shall find me an obedient daughter."

By this time Henry's fury ignited instantly on meeting resistance. He sent a delegation to warn Mary of "the folly and danger of her conduct." He was surprised, he informed her, that she had so far forgotten her filial duty that she was trying "arrogantly to usurp the title of Princess, pretending to be heir apparent." For this she deserved both the King's high displeasure and punishment by law. But if she would mend her ways Henry would forgive her out of his "fatherly pity."

Mary replied that she would obey him like a slave—in everything except denying her legitimate birth. Henry retaliated by breaking up Mary's household and having her removed to Hatfield, the new residence set up for Princess Elizabeth.

The Duke of Norfolk was sent to conduct Mary there. She made a point of objecting to his calling Elizabeth the Princess of Wales: "That is the title which belongs to me by right, and to no one else." But she went with him.

On arriving at Hatfield, Norfolk asked if she would like to have a look at her new baby sister, "the Princess." Mary answered stiffly that she knew of no other princess in England but herself. Anne Boleyn's daughter, she told Norfolk flatly, was no princess at all. She understood that Henry acknowledged Elizabeth as his daughter—just as he acknowledged the bastard Richmond as his son. Because of this, Mary said, just as she called Richmond her brother, she was willing to treat Elizabeth as a sister—but not as Princess of Wales.

The exasperated Duke of Norfolk prepared to take his leave of her, asking if she wanted him to convey any message to Henry.

"None," Mary replied, "except that the Princess of Wales, his daughter, asked for his blessing."

Norfolk pointed out that since the King no longer considered her the Princess of Wales, he would not dare take such a message.

"Then go away, and leave me alone!" Mary stormed. And with that, Chapuys wrote, "she retired to her chamber to shed tears, as she is now continually doing."

At about the same time as the Duke of Norfolk's failure to humble Mary, the Duke of Suffolk was having even less success in intimidating Catherine.

In midsummer, Catherine had been removed farther from London, to Buckden, in Huntingdonshire. Now, toward the closing days of 1533, Suffolk and other commissioners arrived at Buckden with a force of armed men—to take her to Somersham, an even more isolated residence surrounded by marshes.

Catherine received Suffolk in her great chamber, with all her servants assembled. In a ringing voice, she declared her refusal to go to Somersham. It was a notoriously unhealthy place, especially now in the depths of winter. To go there voluntarily, she told Suffolk, would be the same as committing suicide. She would not be guilty of such a sin. If they wanted her there they would have to drag her.

"We find this woman more obstinate than we can express," wrote Suffolk and the other baffled commissioners to Cromwell. "As there is no help for it but to convey the woman by force, which is not in our instructions, we desire to know the King's pleasure."

Afraid to use the force required without specific orders to do so, they also wrote to Henry: "Wish to know the King's pleasure, as she will not remove to Somersham, against all humanity and reason, unless we were to bind her with ropes."

While awaiting Henry's answer, they concentrated for the moment on carrying out the second part of Henry's commission. Summoning all of Catherine's servants, Suffolk ordered them to swear a new oath to serve her as the princess dowager. They refused, saying that this would be in violation of the oath they had previously sworn to serve Catherine as queen.

The furious Duke of Suffolk arrested some of Catherine's household. He discharged all the rest—except for two female attendants and Catherine's Spanish confessor, doctor and apothecary—and replaced them with new servants who swore never to address Catherine as Queen. Catherine reacted in character: Since she *was* the Queen, she could not regard these replacements as servants. She would consider them as guards, she said, and herself as a prisoner. Catherine locked herself in her room, with the few attendants still sworn to her as queen, and refused to come out.

For six days Suffolk waited for fresh orders—which did not come. It was still his duty to get Catherine out of Buckden by any means he could think of, short of actually laying violent hands on her. Suffolk tried a final bluff. He made a show of closing down the entire house. Furniture was carried out of the building. Catherine's things were packed. Baggage was loaded on mules and carts. A carriage was made ready to take Catherine to Somersham. All this was done where she could watch the preparations from her windows. When all was in readiness for the journey, Suffolk and the others stood outside Catherine's locked door. It was time, they shouted through the door, for her to leave.

"If you wish to take me with you," she shouted back, "you will have to break down the door."

The thwarted commissioners returned in defeat to London. Catherine remained at Buckden, still pleading with her nephew Charles to force Pope Clement to render sentence on the question of her marriage to Henry. "I beg you," she wrote, "to show more affection for me, and for my daughter."

Three months later the event so long wished for by Catherine finally took place: on March 24, 1534, Pope Clement announced his decision in the King of England's "great matter"—in Catherine's favor.

The Pope's decision came too late. Henry had already taken measures to insure that such a sentence could not be recognized by anyone in England. The rendering of the decision against Henry served only to drive him to even more drastic measures.

The Duke of Norfolk had failed to humble Mary. The Duke of Suffolk had failed to humble Catherine. But the man Henry chose for the job of humbling the Church in England was Thomas Cromwell. And Cromwell did not fail.

3

A STOCKY, awkward-looking man with a keen, expressive face, Thomas Cromwell viewed his world through practical, cynical eyes, without illusions. His pleasant, attentive manner cloaked a cool, mocking appraisal of his fellow man and a brisk, businesslike approach to the making of history.

Born about the same time as Catherine, the son of a disreputable Putney workman, Thomas Cromwell matured early into a resourceful, self-reliant boy who knew how to get his education from the school of experience. At the age of eighteen, wanderlust and a thirst for more practical learning took him to the Continent. He spent the next ten years there, earning his way in the French army, in a banking firm in Florence, in a merchant-trading company in the Netherlands.

Those ten years taught Cromwell a great deal about how politics and business were actually conducted—as opposed to the moralistic lip service paid to how they were *supposed* to be conducted. There was, for example, moneylending for a profit. In theory, the Church still regarded this as a sin. But even back in the Middle Ages a realistic observer had differentiated what was said about moneylending from what was actually done about it: "He who takes it goes to Hell, and he who does not goes to the workhouse."

Through his business connections in the Netherlands, Cromwell met and married the daughter of an Englishman in the wool trade, thus improving himself financially and in social status. Returning to England, he added the study of law to his acquired skills at banking, politics and business. With his grasp of such a range of practical know-how, he did well for himself as a lawyer, and was elected to the 1523 session of Parliament.

Typical of Cromwell was his sardonic summing-up of how much this session tackled—and how little it accomplished: "I amongst others have endured a Parliament which continued by the space of seventeen whole weeks . . . communed of war, peace, strife, contention, debate, murmur, grudge, riches, poverty, penurey, truth, falsehood, justice, equity, deceit, oppression, magnanimity, activity, force, attempraunce, treasons, murder, felony, conciliation. . . . Howbeit, in conclusion we have done as our predecessors have been wont to do; that is to say, as well as we might, and left where we began."

While a member of this Parliament, Cromwell successfully opposed Cardinal Wolsey's demand for war funds for the King. This did not make an enemy of the Cardinal. Wolsey was astute enough to recognize a uniquely useful man when he spotted him. He took Cromwell into his service. Cromwell showed himself adroit at handling Wolsey's business affairs— which included disbanding some small monasteries under a grant from the Pope, so that Wolsey could use the profit from their sale to establish colleges.

He was also adroit enough to avoid being crushed when Wolsey toppled from power. Gaining a seat in the first session of the Reformation Parliament in 1529, Cromwell soon managed to make King Henry recognize his skills. Within two years he was a member of Henry's Council.

When Thomas Cromwell entered the King's service, he received a piece of cautionary advice from Sir Thomas More: "Master Cromwell . . . in your counsel-giving unto His Grace, ever tell him what he *ought* to do, but never what he is *able* to do. For if a lion knew his strength, hard were it for any man to rule him."

But it was precisely by helping Henry to recognize his own strength—by showing him what he was able to do, and how to do it—that Cromwell rose swiftly to a position as Henry's chief adviser; and as his spokesman in the House of Commons. Cromwell became the architect of the measures by which royal supremacy was substituted for papal supremacy within England; the measures which crushed the independence of the clergy, and diverted the power and wealth of the Church into the hands of King Henry.

Cromwell has been characterized by most writers as a heartless, cold-blooded monster. This is quite wrong. Cromwell was personally courageous, kind and loyal. After Wolsey lost the King's favor, Cromwell was the only one to speak out boldly and publicly against the attack on the fallen Cardinal. After Anne Boleyn gave birth to Elizabeth, Cromwell earned

Catherine's deepest gratitude by his attempts to soften Henry's angry treatment of the obstinate Mary. Chapuys found Cromwell to be most enjoyable company, remarking ruefully on how easy it was to become too talkative with him. His generosity was admitted even by his enemies; his neighbor noted that Cromwell gave away food to as many as two hundred poor people each day.

Cromwell was not a vicious monster. He was worse than that: a brilliant organization manager who believed the State should have absolute power over every aspect of life within its borders. And for Cromwell, as for Henry, "the King in Parliament" was the State.

His hand was clearly evident in the preamble to the Act in Restraint of Appeals to Rome of 1533, which declared that England was "governed by one supreme head and king" toward whom everyone "ought to bear, next to God, a natural and humble obedience." And in 1534, the year in which Pope Clement finally proclaimed Catherine to be Henry's true wife, Cromwell masterminded a number of further measures which rendered the Pope's decision inoperative within England.

The submission of the clergy to the King's authority was formalized in an act of Parliament. The Pope's power to name new bishops was given to the King. The bishops were to swear allegiance to Henry, and refer to the Pope as merely "the Bishop of Rome." It was no longer heresy to speak against "the Bishop of Rome or his pretended power." The papal power of granting dispensation, a source of considerable income, was transferred to Archbishop Cranmer. Another source of papal income was diverted to King Henry: the first fruits of benefices and a tenth of all clerical revenues. It was enacted that the King "is and ought to be the supreme head of the Church of England"—without the qualifying phrase "in as far as the law of Christ allows."

Most important, there was the Act of Succession. This

made official and final the decision that Henry had never been married to Catherine, that he *was* lawfully married to Anne Boleyn, and that on his death the children of Henry and Anne should succeed to the throne. All adult subjects of King Henry, male and female, were obligated to swear an oath that they supported and would maintain this Act of Succession. Anyone who refused to take this oath would be guilty of treason, punishable by death.

This meant that a person might be executed, not only for what he did or said, but also for what he *thought*. It was no longer enough to support the government decree by one's actions; one must also swear to agree with it in one's conscience. This "putting windows in men's souls" was not new. The Church had done it, by means of the Inquisition. This time it was a temporal government that was doing it, by civil law.

This oath doomed Thomas More. He could not swear, against his conscience, to what he did not believe in.

"I am the king's true faithful subject," More defended himself, in refusing to take the oath. "I do nobody harm. I say none harm. I think none harm, but wish everybody good. And if this be not enough to keep a man alive, in good faith I long not to live."

4

HE HAD BEEN in danger since he had resigned as lord chancellor. But until the oath was demanded of him, he had been able to walk a thin edge of safety.

By giving illness as his reason for resigning, More had been able to avoid publicly offending Henry. And his resignation enabled him to keep his views to himself for a time. However, it left him with practically no source of income. He had to reduce his household severely. For his former retainers he found employment with other men who were able to afford them—including Lord Audley, who replaced him as chancellor. "Chancellor More is Chancellor no more," his jester punned. But the jester's joking gave way to bitter weeping when More sent him into the service of London's Lord Mayor.

The Church tried to alleviate his financial distress by awarding him a large sum of money in recognition of his services. More refused to take it. His reduced circumstances worked little hardship on a man who had never ceased to long for a monk's life. He settled easily, almost with relief, into a modest mode of living, hoping to save himself by keeping out of the limelight.

But he was realistic enough to prepare for death. He had the remains of his first wife moved to the Chelsea Church and placed in a tomb intended to hold himself and his second wife also. "Oh, how well could we three have lived together in matrimony," he wrote, "if fortune and religion would have suffered it. But I beseech Our Lord that this tomb and heaven may join us together. So death shall give us that thing that life could not."

He had earned Anne Boleyn's enmity by refusing to attend her coronation. Not long afterward Henry became determined to force More into committing himself. He was no longer to be allowed the safety of silence. A number of attempts were made to trap More into a dangerous position from which he would be able to extricate himself only by openly declaring his support of what Henry was doing. Such support, from such a man, would do much to sway the rest of the kingdom. And to reassure the King himself that what he was doing was right.

The first attempt was through a charge that he had taken bribes while in office. More easily defended himself against the charge, and the King's Privy Council was unable to make it tenable. Next he was charged with having declared against the Council's published support of Henry's marriage to Anne

Boleyn. Again, More was able to refute the charge. He wrote to Cromwell, "I know my bounden duty to bear more honor to my prince, and more reverence to his honorable Council, than . . . to make an answer unto such a book, or to counsel and advise any man else to do it."

It proved more difficult to extricate himself from a charge implicating him in the treason of a nun named Elizabeth Barton, popularly called "the Holy Maid of Kent." After suffering grave illness, this woman had fits and trances during which, she claimed, she experienced divine revelation. She began to make prophecies which she claimed to be the direct word of God, given to her during regular visits to Heaven once each fortnight. Some regarded her as a harmless fanatic. Others took her more seriously. A number of important persons, including Thomas More, consulted with her and corresponded with her, trying to determine if her revelations actually were divine.

It was when King Henry began pushing toward his divorce, and after being carefully instructed by some monks, that Elizabeth Barton's prophecies became dangerous. God had told her, she declared, that if Henry did divorce Catherine and marry Anne, he should cease to be King within a month "and should die a villain's death." Henry went ahead and divorced Catherine. He married Anne. He continued to rule and did not die.

It had been made treason to predict the death of the king. Elizabeth Barton and five monks charged with being her accomplices were tried and found guilty of high treason. The discredited and convicted "Holy Maid of Kent" was made to deliver a public admission that she was an impostor:

"I, dame Elizabeth Barton, do confess that I, most miserable and wretched person, have been the original of all this mischief, and by my falsehood I have deceived all these persons and many more." She and the monks were executed. A bill was introduced into Parliament charging More and others with complicity in her treason.

More defended himself ably. He had spent some time with Elizabeth Barton. But he had refused to listen to any of her prophecies against the King, and had warned her against telling them to anyone else. He had a copy of a letter he had written to her which proved this.

"Of a truth I had a great good opinion of her," he admitted to Cromwell. But her confession that she was lying had made him realize that she was a "wicked woman." He congratulated Cromwell for having proved it: "You have done . . . a very meritorious deed in bringing forth to light such detestable hypocrisy, whereby every other wretch may take warning, and be feared to set forth their own devilish dissembled falsehood, under the manner and color of the wonderful work of God."

More escaped this charge only to face another. The King's commissioners reminded him of what a great benefactor and friend Henry had always been to him. More assured them that he knew it, was most grateful to the King, and devoted to him. In that case, they suggested, More should give public support to Henry's divorce and remarriage. More avoided giving them any answer at all to this. Their tone became threatening: More might be charged with having induced King Henry to write his book against Luther, which had defended the supranational authority of the pope. This book had earned Henry his title of "Defender of the Faith." It had also given the Pope a sword which he now used against Henry.

"My Lords," More told the commissioners, "these terrors be arguments for children, and not for me." He calmly assured them that once King Henry took the trouble to think back, he would remember that it was he himself who had decided to write the book. All More had done, after Henry finished it, was a little rewriting and smoothing of the composition. As a matter of fact, More clearly remembered having regarded what Henry had written as giving a little too much authority to the Pope. He had told Henry so; and advised that this phrasing should be altered.

"Nay, that it shall not," Henry had answered him. "We are so much bounden unto the see of Rome that we cannot do too much honor to it."

The discomfited commissioners withdrew their charges against him. But More realized that his ordeal was not over. "What is postponed is not abandoned," he told his favorite daughter, Margaret.

The Duke of Norfolk warned him against continuing to incur Henry's displeasure. "For by God's body, Master More, the indignation of a prince is death."

"Is that all?" answered More. "Then in good faith the difference between Your Grace and me is but this: that *I* shall die today and *you* tomorrow."

Within the month came the demand from which there was no escape. Thomas More was summoned to appear before the commissioners at Lambeth to take the oath supporting the new Act of Succession.

He had not yet read the oath. He was ready to take it if he could do so without violating his religious belief. But he suspected that it would be phrased in such a way that he would not be able to. With a heavy heart and a sad, strained manner, he left his home, bade farewell to his wife and family at the gate, and boarded the boat which carried him along the river to Lambeth. He never returned.

On the 13th of April, 1534, Thomas More faced the Lambeth commissioners: Archbishop Cranmer, Chancellor Audley, Thomas Cromwell, and the Abbot of Westminster. They showed him the oath. He read it, and found what he had feared. The oath to maintain the Act of Succession itself did not go against his conscience. He believed in the authority of King and Parliament to make the laws of the land. The Act made it law that Anne Boleyn was Queen and Princess Elizabeth heir to the throne. More was ready to swear to abide by and uphold this law.

But there was a preamble to the oath which disclaimed papal authority in spiritual matters. It consequently denied that Catherine had ever been Henry's lawful wife, and maintained that the King—not the "Bishop of Rome"—was the final judge in questions involving the English Church. This, Thomas More could not swear to.

He was careful not to give his exact reasons. All he would tell the commissioners was that he would swear to the Act itself—but not to the preamble, as it was presently written. The commissioners tried to sway him by showing him the names of all the great nobles and bishops who had already taken the oath. More was not swayed. He had not and would not advise anyone else against taking the whole oath. He could only answer for himself. "I . . . leave every man to his own conscience. And me thinketh in good faith that so were it good reason that every man should leave me to mine."

The commissioners finally gave up their attempts at gentle persuasion. More was imprisoned in the Tower of London. He was not the only one. The commissioners that day presented the oath to the leading members of the London clergy. Almost all of them swore to it. But two would not. One of these was Nicholas Wilson, who had been the King's confessor, and Catherine's. The other was the emaciated, seventy-three-year-old Bishop Fisher of Rochester—who had never ceased to uphold Catherine's rights in the divorce controversy. Fisher became the only English bishop to refuse the oath. His scruples were exactly the same as More's. And like More, he was careful not to be precise about what those scruples were.

Wilson preceded More into a Tower cell by a few hours. Bishop Fisher followed them in a few days.

Archbishop Cranmer was deeply troubled about the refusals of such universally admired men as More and Fisher, and their imprisonment. "I do not know why my lord of Rochester [Fisher] and Mr. More were contented to swear to the Act of the King's Succession, but not to the preamble," he wrote to Cromwell. "They would not give a reason for their refusal; but it must either be the diminution of the authority of the bishop of Rome or the reprobation of the King's first pretensed matrimony." He suggested that More and Fisher be allowed, secretly, to swear to the Act—without the preamble.

Cromwell answered Cranmer: "I have shown your letter to the King, who does not agree with you."

The preamble, Henry realized, was essential to the oath. For unless the Pope's authority was repudiated, Henry's marriage to Catherine was legal. Which, in turn, would mean that any children he had through Anne were illegitimate—even if by English law they were recognized as heirs to the throne. Henry's main purpose through all these years of controversy had been to obtain a *legitimate* male heir. It was this legitimacy that More and Fisher must be made to swear to, by including the preamble in their oath.

But they would not so swear—though their imprisonment stretched on into weeks, and then months. For Thomas More, life in a prison cell was a not-unwelcome hardship. It had been with misgivings that he had chosen to have a family rather than spend his life in the cell of a monk. This was very like it. He spent his time in religious and philosophical writing, continuing to do so with a coal after pen and ink were taken from him.

He could still view his situation with a sense of humor. Walsingham, the Lieutenant of the Tower, told him he was sorry he could not give him more cheerful accommodations. More told him: "Assure yourself, Master Lieutenant, I do not dislike my cheer. But whensoever I so do, then thrust me out of your doors."

The King's representatives continued to try to get More

either to accept the oath, or to incriminate himself by denying the statements in the preamble. He would do neither. Silence on the subject remained his only defense. It baffled his questioners, infuriated his King, and frightened his family.

More's wife came to his cell, and tried to persuade him to gain his freedom by giving in to the King. "I marvel," she told him in blunt exasperation, "that you that have been always hitherto taken for so wise a man, will now so play the fool to lie here in this close filthy prison, and be content thus to be shut up among mice and rats—when you might be abroad at your liberty."

She reminded him that he could have his liberty, and the King's favor, "if you would but do as all the bishops and best learned of this realm have done." She reminded him, too, of what was waiting for him in Chelsea: "Your library, your gallery, your garden, your orchard . . ."

"I pray thee, good Mistress Alice," More asked his wife affectionately, "tell me one thing. . . . Is not this house as nigh heaven as mine own?"

5

IN TRUTH, though More would continue to do whatever his conscience would allow to avoid death, he was coming by the start of 1535 to expect death—and even to look forward to it, as an escape from a difficult world. Through his hours of religious contemplation in his cell, he had come to the conclusion that God would be pleased to receive only those who came unto him joyfully.

There was reason by then to expect death. The Act of Supremacy, passed toward the close of 1534, had pronounced King Henry "Supreme Head of the Church of England." To maliciously deny this royal supremacy, which replaced papal supremacy, was to incur the penalty of execution for high treason. From his cell window, More watched Richard Reynolds of the Sion Monastery, and the priors of three Carthusian Charterhouses, being taken to execution for denying that King Henry was Supreme Head of the Church. One by one, all four were hung, cut down while still alive, and disemboweled.

This example sufficed to bend most of the monks of England to the King's will. But the Carthusians of the London Charterhouse, who had a rare reputation for genuine piety, would not bend. Three more of them were arrested, chained by the neck in a standing position, and forced to remain that way for seventeen days and nights. They were finally hung, drawn and quartered like the first four monks—denying to the end that the King was Head of the Church.

Nicholas Wilson, the King's former confessor, broke under fear of a similar fate. He decided to give in to Henry. But his surrender was a minor victory for Henry. It was what More and Fisher did that was important. If they would accept the King's supremacy, others would accept. If they would not, their deaths would frighten others into accepting.

Fisher was lured from his silence by treachery. Solicitor-General Rich conferred with the aged Bishop in private. King Henry, Rich told him in strictest confidence, had begun to wonder if what he was doing was right in the eyes of God. His conscience was troubled. Was he *really* entitled to regard himself as the spiritual leader of the English Church? The King respected Fisher, wanted his opinion on the question, and swore that whatever his answer it would be kept secret and not used against him.

Bishop Fisher was old, and sick, and had been imprisoned in his cell for over a year. In his weakness, he succumbed to the trick. He told Rich his opinion: By the law of God, the King neither was nor could be Supreme Head on earth of the Church of England.

To the Bishop's folly was added the folly of another. Pope

Clement had died shortly after rendering his opinion on the divorce. The new Pope was Paul III. Under the mistaken notion that his authority might yet have some weight in England, he made public his protection of Bishop Fisher by naming him a cardinal. Henry was furious. Fisher would have to wear the cardinal's hat on his shoulders, Henry thundered, because he would have no head on which to wear it.

Fisher was brought to trial. Solicitor-General Rich submitted Fisher's "secret" opinion as evidence against him. Fisher was found guilty of high treason. He was beheaded, stripped of clothes, and left lying in the open, as one witness wrote, "a long, lean body, nothing in a manner but skin and bare bones." His head was mounted on London Bridge, for all to look at and think upon.

Unlike Fisher, Thomas More proved impossible to trap. Cromwell met with him in the Tower and demanded of him a straight answer: Did he or did he not believe that the King was Supreme Head of the Church? More replied calmly that he had hoped the King would never demand an answer to such a question from him. And he did not answer it. "I have in good faith discharged my mind of all such matters," he explained to Cromwell, "and neither will dispute kings' titles nor popes'. . . . But the King's true faithful subject I am, and will be, and daily I pray for him, and for you all that are of his honorable council, and for all the realm. And otherwise than this, I never intend to meddle."

This was no answer at all, of course. But it was all More could give Cromwell. If it was not enough to keep him alive, "my poor body is at the King's pleasure. Would God my death might do him good."

In spite of his frustration, Cromwell was moved by More's words and manner. He spoke gently and kindly to him before he left. But a week later Cromwell was back, again demanding of More a simple yes or no answer to his question. Again More would not give it to him: "I will not meddle with any such matters, for I am fully determined to serve God, and to think upon his passion and my passage out of this world."

A special delegation from the King's Council tried its hand at interrogating More. The delegation's questions, and More's replies, can still be read in the official report of the interrogation:

"1. Whether he would obey the King as Supreme Head?— He can make no answer.

"2. Whether he will acknowledge the King's marriage with Queen Anne to be lawful, and that with Lady Catherine invalid?—Never spoke against it . . . no answer.

"3. Where it was objected to him that he . . . is bound to answer.— He can make no answer."

So, finally, they prepared to snare Thomas More with perjury.

Thomas More was a pitiful figure when he left the Tower to attend his trial. A sick, worn-out man, he limped to Westminster Hall with the aid of a cane. He stood before his judges, leaning heavily on the cane. His condition was so painful to observe that he was given a chair to use for the duration of the trial. Sitting on it, he heard and answered the charges against him.

They charged that in former days, when speaking to the King, he had maliciously resisted Henry's marriage to Anne Boleyn. More admitted advising Henry against the second marriage. The King had asked for his opinion and he had given it, as a loyal, honest subject was required to. That was no crime. The crime lay in the word "malicious"—and he had not spoken maliciously.

It was charged that his refusal to say whether he was for or against the King's supremacy could only be interpreted to mean that he denied Henry that title—which was treason. On the contrary, More said; according to civil law, "silence indicates consent." Did *More's* silence mean consent? He did not say.

His answers could have no practical effect on judges instructed to condemn him. But even at this late hour, Henry did not want More's death. What he wanted was his support. He instructed Chancellor Audley to give More a last chance: If More would only emerge from his silence, publicly admit his errors, and repent of them, Henry would forgive and free him.

But the reason for More's silence went much deeper than a mere obstinate resistance. To his judges, More explained the right of an individual to his private opinions: "You say that all good subjects are obliged to reply. But I say that the faithful subject is more obliged to respect his conscience and his soul than anything else in the world—provided his conscience, like mine, does not raise scandal or sedition."

More would not move from this position. There was only one thing left for Henry's court to do. Even these picked and instructed judges could not find a man guilty of treason on the grounds of silence. More could not be condemned for what he had *not* said. It was Solicitor-General Rich, the man whose testimony had condemned Fisher, who rose to destroy Thomas More.

He recalled an occasion when he had gone to the Tower with two other men—Southwell and Palmer—to punish More for his defiance by removing all books from his cell. While doing so, Rich said, he had tried to argue More into accepting the Act of Parliament which had made the King Supreme Head of the Church.

He had put to More a theoretical case: Suppose Parliament pronounced Rich to be King of the realm? "Would not you then, Master More, take me for King?"

"Yes, Sir, that I would," More had answered; but while the Parliament had a right to meddle with the state of temporal matters, it had no such power in spiritual matters. Then, according to Rich, More had gone on to explain himself thus: "I put you this case. Suppose the Parliament would make a law that God should not be God? Would you then, Master Rich, say that God were not God?"

"No, Sir," Rich reported himself as having answered, "since no Parliament can make any such law."

Upon which, according to Rich, Thomas More had uttered this treasonous statement: "No more could the Parliament make the King Supreme Head of the Church."

On his oath, Rich swore that this was a true account of their conversation in More's cell.

More tartly labeled Rich's testimony a lie: "If this oath of yours, Master Rich, be true, then pray I that I never see God in the face. Which I would not say, were it otherwise, to win the whole world. . . . In good faith, Master Rich, I am sorrier for your perjury than for my own peril."

He flatly denied having made any such statement to Rich. It was ridiculous, More pointed out, to believe that he would so place himself in the power of a man of such unpleasant reputation. "I have known you from your youth hitherto," he reminded Rich, "for we long dwelled together in one parish. Where yourself can tell—I am sorry you compel me to say so—you were esteemed very light of your tongue, a great dicer, and of no commendable fame."

Turning to his judges, More went on: "Can it therefore seem likely unto your honorable lordships that I would in so weighty a cause so unadvisedly overshoot myself as to trust Master Rich, a man of me always reputed of so little truth? . . . That I would utter unto him the secrets of my conscience touching the King's supremacy?"

It did *not* seem likely. In an effort to shore up his testimony Rich called upon the other two men who had been present in More's cell. He expected them to corroborate his evidence. But neither could bring himself to lie to More's face. Both evaded the question, claiming they had been too busy packing More's books to hear what he said to Rich.

But it no longer mattered whether anyone believed Rich or

not. The verdict was predetermined—and was rendered: More was guilty of high treason. Chancellor Audley hastened to pronounce sentence, but More interrupted him: When *he* was chancellor, More recalled, it was customary to ask the condemned man if he had anything to say before sentencing him.

Disconcerted, Lord Audley asked More: *Did* he have anything he wished to say? More did. There was no longer any way to avoid his fate; and so there was no longer anything to lose in breaking his silence—and telling the truth as he saw it:

"Seeing that you are determined to condemn me . . . I will now in discharge of my conscience speak my mind plainly. . . . This indictment is grounded upon an Act of Parliament directly repugnant to the laws of God and his holy Church." The pope was Supreme Head of the Church, he told the court, and no king could rightfully assume that title.

The sentence, of course, was death. Taking More back to his cell to await execution, Kingston, Constable of the Tower, wept bitterly. More consoled him. "Good Master Kingston, trouble not yourself, but be of good cheer. For I will pray for you, and your good wife, that we may meet in heaven together, where we shall be merry for ever and ever."

"I was ashamed of myself," Kingston later told More's son-in-law, "that, at my departing from your father, I found my heart so feeble, and his so strong, that he was fain to comfort me, which should rather have comforted him."

Shortly before the execution, Cromwell came to More on a final attempt at conciliation. The King was still ready to pardon him, Cromwell said, if only he would change his mind. More told Cromwell that he had already changed it: Originally he had intended to shave before his execution, but now he had decided to let his beard share the fate of his head.

More retained a touch of his old humor to the scaffold. Weak and shaky, he asked one of the officers present to help him up the steps: "I pray you, Master Lieutenant, see me safe up. And for my coming down, let me shift for myself."

He went to the executioner with these words: "I die the King's good servant—but God's first."

Thomas More's head replaced Fisher's on London Bridge.

Erasmus learned of it and shrank inside himself. "Thomas More . . ." he mourned, "whose soul was more pure than any snow, whose genius was such as England never had—yea, and never shall again, mother of good wits though England be." Exactly four hundred years later, in 1935, the Roman Catholic Church pronounced More a saint.

Chapter Ten

WIFE IN LIMBO

"HE HAS NO respect or fear for anyone in the world," Charles V said of Henry VIII.

Henry proved the truth of this by making martyrs of the monks, of Fisher and of More. For five years Catherine had clung to an illusion: If only the Pope would render a decision in her favor, Henry would take her back.

But in the early spring of 1534—more than a year before the executions—Catherine bitterly conveyed her disillusion to Chapuys, who passed it on to Emperor Charles: "She hitherto imagined that the Papal sentence once delivered . . . this King would return to the right path. But she now realizes that it is absolutely necessary to apply stronger remedies to the evil. What these are to be, she darst not say."

She dared not say. But Chapuys and the rest of her supporters, in and out of England, knew what she meant. The Pope must push the excommunication of Henry to its logical conclusion: declare Henry deprived of his kingdom and Englishmen no longer his subjects; call upon the other kings of Christendom to join with all true Christians within England in removing Henry from his throne. The new Pope, however, hesitated to go so far. Henry had once been the staunchest supporter of papal authority. Circumstances might yet persuade him to return to the fold—unless the Church cut off the possibility of reconciliation by declaring war on him. Until the deaths of Fisher and More, Paul III—like Clement before him—continued to hold back his ultimate weapon and hope the situation would solve itself.

While the Pope hesitated, Catherine and Henry discovered one point on which they could agree: Henry wanted her to move to a less accessible residence than Buckden; Catherine wanted to move to a healthier place. Kimbolton Manor, which was drier but remote, proved mutually satisfactory.

She settled into Kimbolton, but refused to have any contact with the new servants assigned to "the Princess Dowager." She kept to her quarters with the few who continued secretly to address her as Queen of England. And she continued to defy Henry's wishes. "I do not fear," she declared. "For there is no punishment from God except for neglected duty." In the increasing bitterness of her defiance, Catherine even went so far as to ignore the danger to her daughter. She advised Mary that it was time "to show her teeth to the King."

Princess Mary was by then quite ready to do so. She, too, had become disillusioned. For a time she had hoped that Henry's fatherly affection might yet soften him toward her. Shortly after she had been reduced to being a member of her half sister's household, Henry paid a visit to the residence of his daughters. But he came only to see the four-month-old Elizabeth. Mary was ordered to remain in her quarters while he was there.

When Henry arrived he sent Cromwell and two other men to her. They urged her once more to renounce her title of Princess. Mary told them that nothing could force her to do so. She begged for permission to see her father and kiss his hand.

The request was not granted. Henry spent some time with the baby Elizabeth. Then he went directly outside to his horse. Mary suddenly appeared on her balcony, falling to her knees and raising her clasped hands toward him. Henry glanced up and saw her there. For a moment it seemed as though she had finally succeeded in reaching through his anger to touch his heart. He acknowledged her presence by touching a hand to his hat and bowing to her.

But that was all. Henry rode off without speaking to her. The pressure on Mary to renounce her title continued. She was forced to remain in Elizabeth's household, under the supervision of Anne Boleyn's aunt, Lady Shelton.

By the time another two months had passed, Mary was furious enough "to show her teeth." The occasion was the removal of Elizabeth's household to another residence. Mary absolutely refused to pay court to her half sister by traveling as part of the entourage. The men of the household had their orders. They seized Mary and dragged her out of the house to a waiting carriage. Digging in her heels all the way, she cried out again and again that she was being taken along by force, against her will. Mary was picked up bodily and placed in Lady Shelton's carriage, still struggling, still protesting loudly that her presence in the party of her half sister was involuntary—and in no way an acceptance of Elizabeth as Princess. That title, she repeated for all to hear, belonged to her and no one else.

Word of Mary's behavior shocked the court, and frightened Chapuys. He warned her that she might irritate Henry into treating her more severely. To safeguard her claim to the title, her formal protest was sufficient. She should not go to the extreme of forcing her guards to use violence on her.

Mary heeded his advice. After that she protested with more care and subtlety. The next time she was ordered to accompany Elizabeth's household to a new residence, she said she would obey the order—though under protest. But as soon as the party started out for the Thames, where a barge was waiting to take them the rest of the way, Mary had the driver of her own carriage go on ahead at full speed. Soon Elizabeth and the others were left far behind, out of sight. In that way Mary contrived to make the required journey—but not as a part of her half sister's entourage. "She made such haste," Chapuys was pleased to report to Charles, "that she arrived at Greenwich one hour before the bastard [Elizabeth]. And when she came to the barge she managed so well that she occupied the most honorable seat in it."

Her father finally allowed Mary to live apart from his other daughter—though still under Lady Shelton's supervision. But he refused Mary's request that she be permitted to live with her mother. To put mother and daughter together, Henry felt, would only result in increasing Mary's obstinacy. It might also encourage Catherine to help her daughter's future by shifting from defensive defiance to aggressive subversion.

"The lady Catherine is a proud, stubborn woman," Henry explained to his Council. "If she took it into her head to take her daughter's side, she might take to the field, muster a great assembly of men, and wage against me a war as fierce as any her mother ever waged in Spain."

There is no evidence that Catherine ever contemplated leading an army against her husband. What Henry really feared was that others would make war against him—in Catherine's name. When he executed the Carthusian monks, and then beheaded Fisher and More, the threat of such a war began to loom larger.

2

"All Venice in great murmuration to hear it . . ." an Englishman wrote home of reaction in Italy to the executions of the first four monks for denying Henry's spiritual supremacy. "I never saw Italians break out so vehemently at anything; it seemed so strange, and so much against their stomach."

From Rome, Casale reported a similar reaction to the martyrdom of the monks for their religious convictions. "Some cardinals said they envied such a death. I told them, if that were so, they might go to England and imitate the monks' folly." More ominous was his news that "the French show great astonishment" at the executions. On the surface, France supported Henry in the great divorce controversy. Actually, Francis I was only encouraging him to make sure that Henry and Charles would not unite against him. At the same time, Francis did not want to go so far as to enrage Charles or alienate the Church.

Pope Paul's representatives asked Francis to speak to Henry on behalf of Bishop Fisher, shortly before Fisher's execution. Francis answered that he would do what he could, but doubted his power to sway the English King. Henry, said Francis truthfully, was a difficult friend—unstable, obstinate and proud. "Sometimes he almost treats me like a subject. . . . He is the strangest man in the world, and I fear I can do no good with him. But I must put up with him, as it is no time to lose friends."

When King Henry did execute Fisher and More, it put an end to Pope Paul's hesitancy. The Pope informed Emperor Charles and the other Catholic rulers of the Continent that he intended to declare Henry VIII deprived of his kingdom. He asked for the assistance of these princes in executing justice on Henry—"who has put to death the cardinal bishop of Rochester, and despised the authority of the Holy See in marrying Anne Boleyn, and repudiating his true wife Catherine."

Pope Paul's letter to King Francis was worded with special care. He recognized, wrote the Pope, that Francis had been allied to England. But he reminded Francis of Henry's "wickedness . . . notorious adultery . . . sacrilegious slaughter . . . heresy and schism." Because of these crimes, the Pope went on, he was "compelled to declare Henry deprived of his kingdom and his royal dignity." He earnestly implored the French

King to "bear this calmly, and be ready to execute justice on Henry when required."

King Francis, like most of the important men of France, had been shocked by the deaths of Fisher and More. But as the reaction to these executions pushed the divorce controversy to the brink of war, he tried to maintain his neutrality—hinting of his sympathy to both sides while allying himself with neither. This was evident even in letters which Frenchmen wrote to each other, in which they carefully referred to Catherine as "the old Queen" and to Anne Boleyn as "the new Queen."

As for Emperor Charles, upon whom would fall the main burden of actually depriving Henry of his kingdom, he had too many other problems to assume one more. Charles was fighting a war against the horde of Ottoman Turks smashing at the eastern borders of his Empire, and contending with the Protestants within his Empire. If he now made war on England he might also find himself enmeshed once more with the French.

So Charles and Francis, separately and for separate reasons, delayed the meeting of a council called by the Pope to declare war on Henry. But although few outside of Pope Paul actually wanted it, at this moment war seemed very close.

King Henry was startled—and perhaps a bit shaken—by the violence of Pope Paul's indignation against him. He told Cromwell to send a letter to Casale in Rome, ordering him to do his best to placate the Pope and his court. Casale was to explain to them that Fisher and More had formed a treasonous conspiracy against their king. Henry had mercifully tried to save them by mild persuasion. He had offered them forgiveness if they would only cease their treason and be penitent. But they had continued to defy him, leaving no other choice under the law than to execute them.

But, Henry told Cromwell to write, if this explanation did not suffice, then Casale was to issue a warning to the Pope:

The King of England was ready to defend himself—"having his kingdom and his subjects so secure that he trusts to repel any injury that may be attempted against him."

Henry meant what he said. If war did come, he intended to be prepared to fight it.

3

A BASIC NECESSITY in preparing for war was money. In the Church of England Henry found it. With the help of Cromwell, he had already begun to divert into his own treasury the wealth that had once gone to Rome. But there was another source within the Church, a juicy plum ripe for the plucking: the more than eight hundred monasteries of England and Wales. Henry named Cromwell his vicar-general and set him to work.

Cromwell began with a detailed investigation of the moral and financial state of every religious house in the kingdom. His investigating agents were at pains to send back horrifying reports of the sinful conditions and lack of spiritual value they found in almost every monastery they visited. Of course, they exaggerated to make the point Cromwell wanted their reports to make. Nevertheless, such conditions did exist in too many of the religious houses—both in England and on the Continent.

At about the same time, Pope Paul made an investigation of his own, with an eye to reforming the Church from within. His investigating commission uncovered scandalous situations reaching from the hierarchy of the Church to its lowest ranks. Just as there were cardinals in Rome who did not bother to hide their dalliance with sinful women, so there were unfit priests hearing confessions, nuns who were immoral and money-hungry, monks who were lustful and dishonest.

Only some of the monasteries were actually guilty of such evils. But it became obvious that most of them had lost their original religious character, the spiritual and educational purpose for which they had been founded. In England it became equally obvious that these ills were merely the excuse, not the reason, for the investigations by Cromwell's agents. By the time they were finished, Cromwell knew the cash value of every single religious house, down to the last penny.

Then Cromwell was ready to carry out the wholesale dissolution of the monasteries, beginning with the smaller houses. What came to be known as "the fall of the monasteries" served two purposes: Loot from the monastery coffers and from the sale and rental of monastery property fattened the royal treasury; at the same time, it insured the loyalty of those members of the nobility and gentry who bought or were given the lands of the dissolved monasteries. They must be loyal to King Henry, for fear of losing what they had obtained if Henry's enemies triumphed over him.

While Henry and Cromwell were thus busy preparing for war, Chapuys was also preparing for it—in his own way. He tried to organize a rebel party around Catherine and Mary, a group of important Englishmen who would lead an internal revolt against Henry when Charles and the other Catholic rulers attacked England. And he kept sending Charles warnings that further delay would be fatal to Catherine and Mary.

"The King himself is not evil by nature," Chapuys wrote of Henry. "It is this Anne who has provoked him to this perverse and wicked temper, and alienates him from his former humanity." Because of Anne Boleyn's constant prodding, he informed Charles, King Henry was threatening to kill his first wife and daughter, as he had Fisher and More, for refusing to take the oath of succession.

Charles agreed that Henry was acting despicably: "As long as the King of England remains in his present abominable condition with his concubine, and holds the divorce to be valid . . . it is impossible with honor and good conscience to deal with him." But he doubted that Henry would stoop so low as to murder Catherine and his older daughter: "The ill will of the King of England to the Queen and Princess is cruel and horrible. But it is impossible to believe that he would be so unnatural as to put them to death. . . . He probably intends by threats to make them swear to and approve his statutes."

Charles felt that Catherine and Mary were right to safeguard their status by refusing the oath. But if Henry's anger became too dangerous, he wrote Chapuys, "advise them to take the oath rather than lose their lives—protesting that they take it out of fear. It cannot then prejudice their rights."

Catherine seconded Chapuys' repeated warnings. She wrote to her nephew urging him to action before it became too late, before she and her daughter followed the holy martyrs to the headsman's ax.

On the same day she wrote to Pope Paul:

Most Holy and Blessed Father . . . I have only one satisfaction in thinking of the present state of things. I give unceasing thanks to our Lord Jesus Christ for having given Christendom a vicar like Your Holiness at a time of such great trouble. . . . Your Holiness knows, and all Christendom knows, what things are done here; what great offense is given to God, what scandal to the world, what reproach is thrown upon Your Holiness. If a remedy be not applied quickly, there will be no end to ruined souls and martyred saints. The good will remain constant and suffer, the lukewarm perhaps fall away, and the rest stray like sheep that have lost their shepherd. . . . And so I conclude, waiting for the remedy from God and from Your Holiness. May it come speedily. If not, the time will be past.

When she wrote this letter Catherine had only three more months to live.

4

CATHERINE had suffered from recurring illnesses over a period of many years. The strain upon her of Henry's decision to obtain a divorce was bound to lower her resistance to disease still further. To this was added the sustained fear with which she lived after the executions of Fisher and More. Fear, not only for her own safety, but for that of her daughter. It was too heavy a burden to carry with what little health she had left. During the year of those executions, 1535, she had periodic attacks of severe stomach pains and nausea. In November, her illness became serious, and she was bedridden for some time. By the second week in December, she seemed to be recovering. She was able to sit up, read her mail, and write answers to a few letters with a shaky hand.

The most important letter came to her from Dr. Ortiz, her special representative in Rome. He wrote to inform Catherine of the determined efforts that Pope Paul was making "to remedy" her situation. He urged her to persevere in the knowledge that the Pope's activities would soon bring her justice.

Catherine wrote back from her sickbed:

Doctor . . . the contents of your letter were a consolation and a comfort to me . . . I shall have patience and wait. But you, Doctor, must urgently advise His Holiness that the good work which he has commenced should be promptly executed. For the Devil has so far been only half-tied. Should there be the least hesitation or delay, it will set him loose and at liberty to do mischief.

I cannot, indeed dare not, write to you in clearer terms. You are a prudent and wise man, and will understand what I mean . . .

Catherine.

Twelve days later, on Christmas Day, Catherine suffered a relapse. This time nothing could ease the attacks. They grew steadily worse. Intense nausea made it impossible for her to keep food or liquids in her stomach. Violent cramps and feverish restlessness prevented her from falling asleep for more than a few minutes at a time. Through the days and nights that followed, the fits of vomiting, her inability to eat or drink, and the lack of rest combined to drain her of the last of her strength. Her physician urged that he be allowed to summon other medical men for consultation. Catherine weakly answered that she wanted no more physicians. Her fate was in the hands of God; to His will she committed herself.

It began to seem certain that Catherine's sickbed was becoming her deathbed. She asked for Chapuys, and for her daughter. A message was dispatched to Chapuys, who petitioned Cromwell for the special permission necessary before he could visit Catherine. Cromwell was certain he would be allowed to go. But Chapuys would have to speak about it to King Henry, who had other important matters to discuss with him.

Chapuys went to Greenwich. Henry received him with warm friendliness, embracing him affectionately. Henry was in an ebullient mood. He too had heard that Catherine was sinking, and felt no reason to be sorry for it. She was no longer his wife. She had become England's enemy, the reason his kingdom was now threatened with invasion and revolt. Henry made no attempt to hide his real feelings on this matter. He told Chapuys plainly that once Catherine was dead, Emperor Charles would no longer have any reason to trouble himself about England's affairs.

Charles, said Henry, had plenty of other troubles to concern himself with—and would soon have more. King Francis was once more preparing to make war against Charles in Italy, and was asking for England's help. Henry was revealing this quite openly to Chapuys, he explained, because as an Englishman he was accustomed to being honest and straight-forward. Not being a Frenchman or a Spaniard, he was not used to dealing in guile and trickery. Charles, for his own good, should stop supporting the claims of Catherine and Mary. Instead he should persuade the Pope to revoke the sentence against Henry. It behooved Charles, threatened by another war with France, to protect himself by coming to terms with Henry—and quickly.

Chapuys replied that he was quite sure Emperor Charles was strong enough to fight a war with the French.

"That may very well be," Henry countered pointedly, "but were I now to throw my sword in the balance, you would find the situation very different."

Having planted this warning, Henry gave Chapuys permission to go to Catherine. Chapuys asked if Mary might also be allowed to see her mother. Henry hesitated. He did not want Mary, who was already stubborn enough, to be further stiffened by any last words from Catherine. Finally, Henry said he would need some time to think it over—which was tantamount to refusal.

Chapuys rode swiftly to Kimbolton. He was immediately summoned to Catherine's bedside. "I found her," he wrote later, "so wasted that she could neither stand nor sit up in her bed." He knelt beside her bed and kissed her hand. His devotion warmed Catherine's heart. She thanked him for having taken the trouble to come to her. If it now pleased God to take her, she said, at least it would be a consolation to die in his arms—and not all alone like a beast.

Chapuys did his best to persuade her that she was not dying. He tried to lift her spirits with an assurance that her husband had decided to treat her more kindly in the future. He begged her to take heart: She must do her utmost to recover—not only for her own sake, but also because the welfare of all Christendom depended in great measure on her recovery.

His talk soothed Catherine; her own exhausted her. She longed for some rest, she told him. In the past six days she had not slept for more than two hours. Now that he was here, she felt she might be able to sleep a little.

Chapuys withdrew from her room, only to be summoned back before long. Catherine was too excited to sleep. She wanted to talk further with him—about her own situation and the future of her daughter; about what the Pope was doing, and what Charles must do. Chapuys was afraid that she would overtire herself. Several times he suggested he withdraw and let her rest. But Catherine would not hear of it. For two more hours she continued to talk with him, telling him how much pleasure and consolation he gave her just by being there.

For four days Chapuys remained at Kimbolton, spending hours each day at her bedside. His presence seemed to help Catherine. By the time he left to return to London, Catherine's doctor was convinced that she was out of danger.

So it continued to seem, for two days after Chapuys was gone. Catherine's condition improved a bit each day. She was able to smile. She became strong enough to sit up in bed; to comb and dress her hair without help from her maids. Then, during the night of January 6, 1536, she suddenly felt herself sinking again. This time she sensed there would be no recovery.

She asked her attendants how much longer it was till dawn. They told her that it was only midnight. Catherine waited, stirring fitfully in her bed. From time to time she asked again how much longer she had to wait for dawn. She was waiting to hear a last Mass and receive the Holy Sacrament.

Frightened, they sent for her confessor. He came quickly, but Catherine insisted on waiting for dawn. It arrived at last. Catherine heard Mass, took the Holy Sacrament, and received extreme unction. She prayed for the salvation of her soul. Also that God would forgive her husband for what he had done to her, and inspire him to follow the right path. She had only hours left to live when she dictated her last words to Henry:

My most dear lord, King, and husband—

The hour of my death now approaching, I must, out of the love I bear you, remind you of the health of your soul. This you ought to prefer before all worldly matters and the pampering of your flesh, for which you have cast me into many miseries and yourself into many troubles. But I forgive you all, and pray God to forgive you also.

For the rest, I commend unto you our daughter, Mary, beseeching you to be a good father to her, as I have heretofore desired. I entreat you also to give marriage portions unto my maids; which is not much, there being but three of them. For all my other servants I ask a year's pay lest otherwise they be unprovided for.

Lastly, I make this vow, that mine eyes desire you above all things. Farewell.

Death came to Catherine of Aragon at two o'clock that afternoon. Thirty-five years earlier she had first set foot on English soil, as a gift from Spain. From the first she had been the pawn of scheming kings, contending powers, and a capricious fate. To the last she had borne herself with a regal pride and a woman's strength. England had given her some joy, a short span of glory, and much misery. It had received from her a princess, a victory in battle, and a stumbling block over which it tripped into a religious and political revolution. Now she was gone.

When the news reached Henry he exclaimed in relief: "Thank God, we are now free from any fear of war."

Anne Boleyn had even more cause for happiness, and no reticence about showing it. But this did not last long. Ironically, both her happiness and the reason for it were to die on the same day that Catherine was buried.

Chapter Eleven

SWORD FOR A SLENDER NECK

EVER SINCE that moment in 1533 in which she had given birth to a daughter, instead of a male heir to the throne, Anne Boleyn had been in danger. Even when her power seemed at its height, Anne knew her only security lay in producing a son for Henry—and quickly.

This insecurity, plus the fact that her husband began to give his attention and time to other attractive ladies of the court, left her frightened, suspicious and vengeful, her mood swinging erratically from arrogance to gloom to hysterical gaiety. According to the charges later made against her, but never proved, it also drove her to the beds of various men other than her husband—in a frenzied effort to get pregnant again.

Less than a year after Elizabeth's birth, it was believed that Anne Boleyn *was* pregnant again. Once more her future seemed brilliant. Henry had Parliament pass a law that, if he should die before Anne, she was to become Regent—"absolute governess of her children and kingdom."

In the same year Anne's sister, Mary Boleyn—who had once been Henry's mistress—married for a second time. Anne had reason to compare the good fortune her determination had won her with the depths to which her sister had sunk.

The man Mary Boleyn fell in love with in 1534—Sir William Stafford—had little fortune, and not much rank. But she married him, without first consulting Anne. Neither Anne nor Henry was pleased. Both made this painfully apparent to the newlyweds. After three months of suffering royal disfavor, the unhappy Mary Boleyn wrote to Secretary Cromwell, pleading with him to help her and her husband:

I am sure that it is not unknown to you the high displeasure that both he and I have of both the King's Highness and the Queen's Grace by reason of our marriage, without their knowledge; wherein we both do yield ourselves faulty. . . . But one thing consider, Master Secretary; that he was young, and love overcame reason . . . I might have had a man of greater birth, and a higher; but I assure you I could never have had one that should have loved me so well, nor a more honest man.

She asked Cromwell to beg Henry "to have pity on us; and to speak to the Queen's Grace for us. For Her Grace is so highly displeased with us both that, unless the King be so good as to withdraw his rigor, and sue for us, we are never likely to recover Her Grace's favor—which is too heavy to bear. . . . For God's sake help us; for we have been now a quarter year married, and too late now to change this. . . . But if I were at liberty, and might choose, I assure you I had rather beg my bread with him than to be the greatest queen christened. And I believe verily he would not forsake me to be a king."

If there was in this a veiled hint of resentment toward Anne, it was understandable. As understandable as the possibility that Anne felt a tinge of envy toward her less fortunate sister—who at least had a husband who loved her. By this time, Anne Boleyn was quite aware that the King no longer loved her. Perhaps he had struggled too long and hard to make her his wife, and had found the prize not worth the price. At any rate, the only thing binding Henry to Anne now was his expectation of acquiring a legitimate son through her.

But by midsummer of 1534 it became apparent that Anne was not pregnant. Either the signs of it had been false, or she had aborted in the early months. By September, Chapuys was writing: "Since the King began to doubt that his lady was pregnant, he has renewed and increased the love he formerly had for a very beautiful damsel of the court. And because the said lady [Anne] wished to drive her away, the King has been very angry. . . . To which it is not well to attach too much importance, considering the changeable character of the said King, and the craft of the said lady, who knows well how to manage him."

2

BUT ANNE was becoming less able to manage Henry. When she complained of his attentions to other women, he warned her to mind her own business. Less than a month later Chapuys saw more reason for glee: "Of late days Lord Rochford's wife has been banished from the Court because she conspired with the Concubine [Anne] to procure the withdrawal from Court of the young lady whom this king has been accustomed to favor, whose influence increases daily. Meanwhile, that of the Concubine diminishes, which has already abated a good deal of her insolence."

Early in the following year, the French Admiral, in London to discuss a possible meeting between Henry and King Francis, had an uncomfortable moment with Anne Boleyn during a feast in his honor. He was seated beside Anne when she suddenly "burst into a fit of uncontrollable laughter without any occasion."

The Admiral frowned and said, "What, madam, do you laugh at me?"

Anne explained, with an attempt at lightness, that what had amused her was something that had just been done by her husband, whom she had been watching. Henry had left her, saying he was going to speak to the Admiral's secretary. Instead he had met a certain young lady and become so engrossed with her that he had apparently forgotten his errand.

The following month another French representative recorded a disturbing conversation with Anne. She told him that unless France acted to support Henry against the sentence of the Pope, she feared she would "be ruined and lost. For she sees herself very near that; and in more grief and trouble than before her marriage."

Anne pleaded that France consider her situation, "of which she could not speak as fully as she wished, on account of her fears, and the eyes which were looking at her, her husband's and those of the lords present. . . . She then left. The King went, without her, into the next room where a dance was beginning.

"As far as can be judged," the Frenchman concluded after this interview with Anne, "she is not at her ease . . ."

That was a masterful understatement. But by the closing months of that year, Anne was once more pregnant, and this time there was no question about it. She was beginning the fourth month of her pregnancy when Catherine died. Anne's confidence soared on the wings of Henry's elation. With his first wife gone, Anne's situation and that of her baby daughter were secure—if only she would deliver a boy this time.

There was little likelihood that the Pope would proceed with his sentence against Henry now that the Queen, on whose behalf the sentence had been rendered, was dead. Catherine's death also meant that Henry no longer would have to court King Francis' support.

"The time has come for dealing with the French much more to our advantage than heretofore," Henry exulted on hearing that Catherine was dead. "They will fear my becoming the Emperor's friend and ally now that the real cause of our enmity no longer exists. I shall be able to do anything I like with them."

"On the following day, which was Sunday," Chapuys reported bitterly, "the King dressed entirely in yellow from head to foot, with the single exception of a white feather in his cap. His bastard daughter [Elizabeth] was triumphantly taken to Church to the accompaniment of trumpets and great display. After dinner, the King went to the hall where the ladies were dancing, and there made great demonstration of joy. Then he went into his apartments, took the little bastard and carried her out in his arms, showing her off to everyone present."

Both Henry and Anne believed at first that with Catherine gone it would be easier to handle Princess Mary. Anne tried to woo Catherine's daughter with softness. She was eager, Anne informed Mary, to become her warmest friend and a second mother to her—if only Mary would stop being obstinate, and obey her father's commands like an obedient daughter. If Mary would take the oath of succession she would be allowed to return to court. Anne would exempt Mary from the duty of carrying her train, and allow her to walk by her side as an equal.

Mary replied that she *was* obedient to her father's commands—as long as they were in conformity with her honor. Anne's aunt, Lady Shelton, pleaded with Mary to reconsider Anne's offer of friendship. Mary refused. Anne resorted to indirect threats. She sent her aunt a letter, which was intended to be left open where Mary would be likely to find and read it:

Mrs. Shelton—
My pleasure is that you seek to go no further to move the Lady Mary towards the King's Grace. . . . What I have done myself has been more for charity than because the King or I care

what course she takes, or whether she will change or not change her purpose. When I shall have a son, as soon I look to have, I know what then will come to her. Remembering the word of God, that we should do good to our enemies, I have wished to give her notice before the time, because . . . the King . . . will not value her repentance when she has no longer power to choose. . . .

Mrs. Shelton, I beseech you, trouble not yourself to turn her from any of her willful ways, for to me she can do neither good nor evil.

Your Good Mistress,
Anne R.

"When I shall have a son . . ." Anne knew only too well that her future rested entirely upon that event. She was the first to see that future darkening when disaster struck.

On January 29, Catherine was buried. It was as though she reached out of her grave to inflict the most appropriate revenge: On that same day, Anne Boleyn miscarried. The physicians revealed that the dead child was a boy.

Henry stalked into Anne's room, spoke a few cold words to her, and strode out. He had given her a son—and she had killed him.

In a vain attempt to move Henry to sympathy for her, Anne pleaded that the disaster had been caused by the intensity of her love for him: A report had come to her that Henry had fallen with his horse while hunting, and was seriously injured. She had become so frightened that she had fainted. In falling, she had hurt herself, bringing on the miscarriage.

Few believed her. "Upon the whole," Chapuys recorded, "the general opinion is that the Concubine's miscarriage was entirely owing to a defect in her constitution, and her utter inability to bear male children. Others imagine that it was caused by fear that the King would treat her as he treated his late Queen. This is not unlikely, considering his behavior towards a damsel of the Court, named Miss Seymour, to whom he has lately made very valuable presents."

Anne Boleyn was doomed. And not merely because of "Miss Seymour"—who was Jane Seymour, and destined to be no passing fancy. It had not been merely for love of Anne Boleyn that Henry had discarded Catherine. It was not primarily infatuation with Jane Seymour that caused Henry to rid himself of Anne.

Henry had gone to enormous trouble to make Anne Boleyn his wife and queen. He had set himself against the power of Emperor Charles and the prestige of the Pope. He had risked civil disorder within his own kingdom. And for what? To obtain a legitimate son. He had expected a son from Anne. She had *promised* him a son. What had she given him? Another daughter. And when she finally had a son growing within her, it had died unborn. God's meaning was quite clear—to Henry. What to do about it was equally clear.

Catherine was gone. But the Pope, Emperor Charles, and many others had made it obvious that they would never concede that Anne Boleyn was his legitimate wife; nor that any of her children were legitimate heirs to the throne. Since God had shown that He frowned on this marriage anyway, and would not allow Anne to have male children, there was no point in continuing the struggle while handicapped by Anne.

According to the Pope and Charles, Anne Boleyn was not Henry's legal wife. If that was so, Catherine's death left Henry a widower with a perfect right to marry again. No one would question the legitimacy of any children he had by his next wife. It had been difficult for the King to rid himself of Catherine. Ridding himself of Anne Boleyn would be easier.

Anne Boleyn recovered from her miscarriage. But any small hope she clung to soon vanished. She found herself seeing less and less of her husband in the ensuing weeks. Ignored and abandoned, she was left to dwell upon the rumors that came to her about Jane Seymour.

Jane Seymour was one of Anne's maids of honor. She was neither younger nor prettier than Anne, and not nearly so bright. But she was attractive enough to stir Henry's desire, with a healthy, full-curved figure that promised both pleasure and the ability to bear children. As soon as Jane Seymour attracted the attentions of Henry, the large Seymour family seized this chance to replace the Boleyn family as satellites around the throne. Enemies of Anne Boleyn joined the Seymours in teaching Jane exactly how to handle the King.

They had learned much from Anne Boleyn's example. Carefully schooled and advised each step of the way, Jane Seymour sweetly encouraged Henry's interest in her and flattered his vanity. Modestly, she retreated from his aggressive desire for fulfillment. She was a good girl, who would never give the delights of her body to any man—unless, of course, that man was her husband.

Henry became more and more intrigued with her. It was rumored that Jane Seymour found ways to remind him that his people hated and would never accept Anne Boleyn. It was also rumored that Jane had managed to enlist Princess Mary as a friend and ally in her campaign. What rapidly became more than a rumor was Henry's determination to make Jane Seymour his wife.

3

HENRY knew now what he wanted. He had few of his old scruples left concerning the means he used to get it. It has been written of a group beheading that occurred at one time in England: "The first head to be exhibited produced a deathly hush; the fifth a gust of laughter." It was thus with Henry. It had hurt him at first to act viciously toward Catherine; it had run counter to his image of himself as a decent, good-natured man. By the time Catherine died, he had become accustomed to the use of brutality to solve his problems. It came easily and more naturally to him now.

Henry had learned something else from his experience with Catherine. Divorcing her had solved little. She had remained a painful thorn in his side. It had proved impossible to remove that thorn as long as Catherine lived. And Henry had been obliged to wait many years for her to die a natural death. He did not intend to repeat that mistake with his second wife.

Shortly before Thomas More's execution, his daughter Margaret had visited him in his cell, and he had asked her how Anne Boleyn's affairs were progressing. "Never better," she had told him. More had shaken his head sadly: "Alas, Meg, it pitieth me to remember into what misery, poor soul, she shall shortly come."

Less than a year after More's execution, within three months of Catherine's funeral, the jaws of fate snapped shut around Anne Boleyn. In mid-April of 1536, some of Henry's Privy Council began a purposeful investigation of Anne's past behavior, passing on to Cromwell every shred of court gossip about her misconduct with various men. Circumstantial indications were not difficult to find. The Boleyn family had many enemies who had lusted for this moment. Anne had always enjoyed the company of charming men. She was the kind of woman who stimulated men to open adoration. And she had been reared at the French court where lighthearted flirting was a way of life.

The charge that was being plotted against Anne Boleyn was adultery. No evidence has ever emerged from the secrecy of the investigation and subsequent trials to prove her guilty. But toward the end of April began the arrests of five men accused of misconduct with Anne. Three were gentlemen of the royal household: Sir William Brereton, Sir Francis Weston, and Sir Henry Norris. The fourth was Anne's brother,

Lord Rochford, whose wife had apparently been persuaded to murmur the charge of incest. The fifth was a favorite court musician named Mark Smeaton.

That this musician had been deeply infatuated with Anne Boleyn was apparently true. There was one occasion upon which Anne found him gazing unhappily out of the round window of her presence chamber.

"Why so sad?" she asked him.

Smeaton replied that it was nothing. But Anne knew he was suffering from an impossible love for her. She admonished him, not unkindly, "You may not look to have me speak to you as if you were a nobleman, because you are an inferior."

"No, no, madam," the woebegone musician hastened to assure her, "a look sufficeth me."

Mark Smeaton was taken in chains to the Tower. There he was driven by threat of torture and drawn by hope of pardon into confessing that he had committed adultery with Queen Anne several times. No one has ever been able to prove whether this was a true or false confession.

The other four men arrseted were under threat of similar torture, and in danger of a death more horrible than beheading. They might be executed by burning, or disemboweling. Sir Henry Norris was said to have been offered a pardon, by King Henry himself, if he would confess to the adultery. Norris refused, was put in the Tower. There he made the demanded confession, but later swore his confession was a lie. The other three accused denied the fatal charge.

It has been thought by some historians, without proof, that Anne Boleyn might possibly have committed the adulteries charged—out of her desperate need to produce a male heir for Henry. But Henry himself certainly did not behave like a man learning that he was a cuckold many times over. His gaiety during the secret hearings on these charges was noted by many. "You never saw prince or man," Chapuys wrote to

Charles, "who made greater show of his horns, or bore them with more delight."

On the second day in May, Anne was summoned before the Council at Greenwich. Her uncle, the Duke of Norfolk, presided. The secrecy which shrouds this hearing is parted only by Anne's later words to Sir William Kingston, Constable of the Tower.

"I was cruelly handled at Greenwich, with the King's Council," she told Kingston. When she tried to defend herself against the charges, the Duke of Norfolk only shook his head and said, "Tut, tut, tut!"

Anne Boleyn swore to the Council that she was innocent. In vain. She was turned over to the custody of William Kingston. He took her by barge up the river toward the Tower of London. Kingston had found it impossible to maintain his composure with Thomas More at the end. And he had difficulty in maintaining it now. In fact, Kingston seems to have been much too tenderhearted a man for the job he held.

As he conducted Anne Boleyn ashore under the glowering walls of the Tower, she suddenly halted and asked him pitifully, "Mr. Kingston, shall I go into a dungeon?"

"No, Madam," Kingston comforted her, "you shall go into the lodging you lay in at your coronation."

The irony of it twisted the knife inside Anne. "It is too good for me," she said bitterly. "Jesus have mercy on me." She fell to her knees and wept.

Her weeping gave way to wild laughter, which with some difficulty she finally managed to control.

She asked Kingston if she might have the Sacrament in her quarters, so that she might pray for mercy. "For I am as clear of the company of men as for sin, as I am clear from you, and am the King's true wedded wife."

She wanted to know where Henry was, and her father. And: "Oh, where is my sweet brother?" Kingston answered evasively, and led her toward her quarters.

"Mr. Kingston," she asked, "do you know wherefore I am here?"

Kingston said he did not.

"I hear that I shall be accused with three men; and I can say no more than Nay—without I should open my body."

She looked about her and suddenly cried out: "Oh Norris, hast thou accused me? Thou art in the Tower with me, and thou and I shall die together! And Mark, thou art here, too . . ." Again she tried to calm herself. "Mr. Kingston, shall I die without justice?"

"The poorest subject the King hath," Kingston told her, "hath justice."

His answer sent Anne into another fit of laughter.

In her quarters in the Tower, Anne found that she had been assigned ladies whom she did not like to attend her. "I think it much unkindness in the King," she protested, "to put such about me as I never loved." But they were not there to make her happy. They were there to spy on her, and report all she said—or at any rate what they would claim she had said.

Soon these ladies reported conversations with Anne about two of the accused: Norris and Weston. They said Anne recalled having chided Norris for not getting married, saying to him: "You look for dead men's shoes; for if aught but good came to the King, you would look to have me."

They reported that Anne also told them of chiding Weston—for being in love with a woman other than his wife. To which Weston had answered Anne that he loved *her* better than either his wife or the other woman.

It seems unlikely that Anne Boleyn would have made such dangerous statements to those whom she knew to be her enemies. But there was, by then, no one left to defend her against any accusations. Her uncle, the Duke of Norfolk, had already prejudged her guilty to keep his own influence with the King. Her father was too eager to save his own skin to say a word in protest against what was being done. Only the timid, vacillating Archbishop Cranmer expressed doubts of her guilt.

While everyone else was either attacking Anne or keeping a discreet silence, Cranmer wrote to Henry, hesitantly questioning the charges against Anne Boleyn: "I am in such a perplexity, that my mind is clean amazed; for I never had better opinion in woman than I had in her; which maketh me to think that she should not be culpable. And again, I think Your Highness would not have gone so far, except she had surely been culpable . . ."

He wrote partly out of fear. It was Cranmer who had divorced Henry from Catherine and married him to Anne; her guilt might bring him to grief. So in his letter he was careful to write: ". . . if she prove culpable, there is not one that loveth God . . . but must hate her."

And, too, he wrote the letter out of anxiety about the fate of the Reformation of the English Church. Anne was naturally an ally of the enemies of the Pope. And Cranmer was by now a Protestant in his heart. He feared that Anne's fall might trigger a reaction against the Reformers: "I trust that Your Grace will bear no less entire favor unto the truth of the Gospel than you did before; forsomuch as Your Grace's favor to the Gospel was not led by affection unto her, but by zeal unto the truth."

But Cranmer also wrote to Henry out of loyalty to Anne Boleyn, who had been a true friend to him:

"Now I think that Your Grace best knoweth, that next unto Your Grace I was most bound unto her of all creatures living. Wherefore I most humbly beseech Your Grace to suffer me in that, which both God's law, nature, and also her kindness, bindeth me unto; that is, that I may . . . pray for her, that she may declare herself inculpable and innocent."

Before he could send this letter to Henry, Cranmer was summoned to the Star Chamber to meet with a number of the

King's Council. There they told him of evidence they had against Anne Boleyn, of what the King intended to do about it, and the part Cranmer was ordered to play. Cranmer's small store of courage collapsed. He hurried home and added a new ending to his letter to Henry:

"I am exceeding sorry that such faults can be proved by the Queen, as I heard of their relation. But I am, and ever shall be, your faithful subject.

"Your Grace's most humble subject and chaplain . . ."

It should be remembered that whatever Cranmer did— through all his bending and detouring and backtracking—he believed the cause of the English Reformation depended upon his survival.

Anne Boleyn was all alone now.

4

BY HER SECOND DAY in the Tower, Anne Boleyn fought free of her initial terror. She managed, for a time, to persuade herself that everything would turn out all right. She was more cheerful and ate heartily. When Kingston visited her after supper he was surprised to find her positively merry.

"To be a queen, and so cruelly handled," she sighed, and then gave a light, confident laugh. "But I think the King does it to prove me . . . I shall have justice."

"Have no doubt therein," Kingston assured her.

"If any man accuse me," Anne went on determinedly, "I can but say Nay. And they can bring no witness [against me]." Her false cheerfulness began to crumble as she thought about it, and the nagging doubts returned: "If I die you shall see the greatest punishment for it that ever came to England. And then I shall be in Heaven, for I have done many good deeds in my days."

She never wept before anyone again. But her agony of

spirit under such uncertainty revealed itself in emotional imbalance. She swung erratically from defiance to cheerful hopefulness to fear to stoic acceptance of the inevitable.

"One hour she is determined to die," Kingston wrote, "and the next hour much contrary to that." A copy of a letter from Anne to Henry on the fourth day of her imprisonment was later found among Cromwell's papers. It seems to have been written after someone came to her from Henry with a hint that she might save her life by confessing her guilt. Though it has never been established beyond doubt that she actually sent this letter, it would seem to be an accurate reflection of her mood at the time:

To the King, from the lady in the Tower—

Your Grace's displeasure and my imprisonment are things so strange unto me, that what to write or what to excuse I am altogether ignorant. Whereas you send unto me, willing me to confess a truth, and to obtain your favor . . . I rightly conceived your meaning; and if, as you say, confessing a truth indeed may procure me safety, I shall with all willingness and duty perform your command.

But let not Your Grace ever imagine that your poor wife will ever be brought to acknowledge a fault where not so much as a thought thereof proceeded. And to speak a truth, never Prince had wife more loyal in all duty, and in all true affection, then you have ever found in Anne Boleyn; with which name and place I could willingly have contented myself, if God and Your Grace's pleasure had been so pleased. Neither did I at any time so far forget myself in my exaltation or received queenship, but that I always looked for such an alteration as now I find; for the ground of my preferment being on no surer foundation than Your Grace's fancy, the least alteration I knew was fit and sufficient to draw that fancy to some other subject.

You have chosen me from a low estate to be your queen and companion, far beyond my desert or desire. If then you found me worthy of such honor, good Your Grace, let not any light fancy or bad counsel of mine enemies withdraw your princely favor from me; neither let that stain, that unworthy stain, of a disloyal heart towards Your Good Grace, ever cast so foul a blot on your most dutiful wife, and the infant princess, your daughter [Elizabeth].

Try me, good King, but let me have a lawful trial; and let not my sworn enemies sit as my accusers and my judges; yea, let me receive an open trial, for my truth shall fear no open shames . . .

But if you have already determined of me; and that not only my death, but an infamous slander, must bring you the enjoying of your desired happiness, then I desire of God that he will pardon your great sin therein . . . and that He will not call you to a strict account for your unprincely and cruel usage of me.

Anne Boleyn reminded Henry that they must both ultimately face God—"in whose judgment, I doubt not (whatsoever the world may think of me) mine innocence shall be openly known and sufficiently cleared." Her letter continues:

My last and only request shall be that myself may alone bear the burden of Your Grace's displeasure; and that it may not touch the innocent souls of those poor gentlemen who, as I understand, are likewise in strait imprisonment for my sake.

If ever I have found favor in your sight, if ever the name of Anne Boleyn hath been pleasing to your ears, then let me obtain this request; and so I will leave to trouble Your Grace any further; with mine earnest prayers to the Trinity, to have Your Grace in his good keeping, and to direct you in all your actions.

From my doleful prison in the Tower, the 6th of May.

Anne Boleyn.

If such a letter was actually sent, if Anne's husband ever received it, it had no effect upon the onrush of doom. Henry was already making secret arrangements to marry Jane Seymour before the month was out. The end was not long in coming. Four of the men accused of carnal knowledge of the Queen—Norris, Weston, Brereton and Smeaton—were tried at Westminster just ten days after the arrest of Anne Boleyn. Only Smeaton, the court musician, pleaded guilty. All four were found guilty and sentenced to be executed. This is all that is known of their trial.

Three days later, Anne's brother was tried, pleaded innocent, and was found guilty. He was condemned to death.

On the same day, Anne Boleyn had a separate trial, inside the Tower. She had herself under rigid control when Kingston conducted her from her rooms. With quiet dignity, she faced the special commission headed by the Duke of Norfolk. Her own father had been among the judges who had condemned the first four men for adultery with her. He was excused from duty this day.

There is no record at all of the evidence presented at Anne Boleyn's trial. We have only the charges against her, the fact that she pleaded not guilty, and the sentence.

The main charges against her were adultery, incest and treason. The various counts in the indictment were specific and detailed:

"Indictment . . . that whereas Queen Anne has been the wife of Henry VIII for three years and more, she, despising her marriage, and entertaining malice against the King, and following daily her frail and carnal lust, did falsely and traitorously procure by base conversations and kisses, touchings, gifts, and other infamous incitations, divers of the King's daily and familiar servants to be her adulterers . . .

"That she procured, by sweet words, kisses, touches, and otherwise, Hen. Norris . . . to violate her . . . and they had illicit intercourse at various times . . .

"Also the Queen . . . procured and incited her own natural brother, Geo. Boleyn . . . to violate her, alluring him with her tongue in the said George's mouth, and the said George's tongue in hers . . ."

After listing similar counts of adultery with the other three men, the indictment went on:

"The Queen and these other traitors conspired the death and destruction of the King, the Queen often saying she would marry one of them as soon as the King died, and affirming that she would never love the King in her heart."

Anne firmly denied these charges—and was found guilty. According to one report of the trial, one of the judges became so distressed by the proceedings against her that he pleaded sickness and left. According to another, the Duke of Norfolk shed tears as he pronounced sentence upon his niece:

"Judgment—that the Queen be taken by the Constable back to the King's prison within the Tower; and then, as the King shall command, be brought to the green within the said Tower, and there burned or beheaded, as shall please the King."

Anne was to be beheaded. The more painful alternative of being burned to death was only an implied threat—to insure that she would make no statements in her final days that would be harmful to Henry.

"When the sentence was read to her," Chapuys informed the Emperor three days later, "she received it quite calmly, and said that she was prepared to die, but was extremely sorry to hear that others, who were innocent and the King's loyal subjects, should share her fate and die through her. She ended by begging that some time should be allowed for her to prepare her soul for death. . . . Although the generality of people here are glad . . . still a few find fault and grumble at the manner in which the proceedings against her have been conducted, and the condemnation of her and the rest, which is generally thought strange enough."

There exists another account, written three weeks after the trial, of Anne's behavior on hearing the death sentence pronounced against her by Norfolk: "Her face did not change, but she appealed to God whether the sentence was deserved;

then, turning to the judges, she said she would not dispute with them, but believed there was some other reason for which she was condemned than the cause alleged, of which her conscience acquitted her, as she had always been faithful to the King. But she did not say this to preserve her life, for she was quite prepared to die. Her speech made even her bitterest enemies pity her."

Kingston returned Anne to her quarters to await execution. While she waited, Henry was making sure that there would be no question about the title of his next queen. Henry called upon Cranmer to rule that Anne Boleyn had never been his legal wife. Cranmer did so, with dispatch. The grounds for Cranmer's sentence have never been revealed. The most likely grounds would have been that Henry's marriage to Anne was not acceptable in the eyes of God, because he had previously had carnal knowledge of Anne's sister, Mary Boleyn.

Now Anne knew the worst. If she had never been Henry's true wife, she had never had a right to her title. She would not even have the consolation of dying a queen—and her daughter Elizabeth was officially a bastard.

5

"No PERSON ever showed greater willingness to die," Chapuys wrote of Anne Boleyn after a secret conversation with one of her female attendants. Anne had aimed as high as it was possible for any woman to rise. Against stupendous odds, she had won—only to have the fruit of that victory rot in her hands. The taste of her defeat was too bitter to live with. Kingston confided to Cromwell: "I have seen many men and also women executed, and all have been in great sorrow. To my knowledge, this lady has much joy and pleasure in death."

When the hour of her execution was slightly delayed,

Anne grumbled, "I am very sorry for it. I thought to be dead by this time, and past my pain."

When she learned that Henry had imported a skillful executioner from the Continent to behead her with a sword, she laughed—and, encircling her slender neck with her hands, she said it should prove an easy job: "I hear the executioner is very good—and I have a little neck."

Cranmer was appointed to be her confessor, and spent some time with her in the Tower. Before taking the Sacrament, and again after receiving it, Anne swore "on the damnation of her soul, that she had never been unfaithful to the King." It was said that on the day of Anne's execution, Cranmer wept uncontrollably and declared: "She who has been the Queen of England upon earth will today become a Queen of Heaven."

Two days after her trial, Anne Boleyn watched from her window as her brother and the other four men who had been found guilty of committing adultery with her were led out to Tower Hill to face the headsman. Upon the scaffold, Norris, Weston, Brereton and Anne's brother acknowledged publicly that they deserved to die for their general sinfulness. This was something expected of them, to prevent retaliation against their surviving families. But all four bent to the headsman's ax still holding to their previous claims that they were innocent of the specific charge of adultery with Anne. Only Mark Smeaton, the court musician, had confessed to that charge. He went to his death without repudiating his confession.

When Anne learned of this she cried out, "Did he not exonerate me, before he died, of the public infamy he laid on me? Alas! I fear his soul will suffer for it."

The next day it was her turn to die. She spent the morning in prayer. She swore to the end that she was innocent. But when the time came to leave her chambers it was apparent to all around her that she was looking forward to being finished with the misery life had become for her.

Shortly before noon, wearing an ermine cloak and hood,

Anne Boleyn was led out to the Tower Green where the imported executioner awaited her with his sword. Attended by four young ladies, she walked slowly and appeared exhausted. She kept looking around her, as if aware that she would never see the world again. She appeared a bit dazed, but conducted herself to the last with composure. Upon the scaffold she asked for permission to address the people assembled to watch her execution. Her request was granted.

"Good Christian people, I am come hither to die," she said in a voice that trembled with weakness at first, but slowly gathered strength: "According to the law and by the law I am judged to death, and therefore I will speak nothing against it. I am come hither to accuse no man, nor to speak anything of that whereof I am accused." She smiled upon the gathering as she went on: "I pray God to save the King, and send him long to reign over you, for a gentler or more merciful prince was there never. To me he was ever a good and gentle sovereign lord."

She begged that anyone whom she had not treated well would forgive her, and concluded:

"If any person will meddle with my cause, I require him to judge the best. And thus I take my leave of the world and of you. I heartily desire you all to pray for me. To God I commend my soul."

Anne turned to her female attendants. They removed her cloak. She gave them her hood, and put on a cap, gathering her hair up inside it to bare her neck. She knelt upon the scaffold, carefully arranging the hem of her gown around her feet as she took her last look at the world. Her eyes were covered with a linen cloth. The other ladies drew back, weeping.

"Oh, Christ," Anne Boleyn prayed, "receive my spirit!"

The executioner's sword came down on her bared neck, severing her head from her body.

The instant her head fell, one of her female attendants

snatched it up and wrapped it in a white cloth. She carried the head to the chapel for burial, followed by the other three women carrying the rest of Anne Boleyn's body.

The next day King Henry was betrothed to Jane Seymour. He married her ten days later. The following year Jane Seymour gave Henry another child. This time it was a boy— and he was given the name Edward. Jane Seymour never recovered from the effects of childbirth. She died twelve days later. But her son survived.

At last Henry had what he had been waiting for since the day he married Catherine, twenty-eight years earlier—a legitimate male heir to his throne.

Part Four

THE
FINAL VERDICT

*Men never do evil so completely and cheerfully
as when they do it from religious conviction.*
—Pascal

Chapter Twelve

FATHER AND SON—
THE OUTSTRETCHED HAND

THERE IS NO RECORD of Henry's having fathered any child out of wedlock except the Duke of Richmond. This boy was said to resemble his father in looks and personality. He was liked as Henry had been liked as a boy, and for the same reasons. Whether these similarities went deeper only time could have revealed. Richmond died in the same year that Henry's first wife was buried and his second wife was beheaded.

So it was that when Henry's third wife died the following year, the King was left with just three children: Catherine's daughter Mary, Anne Boleyn's daughter Elizabeth, and Jane Seymour's son Edward.

It was these three children who were to render the final verdict in the controversy over Henry's divorce from his first wife.

Despite what had happened in the past—and what was to happen in the future when it came their turn to rule—Henry's children acquired a strong affection for each other during the remainder of Henry's lifetime. Mary felt no personal enmity toward Anne Boleyn's daughter. Elizabeth was only a baby, not responsible for the harm done by her mother. And now she, too, was declared illegitimate.

Elizabeth was only four years old when she was told that she was no longer to be addressed as Princess. "How haps it," she asked, "yesterday my Lady Princess, and today but my Lady Elizabeth?" Mary was in a position to understand the child's confusion. Seventeen years older than Elizabeth, Mary came to enjoy petting and playing with her pretty, precocious little half sister. She even took pride in the girl's early accomplishments. Two months after the death of Anne Boleyn, Mary was writing to Henry:

"My sister Elizabeth is in good health (thanks to our Lord) and such a child toward, as I doubt not, but Your Highness shall have cause to rejoice in time coming."

When little Elizabeth wearied during the long ritual of Edward's christening, it was Mary who impulsively reached out and took her hand, holding it through the remainder of the ceremony. Edward grew into childhood loving and loved by his two sisters. A natural companionship developed between him and Elizabeth, who were so close in age. The much older Mary treated Edward with an almost motherly warmth, and he delighted in her visits and presents.

All three shared something else: an awed love for their royal father—whose wrath had become terrible but whose charm remained irresistible.

Mary's passive resistance to her father's demands wavered and broke after the death of Anne Boleyn. What Henry called Mary's "Spanish pride" was finally crushed by his punishing anger, and she made her bitter capitulation.

Henry forgave her the past. Once more he treated her with all the warmth of a naturally generous father. Easygoing affection was still a basic part of Henry's character. Only opposition pricked him to cruelty.

Jane Seymour had learned this lesson quickly. At first, she had tried urging her new husband toward a reconciliation with Rome. Henry had responded by reminding her threateningly of how he had disposed of his second wife. His tone frightened her into humble submission. She ceased to meddle, and was rewarded by kindly treatment from her husband in the scant span of life left to her.

Jane Seymour's mistake had been a natural one, shared by others in England and on the Continent. With both Catherine and Anne Boleyn gone, even the Pope assumed that there was nothing left to prevent a reconciliation between Henry and himself. He soon learned his error. Henry considered the divorce between England and Rome as final as he considered the divorce between himself and the late Catherine. Shortly before Jane Seymour's death, he proclaimed his decision—in blood.

The north of England had resented Henry's religious "reformation" from the start. This reformation continued after the beheading of Anne Boleyn. Archbishop Cranmer was introducing innovations with a strong scent of Lutheranism. Cromwell was being called "the hammer of the monks" for the methodical efficiency with which he was wiping out the smaller monasteries and convents. The monastery treasuries were looted; their lands were rented or sold; the religious orders were being disbanded; the buildings themselves were being turned into warehouses and barns, or being torn down to provide building materials for other purposes.

The monks and nuns who were turned out—though most were given government pensions if they did not resist—fanned the seething discontent in the north to the exploding point. The explosion was finally triggered by reports of increased taxation. Thousands of armed men rose up in Lincolnshire and Yorkshire to resist the tax assessors. Their revolt came to be known as the Pilgrimage of Grace when they went on to demand an end to religious innovation and a return to the old ways of the Church. They wanted the monasteries restored, Cromwell and other lowborn officials stripped of authority, Cranmer dismissed along with other "heretic" bishops.

Henry responded with an outburst of vengeful fury against the rebels. He ordered the raising of a "great army to invade their territories . . . to burn, spoil, and destroy their goods, wives, and children with all extremity." Henry had no standing army. In such emergencies he had to depend on his popularity to draw sufficient numbers of fighting men to his banner. That he *was* still popular with most of the realm was proved by the swiftness with which the dukes of Suffolk and Norfolk were able to assemble troops to oppose the rebels.

Most of these rebels did not seek to depose Henry—only to force changes on him. Henry was no more anxious than they for a civil war. When the rebels asked for an opportunity to explain their grievances, he choked down his rage and answered softly: "To show our pity, we are content, if we find you penitent, to grant you all letters of pardon . . ."

The leader of the rebellion, a lawyer named Robert Aske, met with Henry. He accepted the King's vague assurance that he would give thought to the complaints. Bitter controversy broke out among the rest of the rebels over whether they should accept Robert Aske's belief in Henry's good will. The rebellion fell apart. Henry pounced like a tiger, forgetting his promise of a general pardon. Robert Aske and over two hundred other rebels were executed for their treason—beheaded, burned or hung in chains.

This frightening example sufficed. The rebellion known as the Pilgrimage of Grace collapsed. England continued on the course its King had set. Henry destroyed the Canterbury shrine of the martyred St. Thomas Becket, symbol for two and a half centuries of the triumph of Church authority over an earlier King Henry of England. Cromwell went on from the smaller monasteries to sack and abolish the large ones. Cranmer obediently made clear Henry's concept of "our Church of England"—Catholic, but in no way connected to "the Bishop of Rome." The Bible, in plain English, was for the first time placed open in every church in the country by official decree for laymen to read for themselves.

Pope Paul now understood that Henry intended never to return to the Roman Catholic Church. Once more he began pressing the French King and the Spanish Emperor to act together to depose Henry. For a time it seemed that they might actually dissolve their mutual hatred and do so. Charles, who only a short while ago had challenged Francis to a duel to settle their differences, now accepted Francis' offer of safe conduct through France—and journeyed to Paris for a friendly meeting.

Alarmed, Henry cast about for an alliance with some other power on the Continent. Cromwell took over the job of forming such an alliance with the Lutheran princes and other German rulers opposed to the Pope. Cromwell had repeatedly proved his skill at domestic manipulation. But in trying to extend himself into foreign affairs, he blundered. He compounded his blunder by tampering with Henry's personal life.

Cromwell thought he had a key—in Anne of Cleves—to a league with the antipapal powers of the Continent. Since the death of Jane Seymour, Henry was an eligible widower. Anne of Cleves was unmarried. Her brother was the Duke of Cleves, who, like Henry, was Catholic but against the Pope, and at odds with Emperor Charles. Her brother-in-law was the influential Elector of Saxony. Cromwell encouraged Henry to shore up his international situation by marrying Anne of Cleves.

The great portraitist, Holbein, painted a picture of Anne of Cleves which was sent to Henry. It showed a plain, stolid-looking woman. Henry's ambassador at Cleves praised her gentleness, but added that she was exceedingly tall, had few social graces, and spoke no language but German. Cromwell did not see the pit yawning before him. Though he had never met Anne of Cleves, he insisted that neither the portrait nor the ambassador's description did her justice. "Everyone praises her beauty both of face and body," Cromwell assured Henry. "One says she excels the Duchess of Milan [a renowned beauty of the day] as the golden sun does the silver moon."

Henry agreed to marry Anne of Cleves. She journeyed in state from Cleves to England. When she arrived, Henry hurried to have a look at her. Cromwell waited on tenterhooks to learn Henry's reaction. As he later recalled it, Henry returned from his first meeting with his intended bride looking "heavy and not pleasant." Cromwell asked how he had found Anne of Cleves.

"Nothing so well as she was spoke of," Henry answered ominously. "If I had known as much before as I now know, she should not have come within this realm."

He did not like her and did not want her for his wife. But to refuse to marry her at this point would mean offending her brother and brother-in-law—perhaps driving them to join with Charles against him. Depressed, Henry demanded to know if Cromwell could come up with a remedy for this dilemma. Cromwell could not. Reluctantly, Henry married the woman he privately referred to as "the Flanders mare."

On the morning after the wedding night, Cromwell hopefully asked Henry again how he liked Anne of Cleves.

"As you know," Henry told him pointedly, "I liked her before not well. But *now* I like her much worse." He said he had found her so distasteful that he had not been able to consummate their marriage. "I felt her belly and breasts, and . . . I had neither will nor courage to proceed any further. . . . I have left her as good a maid as I found her. My heart will never consent to meddle with her carnally."

The ground upon which Cromwell stood began to give way. It collapsed completely when it turned out that Henry's marriage to Anne of Cleves had been as politically unnecessary as it was personally unpleasant. Cromwell had pushed Henry into the marriage on the assumption that Charles and

Francis were about to join against him, and that he would need other friends on the Continent. But the reconciliation between Charles and Francis did not last long. Soon after their Paris meeting they were at each other's throats again; and once more each was soliciting Henry's aid and good will.

Henry had a wife he did not like and did not need. He turned savagely on the man responsible. Cromwell was thrown in the Tower, charged with encouraging heresy, usurping powers that did not belong to him, meddling in the King's foreign affairs, and treason. Once again, the only voice to be heard defending one upon whom Henry had leveled his anger was the voice of the Archbishop Cranmer.

He wrote Henry saying that he could not help but be amazed to learn that Cromwell was a traitor—"he who loved Your Majesty (as I ever thought) no less than God; he who studied always to set forwards whatsoever Your Majesty's will and pleasure; he that cared for no man's displeasure to serve Your Majesty; he that was such a servant, in my judgment, in wisdom, diligence, faithfulness and experience, as no Prince in this realm ever had; he that was so vigilant to preserve Your Majesty from all treasons . . . who shall Your Grace trust hereafter, if you might not trust him?"

Cranmer's lone voice had no effect at all. Cromwell's usefulness was ended. He had already navigated the Reformation Parliament along the course Henry had desired. He had created for the operation of civil government an integrated network of efficient machinery that was to last for centuries after him. He had funneled into the hands of the King every drop of money and power that could be squeezed out of the destruction of the monasteries.

In doing all this, Cromwell had generated an atmosphere of frustrated resentment and fury. English Catholics hated him as the instrument which had slashed their Church to ribbons. The nobility and gentry hated him—as much as they had ever hated Wolsey—as a lowborn upstart who had acquired power over them. Henry could appease much of this anger, and satisfy this hatred, in the very act of killing Cromwell out of personal anger.

Cromwell understood this quite as well as Henry. Abjectly, he wrote to Henry: "I, a most woeful prisoner, am ready to submit to death when it shall please God and Your Majesty. And yet the frail flesh incites me to call to Your Grace for mercy and pardon for mine offenses . . . Written at the Tower, with the heavy heart and trembling hand of Your Highness's most miserable prisoner and poor slave, Thomas Cromwell." Below, he added a postscript: "Most gracious prince, I cry for mercy, mercy, mercy."

Hoping against hope, Cromwell eagerly supplied Henry with what he needed to get rid of his fourth wife: a detailed account of his own knowledge that the marriage had never been consummated. Anne of Cleves allowed her sense of self-preservation to triumph over pride. She, too, testified that her marriage was not consummated.

With Anne of Cleves offering no objection, Archbishop Cranmer, obedient as ever to the orders of his King, had no difficulty in declaring the marriage invalid. Henry rewarded Anne of Cleves by enabling her to live out the rest of her life in England, treated with dignity and generosity. He rewarded Cromwell by allowing him a quick death under the headsman's ax.

Lord Surrey expressed the satisfaction of most of England: "Now is that foul churl dead, so ambitious for others' blood. Now is he stricken with his own staff."

2

THOMAS MORE had held back, clinging to his allegiance to Rome, and had died. Thomas Cromwell had pushed forward, reaching for an alliance with the Lutherans, and had died.

Henry assigned religious reformers as tutors to his son Edward, who would be the next king of England. Simultaneously he stopped the English Reformation by enacting the Six Articles, which defined England's religion by law as antipapist—but Catholic.

Henry had taken his position in the middle, between the two opposing faiths. For those trying to follow Henry through his every religious twist and turn, it was extremely confusing. And it was fatal not to follow. The King had made himself sole judge not only of what his people did, but also of what they thought, said and believed.

"Junker Harry means to be God," Luther observed sardonically. "For what Junker Harry wills must be accepted as an article of faith as a matter of life and death."

Henry himself made this crystal-clear at one point: On the same day, and in the same place, he burned three Roman Catholics for treason—and three Lutherans for heresy.

There were many other executions in the remainder of King Henry's reign. He no longer waited for opposition to fully declare itself. He struck out at the first hint of it. "It was," Bishop Latimer was later to explain to Henry's son, in speaking of this period, "a dangerous world; for it might soon cost a man his life for a word's speaking." People were killed for treason, for heresy, for the suspicion of having committed either, and for the possibility that they might be contemplating either. And another queen died for adultery.

Katherine Howard was Henry's fifth wife. He was beginning to get old now, in need of a young, innocent virgin to stir his appetite. Katherine Howard, like Anne Boleyn, was a niece of the power-hungry Duke of Norfolk. Norfolk brought her to the King. She was young, pretty and saucy. Henry was stirred. He married her, and was happy with her until the blow fell.

This time there was no question of an innocent girl's being perjured to death. Henry was genuinely stunned when Cranmer unhappily revealed what he had learned: Katherine Howard had not been a virgin when she married Henry. Henry at first refused to believe it. But the proofs were forthcoming, and they were incontrovertible.

Katherine Howard had been indulging her passion for a variety of men without restraint since a startlingly early age. That was bad enough. What was worse, she had continued to indulge—especially with her cousin Culpeper—after she was Henry's bride and queen.

Katherine Howard went to her death. But upon the scaffold she managed a last nose-thumbing in King Henry's direction. "I die a queen," she announced before bending to the headsman's ax, "but would rather die the wife of Culpeper."

3

HENRY lost his composure so completely as to weep before members of his Council. Katherine Howard's unfaithfulness caused him so much grief, noted one Frenchman at court, "that of late it was thought he had gone mad."

Henry did not go mad. But there could be no doubt that the experience with his fifth wife left him considerably shaken. When he married again it was a sensible marriage, to a sensible little widow named Catherine Parr. His sixth and last wife had been married twice before, and would live on after Henry to marry yet again.

To Henry, Catherine Parr became as much a nurse as a wife. For Henry was now racked by disease and continually tormented by an ulcerated leg that would not heal. With the death of Katherine Howard and his marriage to Catherine Parr he aged rapidly. His appearance became a travesty of the physique of his youth. His once-powerful figure became corpulent. His eyes were suspicious little slits sunk deep in fat. The natural exuberance of his personality displayed itself more and more through violent displays of temper.

And yet, Henry was still capable of real affection when assured that he was not opposed. There was the occasion when he attacked Catherine Parr because he heard that she had expressed an opinion contary to his own. She answered sweetly that though she might dispute with others, she would consider it unwomanly to presume to contradict her husband in any matter. Henry's anger dissolved: "Is it so, sweetheart? Then are we perfect friends."

To Henry's three children, Catherine Parr became a warmly protective stepmother. She was loved in return by them. Mary was not far from her in age, and the two became good friends though they had separate households. The two much younger children, Edward and Elizabeth, were taken into Catherine Parr's household, and experienced with her the first real family life they had ever known.

This common attachment to Catherine Parr served to further strengthen the emotional bond between Henry's three children. At one point, when Edward was eight and he and Elizabeth were temporarily sent to separate households, he wrote to her: "Change of place did not vex me so much, dearest sister, as your going from me. . . . It is some comfort in my grief that my chamberlain tells me I may hope to visit you soon." And to Mary he sent a touching excuse for being remiss in his correspondence: "Although I do not write more often, you must not think me ungrateful, for just as I seldom put on my best clothes, yet I love them more than the others."

In his last years the ailing, aging King did his best to leave his children a secure, united realm. He waged successful war against France and Scotland to prevent these two nations from combining against England. He attempted, with less success, to join Scotland to England by arranging a marriage between Edward and Mary Stuart—who was the granddaughter of Henry's older sister, Margaret, and who inherited the throne of Scotland when only one week old.

Henry tried, also, to put an end to the discord within his own realm that he himself had caused. In his last speech before Parliament he chided the assembled members: "Behold then, what love and charity is amongst you, when one calleth another heretic and Anabaptist, and he calleth him again papist, hypocrite and Pharisee?"

But it was too late for soft words to stem the turbulence that the divorce had churned. England raged with religious controversy. Ambitious politicians, realizing that Henry's place upon the throne would soon be taken over by his very young son, used charges of religious deviation as lethal weapons to eliminate other ambitious politicians.

"Who shall Your Grace trust hereafter," Archbishop Cranmer had asked Henry when Cromwell fell from favor, "if you might not trust him?" The answer, of course, was Cranmer himself. Cranmer and Cromwell had been the two chief instruments responsible for working out Henry's divorce. But unlike Cromwell, Cranmer never let power go to his head.

Though his heart had come to be dedicated to the Reformation, Cranmer never took a step in that direction unless Henry took the step first. And he backed up the instant Henry hesitated. If Henry wished the burning of Lutherans, Cranmer saw to it that they were burned. If Henry wanted the banning of reformers' books, Cranmer had them banned.

When Henry finally lay dying—at the age of fifty-five and after ruling England for thirty-seven years—it was Cranmer he sent for. By the time Cranmer reached him, the King was no longer able to speak. But he stretched out his hand, and Cranmer grasped it in his own. Henry VIII squeezed Cranmer's hand tightly—and died.

4

PRINCE EDWARD and his half sister Elizabeth were together when they learned that their father was dead. The two wept piteously, clinging to each other in their grief.

Elizabeth was thirteen. Edward was nine—and King of England. An extremely handsome child, he was also very bright—with a mind already stuffed with more learning than most people today acquire in a lifetime. He had the grave manner of one trained from birth to accept this responsibility, whenever it came. But he *was* still a child. Even while his father had lain dying, a plot had been hatched to determine who would rule in this child ruler's name.

Henry had tried to provide against this in his will by naming sixteen regents to assist and advise young King Edward VI, and manage the affairs of the country for him until he reached the age of eighteen. One of these regents was Edward's uncle the Earl of Hertford, elder brother of Edward's mother. He was a popular figure, a successful general, and a religious man who shared with Cranmer a desire for the final triumph of the Protestant Reformation in England. He was a man of high, unselfish ideals, who envisioned sweeping political and economic changes which would improve the lot of every person in the realm. But as he saw it, the only way to insure the success of these ideals was to seize absolute power for himself.

Proving himself as forceful in politics as he was in war, he seized the power he wanted. He neutralized the authority of the other regents—and had himself proclaimed Duke of Somerset, guardian of the boy king, and protector of the realm.

From the first, Somerset's position as virtual dictator was coveted by two other men. One was John Dudley, who had been named Earl of Warwick—and whose father had been that Dudley whom Henry VIII had executed long before as one of his first official acts as king. The other was Somerset's own young brother, Thomas Seymour.

Warwick approached his objective with the cunning, patient stealth of a stalking lion—and went undetected until he finally had his prey between his teeth. But Thomas Seymour, with enormous ambition and little common sense, pounced for all to see—tripping himself, and almost dragging young Elizabeth down to ruin with him.

Thomas Seymour was a handsome, fun-loving swashbuckler of a man. His easygoing charm had a devastating effect on many women—and misled some politicians into believing that he knew what he was doing. Seymour was also a man who was avid for power—and after Henry died, Somerset had all of it. Seymour did not think this was fair. After all, Jane Seymour had been *his* sister, too; Edward was as much *his* nephew as Somerset's.

Seymour plunged into a many-faceted intrigue to undermine the authority of his older brother. Somerset had made him admiral of the Navy. Seymour used this position to build up a large cash reserve. He allowed pirates the freedom of the Channel, in exchange for a share of their loot. Much of this money he used to gather hidden stores of arms, and to induce various men to be ready to back him up when the time came for him to seize power. As a prelude to this contemplated coup, Seymour insinuated himself into young King Edward's favor.

He began drawing Edward away from Somerset, to himself. This was not difficult to do. Within months of becoming king, Edward lost most of the affection he had for his older uncle. He resented the fact that Somerset had made him a king in name only, never consulting him; not allowing him even the illusion of having anything at all to say about whatever Somerset did in his name. He chafed under Somerset's strict guardianship, his miserly refusal to allow him even enough money to give presents to his friends and tutors.

Seymour flattered the boy by declaring that a king, no matter how young, was qualified to rule. He hinted that if Edward were to help make *him* the protector, instead of Somerset, he would regard the boy's every word as divine wisdom. He secretly lent Edward money, till the boy was deeply in debt to him.

At the same time, Seymour was groping for power from a different direction—through a marriage calculated to improve his status. His first approaches were to Mary and Elizabeth. Rebuffed, he fastened his attentions on Henry's widow, Catherine Parr. She still bore the title of Queen; and Edward was deeply attached to her, as the only mother he had known. She had been infatuated with Seymour before Henry had made her his wife. It was not difficult to reassert his power over her. He waged a swift, irresistible courtship, and they were married.

There were those who were shocked at Catherine Parr's marrying again so soon after Henry's death. One of these was Mary, and she unburdened her feelings about the marriage to Elizabeth. Mary was thirty-one years old at the time; Elizabeth had just turned fourteen. But Elizabeth's was the cooler head. Her answering letter to Mary was masterful.

She began by soothing Mary with the assurance of shared feelings: "You are very right in saying that, our interests being common, the just grief we feel in seeing the ashes, or rather the scarcely cold body of the King, our father, so shamefully dishonored by the Queen, our stepmother, ought to be common to us also."

But then, with the worldly wisdom of a born politician, Elizabeth went on to advise Mary to accept the marriage without complaint: ". . . neither you nor I, dearest sister, are in such a condition as to offer any obstacle thereto without running heavy risk of making our own lot much worse than it is. . . . We have to deal with too powerful a party, who have got all authority into their hands, while we, deprived of power, cut a very poor figure at court. I think, then, that the best course we can take is that of dissimulation. . . . Let us console ourselves by making the best of what we cannot remedy. If our silence do us no honor, at least it will not draw down upon us such disasters as our lamentations might induce."

It was a remarkable letter to have been written by a four-teen-year-old girl. That she *was* remarkable had already been realized by everyone close to her for quite some time. "Her mind has no womanly weakness," her tutor had written to a scholar friend, "her perseverance is equal to that of a man, and her memory long keeps what it quickly picks up." At a very young age Elizabeth understood that she was brilliant—and took pride in it. When Edward asked for a portrait of her, she sent it with these words: "For the face, I grant I may well blush to offer, but the mind I shall never be ashamed to present."

She had no reason to be ashamed of her face, either, or the rest of her appearance. She was tall, with a good figure, red-gold hair, and beautiful long-fingered hands. She had her father's passion and political cunning, her mother's patient determination, and an ingrained sense of caution learned from the fate of her mother and others since.

But she was still an adolescent, experiencing a surge of new emotions that she did not yet have the experience to cope with. She had been living with Catherine Parr, and continued to do so after her stepmother married the dashing Thomas Seymour. He was more than twenty years older than Elizabeth. But living under the same roof with him, in daily contact with him, the young girl began to show signs of being no more immune to his overpowering charm than women older than herself. Seymour was quick to recognize these signs in her, and to take advantage of them.

Even for that lusty, bawdy age, Seymour's familiar attentions to Elizabeth were remarked upon. He took to slapping her buttocks when he passed her. In the mornings he would go to her bedroom in his nightgown, awaken her by drawing her bed curtains and tickling her. He would pretend he was about to climb under the covers with her, and Elizabeth would scurry away across the bed, giggling. That Elizabeth was thrilled and excited by his boisterous playfulness with her became obvious. The mere mention of his name would be enough to make her blush.

For a time, Catherine Parr apparently accepted Seymour's attitude that these romps were only innocent horseplay. One morning she even joined her husband in it, both of them climbing into bed with Elizabeth and tickling her into a fit of laughing hysterics. On another occasion, in the garden, the Queen held a struggling, squealing Elizabeth while Seymour cut the girl's gown to ribbons. But after Catherine Parr became pregnant, she began to be disturbed about her husband's behavior with her stepdaughter. Elizabeth moved out of the Seymour household, and she and her stepmother remained good friends—at a distance.

Catherine Parr died in childbirth. Thomas Seymour was free to marry again—and Elizabeth and her governess began to make a joking game of the possibility that Elizabeth might be his next wife.

But Edward had cooled toward his younger uncle. Seymour tried to use the boy King's debt to him to pressure Edward into writing a statement that he wanted Seymour to replace Somerset as protector. Edward balked at this, became wary, and saw to it that Seymour never had another chance to talk to him without the presence of witnesses.

Frustrated, needing a private interview with Edward to push his secret plot to its conclusion, Seymour committed the ultimate folly. He obtained duplicate keys to all the door locks in Edward's quarters, and slipped inside one night after the boy was asleep. He managed to get past the guards, down the dark corridors, and through the door to Edward's bedroom. But as he stepped inside, he tripped over Edward's pet dog. It leaped at him, barking furiously. In a panic, Seymour snatched out his pistol and shot the dog. The noise woke everyone within hearing. After that, Seymour's fate was in the hands of others.

There was Edward, his face freezing as he looked from his

slain dog to the dazed slayer. There were armed guards, taking Seymour away. There was an investigation which dug out his secrets—his dealings with the Channel pirates, his hidden arms, his conspiracy to seize power. And his "bedroom romps" with Elizabeth.

Elizabeth found herself under a heavy cloud of suspicion. She was questioned harshly and repeatedly about what she knew of Seymour's plans, about whether she had actually been indiscreet with him, about whether she had planned to marry him without permission. Elizabeth finally managed to escape from under the cloud—by an adroit combination of indignant denials, tears, refusals to answer further questions, and demands that those slandering her name be punished. But Seymour went to the headsman's block.

Some have believed that the vicious migraines which were to plague Elizabeth for most of her life began with Seymour's death. But his fate was a lesson in caution that she added to those learned in earlier childhood. Whatever she felt when Seymour was executed, all that she said was: "This day died a man of much wit and very little judgment."

5

SOMERSET did not long survive his impetuous younger brother. For working carefully behind his back all the time was John Dudley, Earl of Warwick—a devious, cold-blooded self-seeker. It was Warwick who, without drawing attention to himself, helped to magnify all of England's discontents and focus them on Somerset.

There was a great deal of discontent for him to work with. Somerset had dedicated himself to completing the English Reformation which Henry had unwittingly begun in his struggle for the divorce. As Catholic bishops died, they were replaced by reformers. Cranmer, with Somerset's backing, introduced Protestant innovations into the churches to take the place of the old Catholic rituals. The most effective of these was the Book of Common Prayer, most of it written by Cranmer himself, which made church services something no pious Catholic could swallow. But at the same time, Somerset urged and practiced tolerance toward Catholics and other religious dissenters, and by repealing the heresy and treason laws which had taken so many lives under Henry, he gave Englishmen back the freedom of thought and speech they had lost.

Somerset's religious changes infuriated Catholics. His leniency alarmed reformers. Upon religious controversy was being heaped the growing economic misery of the common people. The export of wool was England's most profitable business, and the moneyed classes were enclosing more and more of the country's land as pasture for their sheep. In the process, they were evicting thousands of tenant farmers, who had nowhere else to go with their starving families—and no other way to earn a living. Finally, the peasants of the eastern counties rose in rebellion against the enclosure system.

About the same time, a religious rebellion broke out in another part of the country, where the peasants revolted against the substitution of the liturgy of Cranmer's prayer book for the old Catholic Mass. "We will have the religion of our fathers!" was their cry—and it proved contagious.

Somerset appealed for arbitration of the disputes by peaceful, legal means. "Content yourselves, good people," he pleaded. "Do not with this rage and fury drive yourselves to the sword. . . . If anything be to be reformed in our laws, the Parliament is near at hand."

But the moderate voice calling for patient compromise was no more listened to in that age than it has been in any age since. Both rebellions—the economic one and the religious one—spread and gathered momentum across the country. And it was Warwick who gained the major credit for breaking the back of the combined rebellions—when the army of imported mercenaries he led slaughtered thirty-five hundred peasants in a single day. With that convulsive display of strength, Warwick became the man around whom everyone who disliked Somerset rallied.

Warwick had carefully laid the groundwork for this day. Catholics were certain he was a Catholic. Reformers assumed he was one of them. The combination assembled by Warwick proved too much for Somerset. He fell from power—after just two and a half years as protector—and was eventually executed on Warwick's trumped-up charge of treason.

Now that he had the reins of government secure in his own hands, Warwick decided that the future belonged to the Protestant reformers—and proceeded to push the English Reformation to extremes which Somerset had never contemplated. Cranmer's prayer book had offended many reformers by not going far enough. Under Warwick, Cranmer revised his book to eliminate what offended the new reigning spirit.

His second prayer book left the Catholics nothing—not even a memory. The Forty-two Articles of religion laid down the specifics of the Protestant creed for all Englishmen. And Warwick's new Act of Uniformity made it a crime for any English man or woman to fail to attend the new services every Sunday.

Like Somerset, Warwick did all this in the name of King Edward. Unlike Somerset, Warwick actually allowed the boy a leading role in what was being done, encouraging him to assert his authority as king.

For one thing, Edward was growing older—and harder to ignore. For another, Edward agreed with what Warwick was doing. This was partly due to Warwick's ability to dominate the boy's mind. It was also the result of his upbringing. He had been reared by reformers. The new faith was ingrained in him—and he believed in his heart that to tolerate any other faith was an offense to God that would be punished by damnation.

His sister Elizabeth had received the same religious upbringing. But Elizabeth had a mind of her own—and was much more flexible in her faith. She considered with indifference arguments about the fine points of exactly how one should worship God. "There is only one Christ Jesus," she was to say later, "and one faith. The rest is a dispute about trifles."

Edward was more like his other sister. He was as rigid in his Protestantism as Mary was in her Catholicism. Each regarded the faith of the other with genuine horror. Warwick's Act of Uniformity was the law of the land. The love Edward and Mary felt for each other could not prevent them from being drawn inevitably into a religious clash of wills. This was their legacy from Henry's divorce. Their love only made the clash more bitter.

Chapter Thirteen

THE DAUGHTERS—
AND THE BURNING HAND

MARY HAD CONTINUED to worship as a Catholic after her brother became king—pointing out that their father, in spite of his break with Rome, had died a Catholic. So long as the tolerant Somerset ruled, Edward looked the other way, allowing his sister to observe her religion unmolested. But once Warwick gained ascendency over the kingdom—and over Edward's mind—increasing pressure was applied to force Mary to conform to Cranmer's prayer book and Warwick's Act of Uniformity. She was asked to stop celebrating the Mass in her household.

The controversy over Henry's divorce from Catherine of Aragon was far from ended. It still raged—partly out in the open, but mostly, for the moment, in the form of underground crosscurrents. These currents now began to swirl with increasing turbulence around the focal issue of "Mary's Mass."

Mary refused to give up her Mass. "And as for your new books," she told one of the Protestant bishops, "I thank God I never read any of them. I never did, nor ever will do."

Edward was deeply troubled, caught between his fondness for Mary and his fear of God's displeasure. He asked Elizabeth to try to persuade Mary to comply with the new religious laws. The supple Elizabeth did not refuse to do so; she merely neglected to get around to it.

The King's Council, prodded by Warwick, commanded Mary to cease attending Mass. Mary ignored the command—and she had powerful support for her resistance from abroad. Emperor Charles was angered by the pressure being applied to make Mary act against her conscience. On his command, the new Spanish ambassador complained of it to some of the King's Council. He was given a curt answer: "You talk much of the Lady Mary's conscience. You should consider that the King's conscience will receive a stain if he allows her to live in error."

This last worried Edward more and more as time went by. This—and the thought that Mary's example might lead the rest of his people into sinful error. He was thirteen now, fully convinced that he was God's spokesman in England. When next Mary visited the court, Edward sternly asked her to explain rumors that she was still regularly attending Mass in her household.

Mary started to answer, stared at her young brother, and suddenly began to cry. Edward lost his composure. Tears flowed from his own eyes. "I think no harm of you,"—he wept, and the interview between them was abruptly broken off.

Later, Mary made the mistake of telling certain members of the court: "I hope His Grace will not be wroth with me, until he is of an age to judge for himself."

Warwick made certain that the boy felt the full sting of her remark. Edward's attitude toward his elder sister began to harden. His Council sent Mary another letter warning that she must conform to the religion of England. Edward wrote a postscript to it:

"Truly, sister, I will not say more and worse things, because my duty would compel me to use harsher and angrier words. But this I will say with certain intention, that I will see my laws strictly obeyed, and those who break them shall be watched and denounced."

The anger of her young brother, and the knowledge of what it could lead to, kept Mary in a constant state of tension. The illnesses and attacks of nerves, to which she had become subject after her father had divorced her mother, now began to recur with increasing frequency. But she continued to hear Mass in her household—relying on Emperor Charles to save her from being punished for it. Her defiance, and rumors that Charles was planning to spirit her out of England, resulted in a formal summons for her to appear before the King and his Council.

Mary rode into London with a large number of noblemen—each wearing a black rosary. Thousands of Londoners massed in the streets to cheer her. At the palace, she faced Edward and his Council resolutely. The Council reminded her that she was a subject of the King; in religion, as in all else, a subject was bound to bow to the King's knowledge and understanding.

Mary looked at her young brother. "Although Your Majesty is of great understanding, yet riper age and experience will teach you much more yet." Again she had touched his sore spot.

"You also may have something to learn," he snapped. "No one is too old for that."

Mary answered that she could not learn to change her religion—which had been the religion of their father, and the religion their father had wanted the nation to continue in. She charged that Edward's advisers were pushing him in a direction contrary to what Henry had wished for his realm, a direction that was harmful to Edward.

At this Warwick burst out: "How, now, my Lady! It

seems Your Grace is trying to show us in a hateful light to the King our master."

"I have not come hither to do so," Mary flung at him, "but you press me so hard I cannot dissemble." Turning back to her brother, she requested again that she be allowed to continue hearing Mass in her household. "I had hoped that . . . because I am Your Majesty's sister, Your Majesty would have allowed me to continue in the old religion. There are two things only, soul and body. My soul I offer to God, and my body to Your Majesty's service. May it please you to take away my life rather than the old religion."

"I desire no such sacrifice," Edward protested. Nor did he. But he was quite determined by now to make her give up her Mass. Mary was given permission to withdraw. Then Edward left to his Council the task of deciding how to force Mary to capitulate.

Before they could come to a decision, their deliberations were cut short by the arrival of a threatening note from Charles. Edward recorded in his diary: ". . . the Emperor's ambassador came with a short message from his master of war if I would not suffer his cousin the Princess to use her Mass." Warwick and the rest of the Council decided to be practical and let Mary have her Mass—for the time being. But they knew King Edward was too convinced he was following God's will to be swayed from it by the threat. So they sent for Archbishop Cranmer and bishops Ridley and Ponet. These leaders of the Church were told that if Charles invaded England he might very well conquer it; then everything they had done would come to naught. In such a situation was it permissible to let Mary continue in her false religion?

Cranmer and the others decided: ". . . although to give licence to sin was sin, yet . . . to suffer and wink at it for a time might be borne." They all met with Edward, informed him of the situation, and of the bishops' decision.

Edward stared at Cranmer, unable to believe he would

agree to such a thing. "Is it lawful by Scripture," he demanded, "to sanction idolatry?"

"There were," the discomfited Cranmer pointed out, "good kings in Scripture, Your Majesty, who allowed the hill altars, and yet were called good."

"We follow the example of good men when they have done well," Edward answered sharply. "We do not follow them in evil. David was good, but David seduced Bathsheba and murdered Uriah. Are we to imitate David in such deeds as these?"

But the combined insistence of Warwick and Cranmer, of the Council and the bishops, prevailed. Edward gave in to them, but with bitter tears and the fear that he was offending God by doing so. For a time Mary was allowed her Mass. Then the struggle of faiths between her and Edward began again. Finally they reached an uneasy sort of truce. Mary had her Mass, but in strict secrecy, with only a few intimates attending.

"When the King comes to such an age as he may be able to judge these things himself," Mary said at one point in their struggle, "His Majesty shall find me ready to obey his orders in religion."

But Edward never reached that age. Illnesses that the doctors could not diagnose properly began to drain his strength during his fifteenth year. By the time he was sixteen he was bedridden. It became increasingly certain that he would not live to reach the age of seventeen. In a last attempt to save him, he was apparently given overdoses of arsenic. For a time he rallied. Then his hair and fingernails began to fall out, and he sank rapidly.

Warwick became desperate. According to Henry VIII's will, when Edward died, the Catholic Mary would inherit the throne. Warwick was quite sure that if this happened, his life would be forfeit. He considered using his power to have Mary bypassed. Elizabeth was next in line. She was a Protestant, but too self-assured and independent. Warwick wanted someone he could manage. So he decided on sixteen-year-old Jane Grey. The royal Tudor blood flowed in her. She was descended from the marriage of Henry's younger sister to the Duke of Suffolk. Warwick could handle her family. To make sure he would be able to control Jane Grey herself, he arranged to marry her to his own son, Guilford.

Jane Grey balked. She did not like Warwick's son. Her parents beat her until she consented, and became Warwick's daughter-in-law. Then Warwick hurried to persuade the dying Edward to declare Jane Grey his successor. The boy was not easy to persuade. He still cared for both his sisters, despite his religious quarrel with Mary. And he did not want to go against his father's will.

Warwick used terror to get his way. He explained that if Mary became queen, England would certainly revert to the Roman Catholic Church. If Elizabeth became queen, she would probably marry a foreign Catholic prince—and the result for England would be the same. Edward, about to meet his Maker, would be responsible—and would be damned. Frightened, Edward on his deathbed signed a "Device for the Succession" which bestowed the crown on Jane Grey.

By threats, promises and persuasions, Warwick induced every important figure he could reach to sign this document. Not until all the other signatures were obtained was Archbishop Cranmer approached. Cranmer was horrified. Though he had at least as much to fear from a Catholic queen as Warwick, this subversion of Henry's will and Mary's right was too unsavory for Cranmer to swallow. He went to Edward's bedside and begged to be excused from signing.

Edward insisted. He reminded Cranmer that everyone else had submitted to his wishes. "Be not more repugnant to my will than the rest. . . . You alone must not stand out." Edward was close to death, and Cranmer loved him. He could not refuse this boy whom he had helped raise and educate, this king whose will he considered the supreme authority. Cranmer signed.

Edward died in agony on July 6, 1553. Warwick proclaimed a sobbing, protesting Jane Grey Queen of England—and tried to lay his hands on Mary and Elizabeth. Both princesses were some distance away at the time. Warwick sent word that Edward was dying and wished to see them. The wary Elizabeth stayed where she was. Mary started for London, but was warned in time. She fled to Framingham Castle, and proceeded to prove that she was indeed Queen Isabella's granddaughter. Boldly pronouncing that the throne was now hers by right of inheritance, she called upon Englishmen to arm themselves and rally to her cause. They did, by the thousands.

Warwick set out with an army to seize her. His troops deserted in large numbers to join Mary. Even London, predominantly Protestant by now, rang with shouts of "God save Queen Mary!" and "Death to the Traitors!" The Council, which Warwick had thought committed to him, read the temper of the people and declared for Mary.

With Elizabeth at her side, Mary rode into London amid a joyous tumult of cheering citizens and ringing bells. Warwick, imprisoned, swore he would be a good Catholic in the future if only Mary would forgive him. "An old proverb there is," he whimpered, "and that most true . . . a living dog is better than a dead lion. Oh that it would please Her Good Grace to give me life—yes, the life of a dog." But the choice was no longer his to make. He had lived as a lion; he died a dog.

Catherine of Aragon's daughter, a strained and sickly woman of thirty-seven, was Queen of England.

2

THE QUEEN who was to become known to history as "Bloody Mary" began her reign with a generous display of good will. She pardoned most of those who had been her enemies, upon their admission of guilt and submission to her mercy. Protestants were given time to recant or escape from the country. The Spanish ambassador, who became closer to her than anyone in her own government, urged her to execute Jane Grey. Mary was content merely to keep the nine-day queen and her husband confined in the Tower.

Mary felt she could afford to be generous. Her suffering was over. Her tormented resistance and unswerving faith had been rewarded. She had won—and through her, her dead mother had won. The first session of her Parliament declared that Henry's divorce from Catherine of Aragon was invalid. They had never been legally divorced—and Mary was Henry's legitimate daughter; his only legitimate daughter. The same Parliament wiped out all of the Reformation Acts accomplished under Edward. England was once more a Catholic nation. Any other faith, any deviation in forms of worship, was heresy.

Elizabeth became subject to the pressures Mary had felt under Edward. Their brother had insisted that Mary conform to his new laws and stop going to Mass. Now Mary insisted that Elizabeth conform to her new laws and go to Mass. Unlike Mary, Elizabeth saw nothing to be gained by a prolonged display of open resistance.

Pointing out that it was not her fault she had been raised as a Protestant, Elizabeth told Mary that as a loyal subject, she was willing to be taught the error of her ways. Mary sent her Catholic books and instructors. Elizabeth read the books, listened to the priests—and began attending Mass.

At first Mary was delighted. She bestowed gifts upon Elizabeth as tokens of her pleasure. But the Spanish ambassador and others feared Elizabeth would become a rallying point for Protestant reaction. They began hinting that her conversion was only a piece of cynical playacting.

Mary questioned her sister about this. Elizabeth looked her

in the eye and swore she was a sincere convert to Catholicism. Mary wanted to believe her. But rumors about Elizabeth's bored attitude during Mass continued. It eased Mary's troubled mind, and relieved her sister from being constantly spied upon, when Elizabeth moved to another residence some distance from court. The separation enabled them to retain their sisterly affection—for a time.

What finally turned them into enemies began with England's belief—shared by Mary—that a woman on the throne needed a husband to help her rule. It was also imperative that she marry soon if she was to have sons to inherit the throne. The most eligible Englishman was twenty-eight-year-old Edward Courtenay, whom Mary had released from the Tower after fourteen years of imprisonment. He was the great-grandson of King Edward IV. Henry had imprisoned him, after beheading his father for treason, as a possible threat to the Tudor succession. Somerset and Warwick had left him in the Tower as an enemy of the Reformation.

It was obvious that Edward Courtenay had all the qualifications. He was an Englishman, of royal descent, handsome, and a Catholic. Mary's government and people decided that he was the man she should take as her husband. Mary decided otherwise.

Emperor Charles secretly offered her his own son Philip, who was soon to become King of Spain and ruler of the Netherlands. To Mary, it must have seemed like a dream come true. Her mother had come to England from Spain to join the two countries. The alliance had failed. England had been torn from the bosom of the Roman Catholic Church. Now her daughter, half English and half Spanish, could achieve all her mother had longed for. By marrying Philip, and bearing his son, she would unite Catholic England and Catholic Spain—and help to reunify Christian Europe under the true Church.

The Spanish envoy skillfully added personal incentive in

describing Philip to Mary: Though only twenty-seven, Philip was mature beyond his years—being already a widower with a son. Praises were sung of his character, his looks, his intellect. By the time a portrait arrived from Spain, Mary was primed. She studied the picture of the slim, elegant, liquideyed Philip—and fell passionately in love.

When she revealed her choice to her own people, they were aghast. A husband would naturally exercise great influence over his wife's decisions. They were sure that if Philip became Mary's husband, England would be maneuvered into acting in Spain's best interests, not its own. Her Council and her most trusted bishops pleaded with her to choose an Englishman instead.

Mary responded with womanly tears and royal rage. She was Queen, and would marry whom she wished to marry. Philip was the man she wanted. Philip was the man she intended to have. Parliament formally petitioned her to forget Philip and marry one of her English subjects. Hot with resentment and defiance, she faced a deputation from Parliament and warned them she would die if she married a man she didn't love: "I will not live three months, and I will bear no children. And so, Mr. Speaker, you will defeat your own ends."

God, she declared, was guiding her into this marriage. England would have to accept God's decision and her choice: Philip of Spain.

England did not want to accept it. A dead dog was thrown through one of Mary's windows, with a noose around its neck and a note demanding the hanging of all papists. When the representatives of Emperor Charles rode into London in January to work out the treaty of marriage between Mary and Philip, they were bombarded with snowballs.

In Kent, Sir Thomas Wyatt—son of the poet who had been in love with Anne Boleyn—measured the depth of English anger. He decided the time had come for rebellion. His plan was to remove Mary from the throne and substitute Elizabeth, who would be married to the popular Edward Courtenay. Whether Elizabeth had any knowledge of this is not known. Courtenay joined Wyatt in the plot. They received support from the French, who feared an alliance of Charles and England against France.

Wyatt's call to revolt against the Queen who wanted to marry a foreigner drew thousands of incensed Englishmen to his standard. He marched on London. Courtenay, who was supposed to lead another force to join him, lost his nerve and failed to appear. But Wyatt had a formidable army behind him when he neared London. It seemed probable that the people of London would join him to depose Mary—and keep Philip out of England.

Most of Charles's representatives fled for safety. The Council began planning an escape for Mary. But Mary no longer trusted them, was no longer sure she could trust anyone around her. She knew only that she was Queen of England— and would live or die acting like one. So she went to Guildhall, and there delivered a passionate call to the citizens of London.

"I am come to you in mine own person," she cried out to the Londoners, "to tell you that which already you see and know . . . how traitorously and rebelliously a number of Kentishmen have assembled themselves against both us and you! . . . Now, loving subjects, what I am, ye right well know. I am your Queen . . . to whom, at my coronation, you promised your allegiance and obedience . . ."

She had touched them in their pride, as Londoners faced with an armed invasion from another part of the country. She next reminded them of their duty to prove themselves loyal Englishmen by supporting their rightful leader. And she went on to tell them of her feeling for them:

"I cannot tell how naturally the mother loveth the child, for I was never the mother of any. But certainly if a prince may love her subjects as the mother doth love her child, then I . . . do as earnestly and tenderly love and favor you."

As for their fears that she might marry a man they did not care for: "I am not so bent to my will . . . that rather for mine own pleasure I would choose where I lust." She assured them that she would not marry Philip if Parliament decided against it. Actually, she had by then made fairly certain that Parliament would agree to Philip, under the marriage treaty which had been worked out. But this, of course, was no time to explain such fine points.

"Good subjects, pluck up your hearts!" Mary exhorted the Londoners. "Like true men, stand fast with your lawful sovereign against these rebels, and fear them not. For I do not, I assure you. I am minded to live and die with you."

London did stand fast with her. Faced with this unexpected resistance, many of Wyatt's troops deserted him. Wyatt, with the rest of his rebels hacked apart and trapped in the streets of London, surrendered.

Mary had won. The rebellion was over. But it had opened the deep wounds Mary had sustained in her tormented struggles with her father and brother. She was never quite the same again.

She brooded over the injuries England had inflicted on her and her mother. She remembered that through all the strife only her cousin Charles had offered any comfort. From now on she placed her trust in the Spanish. For the English she was to feel an increasing suspicion. Her sister Elizabeth was among the first to learn how frightening Mary's suspicion could be.

3

MARY SENT an armed guard to fetch Elizabeth to London for questioning about her supposed role in Wyatt's rebellion. On the day that Elizabeth started for London, Mary revealed that her days of leniency were over. She finally beheaded luckless little Jane Grey for treason. Elizabeth knew that she too was suspected of treason—of having been in communication with Wyatt and the French at the start of the rebellion. She rode into London proud and erect, but with a face as pale as the white gown she wore. She was greeted by the sight of the mangled corpses of drawn and quartered rebels.

Her feeling of impending doom grew heavier when she learned that Mary did not want to see her. It was the Queen's investigators who met with her instead—and they questioned her relentlessly for three weeks. Elizabeth stubbornly denied having had anything to do with the rebellion. She wrote to her sister, asking for a chance to speak to her in person, begging that she not be condemned unheard. Mary refused to talk to her until she cleared herself of the charges.

Elizabeth's questioners could not prove her guilty, but did not believe her innocent. They advised her to throw herself on the Queen's mercy and plead for pardon. Elizabeth retorted that she could not ask pardon for what she had not done.

Mary's anger seemed to increase with the failure to find evidence against her sister. She ordered her imprisoned in the Tower.

Elizabeth's barge reached the Traitor's Gate in a downpour. She remembered what had become of her mother after her last arrival at this place. In a fit of fright and temper, she sat down on the wet stone steps and would not go inside. The men escorting her pleaded for Elizabeth to come in out of the rain. She refused to move. Then one of the men began to cry. In disgust, Elizabeth rose to her feet, told him to act like a man, and marched inside.

In years to come, Elizabeth was to reveal that during her long confinement in the Tower her thoughts turned more and more to the fate of Anne Boleyn. She became convinced that her fate was to be the same, and decided to follow her mother's example: to ask that her beheading be done with a sword.

Elizabeth was still in the Tower when Prince Philip arrived in England and married Queen Mary. The terms of the marriage treaty to which Parliament agreed declared that Philip was to have no part in the government of England. But even the most loyal English Catholics realized that Mary would certainly be influenced by a husband she cared for. And she did care for Philip. She had fallen in love with his portrait. Her love deepened when she finally met him. It grew deeper still when she believed herself pregnant by him.

Philip, for his part, was interested in Mary only as a means of riveting England to the Spanish Empire. It was Philip who, for the good of Spain, persuaded Mary to release Elizabeth from the Tower. He feared that Mary, sickly and old for childbearing, might die without leaving a male heir to the throne. In that case, there would be only two persons with a strong claim to the throne. One was Elizabeth. The other was Mary Stuart, Queen of Scots, granddaughter of Henry's elder sister. Mary Stuart was betrothed to the French dauphin, had been brought up in France, and could be expected to join England to France. So Philip preferred Elizabeth, whom he believed he could bend to his own purposes. He did not understand Elizabeth's character, and was to learn to understand it only after many years of painful lessons.

Mary's apparent pregnancy was the happiest period of her life. She was married to a man she loved and admired. She carried his child inside her. The Pope had accepted England back into the bosom of the Roman Catholic Church.

The destruction of her happiness began with a cruel disappointment.

The time for her to deliver her child came—and went. There was no child. She had never been pregnant. The symptoms had been misread; or a false pregnancy had been induced by Mary's obsessive desire to have Philip's baby—or a combination of both. In any case, the result was to make Mary look ridiculous.

In shame and despair, Mary hid herself from her people. Philip, irritated at having tarried so long in England for what had turned out to be no reason at all, sailed for the Continent to tend to Spanish affairs. Mary, abandoned in her misery, brooded over what had happened to her. God, it seemed, was punishing her for allowing Protestant heretics to remain alive in her kingdom. She had already re-enacted the laws against heresy. Now she ordered them to be carried out—with the most pious severity. The burnings began.

4

"MEN never do evil so completely and cheerfully as when they do it from religious conviction," Pascal wrote. This could be applied to Edward's reign, as well as Mary's. But there was a great difference in the lengths to which the faith-inspired evil was carried. In Edward's reign, two men were burned for heresy—both of them foreigners. Mary burned some three hundred people to death solely because their religion differed from hers.

Most of these heretics who died by fire were common people. Many were women. One of these women was pregnant, and the flames consumed the child in her womb along with her.

Only a tiny percentage of those burned had any connection with politics. Among this small number were Cranmer and two powerful Reformation preachers, Latimer and Ridley. Mary had kept Cranmer in prison for a number of years. The Pope had excommunicated him and stripped the title of Archbishop from him. But for a time he was allowed to go on living, so that he might recant.

Cranmer did recant. He did so partly because Queen Mary

ordered him to. He had always believed that the monarch of the kingdom was the supreme authority, and must be obeyed. But he obeyed, also, because he had little physical courage. He shuddered at what might be inflicted upon him. It was to save himself from being burned to death that the sixty-six-year-old Cranmer denounced all that he had done and said and stood for.

He wrote one recantation after another. In them he found himself guilty of having wrongfully declared Henry and Catherine of Aragon divorced, when they were actually husband and wife by law and in the sight of God. He wrote that the Catholic faith was the true one, and that he believed in the Roman Catholic Church and all its Sacraments. As for his prayer books and other works, he swore that his writing "troubleth my conscience more than any other thing that ever I did." They were "contrary to the truth of God's word . . . I renounce and condemn, and refuse them utterly as erroneous."

Nothing his enemies could have done would have made of him a lower object than he had now made of himself by such groveling. If Mary had been clever, she would have let Cranmer live on this way. But she was not trying to be clever. She was trying to do what she considered right in the eyes of God. So she ordered the burning of Cranmer.

In his convulsive efforts to go on living, Cranmer had become something to be pitied and disdained. But in his death, and his manner of dying, he became something else again. He was still afraid as he went to the stake and the waiting faggots. But he no longer had anything to lose by telling the truth.

For all to hear, he took back all his recantations, "written for fear of death, and to save my life if it might be." From the stake he renounced "all such bills which I have written or signed with mine own hand since my degradation. . . . And as for the Pope, I refuse him as Christ's enemy and Antichrist, with all his false doctrine!"

Since he said these things as he was preparing to meet his Maker, their sincerity was obvious—and their effect enormous. The horrified priests hastened to silence him for all time. But the damage had been done. It was increased even more by his final act, for which he was to be remembered more than for anything else he had done in his entire lifetime.

Holding up the right hand with which he had signed his false recantations, the weak and timid Cranmer declared firmly: "Forasmuch as my hand offended in writing contrary to my heart, therefore my hand shall first be punished." As the fire licked up around him, he thrust his right hand down into the flames—and held it there as long as he was conscious, so that it burned before the rest of him.

Mary had hoped to burn the Protestant heresy out of England. The burnings had the opposite effect. In the reigns that followed hers, children were to be brought up on stories about the martyrdom of Cranmer and the others. In her own reign their example stiffened her people against her, and made hot reformers out of many who had been only lukewarm Protestants.

When the slow fires ate at the flesh of the Protestant bishops, Latimer and Ridley, the eighty-year-old Latimer called out to his fellow sufferer: "Be of good cheer, Master Ridley, and we shall this day light such a candle in England, by God's grace, as I trust shall never be put out!"

So it was to be. And Mary compounded her mistake out of love for her husband. Philip of Spain returned to England only once, for a short time. He came solely for the purpose of inducing England to join him in making war on France. Mary must have known such a war at that time was not in the best interests of her kingdom. But she could not resist Philip.

Mary declared war on France and prepared to invade it. Philip sailed away and never returned. Spain got what it wanted from this war. England gained nothing—and lost its last foothold on the Continent: Calais, which the English had held for over two hundred years.

All England felt this blow to its pride—sustained only to help Spanish ambition. "When I am dead and opened," Mary groaned, "you will find Calais lying in my heart."

In all, Mary could not have done her religion more harm if she had deliberately set out to destroy English Catholicism. She had proved to the English that what they had feared was true; in their minds Catholicism was linked to foreign domination, Spain, and the terrors of the Inquisition. To be English came increasingly to mean being anti-Rome, anti-Spain, anti-Catholic.

Hated by her people and abandoned by her husband, the unfortunate Mary finished her life a bitter, unbalanced wreck of a woman, wandering the corridors of her palace alone. Sir Walter Raleigh was to sum up both mother and daughter thus, in describing Catherine of Aragon: ". . . the mother of many troubles in England, and the mother of a daughter that, in her unhappy zeal, shed a world of innocent blood, lost Calais to the French, and died heartbroken, without increase."

Mary died on November 17, 1558. She had reigned for five years. Now Anne Boleyn's daughter was Queen.

5

QUEEN ELIZABETH began her reign by settling the religion of her people in the Church of England. In doing so, Anne Boleyn's daughter rendered the final decision in the controversy over Henry's divorce from Mary's mother. Though the controversy was to plague much of her reign, papal authority was never again to gain so much as a toehold within the kingdom. For Elizabeth ruled brilliantly, with all of her father's domineering energy and flair for popular politics; and with much more tolerance. By the time she died, in 1603, England was utterly and irrevocably committed to the Reformation.

This was partly due to the sheer length of her reign. She was twenty-five when she mounted the throne of England. She was seventy when she died, and by then few Englishmen could remember when England had been Roman Catholic. But she started with a great advantage. She caught the English people, as it were, on the rebound from her sister's violent Catholicism. And she was perfectly in tune with the feelings of her fellow Englishmen in refusing to carry them too far in the other direction—to the extreme Protestantism of the bloodthirsty Luther or the unbending Calvin.

The "Elizabethan Settlement" of the religion of the realm, worked out together with her first Parliament, established the Anglican Church—as Protestant and nationalistic. Mary's religious legislation was swept away. Most of the statutes of the English Reformation were restored. The authority of Cranmer's prayer book was revived, along with most of his Forty-two Articles of religion. Mary's Catholic bishops were replaced by reformers who emerged from hiding and flocked back from exile. As a sop to Catholic Europe, Elizabeth was not titled Supreme Head of the Church, but merely its "Governor."

Elizabeth was quite willing to offer a concession to papal Rome—if it did not infringe on her authority as Queen, or on the status of England as a separate, inviolate nation. She was not interested in making a stand on faith. Religion did not dominate her, as it had dominated her brother and sister. She identified herself with the English Reformation—but she was not anti-Catholic, except when it served a political purpose.

It was Elizabeth's position that the government should not interfere with any individual's private conscience—that she intended to "make no windows into men's souls." From the start she had to contend with plots against her crown and her person, and these plots were usually Catholic. Elizabeth dealt cold justice without mercy to those suspected of plotting against her. Yet she also repealed the laws which decreed burning for heretics, laws which Mary had used against reformers. Under Elizabeth—though England was Protestant by law—an English Catholic could hold to his faith in safety, so long as his faith did not incline him to anything against the Queen or her government.

But Catholics at home and abroad believed that Elizabeth had no right to wear the crown. According to their Church, Henry and Catherine of Aragon had never been divorced. Therefore Henry had never been legally married to Anne Boleyn—and Anne Boleyn's daughter was not a legitimate heir to the throne. It would be a righteous act in support of divine law to depose Elizabeth, and replace her with a legitimate heir like Mary Stuart.

But as her father had done, Elizabeth for years prevented invasion from abroad by playing those old enemies, France and Spain, against each other. She dealt as successfully with religion-inspired rebellion within her nation. The failure of these internal conspiracies was in some measure due to the fact that many English Catholics refused to support them. Some put national loyalty above religious conviction. Others simply preferred not to get involved, and remained neutral. Those who did join in the conspiracies against Elizabeth were detected and crushed before their plans reached fruition.

All the intrigues of the Catholic Church against Elizabeth failed—even after the Pope gave the intriguers a powerful new weapon by formally pronouncing Elizabeth excommunicated and deprived of her realm. Elizabeth, on the other hand, succeeded spectacularly in her counterblows against the Catholic Church.

She aided the rebellion of the Netherlands against Spanish misrule—which also developed into a victorious revolt against the Roman Catholic Church. She encouraged and materially supported the Protestant revolution in Scotland which was led by the fanatic John Knox. This Scottish Reformation scored a swift and complete triumph. The Catholic Queen of Scots, Mary Stuart, had to leave her son and heir behind, and escape from her country. She made the mistake of running to England, from which there was to be no escape. For almost twenty years she remained in England, in an ambiguous position as Elizabeth's guest—and prisoner.

From the time Mary Stuart entered England, there were those among Elizabeth's advisers who urged that she be killed as a threat to the throne. Elizabeth resisted this advice for nineteen years—not entirely out of mercy toward her trapped guest. So long as the France-oriented Mary Stuart remained alive as next in line to the English throne, Philip of Spain would hesitate to act to depose Elizabeth. Elizabeth took advantage of this to "singe the King of Spain's beard" with relative impunity. She permitted Drake and other English "sea dogs" to enrich her treasury by acts of piracy against Spanish ships and possessions. While Mary Stuart lived, King Philip did not do much in retaliation.

But conspiracies against Elizabeth continued to take root in her realm. They aimed at triggering an English counter-reformation by assassinating Elizabeth.

Catholic Mary Stuart, as next in line for the succession, increasingly became the rallying point for these conspiracies.

Finally, reluctantly, Elizabeth gave in to her advisers. Mary Stuart was executed in 1587, calmly declaring herself a martyr dying for her religion.

The execution of Mary Stuart rang down the final curtain on "the divorce." Macaulay, the historian, contemplating "the instruments by which England was delivered from the yoke of Rome," summed it up this way: "The work which had been begun by Henry, the murderer of his wives, was continued by Somerset, the murderer of his brother, and completed by Elizabeth, the murderer of her guest." Whether or not the choice of the word "murderer" was correct in each case, he was right about the result.

After Mary Stuart was gone, King Philip decided the time had come for him to enforce the papal excommunication of Elizabeth—and take over England. He sent his "Invincible Armada" against England—hundreds of ships packed with thousands of troops, the most experienced fighting men in Europe.

With the threat of Spanish invasion looming ever closer, Elizabeth went among the English fighting men who had gathered at Tilbury, heartening them for the coming battle. She had been warned not to risk going among these troops, some of whom might be more loyal to the old Church than to her. She had scorned this advice. The ringing words she spoke to her soldiers that perilous day carry to us through the centuries the sound of the Elizabethan spirit:

"My loving people, we have been persuaded by some that are careful for our safety to take heed how we commit ourselves to armed multitudes, for fear of treachery. But I assure you I do not desire to live to distrust my faithful and loving people. Let tyrants fear. I have always . . . placed my chiefest strength and safeguard in the loyal hearts and good will of my subjects. And therefore I am come amongst you . . . to lay down—for my God and for my kingdom and for my people—my honor and my blood, even in the dust. I know I have the body of a weak and feeble woman. But I have the heart and stomach of a king."

The Spanish Armada was beaten—by this spirit, by the winds, and by the navy which Henry VIII had established and Elizabeth had nurtured. With the defeat of the Armada began the decline of Spanish power and the rise of the English empire. And with it ended the last serious threat to England's separation from the Church of Rome.

Nothing would happen in the future to alter this. Not even when Elizabeth died and Mary Stuart's son, James, became

King of England and united it with Scotland. For James had not known his Catholic mother, and had been raised by John Knox as a Protestant. And when a subsequent Stuart king showed signs of turning back to Rome, the English deposed him. England had rendered its own verdict.

The dream of a united Catholic Europe under Spain, which had long ago brought Catherine of Aragon to England, was dead. The English Reformation, which had begun with Henry's first lustful glance in Anne Boleyn's direction, was victorious.

"The divorce" was final.

Bibliography

Since this book obviously is intended for the general reading public, it is presented without the detailed mass of footnotes on sources common to the works of specialists in the period which are intended for their fellow specialists—what J. E. Neale has called "the elaborate scaffolding of documentary authority." Should any scholar specializing in the early Tudor period fail to recognize the documentary authority for any quotation or incident in this book, I will be pleased to supply it upon request. The general reader should bear in mind that every personality and event connected with "the divorce" aroused—and continues to arouse—passionate historical and religious controversy. Even some of the best books on the subject are affected by their authors' personal bias, and editorial slanting is as common as violent disagreements between equally eminent authorities in this field.

For anyone wishing to dip deeper into the period covered in this book, there is no more fascinating historical material than the *Letters and Papers, Foreign and Domestic, of the Reign of Henry VIII* (eds. J. Brewer, J. Gairdner, and K. H. Brodie, 21 vols. London, 1862–1910).

Of almost equal importance as source material, and of interest for browsing, are the volumes of the *Calendar of Letters, Despatches, and State Papers relating to the Negotiations between England and Spain preserved in the Archives at Simancas and elsewhere* (ed. G. A. Bergenroth, et al., 13 vols., London, 1862–1954) and the *Calendar of State Papers and Manuscripts relating to English Affairs, Preserved in the Archives of Venice, and in other Libraries of Northern Italy* (ed. R. Brown, et al., 9 vols., London, 1864–1898).

For those interested in reading more about the lives of specific personalities, I would especially recommend these biographies:

Chambers, R. W., *Sir Thomas More*. New York, 1935.

Chapman, Hester W., *The Last Tudor King*. London, 1858 (on Edward VI).

Ferguson, Charles W., *Naked to Mine Enemies*. New York, 1958 (on Thomas Wolsey).

Mattingly, Garrett, *Catherine of Aragon*. London, 1942.

Neale, J. E., *Queen Elizabeth I*. London, 1939.

Pollard, A. F., *Henry VIII*. London, 1925.

Prescott, H. F. M., *Mary Tudor*. London, 1953.

Ridley, Jasper, *Thomas Cranmer*. Oxford, 1962.

In addition to these modern works, no one should miss two delightful contemporary biographies:

The Life of Thomas More, by his son-in-law, William Roper. London, 1822.

The Life of Cardinal Wolsey, by his servant, George Cavendish. London 1885.

Unfortunately, there is no biography of Anne Boleyn which ranks with any of the above works.

A list of some of the other works consulted during the preparation of *The Divorce* follows. For information on further writings related to the subject, one should turn to Conyers Read's *Bibliography of British History, Tudor Period, 1485–1603*. Oxford, 1933.

Acton, John E., Lord, *Lectures on Modern History*. London, 1950.

Allen, J. W., *A History of Political Thought in the Sixteenth Century*. London, 1951.

Armstrong, Edward, *The Emperor Charles V*. London, 1910.

Bacon, Francis, *The Life of Henry VII*, Harmondsworth, 1950.

Belloc, Hilaire, *Wolsey*. London, 1930.

Bindoff, S. T., *Tudor England*. London, 1950.

Black, J. B., *The Reign of Elizabeth, 1558–1603*. Oxford, 1936.

Brandi, C., *Emperor Charles V*. London, 1939.

Brewer, J. S., *The Reign of Henry VIII From His Accession to the Death of Wolsey*. London, 1884.

Bridgett, T. E., *Life and Writings of Sir Thomas More*. London, 1892.

Byrne, M. St. Clare, ed., *The Letters of King Henry VIII*. London, 1936.

Calendar of Entries in the Papal Registers; Papal Letters Relating to Great Britain and Ireland. London, 1960.

Calendar of State Papers and Manuscripts (*Milan*). London, 1912.

Calendar of State Papers (*Domestic Series*) *of the reigns of Edward VI, Mary, Elizabeth*. London, 1856.

Calendar of State Papers (*Foreign Series*) *of the reign of Edward VI . . . Mary . . . Elizabeth*. London, 1861; 1863–1950.

Cambridge Modern History, vols. 1, 2 & 3. New York, 1907.

Campbell, John lord, *Lives of the Lord Chancellors of England*, vols. 1 & 2. London, 1845–47.

Campbell, W. E., *Erasmus, Tyndale, and More*. London, 1949.

Cardinal, Edward V., *Cardinal Lorenzo Campeggio*. Boston, 1935.

Chamberlin, Frederick C., *Private Character of Henry the Eighth*. New York, 1931.

Claremont, Francesca, *Catherine of Aragon*. London, 1939.

Coulton, George G., ed., *Life in the Middle Ages*. Cambridge, England, 1930.

Crabitès, Pierre, *Clement VII and Henry VIII*. London, 1936.

Creighton, Mandell, *Cardinal Wolsey*. London, 1888.

Deans, R. Storry, *Trials of Five Queens*. London, 1909.

Dickens, Arthur Geoffrey, *Thomas Cromwell and the English Reformation*. London, 1959.

Dixon, William Hepworth, *History of Two Queens*. London, 1873.

Duboys, A., *Catherine of Aragon and the Sources of the English Reformation*. London, 1881.

Einstein, Lewis, *Tudor Ideals*. New York, 1921.

Elton, G. R., *England Under the Tudors*. London, 1955.

Elton, G. R., *Star Chamber Stories*. London, 1958.

Elton, G. R., *The Tudor Revolution in Government*. Cambridge, England, 1953.

Fabyan, Robert, *The New Chronicles of England and France*. London, 1811.

Farrow, John, *The Story of Thomas More*. London, 1956.

Fisher, H. A. L., *History of England (1485–1547)*. London, 1934.

Fitzpatrick, Benedict, *Frail Anne Boleyn*. New York, 1931.

Fortescue, John, Sir, *A History of the British Army*. London, 1899–1930.

Foxe, John, *Acts and Monuments* (Book of Martyrs). London, 1841.

Friedmann, Paul, *Anne Boleyn*. London, 1884.

Froude, James Anthony, *The Divorce of Catherine of Aragon*. New York, 1881.

Froude, James Anthony, *History of England from the Fall of Wolsey to the Defeat of the Spanish Armada*. London, 1856–70.

Froude, James Anthony, *Life and Letters of Erasmus*. New York, 1894.

Gairdner, James, *The English Church in the Sixteenth Century*. London, 1912.

Gairdner, James, *Henry the Seventh*. London, 1899.

Gairdner, James, ed., *Memorials of King Henry VII.* London, 1858.

Gairdner, James, ed., *Letters and Papers Illustrative of the Reigns of Richard III and Henry VII.* London, 1861–1863.

Garvin, K., *The Great Tudors.* London, 1935.

Gasquet, Francis Cardinal, *Henry VIII and the English Monasteries.* London, 1888.

Glenne, Michael, *Henry VIII's Fifth Wife, the Story of Catherine Howard.* New York, 1948.

Gordon, M. A., *Life of Queen Katharine Parr.* Kendal, England, 1952.

Hackett, Francis, *Francis the First.* New York, 1935.

Hackett, Francis, *Henry the Eighth.* New York, 1929.

Haggard, Andrew C. P., *Two Great Rivals: François I and Charles V.* London, 1910.

Hall, Edward, *Hall's Chronicle.* London, 1809.

Harbison, E. Harris, *The Age of Reformation,* Ithaca, 1955.

Harleian Miscellany. London, 1808–11.

Harpsfield, Nicholas, *The Life and Death of Sir Thomas More, Knight.* London, 1932.

Harpsfield, Nicholas, *A Treatise on the Pretended Divorce between Henry VIII and Catherine of Aragon.* Westminster, 1878.

Herbert of Cherbury, Lord, *The Life and Reign of Henry VIII.* London, 1649.

Holinshed, Raphael, *Chronicles.* London, 1807.

Hope, A., *The First Divorce of Henry VIII.* London, 1894.

Hughes, Philip, *The Reformation.* London, 1960.

Hughes, Philip, *The Reformation in England.* New York, 1951–1954.

Huizinga, J., *Erasmus of Rotterdam.* London, 1952.

Huizinga, J., *The Waning of the Middle Ages.* London, 1948.

Innes, Arthur, *Cranmer and the Reformation in England.* New York, 1900.

Innes, Arthur, *Ten Tudor Statesmen.* London, 1934.

Jacob, E. F., *The Fifteenth Century.* Oxford, 1961.

Jenkins, Elizabeth, *Elizabeth the Great.* London, 1958.

Jerrold, Walter, *Henry VIII and His Wives.* London, 1925.

Knox, John, *Collected Works,* ed. David Laing. Edinburgh, 1854.

Latimer, Hugh, *Sermons.* Everyman's Library, London, 1926.

Lingard, John, *History of England.* London, 1855.

Mackie, J. D., *The Earlier Tudors, 1485–1558.* Oxford, 1952.

Maitland, Frederick William, *The Constitutional History of England.* Cambridge, England, 1920.

Maynard, Theodore, *Bloody Mary.* Milwaukee, 1955.

Maynard, Theodore, *The Life of Thomas Cranmer.* Chicago, 1956.

McCabe, L., *Henry VIII, His Wives, and the Pope.* London, 1935.

Bibliography

Merriman, Roger B., *Life and Letters of Thomas Cromwell.* Oxford, 1902.

Morris, Christopher, *Political Thought in England: Tyndale to Hooper.* London, 1953.

Muller, James Arthur, *Stephen Gardiner and the Tudor Reaction.* New York, 1926.

Murray, J., *Charles V.* London, 1867.

Neale, J. E., *Elizabeth and Her Parliaments.* London, 1953.

Nichols, J. G., ed., *Chronicle of Queen Jane and of Two Years of Queen Mary.* London, 1850.

Nichols, J. G., ed., *Literary Remains of Edward VI.* London, 1857.

Orliac, J., *Francis I.* London, 1932.

Paul, Leslie A., *Sir Thomas More.* London, 1953.

Pickthorn, Kenneth, *Early Tudor Government: Henry VIII.* Cambridge, England, 1934.

Pocock, Nicholas, *Records of the Reformation: the Divorce 1527–1533.* Oxford, 1870.

Pollard, A. F., *Factors in Modern History.* London, 1907.

Pollard, A. F., *The History of England from the Accession of Edward VI to the Death of Elizabeth.* London, 1911.

Pollard, A. F., *The Reign of Henry VII from Contemporary Sources.* London, 1913–1914.

Pollard, A. F., *Thomas Cranmer and the English Reformation.* New York, 1904.

Pollard, A. F. ed., *Tudor Tracts.* London, 1903.

Pollard, A. F., *Wolsey.* London, 1929.

Powicke, F. M., *The Reformation in England.* London, 1941.

Ranke, Leopold, *History of the Popes.* London, 1878.

Read, Conyers, *The Tudors.* New York, 1936.

Rival, Paul, *The Six Wives of King Henry VIII.* New York, 1936.

Rowse, A. L., *The England of Elizabeth.* London, 1951.

Salzman, L. F., *England in Tudor Times.* London, 1926.

Schenk, W., *Reginald Pole, Cardinal of England.* London, 1950.

Sergeant, Philip W., *Anne Boleyn: A Study.* London, 1934.

Simpson, H., *Henry VIII.* London, 1934.

Smith, H. Maynard, *Henry VIII and the Reformation.* New York, 1948.

Stephen, Leslie, et al., eds., *Dictionary of National Biography.* Oxford, 1882–1949.

Stone, J. M., *The History of Mary I, Queen of England.* London, 1901.

Stow, John, *The Annales of England.* London, 1592.

Strickland, Agnes, *Lives of the Queens of England.* London, 1852.

Strype, John, *Ecclesiastical Memorials.* Oxford, 1822.

Strype, John, *Annals of the Reformation.* Oxford, 1824.

Tawney, R. H., *The Agrarian Problem in the Sixteenth Century.* London, 1912.

Tawney, R. H., *Religion and the Rise of Capitalism.* New York, 1926.

Taylor, Henry Osborn, *Thought and Expression in the Sixteenth Century.* New York, 1920.

Temperly, Gladys, *Henry VII.* Boston, 1914.

Trevelyan, George M., *English Social History.* London, 1942.

Trevor-Roper, H. R., *Men and Events: Historical Essays.* New York, 1957.

Trovillion, Violet and Hal W., *Love Letters of Henry VIII.* Herrin, Ill., 1936.

Tytler, P. F., *England Under the Reigns of Edward VI and Mary.* London, 1839.

Waldman, Milton, *Queen Elizabeth.* London, 1952.

White, Beatrice, *Mary Tudor.* London, 1935.

Williamson, James A., *The Tudor Age.* New York, 1961.

Wood, Anthony, *Athenae Oxonienses.* Oxford, 1848.

Wood, Mary A. E., *Letters of Royal and Illustrious Ladies.* London, 1846.

Wood, Mary A. E., *Lives of the Princesses of England.* London, 1849–55.

Wriothesley, Charles A., *A Chronicle of England During the Reigns of the Tudors.* London, 1875, 1877.

Zweig, Stefan, *Erasmus of Rotterdam,* tr. Eden and Cedar Paul. New York, 1934.

Index

Index

Index

DATE DUE